Calcium and Ion Channel Modulation

Roger Otto Eckert
1934–1986

Calcium and Ion Channel Modulation

Edited by
Alan D. Grinnell
UCLA School of Medicine
Los Angeles, California

David Armstrong
National Institute of Environmental Health Sciences
Research Triangle, North Carolina

and
Meyer B. Jackson
University of California, Los Angeles
Los Angeles, California

Plenum Press • New York and London

Library of Congress Cataloging in Publication Data

Calcium and ion channel modulation/edited by Alan D. Grinnell, David Armstrong,
and Meyer B. Jackson.

    p.   cm.

"Proceedings of a symposium on ion channel modulation, honoring Roger Eckert,
held February 26–March 1, 1987, in Los Angeles, California"—T.p. verso.

"Publications [of Roger Otto Eckert]": p.

Includes bibliographies and index.

ISBN 0-306-42834-2

1. Calcium channels—Congresses. 2. Ion channels—Congresses. 3. Eckert, Roger—
Congresses. I. Grinnell, Alan, 1936–   . II. Armstrong, David, 1951-   . III.
Jackson, Meyer B. IV. Eckert, Roger.

QP535.C2C2619   1988

574.19′214—dc19   88-5928

                                                     CIP

Proceedings of a symposium on Ion Channel Modulation,
honoring Roger Eckert, held February 26–March 1, 1987,
in Los Angeles, California

© 1988 Plenum Press, New York
A Division of Plenum Publishing Corporation
233 Spring Street, New York, N.Y. 10013

ACKNOWLEDGEMENTS

We would like to express our gratitude to everyone who helped us in the preparation of this volume. At UCLA Vice Chancellor Albert A. Barber of the Office of the Chancellor, Dean J. D. O'Connor of the College of Letters and Science, Dean Kenneth Shine and Associate Dean Irv Zabin of the School of Medicine, and Drs. Paul Boyer and Sue Huttner of the University of California Biotechnology Research and Education Program all helped us obtain financial support for the preparation of this volume and the Symposium on which it was based. Pan American World Airways generously helped with European travel arrangements.

In the preparation of this volume, as in the planning and staging of the symposium, we have depended especially on the organizational skills and efficiency of Carol Gallion. We are also indebted to Sandra Nath-Singh and Jill Sturm in the Jerry Lewis Neuromuscular Research Center, who carried out many essential tasks with dedication and good humor, to Terry Zeyen and Alan Strozer in the Department of Biology, who patiently prepared camera-ready manuscripts for many of the authors, to Lisa DiMeglio, the talented photographer of the Brain Research Institute who was responsible for the pictures taken during the symposium, to Taras Momdjian for carefully assembling all these components into camera-ready form, and to the Eckert family and other symposium participants for photographs that have been included in the volume.

Finally, we would like to express our gratitude to Roger's family and to his former collaborators in other fields who travelled to Los Angeles at their own expense and sat patiently through three long days of esoteric presentations. Their gracious participation added immeasurably to the value of the final product as a tribute to Roger. As organizers of the symposium and editors of this volume, we feel privileged to have known and worked with Roger, and grateful to the many participants who added luster to the meeting and to the quality of this volume.

Alan Grinnell
David Armstrong
Meyer Jackson

UCLA, January, 1988

Roger's family: his mother, Carla Heims and his four sons, Glenn, Brian, Kevin and Tim

Some of Roger's past and present collaborators and students present at meeting: front row: D. Kalman, Anton Hermann, Joy Umbach, Dieter Lux, Kathy Dunlap, Bob Zucker; second row: J. Chad, George Augustine, Paul Brehm, D. Armstrong, Erwin Neher, Akira Murakami; third row: Miles Epstein, Yutakah Naitoh, C. Erxleben, V. Brezina, Doug Tillotson, Joachim Deitmer

CONTENTS

III.   ION CHANNEL MODULATION BY NEUROTRANSMITTERS
AND SECOND MESSENGERS

INTRODUCTION

Cellular neurobiology has been transformed in the past decade by new technologies and fundamental discoveries. One result is an enormous increase in our understanding of how ion channels function in nerve and muscle cells and a widening perspective on the role of ion channels in non-neuronal cell physiology and development. Patch clamp techniques now permit direct observation of the transitions between functional conformations of individual ion channels in their native membrane. Recombinant DNA techniques are being used to determine the primary structure of ion channel proteins and to test hypotheses about channel conformations, sites of grating and modulation, and the basis of ion selectivity. At the same time, biochemical techniques have revealed intricate signalling systems inside cells, involving second messengers such as calcium, phospholipids and cyclic nucleotides, which interface with the external milieu through GTP binding proteins and regulate cell metabolism by altering protein phosphorylation. This panorama of second messenger systems has greatly increased our application for their potential role in regulating ion channel function.

We now recognize that ion channels are much more complicated than we once thought, and more interesting. They are not simply isolated macromolecules in the membrane, gated directly by depolarization or transmitter binding to open briefly at a fixed conductance and then close or inactivate. Instead, individual channels now appear to have many open and closed states that are regulated independently by voltage and transmitters. In addition, transmitters can influence channel gating in indirect ways by stimulating or inhibiting other biochemical pathways. The biochemical control of ion channel function not only provides a way of integrating ion fluxes across the plasma membrane with cell metabolism; recent discoveries suggest that it may also play a fundamental role in the brain's remarkable ability to alter connections and learn from experience.

One of the early revelations of channel modulation was the demonstration by Roger Eckert and his colleagues that voltage-activated Ca^{++} channels in _Paramecium_ and _Helix_ neurons are inactivated by intracellular Ca^{++} accumulation. It now appears that this phenomenon may involve a number of intermediate biochemical steps which culminate in channel inactivation by dephosphorylation of the channels. This phosphorylation-dependence of voltage-activated Ca^{++} channel activity provides a simple framework for understanding the modulation of calcium influx and its integration with cell physiology by external signals, intracellular calcium levels and the metabolic state of the cell. Comparable mechanisms are being discovered for the regulation of a wide variety of other channels affecting membrane potential, endogenous activity, neurosecretion, and the responses of cells to sensory and synaptic inputs.

Thus, ion channel modulation is an area of great excitement and progress, both of which we have tried to capture here. The value and significance of this volume, at least to the contributors, are much enhanced by the fact that these papers are presented in memory of Roger Eckert, whose tragic death on June 16, 1986, cut short a distinguished career in the field. That purpose strongly influenced our choice of

participants and topics. Consequently, the papers, which were presented
initially at a symposium on Ion Channel Modulation at UCLA February 27 -
March 1, 1987, are largely electrophysiological in approach and are
focused broadly on calcium channels and the role of intracellular calcium
in ion channel modulation.

The first section begins with a review by Byerly and Hagiwara, which
focuses on the many mechanisms by which calcium channel activity may be
modulated. That is followed by two papers on calcium channels in
unicellular organisms. Deitmer reviews his elegant experiments on the
distribution and physiological specialization of calcium channels in
Stylonychia, and Ehrlich and Forte describe their preliminary results on
calcium channels, purified from Paramecium, which they have successfully
incorporated into artificial membranes. Mironov discusses calcium
channel selectivity and its modulation by divalent cations. Morad and
Lux describe a dramatic transformation of calcium channel permeability at
higher concentrations of hydrogen ions. Fox et al. review the multiple
types of calcium channels in neurons from the chick dorsal root ganglion
and suggest a central role in neurotransmitter release for the "N"
channels. Finally, Umbach and Gundersen describe the isolation from
Torpedo electric lobes and expression in Xenopus oocytes of a messenger
RNA species which codes for such an omega toxin-sensitive calcium
channel.

The second section turns to some of the effects of intracellular
calcium ions on membrane excitability and other physiological processes,
particularly neurosecretion. Dunlap and Brehm, both former students of
Roger, describe their work in Woods Hole on the role of calcium in the
sensory transduction of light by Obelia, a hydrozoan coelenterate. Fain
and Schroder introduce a new technique for determining the distribution
of total calcium in cells and discuss the implications of that
distribution for the calcium economy of vertebrate rods. Hermann et al.
report the results they obtained with charybdotoxin on calcium-activated
potassium channels in Roger's lab at UCLA. The next two papers by
Tillotson and Nasi and by Smith et al. use two different calcium
indicators to provide complementary pictures of the intracellular
distribution of the calcium entering molluscan nerve cells during
excitation. Then Augustine et al. and Kriebel describe in detail many of
the physiological parameters of transmitter release which ultimately must
be accounted for by the amplitude and kinetics of calcium ion movements
inside the cell.

There is widespread agreement now that many of the events underlying
behavioral plasticity occur at the synapse, and the third section
presents one of the central themes of the symposium: ion channel
modulation by protein phosphorylation and intracellular messengers.
Kostyuk reviews the evidence implicating cyclic AMP-dependent
phosphorylation in the regulation of voltage-activated calcium channels.
Chad describes his experiments on dialyzed molluscan neurons in Roger's
lab that led them to postulate a molecular mechanism for calcium-
dependent inactivation of the phosphorylation-dependent channels, and
Armstrong and Kalman discuss some of the physiological repercussions of
that phosphorylation dependence on calcium channel function in mammalian
cells. Perozo et al. present compelling evidence that phosphorylation
also modulates potassium channel function in that bastion of membrane
biophysics, the squid giant axon, and Kramer et al. illustrate some of
the interactions between calcium and cyclic AMP which produce spontaneous
electrical activity in intact molluscan neurons.

After completing her Ph.D. with Roger, Kathy Dunlap became one of the
pioneers in an important new area of ion channel modulation with the
demonstration that neurotransmitter-receptor interactions also modulate
the activity of voltage-activated calcium channels elsewhere in the
membrane. She and her present collaborators review their experiments on
the molecular events underlying that modulation and its significance for

neurotransmitter release. Ewald et al. extend the description of this
phenomenon in dorsal root ganglion neurons to neuropeptide Y, also linked
to diacylglycerol production through a GTP binding protein. Brezina and
Erxleben review the work they began in Friday Harbor with Roger on
FMRFamide, an endogenous molluscan neuropeptide which counteracts many
of the excitatory synaptic effects of serotonin by modulating
voltage-activated calcium and potassium channels in the cell membrane.
Yakel et al. provide further evidence for potassium channel modulation by
neurotransmitters via GTP binding proteins in cell cultures from the
mammalian CNS.

The fourth section introduces a new context for ion channel
modulation: embryonic development. Not only may ion channels be involved
in transducing the cell interactions which trigger and inform differentiation, ion channels also provide a convenient and important functional
assay of the development of nerve and muscle. In that context Jaffe
implicates a GTP binding protein mediated release of inositol trisphosphate in the stimulation of cortical vesicle exocytosis by
fertilization. Lux describes the distribution of voltage-activated
calcium channels in embryonic neurons, and Kater illustrates how those
channels might regulate neuronal architecture through their effects on
growth cone behavior. Chow et al. discuss some of the inductive
interactions between nerve and target muscle cells that influence the
development of the synapse and neurotransmitter release; while Brehm
et al. analyze the role of innervation in regulating the expression and
distribution of acetylcholine receptors on embryonic skeletal muscle.
Finally Boyle and Kaczmarek describe a very profound effect of estrogen
on the expression of messenger RNA species encoding potassium channels,
which may have important implications for memory as well as development.

In the last section several new experimental and conceptual
approaches to ion channel function and modulation are grouped together.
Neher describes the apparatus he developed with Roger in Gottingen to
facilitate rapid internal perfusion of patch pipettes. Orkand reports
the evidence for a novel ion channel for bicarbonate in neuroglia, and
Westbrook and Mayer review the properties of glutamate channel block by
magnesium and its implications for cell pathology and plasticity in the
CNS. Connor evaluates recent approaches for imaging calcium inside cells
and DeRiemer et al. provide a preliminary catalogue of the ion channels
in isolated synaptic vesicles and synaptosomes. We have chosen to end
the volume with a paper from Kung's lab that brings Roger's research
interests full circle by applying the patch clamp technique to ion
channel function in three unicellular organisms, for which very powerful
genetic techniques have already been developed: paramecium, yeast and E.
coli!

We recognize that any collection of papers on ion channel modulation
can present only a limited snapshot of the field. However, we feel that
the work reported in this volume includes a significant fraction of what
is most exciting about the subject, and we consider it a tribute to Roger
that the contributors, almost without exception, were his former students
and close collaborators and friends.

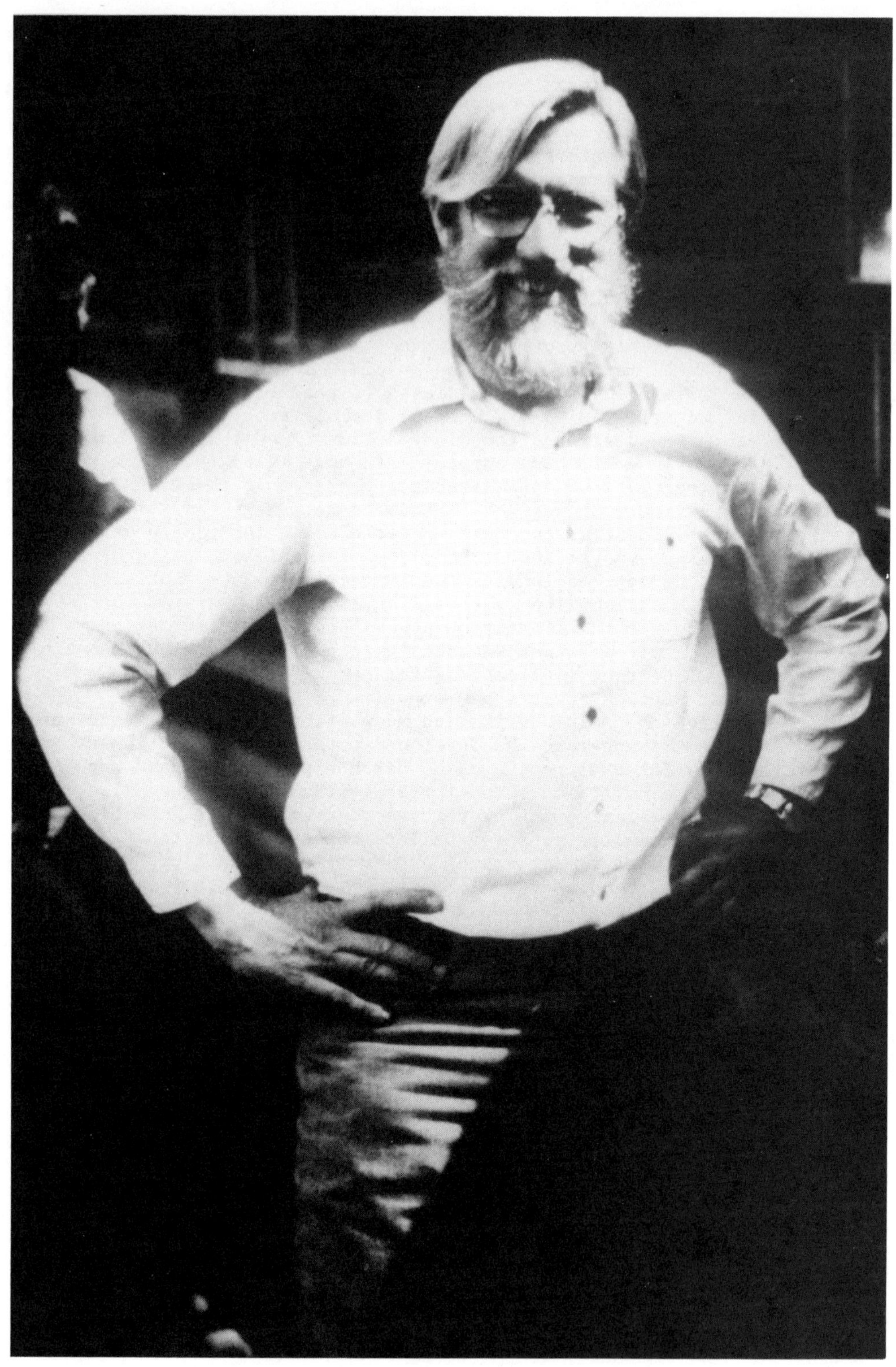

Roger at Friday Harbor, 1984

ROGER OTTO ECKERT
1934-1986

The study of ion channel modulation and the field of cellular neurobiology in general lost one of its luminaries on June 16, 1986 with the death of
Roger Eckert. Roger's many scientific contributions have provided valuable
insights into the properties of ion permeation channels, the metabolic and
ionic mechanisms which regulate their function, and their role in coupling
extracellular signals to cellular responses across the membrane. It is the
respect of his peers for the excellence of his work and the affection of
those who knew him and worked with him that have given rise to this volume.
    Roger Otto Eckert (b. Dec. 12, 1934) was awarded the Ph.D. degree by
Columbia University in 1960 for his work on crayfish stretch receptor
reflexes in the laboratory of E.S. Hodgson. He then undertook postdoctoral training at Harvard with John Welsh and at the Woods Hole Marine
Biological Laboratory in Steve Kuffler's summer course on nerve and
muscle cells in Woods Hole. Among his other projects during that period,
Roger was one of the first to demonstrate electrical coupling between
nerve cells, specifically the giant Retzius cells of the leech. For many
years Roger was a regular summer denizen of Woods Hole, and in later years of
the Friday Harbor Laboratories on San Juan Island in Puget Sound. Those who
accompanied him know how much he loves the summer atmosphere in those
laboratories and the opportunity to concentrate on research.
    In 1962 Roger accepted a faculty position at Syracuse University.
There he initiated an important series of investigations on the dinoflagellate, _Noctiluca_, in which he identified and characterized the bioelectric mechanisms controlling bioluminescence. That study led to one
of the first concise descriptions of the relationship between excitation
and luminescence. In the late 1960's Roger and Yutakah Naitoh and their
collaborators embarked on what is now a classic study of the bioelectric
control of behavior in a unicellular organism. They showed that the membrane around each _Paramecium_ contains both mechanically-activated and
voltage-activated calcium and potassium channels, like those in neuronal
membranes. Furthermore, they demonstrated that the animal's direction of
movement was controlled by the spatial segregation and function of those
channels. The work on _Paramecium_ continued in Los Angeles after 1968
when Roger joined the faculty of the University of California as
Professor of Biology. The ability to find mutants with sensory and loco-
motor dysfunctions made it possible to begin to dissect the electrical
properties of the membrane and to reveal general principles of bioelectric control of ciliary movement. That approach ultimately led,
through the work of his former student, Ching Kung, to one of the first
successful genetic approaches to ion channel function.
    The work on _Paramecium_ in Roger's lab also led to the first clear
demonstration that a certain class of voltage-activated calcium channels,
widely distributed throughout the animal kingdom, only inactivate when
channel activation results in the accumulation of calcium ions intracellularly. Thus, unlike voltage-activated sodium channels, inactivation
of these calcium channels is only voltage-dependent to the extent that
calcium ion entry depends on voltage. This novel finding, demonstrated
simultaneously on molluscan neurons in Roger's lab, introduced a previously unsuspected mechanism for channel inactivation that has helped

open the new and important chapter in neurobiology that is the subject of
this volume.

After that discovery, Roger and his students concentrated increasingly on the role of intracellular calcium in modulating ion channel
function. They studied its role in potassium channel activation and in
calcium channel inactivation; they measured its distribution with dyes
and developed computer models of its diffusion throughout the cell.
Ultimately they began to reveal some of the molecular reactions responsible for its actions. In fact it now appears that one class of voltage-
activated calcium channels is also phosphorylation-dependent, not responding to depolarization at all unless the channels have been phosphorylated by the cyclic AMP dependent-protein kinase. That finding led in
turn to the idea that the rise in internal calcium concentration produced
by channel activation shuts itself off by stimulating an endogenous
calcium-dependent phosphatase to dephosphorylate the channel and inactivate it. Although Roger pursued those problems with his characteristic
thoroughness, he never lost sight of their significance for neuronal
function. He and his students were carrying out parallel investigations
on the mechanisms by which neuropeptides activate these internal messenger systems to alter neuronal function in the _Aplysia_ nervous system.

Roger developed his interest in molluscan neurons as an experimental
preparation for studying ion channel function during several sabbatical
visits in the 1970's to Professor Dieter Lux's laboratory at the Max
Planck Institut fur Psychiatrie in Munich. Roger was very much an inter-
national scientist; the attendance at the Memorial Symposium in his honor
was evidence of that. His past students and postdoctoral collaborators
now hold important positions in Germany, England and Japan as well as the
United States. He himself spent significant lengths of time in outstanding German laboratories on several occasions. In fact, he was on sabbatical in Gottingen, working with Dr. Erwin Neher, until shortly before his
death.

With his many important contributions to our understanding of the
integration of ion channel activity, cellular metabolism and cell behavior, Roger was a central figure in the field of membrane biophysics.
His outstanding accomplishments were recognized by numerous awards, including the prestigious Alexander von Humboldt Senior Scientist Award on
two occasions (1974, 1985), a Jacob Javits Award (1984) crowning 20 years
of uninterrupted funding by both the NIH and NSF, and a Fogarty Inter-
national Award (1985). With evidence accumulating for a molecular mechanism to explain calcium dependent inactivation of voltage-activated
calcium channels, these awards were clearly just the beginning.

Roger was also a conscientious teacher, well known for the precision
of his drawings. He was widely acclaimed for his superb textbook with
David Randall, entitled _Animal Physiology: Mechanisms and Adaptations_,
published by W.H. Freeman and Co. That book, the third edition of which
is now in press, has been translated into four languages and has been
read by more than 50,000 students worldwide. Roger was also an exacting
mentor of the students in his laboratory. He had high standards and pursued them diligently, insisting on excellence from himself and his students. Meeting Roger's expectations paid off not only in his loyal support but in a lifetime of productive research habits, and there may be no
better memorial to his scientific accomplishments than the fact that so
many of his students were invited to participate in this symposium as
much for their pioneering work on ion channel modulation as for their
former association with his laboratory.

Roger is survived by his mother and his four sons: Kevin, Glenn,
Brian, and Tim, in all of whom he took great pride. He is also survived
by many friends and former collaborators who will miss his eagerness to
discuss science, his thoughtful candor and his irrepressible sense of
humor.

<u>Publications</u>

1. Eckert, R. (1961) Reflex relationships of the abdominal stretch receptors of the crayfish. I. Feedback inhibition of the receptors. <u>J. Cell. Comp. Physiol</u>. <u>57</u>:149-62.

2. Eckert, R. (1961) Reflex relationships of the abdominal stretch receptors of the crayfish. II. Stretch receptor involvement during the swimming reflex. <u>J. Cell. Comp. Physiol</u>. <u>57</u>:163-74.

3. Eckert, R. (1963) Electrical interaction of paired ganglion cells in the leech. <u>J. Gen. Physiol</u>. <u>46</u>:573-578.

4. Eckert, R. (1965) Bioelectric control of bioluminescence in the dinoflagellate <u>Noctiluca</u>. I. Specific nature of triggering events. <u>Science</u> <u>147</u>:1140-1142.

5. Eckert, R. (1965) Bioelectric control of bioluminescence in the dinoflagellate <u>Noctiluca</u>. II. Asynchronous flash initiation by a propagated triggering potential. <u>Science</u> <u>147</u>:1142-1145.

6. Eckert, R. (1966) Subcellular sources of luminescence in <u>Noctiluca</u>. <u>Science</u> <u>151</u>:349-352.

7. Eckert, R. (1966) Excitation and luminescence in <u>Noctiluca miliaris</u>. <u>Bioluminescence in Progress</u>, F.A. Johnson (ed.), Princeton University Press, pp. 269-300.

8. Eckert, R. and Reynolds, G.T. (1967) The subcellular origin of bioluminescence in <u>Noctiluca miliaris</u>. <u>J. Gen. Physiol</u>. <u>50</u>:1429-1458.

9. Eckert, R. (1967) The wave form of luminescence emitted by <u>Noctiluca</u>. <u>J. Gen. Physiol</u>. <u>50</u>:2211-2237.

10. Eckert, R. and Sibaoka, T. (1967) Bioelectric regulation of tentacle movement in a dinoflagellate. <u>J. Exptl. Biol</u>. <u>47</u>:433-446.

11. Sibaoka, T. and Eckert, R. (1967) An electrophysiological study of the tentacle-regulating potentials in <u>Noctiluca</u>. <u>J. Exptl. Biol</u>. <u>47</u>:447-458.

12. Eckert, R. and Sibaoka, T. (1968) The flash triggering action potential of the luminescent dinoflagellate <u>Noctiluca</u>. <u>J. Gen. Physiol</u>. <u>52</u>:258-282.

13. Naitoh, Y. and Eckert, R. (1968) Electrical properties of <u>Paramecium caudatum</u>: Modification by bound and free cations. <u>Zeitschr. vergl. Physiol</u>. <u>61</u>:427-452.

14. Naitoh, Y. and Eckert, R. (1968) Electrical properties of <u>Paramecium caudatum</u>: All-or-none electrogenesis. <u>Zeitschr. vergl. Physiol</u>. <u>61</u>:453-472.

15. Naitoh, Y. and Eckert, R. (1969) Ionic mechanisms controlling behavioral responses of <u>Paramecium</u> to mechanical stimulation. <u>Science</u> <u>164</u>:963-965.

16. Naitoh, Y. and Eckert, R. (1969) Ciliary orientation: Controlled by cell membrane or by intracellular fibrils? <u>Science</u> <u>166</u>:1633-1635.

17. Eckert, R. and Naitoh, Y. (1970) Passive electrical properties of
Paramecium and problems of ciliary coordination. J. Gen. Physiol.
55:467-483.

18. Eckert, R. and Murakami, A. (1971) Control of ciliary activity. In
Contractility of Muscle Cells and Related Processes, R.R. Podolsky (ed.),
Prentice-Hall, Englewood, N.J..

19. Kung, C. and Eckert, R. (1972) Genetic modification of electric
properties in an excitable membrane. PNAS 69:93-97.

20. Naitoh, Y. and Eckert, R. (1972) Electrophysiology of ciliate
protozoa. In Experiments in Physiology and Biochemistry, Vol. V G.A.
Kerkut (ed.), Academic Press, London.

21. Murakami, A. and Eckert, R. (1972) Cilia: Activation coupled to
mechanical stimulation by Ca influx. Science 175:1375-1377.

22. Eckert, R. and Naitoh, Y. (1972) Bioelectric control of locomotion
in the ciliates. J. Protozool. 19:237-243.

23. Naitoh, Y., Eckert, R. and Friedman, K. (1972) A regenerative
calcium response in Paramecium. J. Exptl. Biol. 56:667-681.

24. Eckert, R., Naitoh, Y. and Friedman, K. (1972) Sensory mechanisms in
Paramecium. I. Two components of the electric response to mechanical
stimulation of the anterior surface. J. Exptl. Biol. 56:683-694.

25. Eckert, R. (1972) Bioelectric control of ciliary activity. Science
176:473-481.

26. Eckert, R. (1972) Electric mechanisms controlling locomotion in the
ciliated protozoa. In Behavior of Microorganisms, A. Perez-Miravete
(ed.), Plenum, London, pp. 117-126.

27. Eckert, R. and Murakami, A. (1972) Calcium dependence of ciliary
activity in the oviduct of the salamander Necturus. J. Physiol.
226:699-711.

28. Naitoh, Y. and Eckert, R. (1973) Sensory mechanisms in Paramecium.
II. Ionic basis of hyperpolarizing mechanorecptor potential. J. Exptl.
Biol. 59:53-65.

29. Friedman, K. and Eckert, R. (1973) Ionic and pharmacological
modification of input resistance and excitability in Paramecium. Comp.
Biochem. Physiol. 45A 101-114.

30. Machemer, H. and Eckert, R. (1973) Electrophysiological control of
reversed ciliary beating in Paramecium. J. Gen. Physiol. 61:572-587.

31. Epstein, M. and Eckert, R. (1973) Membrane control of ciliary
activity in the protozoan Euplotes. J. Exptl. Biol. 58:437-462.

32. Lux, H.D. and Eckert, R. (1974) Inferred slow inward current in
snail neurones. Nature 250:574-576.

33. Murakami, A., Machemer-Rohnisch, S. and Eckert, R. (1974)
Stimulation of ciliary activity by low levels of extracellular adenine
nucleotides in the amphibian oviduct. Exptl. Cell Res. 85:154-158.

34. Naitoh, Y. and Eckert, R. (1974) The control of ciliary activity in _Protozoa_. In _Cilia and Flagella_, M.A. Sleigh (ed.), Academic Press, pp. 305–352.

35. Eckert, R. and Lux, H.D. (1975) A non-inactivating inward current recorded during small depolarizing voltage steps in snail pacemaker neurons. _Brain Res._ _83_:486–489.

36. Eckert, R. and Machemer, H. (1975) Regulation of ciliary beating frequency by the surface membrane. In _Molecules and Cell Movement_ (Symp. Soc. Gen. Physiol.), R. Stephen and S. Inoue (eds.), Raven Press, New York, pp. 151–163.

37. Machemer, H. and Eckert, R. (1975) Ciliary frequency and orientational responses to clamped voltage steps in _Paramecium_. _J. Comp. Physiol._ _104_:247–260.

38. Eckert, R., Naitoh, Y. and Machemer, H. (1976) Calcium in the bioelectric and motor functions of _Paramecium._ Soc. Exptl. Biol. Symp. _30_:233–255.

39. Eckert, R. and Lux, H.D. (1976) A voltage-sensitive persistent calcium conductance in neuronal somata of _Helix_. _J. Physiol._ _254_:129–151.

40. Schmidt, J.A. and Eckert, R. (1976) Calcium couples flagellar reversal to photostimulation in _Chlamydomonas reinhardtii_. _Nature_ _262_:713–715.

41. Machemer, H. and Eckert, R. (1976) Electromechanical coupling of ciliary activity in _Paramecium_. A film analysis. In _Fortschritte der Zoologie_, _24_:211–215.

42. Eckert, R., Tillotson, D. and Ridgway, E.B. (1977) Voltage-dependent facilitation of Ca^{2+} entry in voltage-clamped aequorin-injected molluscan neurons. _Proc. Natl. Acad. Sci. USA._ _74_:1748–1752.

43. Eckert, R. and Lux, H.D. (1977) Ca-dependent depression of a late outward current in snail neurons. _Science_ _197_:472–475.

44. Eckert, R. (1977) Genes, channels and membrane currents in _Paramecium_. _Nature_ _268_:104–105.

45. Brehm, P., Dunlap, K. and Eckert, R. (1978) Ca-dependent repolarization in _Paramecium_. _J. Physiol._ _274_:639–654.

46. Eckert, R. and Tillotson, D. (1978) Potassium activation associated with intraneuronal free Ca. _Science_ _200_:437–439.

47. Eckert, R. and Randall, D. (1978) _Animal Physiology_, 1st edition, W.H. Freeman and Co., San Francisco.

48. Brehm, P. and Eckert, R. (1978) An electrophysiological study of the regulation of ciliary beat frequency in _Paramecium_. _J. Physiol._ _283_:557–568.

49. Brehm, P. and Eckert, R. (1978) Calcium entry leads to inactivation of calcium channel in _Paramecium_. _Science_ _202_:1203–1206.

50. Eckert, R. (1979) _Paramecium_ as a model for genetic approaches to neurobiology. In _Mutations, Biology and Society_, D.N. Walcher, N. Kretchmer and H.L. Barnett (eds.), Masson, New York.

51. Eckert, R. and Tillotson, D. (1979) Ca entry and the Ca-activated potassium systems of molluscan neurons: Voltage clamp studies on Aequorin-injected cells. In <u>Detection and Measurement of Free Ca in Cells</u>, C.C. Ashley and D.K. Campbell (eds.), Elsevier, Amsterdam.

52. Eckert, R. and Brehm, P. (1979) Ionic mechanisms of excitation in <u>Paramecium</u>. <u>Ann. Rev. Biophys. Bioeng.</u> 8:353-383.

53. Brehm, P., Eckert, R. and Tillotson, D.L. (1980) Calcium-mediated inactivation of calcium current in <u>Paramecium</u>. <u>J. Physiol.</u> <u>306</u>:193-203.

54. Eckert, R. and Tillotson, D.L. (1981) Calcium-mediated inactivation of the Ca conductance in caesium-loaded giant neurones of <u>Aplysia californica</u>. <u>J. Physiol.</u> <u>314</u>:265-280.

55. Eckert, R., Tillotson, D.L. and Brehm, P. (1981) Calcium-mediated control of Ca and K currents. <u>Fedn. Proc.</u> <u>40</u>:2226-2232.

56. Eckert, R. (1981) Calcium-mediated inactivation of voltage-gated Ca channels. In <u>The Mechanisms of Gated Calcium Transport Across Biological Membranes</u>, S.T. Ohnishi and M. Endo (eds.), Academic Press, New York.

57. Eckert, R. and Ewald, D. (1982) Residual calcium ions depress activation of calcium-dependent current. <u>Science</u> <u>216</u>:730-733.

58. Eckert, R. and Randall, D. (1983) <u>Animal Physiology</u>: <u>Mechanisms and Adaptations</u>, 2nd Ed., W.H. Freeman and Co., San Francisco.

59. Eckert, R., Ewald, D. and Chad, J. (1983) Calcium-mediated inactivation of calcium current in neurons of <u>Aplysia californica</u>. In <u>The Physiology of Excitable Cells</u>, A. Grinnell and W. Moody (eds.), Alan Liss Inc., New York.

60. Eckert, R. and Ewald, D. (1983a) Calcium tail currents in voltage-clamped intact nerve cell bodies of <u>Aplysia californica</u>. <u>J. Physiol.</u> <u>345</u>:533-548.

61. Eckert, R. and Ewald, D. (1983b) Inactivation of calcium conductance characterized from tail current measurements in neurones of <u>Aplysia californica</u>. <u>J. Physiol.</u> <u>345</u>:549-565.

62. Ewald, D. and Eckert, R. (1983) Cyclic AMP enhances calcium-dependent potassium current in <u>Aplysia</u> neurons. <u>Cell. Mol. Neurobiol.</u> <u>3</u>:345-353.

63. Augustine, G.J. and Eckert, R. (1984) Divalent cations differentially support transmitter release at the squid giant synapse. <u>J. Physiol.</u> <u>346</u>:257-271.

64. Chad, J., Eckert, R. and Ewald, D. (1984) Kinetics of calcium-dependent inactivation of calcium current in voltage-clamped neurones of <u>Aplysia californica</u>. <u>J. Physiol.</u> <u>347</u>:279-300.

65. Eckert, R. and Chad, J. (1984) Inactivation of calcium channels. <u>Prog. Biophys. Molec. Biol.</u> <u>44</u>:215-267.

66. Chad, J. and Eckert, R. (1984) Calcium domains associated with individual channels can account for anomalous voltage relations of Ca-dependent responses. <u>Biophys. J.</u> <u>45</u>:993-999.

67. Deitmer, J.W. and Eckert, R. (1985) Two components of Ca-dependent potassium current in identified neurones of _Aplysia californica_. _Pflügers Arch._ _403_:353-359.

68. Eckert, R. and Chad, J. (1985) Calcium-dependent regulation of Ca channel inactivation. in _Neural Mechanisms of Conditioning_, D. Alkon and C. Woody (eds.), Plenum, New York.

69. Eckert, R. and Chad, J. (1985) Mechanism for calcium-dependent inactivation of Ca current. Exptl. Brain Res. _14_:35-50.

70. Chad, J. and Eckert, R. (1986) An enzymatic mechanism for calcium current inactivation in dialyzed _Helix_ neurones. _J. Physiol._ _378_:31-51.

71. Eckert, R., Chad, J. and Kalman, D. (1986) Enzymatic regulation of the calcium current. _J. Physiol._ (Paris) _81_:318-384.

72. Brezina, V., Eckert, R. and Erxleben, C. (1986) Modulation of K conductances by endogenous neuropeptide in neurones of _Aplysia Californica_. _J. Physiol._ _382_:267-290.

73. Brezina, V., Eckert, R. and Erxleben, C. (1987) Suppression of calcium current by an endogenous neuropeptide in neurones of _Apylsia californica_. _J. Physiol._ _388_:565-595.

74. Armstrong, D. and Eckert, R. (1987) Voltage-activated calcium channels that must be phosphorylated to respond to membrane depolarization. _Proc. Natl. Acad. Sci._ (USA) _84_:2518-2522.

75. Neher, E. and Eckert, R. (1987) Fast patch pipette internal perfusion with minimum solution flow. In _Calcium and Ion Channel Modulation_, Grinnell, A., Armstrong, D. and Jackson, M., editors, Plenum Press, New York.

Alan Grinnell and
Dick Orkand, standing;
Roger Eckert, seated,
circa 1970

Above: The Eckert lab, 1985, Daniel Kalman, Behnam Varasteh,
Vladimir Brezina, Christian Erxleben, Roger, John Chad,
Anton Hermann, David Armstrong.

SECTION 1

VOLTAGE-ACTIVATED CALCIUM CHANNELS

CALCIUM CHANNEL DIVERSITY

Lou Byerly* and Susumu Hagiwara[+]

*Neurobiology Section
Department of Biological Sciences
University of Southern California
Los Angeles, California 90089-0371

[+]Department of Physiology
Ahmanson Laboratory of Neurobiology
Brain Research Institute and
Jerry Lewis Neuromuscular Research Center
University of California, School of Medicine
Los Angeles, California 90024

INTRODUCTION

In recent years a great deal of evidence has accumulated that demonstrates the large amount of diversity that exists between Ca channels. By Ca channels we mean membrane pores that are opened by depolarization, allow Ca^{2+} to flow down its electrochemical gradient when the channel is open and show a standard selectivity between different divalent cations. Ca^{2+}, Ba^{2+} and Sr^{2+} can pass through Ca channels, while Co^{2+} and Mg^{2+} cannot. Cd^{2+}, Co^{2+}, Ni^{2+} and Mn^{2+} block the channel, but some of the blockers, e.g., Mn^{2+}, can themselves carry current through the channel. This definition of Ca channels excludes several interesting channels which have been shown to allow Ca^{2+} to enter cells.

In accord with the theme of this symposium we are going to focus on the diversity of Ca channels that is indicated by modulation. This subject is fundamentally connected with two other phenomena found with Ca currents, "washout" and inactivation. Roger Eckert's laboratory led the world in demonstrating the relation between Ca current inactivation and accumulation of intracellular Ca^{2+} . As will be demonstrated by the following talks of this symposium, Roger and his colleagues became convinced that inactivation was a part of the broader subject of modulation of Ca channels by intracellular messengers and that it was closely associated with the annoying tendency of Ca currents to disappear when cells were dialyzed internally. Much of the conflicting data and confusion in this field can probably be credited to the diversity of Ca channels. We will take advantage of the many cells throughout the animal kingdom in which more than one type of Ca channel coexist to establish the existence of two fundamental classes of Ca channels. Then we will propose a number of questions related to modulation which may define important subclasses of the two fundamental classes.

CELLS WITH MULTIPLE TYPES OF CA CURRENT

When we reviewed Ca currents seven years ago, the only clear example
of two different types of Ca channels in one cell was the starfish egg
(Hagiwara et al., 1975). However, since then, especially in the last
three years, the co-existence of multiple types of Ca channels has been
demonstrated for a large number of cells. Comparative studies of channels
in the same membrane are especially valuable in establishing intrinsic
difference between channels, because it is clear that the channels are
exposed to the same environment and studied by the same techniques. In
this sense, these cells which contain more than one type of Ca channel are
natural occurrences of the much more difficult experiments now being begun
in which intrinsic channel properties are compared by incorporating
different purified channels into artificial bilayers (Rosenberg et al.,
1986) or by injecting mRNA for different channels into _Xenopus_ oocytes
(Dascal et al., 1986). These studies on cells with more than one type of
Ca current reveal that there are two classes of channels which are found
throughout the animal kingdom. We will refer to them as Type I and Type
II, as was done in the first such study on starfish egg and in several of
the subsequent studies. We will define Type I in all examples to be the
type of channel that activates at more negative potentials. Figure 1
lists a number of examples of cells in which two different Ca currents
have been found and studied in sufficient detail to characterize the
channels. The examples range from the ciliate Stylonychia (Deitmer, 1984
and 1986) to vertebrate nerve, muscle and secretory cells. The list is
not complete; many additional types of vertebrate neurons and several
types of smooth muscle have also been found to have two types of Ca
current. Also the references given are quite incomplete; in each case

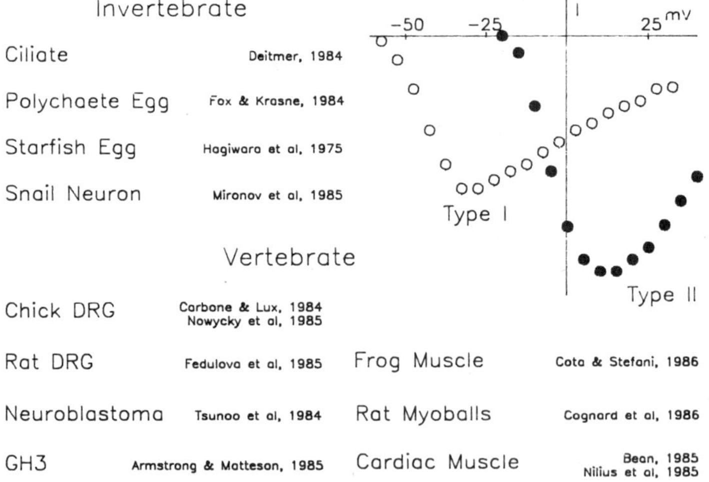

Fig. 1. A partial list of cells in which both Type I and Type II Ca
channels coexist. One or two references which characterize
the two types of channels are given for each cell.
Idealized data illustrate typical current-voltage
relationships for the two types of whole-cell Ca currents.

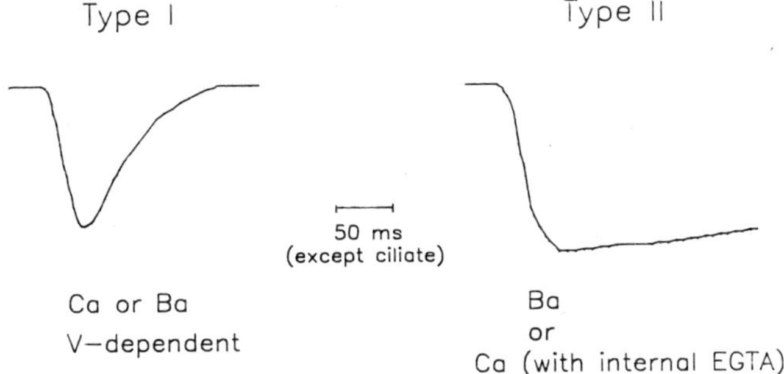

Fig. 2. Idealized current records illustrate the typical inactivation observed for Type I and Type II Ca currents when the membrane potential is stepped to and held at a level that activates the Ca current. The inactivation of the Type I current is the same for Ca^{2+} or Ba^{2+} as current carrier; consequently the inactivation of the Type I channel is probably voltage−dependent. The rate of inactivation of the Type II current is usually slower for Ba current than for Ca current, and usually internal EGTA greatly reduces the rate of inactivation for Type II Ca current. The slower inactivation of Type II channels relative to Type I holds for all the cells of Fig. 1 except snail neurons and frog muscle. Typically the time constant for inactivation of Type I currents is on the order of 50 ms, except for the ciliate where it is a few milliseconds.

only one to two studies are cited. A characterization of typical current−voltage curves for Type I and Type II currents is also given in Figure 1. Typically Type I currents activate at potentials about 40 mV more negative than do Type II currents. Not surprisingly, the absolute voltages for activation vary greatly among the examples and are very dependent on the external concentration of divalent cations. Nowycky and collaborators (1985) have identified three types of Ca currents in chick dorsal root ganglion neurons. We take their T type for Type I and their L type for Type II channels; their N type seem to be basically Type I although they activate at more positive potentials than do the T Type.

The interesting point is that these two classes of Ca current (Type I and Type II) have common properties throughout much of the broad range of examples given. The second property (activation potential being the first) that distinguishes Type I and Type II currents is inactivation. As illustrated in Figure 2, Type I currents typically inactivate (during a maintained depolarization) at least ten times faster than Type II currents. There are two exceptions: Type I Ca currents in frog skeletal muscle exhibit no decline during a 2 s depolarization (Cota and Stefani, 1986), and the inactivation of Type I snail Ca current has about the same

Type I   Demonstrated for:

$I_{Ba} \leq I_{Ca}$ Polychaete Egg   Exceptions:
                        Starfish Egg   Snail Neuron
                        DRG Neurons   Neuroblastoma
Type II   GH3

$I_{Ba} > I_{Ca}$ Cardiac Muscle

Fig. 3.   Typically Ca^{2+} passes through Type I channels as well as
          (or better than) Ba^{2+} does, while Ba^{2+} passes through
          Type II channels better than Ca^{2+} does. This
          generalization has been shown to hold for polychaete egg
          cells, starfish egg cells, DRG neurons, GH3 cells and
          cardiac muscle. In snail neurons and neuroblastoma cells
          Ba^{2+} passes through both types of Ca channel better than
          does Ca^{2+} .

time course as the Type II current (Mironov et al., 1985). The prolonged
nature of the Type II current is observed under conditions which reduce
the accumulation of intracellular Ca^{2+} ; either intracellular Ca^{2+} is
buffered by EGTA or Ba^{2+} is substituted for Ca^{2+} as current carrier.
In the microelectrode studies of the ciliate Stylonychia, the Type II
current when carried by Ca^{2+} inactivates almost as rapidly as Type I
current, but when Ba^{2+} is substituted for Ca^{2+}, the rate of
inactivation of the Type II current is greatly reduced (Deitmer, 1984).
In contrast, the inactivation of Type I current is the same for Ca^{2+} and
Ba^{2+} in all cases where Ca^{2+} has been replaced by Ba^{2+} . This and
other evidence suggest that the inactivation of Type I Ca current is
voltage-dependent. The Type I current usually inactivates with a time
constant on the order of 50 ms, except for ciliate where it inactivates
faster. It is interesting that the inactivation of Ca current in Ascidian
egg is voltage-dependent (like Type I), but in the cleavage-arrested
embryo the Ca current is current-dependent (like Type II) (Hirano and
Takahashi, 1984). Also the selectivity of the Ca current changes in a way
(see below) that supports the idea that Type I Ca channels in the Ascidian
egg are replaced by Type II Ca channels in the embryo.

The selectivity between Ba^{2+} and Ca^{2+} is also characteristic of
the two types of Ca currents. Selectivity is usually measured by
determining permeabilities from reversal potentials for currents carried
by different species of ions (Hille, 1975), but it has only been possible
to determine reversal potentials for the Ca channel in a few preparations
due to contamination by other currents. However, the relative sizes of
currents carried by Ca^{2+} and by Ba^{2+} after equimolar replacement are
easily measured and indicate a consistent difference in selectivity
between Type I and Type II channels. For the types of cells listed in
Figure 3, the Ba currents carried by Type I channels are less than or
roughly equal to the currents carried by equimolar Ca^{2+}, while the Ba
currents carried by Type II channels are consistently larger than the Ca
currents. Only two exceptions have been demonstrated to this generali-

zation. In neuroblastoma cells the ratio of Ca current to Ba current is
0.6 for Type I and 0.3 for Type II (Yoshii et al., 1985), so that Ba
carries larger currents than does Ca through both channels but still
follows the rule that the relative permeability of Ba to Ca is greater for
Type II channel. Mironov and coworkers (1985) reported that the ratio of
Ba currents to Ca currents was 1.7 for both types of channels in snail.

K_d for Neuroblastoma   Tsunoo et al, 1985

	Co	Cd
Type I	160uM	160uM
Type II	560uM	7uM

Co Block

    Type II > Type I

Ciliate
Polychaete Egg
Starfish Egg

    Type II ≃ Type I

Rat DRG
Neuroblastoma
Cardiac Muscle (Atrial)

Cd Block

    Type II > Type I

Ciliate
Polychaete Egg
Chick DRG
Neuroblastoma
Cardiac Muscle (Ventricular)

    Type II ≃ Type I

Rat DRG

Fig. 4. The two types of Ca channels have different sensitivities
to the divalent blockers. Co^{2+} blocks Type II channels
better (at lower concentrations) than it does Type I
channels in ciliate and invertebrate egg cells, but it
blocks both types about the same in vertebrate cells.
Cd^{2+} blocks Type II channels better than Type I channels
in almost all the cells on which it has been tested.

The two types of Ca channels have also been distinguished by their
sensitivity to the inorganic blockers Cd^{2+} and Co^{2+} . In most cases
where a difference has been noted the Type II channel is found to be more
sensitive (see Figure 4). For the invertebrate cells, ciliate, polychate
egg and starfish egg, Co^{2+}, as well as Cd^{2+} (if tried), blocks Type II
at lower concentrations than required to block Type I. However, for the
vertebrate cells only Cd^{2+} is found to be a more effective blocker for
Type II. Since it has been shown that the Cd^{2+} block of snail Ca
currents is voltage-dependent being much more effective at positive
potentials (Byerly et al., 1984), it is not clear if the insensitivity of
vertebrate Type I channels is due to intrinsic channel properties or due
to the fact that the Type I channel is usually studied at more negative
potentials. The block by Co^{2+} is not voltage-dependent; so the Type II
channel of egg cells and ciliate is more sensitive to inorganic blockers

Dihydropyridine (DHP) Sensitivity	Stability During Internal Perfusion
Only Type II channels are sensitive to DHP	Only Type II channels exhibit "Washout"
Chick DRG GH3 Cardiac Muscle Rat Myoballs	Snail Neurons Rat DRG Neurons GH3 Cardiac Muscle (Ventricular)
Exception: Snail Neurons	

Fig. 5.   Only Type II Ca channels are sensitive to dihydropyridines and exhibit washout. In the vertebrate cells listed the Type II channels are much more sensitive to dihydropyridines than are the Type I channels. Dihydropyridines do not appear to have selective effects on the Ca current of snail neurons or other invertebrates. Type II Ca channels are much less stable than Type I channels when exposed to internal perfusion; this has been demonstrated for snail neurons, rat DRG neurons, GH3 cells and ventricular cardiac muscle.

than the Type I channel. At present that generalization does not appear to also apply to vertebrate cells. As the dissociation constants (K_d) given in Figure 4 show, in neuroblastoma Cd^{2+} blocks Type II channels much more effectively than Type I channels, but Co^{2+} actually blocks Type I a little better than Type II.

Type I and Type II Ca channels in vertebrate cells can also be distinguished by their sensitivity to dihydropyridines (DHP). Only Type II channels are found to be DHP-sensitive (See Figure 5); no effects have been noted on Type I channels. Both DHP antagonist (nifedipine, nitredipine, nimodipine) and agonist (Bay K8644) are found to be effective at micromolar concentrations on Type II channels of vertebrate muscle (Bean, 1985; Nilius et al., 1985; Cognard et al., 1986). Bay K8644 strongly increases open probability of Type II channels in chick DRG neurons (their L type; Nowycky et al., 1985), but neuronal Ca channels show less sensitivity to DHP antagonist than do muscle Ca channels (McClesky et al., 1986). DHP-sensitivity does not distinguish between Type I and Type II channels in invertebrate cells, largely because it is not clear that dihydropyridines have any selective effect on Ca channels in invertebrate cells. Mironov and collaborators (1985) found that nifedipine blocked the Type I and Type II Ca channels of snail neurons about equally, with a K_d of 60 um. At such high concentrations it is doubtful that the effect is specific. In _Aplysia_ bag cells 1–10 um nifedipine blocks K as well as Ca currents (Nerbonne and Gurney, 1987). Other investigators have found no effects of high levels of dihydropyridine on molluscan Ca currents (Byerly et al., 1984; Augustine

et al., 1987). We are not aware of studies on any invertebrate cell which
demonstrate a specific action of dihydropyridine on Ca channels.
Nifedipine (10 um) has no effect on Ca currents of _Drosophila_ neurons
(Hung-Tat Leung and Lou Byerly, unpublished results) or of Paramecium
(Ehrlich et al., 1987).

When the internal side of the membrane is dialyzed against a
low-calcium saline solution, Type II Ca currents "wash out" while Type I
currents are much more stable (Figure 5). This difference in stability is
found for all internally dialyzed cells which have both types of Ca
current, with the possible exception of DRG neurons. Carbone and Lux
(1984) did not see a washout of Ca currents in excised outside-out patches
from chick and rat DRG, but washout of Ca current was seen in whole cell
studies of chick (Forscher and Oxford, 1985) and rat (Fedulova et al.,
1985) DRG cells. For cardiac ventricular cells Type II Ca channels
disappear irreversibly a few minutes after excision of inside-out patches,
while Type I channels showed no decrease in activity for many minutes
after excision (Nilius et al., 1985). Since this washout of Ca currents
can be greatly reduced by adding ATP and Mg^{2+} (and cAMP in some cases)
(Fedulova et al., 1985; Forscher and Oxford, 1985; Byerly and Yazejian,
1986; Chad and Eckert, 1986), the concept emerged that Type II Ca channels
are "metabolically dependent" and Type I channels are "metabolically
independent" (Fedulova et al., 1985).

HOW GENERAL ARE CERTAIN PROPERTIES OF TYPE II CHANNELS?

Type I and Type II Ca channels do share the same basic permeation
mechanism. Both show the same selectivity between divalent cations that
is commonly accepted for the definition of Ca channels (see
Introduction). The selectivity of Ca channels for divalent cations over
monovalent cations appears to depend on the high affinity binding of the
permeant divalent cation to the channel. This has been demonstrated both
for the predominantly Type II Ca channels of snail neurons (Kostyuk et
al., 1983), cardiac muscle (Hess and Tsien, 1984), and skeletal muscle
(Almers and McClesky, 1984) and for the apparently Type I Ca channel of B
lymphocytes (Fukushima and Hagiwara, 1985). Presumably the small
differences between Type I and Type II channels in relative permeability
of Ca^{2+} and Ba^{2+} and of sensitivity to Co^{2+} block can be explained
by small differences in binding affinities.

However, Type I channels do not appear to share with Type II the
three properties on which we wish to focus in this paper, Ca-dependent
block (or inactivation), modulation by phosphorylation, and washout.
Therefore, we will restrict our attention to Type II Ca currents and
examine how general each property is. The correlation between the
presence of these properties in different Type II channels should be
helpful in understanding the relationships between the three properties.
The mechanisms underlying the three phenomena are not known for certain.
For example, it is not known what actually gets phosphorylated when Ca
channels are modulated by internal phosphorylation.

Washout

As discussed in the previous section, washout seems to be a common
property of Type II Ca channels whenever the internal environment of the
channel is dialyzed against simple saline solutions. The phenomena has
been well documented for snail neurons (Doroshenko et al., 1982; Byerly
and Yazejian, 1986; Chad and Eckert, 1986), rat and chick DRG neurons
(Fedulova et al., 1985; Forscher and Oxford, 1985), secretory cells
(Fenwick et al., 1982; Armstrong and Matteson, 1985), cardiac muscle

	Positive ●		Negative ○	
Invertebrate	Blocked by Intracellular Ca^{2+}	Washout	DHP Sensitive	Increased by cAMP–dependent Phosphorylation
Protozoa	●	?	○	?
Polychaete Egg	○	?	?	?
Insect Nerve	?	●	○	?
Arthropod Muscle	●	?	?	?
Molluscan Nerve	●	●	◐	◐
Vertebrate				
Neuronal	●	●	●	◐
Secretory	●	●	●	●
Cardiac Muscle	●	●	●	●
Skeletal Muscle	?	?	●	●
Smooth Muscle	●	?	●	○

Fig. 6. Type II Ca channels throughout the animal kingdom are
characterized according to four properties, sensitivity to
(1) block by intracellular Ca^{2+}, (2) washout, (3)
dihydropyridines (DHP) and (4) enhancement by cAMP–dependent
phosphorylation. Filled circles indicate that the property
has been observed in Type II Ca channels from that class of
cells. Open circles indicate that there is evidence that
the class of cells do not have this property. Half–filled
circles indicate that the property is true for some cells of
that class, but not for others. Question marks indicate
that we are not aware of studies of this property in this
class of cells.

(Nilius et al., 1985), and has been described by numerous other
laboratories. However, there are several laboratories which do not
observe the washout phenomena with Type II Ca channels. While some of
these might be due to limited dialysis of the internal environment, that
certainly can't explain the absence of washout of the Type II Ca current
found in the squid giant axon (Dipolo et al., 1983; Francisco Bezanilla,
personal communication). Nevertheless, we indicate in Figure 6 that
washout is a property of Type II Ca channels in molluscan neurons,
vertebrate neurons, secretory cells and cardiac muscle. Washout has not
been reported in other invertebrate cells because very few of these have
been internally dialyzed. The Type II Ca current of _Drosophila_ neurons
does appear to washout (Hung-Tat Leung and Lou Byerly, unpublished data).
We are unaware of any reports on the stability of Ca current in skeletal
or smooth muscle.

Dihydropyridine agonists appear to delay washout of Type II Ca
channels in excised patches from GH3 cells (Armstrong et al., 1987).
Also, the channels for which increases of activity by cAMP–dependent
phosphorylation have been best established are sensitive to

dihydropyridines. Therefore, we also indicate in Figure 6 the
distribution of DHP-sensitivity among the range of preparations whose Ca
channels have been studied in detail. As discussed previously, selective
effects by dihydropyridines and enhancement by dihydropyridines agonists
in particular have been only demonstrated for vertebrates.

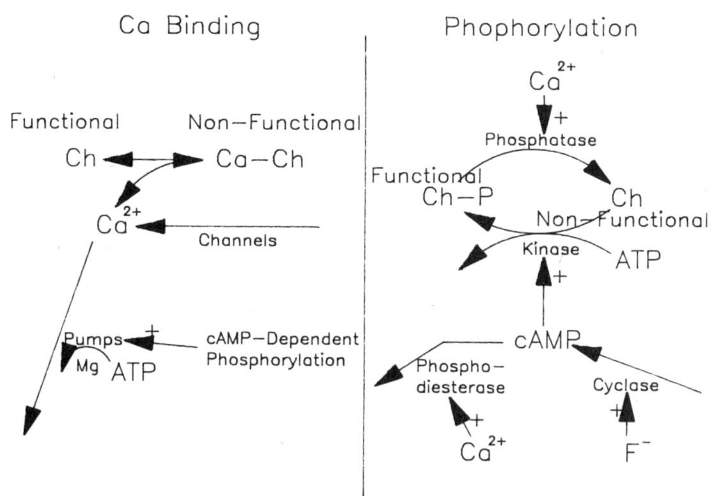

Fig. 7.   Two models for the modulation of Ca channels activity
         underlie much of the thinking about inactivation, washout,
         and control by external hormones and neurotransmitters. In
         the Calcium Binding model (shown on left) intracellular
         Ca^{2+} binds directly to the channel (Ch) and blocks it.
         The concentration of intracellular Ca^{2+} depends on the
         balance between influx of Ca^{2+} through channels and
         removal of Ca^{2+} by pumps, which require ATP and may be
         controlled by cAMP-dependent phosphorylation. In the
         cAMP-dependent Phosphorylation model (shown on right) the
         channel must be phosphorylated to function. Intracellular
         Ca^{2+} acts on enzymes which control the extent of
         phosphorylation of the Ca channels. Arrows indicate fluxes;
         an arrow with a plus sign indicates positive modulation
         (enhancement of activity) of the enzyme at the point.

Block by Intracellular Calcium

The block of Ca channels by intracellular Ca^{2+} has been best
demonstrated in molluscan neurons, both by maintained increases in
[Ca^{2+}] i (Plant et al., 1983; Byerly and Moody, 1984), and by Ca current
inactivation (Tillotson, 1979; Eckert and Tillotson, 1981). The original
demonstration of the block of Ca spikes by intracellular Ca^{2+} was done
on barnacle muscle (Hagiwara and Nakajima, 1966). Ca-dependent

inactivation has also been well demonstrated for paramecium (Brehm and
Eckert, 1978), stick insect muscle (Ashcroft and Stanfield, 1982), and
cardiac muscle (Tsien and Marban, 1982). Evidence for Ca-dependent
inactivation has also been published for chick dorsal root ganglion cells
(Dunlap and Fischbach, 1981), GH3 cells (Kalman et al., 1987) and
guinea-pig smooth muscle cells (Ganitkevich et al., 1986). One reason for
the lack of more evidence is that many cells have only been studied by
techniques which involve dialyzing the inside of the cell against
Ca-buffered solution (high EGTA). There may be Type II Ca channels that

Vertebrate Cells					Molluscan Neurons			
Type	Effect	Internal	External		Type	Effect	Internal	External
Cardiac Muscle	+	cAMP	NE		Aplysia	+	cAMP	5HT
Reuter & Scholz, 1977; Bean et al, 1984					Pellmar, 1981			
GH3	+	cAMP			Helix	+	cAMP	
Armstrong & Eckert, 1987					Doroshenko et al, 1982; Chad & Eckert, 1986			
Neuro-blastoma	+	cAMP			Aplysia Bag Cells	+	Kinase C	
Tsunoo et al, 1984					Kaczmarek, 1986			
Smooth Muscle	−	cAMP			Helix			
					D2	−	Ca	CCK8
Saida & van Breemen, 1985					D2,E2	−	?	DA
					D6,D7	+	cGMP	5HT
Rat SCG Neurons	−	?	NE		D6,D7	0	cAMP	
Horn & McAfee, 1980					Gerschenfeld et al, 1986			
					Aplysia Sensory	0	cAMP	5HT
Chick DRG Neurons	−	Kinase C	NE				Klein & Kandel, 1980	
					Lymnaea	0	cAMP	
Rane & Dunlap, 1986					Byerly & Yazejian, 1986			

Fig. 8. The activity of Type II Ca channels are modulated by
external and internal messengers. The results of studies of
such modulation on a number of vertebrate (left) and
molluscan (right) cells are summarized in four columns. The
first column gives the cell type. The second column
indicates if the messenger increases (+), decreases (−) or
does not affect (zero) the whole-cell Ca current. The third
column gives the intracellular messenger and the fourth
column gives the extracellular messenger. In some cases
only the external or only the internal messenger is known.

are not blocked by intracellular Ca^{2+}. In the polychate egg the rate of
inactivation of the Type II Ca current decreases as the Ca current gets
larger due to increased bath Ca concentration (Fox and Krasne, 1984).
Type II muscle Ca channels incorporated into bilayers function with high
concentrations of Ba^{2+} on the intracellular side (Coronado and Affolter,
1986; Ehrlich et al., 1986; Rosenberg et al., 1986). Under these
situations the concentration of Ca^{2+} must be well above the micromolar
levels that block molluscan Ca channels.

What is the mechanism by which a rise in $[Ca^{2+}]_i$ blocks Ca

channels? The simplest model is shown on the left of Figure 7. It assumes
that Ca binds directly to the channel and prevents it from activating. A
second model has been proposed in recent years (Doroshenko et al., 1982;
Chad and Eckert, 1986) in which Ca^{2+} is proposed to control the level of
phosphorylation of the Ca channel. In this model (right hand side of
Figure 7) the activity of the Ca channel is assumed to require
cAMP-dependent phosphorylation of the channel, or of some closely related
protein. Doroshenko and colleagues suggested that intracellular Ca^{2+}
activated a phosphodiesterase, thus reducing the rate of phosphorylation
of the channel. Chad and Eckert proposed that Ca^{2+} activated a
calcineurin-like phosphatase, increasing the rate of dephosphorylation of
the channel. If this phosphorylation mechanism is generally involved in
block of Ca channels by intracellular Ca^{2+}, it would imply that the
enhancement of channel activity by cAMP-dependent phosphorylation is a
very common property of Type II Ca channels.

Modulation by Phosphorylation

     The enhancement of Ca channel activity by cAMP-dependent
phosphorylation is most dramatic and best demonstrated in cardiac muscle
(Reuter and Scholz, 1977; Bean et al., 1984; Kameyana et al., 1986).
There is also evidence for upward modulation of the Ca current by
cAMP-dependent phosphorylation in GH3 cells (Armstrong and Eckert, 1987),
neuroblastoma cells (Tsunoo et al., 1984), molluscan neurons (Pellmar,
1981; Doroshenko et al., 1982) and skeletal muscle (Stefani et al.,
1987). Based on the evidence that manipulations that should enhance
cAMP-dependent phosphorylation reduce washout, it has been proposed that
the washout of Ca currents is basically due to dephosphorylation of Ca
channels (Doroshenko et al., 1982; Chad and Eckert, 1986; Armstrong and
Eckert, 1987). However, other laboratories have found that washout is
greatly slowed (and even reversed) by the addition of ATP and Mg alone
(Forscher and Oxford, 1985; Byerly and Yazejian, 1986). Byerly and
Yazejian (1986) found that the snail Ca current was insensitive to a
number of manipulations intended to enhance cAMP-dependent phosphorylation
(including addition of the catalytic subunit of cAMP-dependent protein
kinase). In fact, by making different assumptions as to what is lost
during internal dialysis and how well the concentration of internal Ca^{2+}
is controlled, it is possible to explain most of the washout data by
either of the models for Ca channel modulation shown in Figure 7.

     The enhancement of Ca currents by cAMP-dependent phosphorylation does
not seem to be a universal phenomenon. As summarized in Figure 8,
cAMP-dependent phosphorylation has no effect, or even reduces, Ca currents
in some cells. Saida and von Breemen (1985) concluded that cAMP-dependent
phosphorylation inhibits the opening of voltage-dependent Ca channels of
artery smooth muscle. The effects of elevating cAMP or applying catalytic
subunit of cAMP-dependent protein kinase have been extensively studied in
a number of molluscan cells in which an enhancement of the Ca current is
not one of the effects; these include Aplysia sensory neurons (Klein and
Kandel, 1980), Aplysia bag cells (Kaczmarek, 1986), and identified neurons
(D6 and D7) of Helix (Gerschenfeld et al, 1986). Since Ca current washout
is a universal phenomenon with snail neurons (Doroshenko et al., 1982;
Byerly and Hagiwara, 1982), it is difficult to believe it is always due to
the loss of cAMP-dependent phosphorylation.

     In Figure 6 we have tried to indicate how general the evidence is for
an enhancement of Type II Ca currents by cAMP-dependent phosphorylation.
Such an enhancement is well demonstrated for cardiac muscle and seems
clear for GH3 cells. However, it is not clear that it is generally true
for vertebrate or molluscan neurons. There is considerable evidence for a
serotonin-induced increase in voltage-dependent Ca current, which is

mediated by cAMP-dependent phosphorylation in some molluscan neurons (see
Pellmar, 1981; Lothshaw et al., 1986). However, as discussed above, such
an enhancement of the Ca current has not been reported in other molluscan
neurons where the effects of cAMP-dependent phosphorylation have been
extensively studied. Fedulova and coworkers (1985) found the washout of
the Type II Ca current of rat DRG neurons to be reduced by addition of
intracellular cAMP, ATP and Mg^{2+}, and Tsunoo and colleagues (1984)
reported that dibutyrl-cAMP enhanced the Type II current of neuroblastoma
cells. However, in studies of chick DRG neurons, where the effects of
cAMP have been studied (Forscher and Oxford, 1985; Dunlap, 1985), there
does not appear to be a cAMP-dependent enhancement of Ca current.

Recently Ca currents have been demonstrated to be modulated by other
types of phosphorylation. The downward modulation of chick DRG cells by
norepinephrine appears to involve phosphorylation by protein kinase C
(Rane and Dunlap, 1986). Phosphorylation by the protein kinase C induces
a new voltage-dependent Ca channel in _Aplysia_ bag cells (Strong et al.,
1987). Application of serotonin enhances a Ca current in identified _Helix_
neurons, an effect that appears to be mediated by cGMP-dependent
phosphorylation (Paupardin-Tritsch et al., 1986). The apparently
cAMP-independent washout of Ca current studied by Byerly and Yazejian
(1986) might be due to the loss of phosphorylation mediated by protein
kinase C or cGMP-dependent kinase.

CONCLUSIONS

In reviewing the data from cells in whose membranes two (or more)
types of Ca channels coexist, two classes of Ca channels can be identified
with common properties throughout the animal kingdom. The Type I channels
activate at more negative potentials, have voltage-dependent inactivation,
carry roughly equal Ba and Ca currents, are blocked less by Cd^{2+}, are
insensitive to dihydropyridines, and do not wash out. Type II Ca channels
activate at more positive potentials, inactivate much more slowly when
internal Ca^{2+} accumulation is prevented, carry larger Ba currents than
Ca currents, and are more sensitive to block by Cd^{2+} . Many Type II
channels exhibit washout and some are sensitive to dihydropyridines.
There is considerable evidence for subtypes within Type II channels.
Selective action of dihydropyridines may be limited to vertebrate Type II
channels. Block by intracellular Ca^{2+} does not occur for all Type II
currents, and the various types of modulation caused by different kinases
suggests subtypes of Type II Ca channels. There is still a great deal of
largely unexplored diversity even within the subtypes we have suggested.
For example the Type II Ca channels of cardiac muscle and skeletal muscle,
which are both sensitive to dihydropyridines, have been shown to differ in
both kinetics and single-channel conductance when inserted into identical
planar bilayer environments (Rosenberg et al., 1986). Also, DHP-
sensitive channels within the same skeletal muscle myoball have been found
to differ in inactivation kinetics and voltage dependence (Cognard et al.,
1986).

Block by intracellular Ca^{2+} and washout of Type II currents have
been proposed to result from dephosphorylation of a cAMP-dependent site in
snail neurons (Doroshenko et al., 1982; Chad and Eckert, 1986), rat DRG
neurons (Fedulova et al., 1985) and GH3 cells (Armstrong and Eckert,
1987). Assuming this mechanism for Ca-dependent inactivation and for
washout does exist in these cells, how general is it? Is washout in all
cells due to a loss of cAMP-dependent phosphorylation? Likewise, are all
Ca channels that exhibit Ca-dependent inactivation modulated by cAMP-
dependent phosphorylation? If the answers to both of these questions are
yes, then upward modulation by cAMP-dependent phosphorylation is a
property of almost all Type II Ca channels (see Figure 6). However, the

direct demonstration of upward modulation by cAMP-dependent
phosphorylation is much more limited, and in a number of cells there seems
to be pretty strong evidence that the function of Type II Ca channels is
not affected by cAMP-dependent phosphorylation. Are Ca channels that are
modulated by the protein kinase C or the cGMP-dependent kinase also
modulated by cAMP-dependent phosphorylation? If not, then a very
interesting way of classifying Type II Ca channels would be according to
the sensitivity of their activity to various types of phosphorylation, as
determined by experiments where the internal side of the channel is
clearly exposed to an environment appropriate for each type of
phosphorylation. (Of course, the problem of washout will have to be
solved universally before these experiments can be done.) Perhaps Type II
Ca channels can be divided into subtypes such as those not modulated by
intracellular phosphorylation, those modulated by only one type of
phosphorylation, and those modulated by more than one type of
phosphorylation. Hopefully, the following papers in this volume will
begin to give answers to the questions above.

REFERENCES

Almers, W. and McClesky, E.W., 1984, Non-selective conductance in calcium
    channels of frog muscle: calcium selectivity in a single-file pore,
    J. Physiol., 353:585-608.
Armstrong, C.M. and Matteson, D.R., 1985, Two distinct populations of
    calcium channels in a clonal line of pituitary cells, Science,
    227:65-67.
Armstrong, D. and Eckert, R., 1987, Voltage-activated calcium channels
    that must be phosphorylated to respond to membrane depolarization,
    Proc. Natl. Acad. Sci. USA, 84:2518-2522.
Armstrong, D., Erxleben C., and Kalman D., 1987, Calcium channels
    modulated by Bay K 8644 appear less susceptible to dephosphorylation,
    Biophys. J., 51:233a.
Ashcroft, F.M. and Stanfield, P.R. 1982, Calcium inactivation in skeletal
    muscle fibres of the stick insect, Carausius morosus, J. Physiol.,
    330:349-372.
Augustine, G.J., Charlton, M.P., and Smith, S.J., 1987, Calcium action in
    synaptic transmitter release, Ann. Rev. Neurosci., 10:633-693.
Brehm, P., and Eckert, R., 1978, Calcium entry leads to inactivation of
    calcium channel in Paramecium, Science, 202:1203-1206.
Bean, B.P., 1985, Two kinds of calcium channels in canine atrial cells,
    J. Gen. Physiol., 86:1-30.
Bean, B.P., Nowycky, M.C., and Tsien, R.W., 1984, β-adrenergic
    modulation of calcium channels in frog ventricular heart cells,
    Nature, 307:371-375.
Byerly, L., Chase, P.B., and Stimers, J.R., 1984, Calcium current
    activation kinetics in neurones of the snail Lymnaea stagnalis, J.
    Physiol., 348:187-207.
Byerly, L., and Hagiwara, S., 1982, Calcium currents in internally
    perfused nerve cell bodies of Lymnaea stagnalis, J. Physiol.,
    322:503-528.
Byerly, L. and Moody, W.J., 1984, Intracellular calcium ions and calcium
    currents in perfused neurones of the snail, Lymnaea stagnalis, J.
    Physiol., 352:637-652.
Byerly, L., and Yazejian, B., 1986, Intracellular factors for the
    maintenance of calcium currents in perfused neurones from the snail,
    Lymnaea stagnalis, J. Physiol., 370:631-650.
Carbone, E., and Lux, H.D., 1984, A low voltage-activated fully
    inactivating Ca channel in vertebrate sensory neurones, Nature,
    310:501-502.
Chad, J.E. and Eckert, R., 1986, An enzymatic mechanism for calcium
    current inactivation in dialysed Helix neurones, J. Physiol.,
    378:31-51.

Cognard, C., Lazdunski, M., and Romey, G., 1986, Different types of Ca^{2+} channels in mammalian skeletal muscle cells in culture, Proc. Natl. Acad. Sci. USA, 83:517-521.

Coronado, R., and Affolter, H., 1986, Insulation of the conduction pathway of muscle transverse tubule calcium channels from the surface charge of bilayer phospholipid, J. Gen. Physiol., 87:933-953.

Cota, G., and Stefani, E., 1986, A fast-activated inward calcium current in twitch muscle fibres of the frog (Rana montezume), J. Physiol., 370:151-163.

Dascal, N., Snutch, T.P., Lübbert, H., Davidson, N. and Lester, H.A., 1986, Expression and modulation of voltage-gated calcium channels after RNA injection in Xenopus oocytes. Science, 231:1147-1150.

Deitmer, J.W., 1984, Evidence for two voltage-dependent calcium currents in the membrane of the ciliate Stylonychia, J. Physiol., 355:137-159.

Deitmer, J.W., 1986, Voltage dependence of two inward currents carried by calcium and barium in the ciliate Stylonychia mytilus, J. Physiol., 380:551-574.

DiPolo, R., Caputo, C, and Bezanilla, F., 1983, Voltage-dependent calcium channel in the squid axon, Proc. Natl. Acad. Sci. USA ., 80:1743-1745.

Doroshenko, P.A., Kostyuk, P.G., and Martynyuk, A.E., 1982, Intracellular metabolism of adenosine 3'-5' cyclic mono-phosphate and calcium inward current in perfused neurones of Neuroscience, Helix Pomatia, 7:2125-2134.

Dunlap, K., 1985, Forskolin prolongs action potential duration and blocks potassium current in embryonic chick sensory neurons, Pflugers Arch, 403:170-174.

Dunlap, K., and Fischbach, G.D., 1981, Neurotransmitters decrease the calcium conductance activated by depolarization of embryonic chick sensory neurones, J. Physiol., 317:519-535.

Eckert, R., and Tillotson, D.L., 1981, Calcium-mediated inactivation of the calcium conductance in calcium-loaded giant neurones of Aplysia californica, 314:265-280.

Ehrlich, B.E., Forte, M., Jacobson, A.R., Sayre, L.M., 1987, Block of Paramecium calcium channels by the calmodulin antagonist W-7 and its analogues,, Biophys. J.:51, 31a.

Ehrlich, B.E., Schen, C.R., Garcia, M.L., and Kaczorowski, G.J., 1986, Incorporation of calcium channels from cardiac sarcolemmal membrane vesicles into planar lipid bilayers, Proc. Natl. Acad. Sci., USA, 83:193-197.

Fedulova, S.A., Kostyuk, P.G., and Veselovsky, N.S., 1985, Two types of calcium channels in the somatic membrane of new-born rat dorsal root ganglion neurones, J. Physiol., 359:431-446.

Fenwick, E.M, Marty, A., and Neher, E., 1982, Sodium and calcium channels in bovine chromaffin cells, J. Physiol., 331:599-635.

Forscher, P., and Oxford, G.S. 1985, Modulation of calcium channels by norepinephrine in internally dialyzed avian sensory neurons, J. Gen. Physiol., 85:743-763.

Fox, A.P., and Krasne, S., 1984, Two calcium currents in Neanthes arenaceodentatus egg cell membranes, J. Physiol., 356:491-505.

Fukushima, Y., and Hagiwara, S., 1985, Currents carried by monovalent cations through calcium channels in mouse neoplastic B lymphocytes, J. Physiol., 358:255-284.

Ganitkevich, V. Ya., Shuba, M.F., and Smirnov, S.V., 1986, Potential-dependent calcium inward current in a single isolated smooth muscle cell of the guinea-pig taenia caeci, J. Physiol., 380:1-16.

Gerschenfeld, H.M., Hammond, C., Paupardin-Tritsch, D., 1986, Modulation of the calcium current of molluscan neurones by neurotransmitters, J. Exp. Biol., 124:73-91.

Hagiwara, S., and Nakajima, S., 1966, Effects of the intracellular Ca ion concentration upon the excitability of the muscle fiber membrane of a barnacle, J. Gen. Physiol., 49:807-818.

Hagiwara, S., Ozawa, S., Sand, O., 1975, Voltage clamp analysis of two inward current mechanisms in the egg cell membrane of a starfish, J. Gen. Physiol., 65:617-644.

Hess, P., and Tsien, R.W., 1984, Mechanism of ion permeation through calcium channels, Nature, 309:453-456.

Hille, B., 1975, Ionic selectivity of Na and K channels of nerve membranes, in : "Membranes: A Series of Advances", G. Eisenman, ed., Marcel Dekker, New York.

Hirano, T., and Takahashi, K., 1984, Comparison of properties of calcium channels between the differentiated 1-cell embryo and the egg cell of ascidians, J. Physiol., 347:327-344.

Horn, J.P., and McAfee, D.A., 1980, Alpha-adrenergic inhibition of calcium-dependent potentials in rat sympathetic neurones, J. Physiol., 301:191-204.

Kaczmarek, L.K., 1986, Phorbol esters, protein phosphorylation and the regulation of neuronal ion channels, J. Exp. Biol., 124:375-392.

Kalman, D., Erxlebel, C., and Armstrong, D., 1987, Inactivation of the dihydropyridine-sensitive calcium current in GH3 cells is a calcium-dependent process, Biophys. J., 51:432a.

Kameyama, M., Hescheler, J., Hofmann, F. and Trautwein, W., 1986, Modulation of Ca current during the phosphorylation cycle in the guinea pig heart, Pflugers Arch, 407:123-128.

Klein, M., and Kandel, E.R., 1980, Mechanism of calcium current modulation underlying presynaptic facilitation and behavioral sensitization in Aplysia, Proc. Natl. Acad. Sci. USA, 77:6912-6916.

Kostyuk, P.G., Mironov, S.L., and Shuba, Ya.M., 1983, Two ion-selecting filters in the calcium channel of the somatic membrane of mollusc neurons, J. Membrane Biol., 76:83-93.

Lotshaw, D.P., Levitan, E.S., and Levitan, I.B., 1986, Fine tuning of neuronal electrical activity: modulation of several ion channels by intracellular messengers in a single identified nerve cell, J. Exp. Biol., 124:307-322.

McClesky, E.W., Fox, A.P., Feldman, D., and Tsien, R.W., 1986, Different types of calcium channels, J. Exp. Biol., 124:177-190.

Mironov, S.L., Tepikin, A.V., and Grishchenko, A.V., 1985, Two calcium currents in the somatic membrane of mollusc neurones, Neirofiziologiya, 17:627-633.

Nerbonne, J.M., and Gurney, A.M., 1987, Blockade of Ca^{2+} and K + currents in bag cell neurons of Aplysia californica by dihydropyridine Ca^{2+} antagonists, J. Neuroscience, 7:882-893.

Nilius, B. Hess, P., Lansman, J.B., and Tsien, R.W., 1985, A novel type of cardiac calcium channel in ventricular cells, Nature, 316:443-446.

Nowycky, M.C., Fox, A.P., and Tsien, R.W., 1985, Three types of neuronal calcium channel with different calcium agonist sensitivity, Nature, 316:440-443.

Paupardin-Tritsch, D., Hammond, C., Gerschenfeld, H.M., Nairn, A.C. and Greengard, P., 1986, cGMP-dependent protein kinase enhances Ca^{2+} current and potentiates the serotonin-induced Ca^{2+} current increase in snail neurones, Nature:323, 812-814.

Pellmar, T.C., 1981, Ionic mechanism of a voltage-dependent current elicited by cyclic AMP, Cell. Mol. Neurobiol., 1:87-97.

Plant, T.D., Standen, N.B., and Ward, T.A., 1983, The effects of injection of calcium ions and calcium chelators on calcium channel inactivation in Helix neurones, J. Physiol., 334:189-212.

Rane, S.G., and Dunlap, K., 1986, Kinase C activator 1,2-oleoylacetylglycerol attenuates voltage-dependent calcium current in sensory neurons, Proc. Natl. Acad. Sci. USA, 83:184-188.

Reuter, H., and Scholz, H., 1977, The regulation of the calcium conductance of cardiac muscle by adrenaline, J. Physiol., 264:49-62.

Rosenberg, R.L., Hess, P., Reeves, J.P., Smilowitz, H., and Tsien, R.W., 1986, Calcium channels in planar lipid bilayers: insights into mechanisms of ion permeation and gating, _Science_, 231:1564–1566.

Saida, K., and van Breemen, C., 1985, Cyclic nucleotides and calcium movements, in : "Calcium Entry Blockers and Tissue Protection," T. Godfraind, P.M. Vanhoutte, S. Govoni, and R. Paoletti, eds., Raven, New York.

Sakmann, B., Methfessel, C., Mishina, M., Takahashi, T., Takai, T., Kurasaki, M., Fukuda, K., and Numa, S., 1985, Role of acetylcholine receptor subunits in gating of the channel, _Nature_, 318:538–543.

Stefani, E. Toro, L., and Garcia, J., 1987, Alpha- and beta-adrenergic stimulation of fast and slow Ca ++ channels in frog skeletal muscle, _Biophys. J._, 51:425a.

Strong, J.A., Fox, A.P., Tsien, R.W., and Kaczmarek, L.K., 1987, Stimulation of protein kinase C recruits covert calcium channels in _Aplysia_ bag cell neurons, _Nature_, 325:714–717.

Tillotson, D., 1979, Inactivation of Ca conductance dependent on entry of Ca ions in molluscan neurons, _Proc. Natl., Acad. Sci. USA_, 76:1497–1500.

Tsien, R.W., and Marban, E., 1982, Digitalis and slow inward current in heart muscle: evidence for regulatory effects of intracellular calcium on calcium channels, in : "Advances in Pharmacology and Therapeutics II, Vol. 3", H. Yoshida, Y. Hagihara, and S. Ebashi, eds., Pergamon, Oxford and New York.

Tsunoo, A., Yoshii, M., and Narahashi, T., 1984, Two types of calcium channels in neuroblastoma cells and their sensitivities to cyclic AMP, _Soc. Neurosci. Abstr._, 10:527.

Tsunoo, A., Yoshii, M., and Narahashi, T., 1985, Differential block of two types of calcium channels in neuroblastoma cells, _Biophys. J._, 47:433a.

Yoshii, M., Tsunoo, A., and Narahashi, T., 1985, Different properties in two types of calcium channels in neuroblastoma cells, _Biophys. J._, 47:433a.

MULTIPLE TYPES OF CALCIUM CHANNELS: IS THEIR

FUNCTION RELATED TO THEIR LOCALIZATION?

Joachim W. Deitmer

Institut für Zoologie I
Universität Düsseldorf
D-4000 Düsseldorf
Federal Republic of Germany

INTRODUCTION

Calcium influx through voltage-dependent Ca channels leads to a transient increase in intracellular Ca, which can initiate a number of biological processes, such as muscle contraction, synaptic transmission, secretion, ciliary motility, enzyme activation, growth and development. The electrical activity promoted by Ca channels is the most variable and widespread form of excitation. Ca channels are ubiquitous, from unicellular organisms to mammals, and Ca-dependent electrogenesis results in membrane depolarizations of wide range in amplitude and duration. Currents through Ca channels are diverse in their properties, and they can have great impact on the Ca homeostasis in cells. During the influx of Ca ions through open channels the initially low intracellular Ca activity of about 0.1 μM may increase several orders of magnitude within milliseconds.

In recent years different types of voltage-dependent Ca channels have been reported to occur in membranes of the same cell. After a first indication of two Ca-dependent inward currents in starfish egg cells (Hagiwara, Ozawa & Sand, 1975), two or even three kinds of Ca channels have been found in various nerve and muscle cells as well as in other excitable cells (for ref. Deitmer, 1986a; McCleskey et al., 1986). It seems likely now that the occurrence of different Ca channels in one cell is much more widespread than previously thought. The coexistence of multiple kinds of Ca channels has lead to questions concerning their functional significance, in particular with respect to their role in cellular Ca homeostasis and Ca-induced processes. Why does a cell employ different types of membrane potential-controlled Ca channels as pathways for Ca influx across the membrane? What are the characteristics of these Ca channels, and which processes does the Ca influx through the different types of channels initiate or control?

The multiple types of Ca channels have been distinguished by several properties, but two of those Ca channels are found together in a wide variety of cell types. Each of these two types of Ca channels has been referred to by different terms, and the nomenclature is not yet unified. One channel has been termed 'low-threshold', 'transient' (T-type) or just 'channel I', the other has been called 'high-threshold', 'slow',

'long-lasting' (L-type), 'classical' or 'channel II'. In the present
paper the main properties of these voltage-dependent Ca channels as they
occur in the ciliated protozoan <u>Stylonychia</u> are reviewed and discussed in
particular with respect to their localization and function in the membrane.

MATERIAL AND METHODS

The experiments were performed on the hypotrichous ciliate
<u>Stylonychia</u> <u>mytilus</u> syngen, which was cloned and cultured in Pringsheim
solution at 18°C. The cells used were 240-320 μm in length, 40-60 μM
in width, and 10-20 μm in dorsoventral extension. For experimentation,
cells were washed and equilibrated in the normal experimental solution
containing 1 mM $CaCl_2$ or 1 mM $BaCl_2$, 1 mM KCl, and 1 mM Tris-HCl or
1 mM HEPES to give a pH of 7.4 (±0.1). The experiments were carried out
with the experimental chamber held at a temperature of 17-18°C. The cell
to be investigated was held stationary by a fine glass micropipette gently
pressing the cell against a cover slip, letting the ciliary organelles
freely project into the fluid.

Electrical recording was achieved by two microelectrodes which
penetrated into the cell. The electrodes were filled with 1 M KCl for
membrane potential recording, and with 2 M K citrate for current
injection. A third microelectrode, filled with 1 M KCl was placed outside
the cell for differential recording of the membrane potential. The
resistance of the electrodes was 10-50 MΩ in normal bathing solution.
The voltage clamp was performed by means of a feed-back system using a
high-gain differential amplifier (AD 171K). The membrane currents were
monitored by a current-voltage converter connected to the bath via a 1 M
KCl-agar bridge. Further details have been described elsewhere (Deitmer,
1984, 1986b).

The voltage was usually clamped to a stable steady-state within
0.1-0.3 ms during a 10 mV step. The time-to-peak of the early inward
current was 1.5 to 4.5 ms or more. Thus, the clamp was sufficiently fast
to record the peak inward current free of capacitive transients. Leak
currents were small due to the very high input resistance of the cell
(50-100 MΩ); they were not subtracted from the current records. Series
resistance was not compensated for, as it was negligible due to the
relatively small current amplitudes (up to 50 nA).

It was discussed previously that the voltage clamp during holding and
command mode is controlled adequately in <u>Stylonychia</u> (Deitmer, 1984,
1986b). In particular along the cilia, which represent long and thin
cables, this might be of considerable problem. When the resistivity of
the cytoplasm and the axoneme of a cilium 55 μm long and 0.25 μm
diameter, and the interciliary space of the compound ciliary organelles
are taken into account, a length constant of approximately 1.5 mm is
calculated. This assumes a mean input resistance of 70 MΩ (in Ca
solution), a total cell-surface area of 0.3 mm^2 (Machemer & Deitmer,
1987), and a homogenous resistance value for ciliary and somatic
membrane. If the ciliary membrane resistance is much larger, as suggested
by experiments with removed cilia (Machemer & Ogura, 1979; Deitmer, 1984),
the quality of the ciliary space clamp would improve, and with a cellular
input resistance of 100-200 MΩ in Ba solution, the length constant would
increase to several millimeters. This is two orders of magnitude larger
than the length of an average cilium in <u>Stylonychia</u>. From these
calculations, voltage-clamp control within 1 mV should be possible.

RESULTS

The action potential of _Stylonychia_ (Fig. 1) has two components, a
large, graded peak, and a second, smaller, all-or-none 'shoulder'. Both
components of the action potential are dependent on extracellular Ca (De
Peyer 7 Machemer, 1977; De Peyer & Deitmer, 1980). Action potentials are
generated spontaneously by the cell, and they can be elicited by brief
intracellular current pulses (Fig. 1 A) or by a mechanical stimulus to the
anterior part of the cell (Fig. 1 B). In a free-swimming cell, an action
potential initiates a backward movement of the cell due to activation of
ciliary organelles.

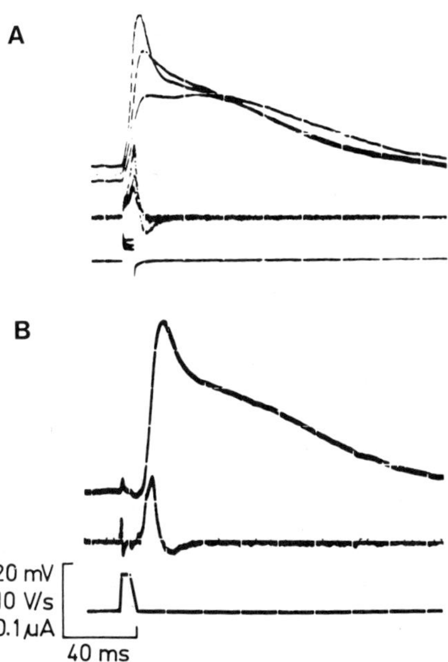

Fig. 1. Action potentials (upper traces) as evoked by intracellular
         current pulses (A, superimposed recordings), or by a
         mechanical stimulus to the cell anterior (B). The middle
         traces give the first derivative of the potential changes,
         dV/dt.

In _Stylonychia_, the ciliary organelles are bundles of 30–80 single
cilia of up to 55 μm in length. The membranelles form a row of
velum-shaped bundles of cilia running at the cell anterior toward the oral
grove; their activity provides a water current which presumably helps
feeding the cell. The ventral and marginal cirri serve cellular
locomotion. High-speed cinematography of cirri and membranelles under
voltage-clamp control revealed that the two types of organelles,
membranelles and cirri, display different motor responses to the same
voltage stimuli (Deitmer et al., 1984). Membranelles beat continuously at
a frequency around 40 Hz irrespective of the voltage pulse applied, while

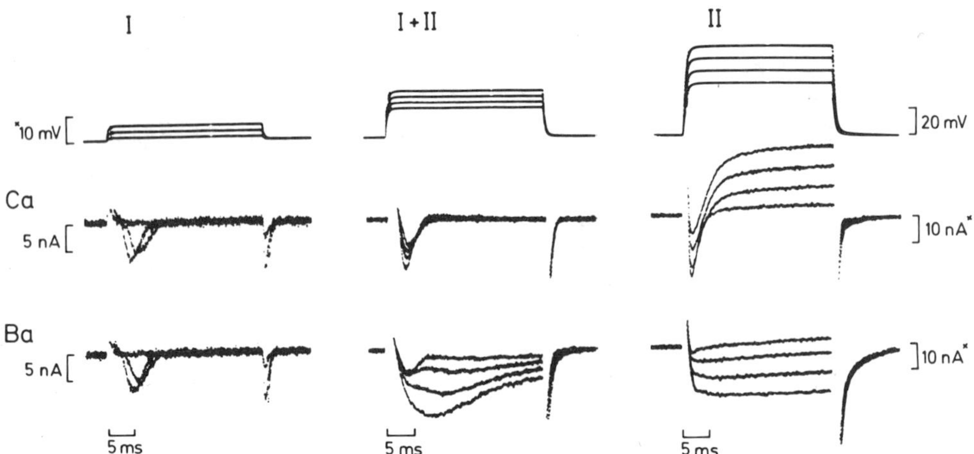

Fig. 2. Ca and Ba currents through the two voltage- dependent
channels (I, II). Left column: depolarizations by 3 and
5 mV from the holding potential of -50 mV elicits current
I. Middle column: depolarizations by 10-16 mV activates
both current I and current II. Right column:
depolarizations by 40-70 mV evoke predominantly current II.

ventral and marginal cirri respond to membrane polarization with a change
in the frequency and direction of their power stroke. In contrast to
membranelles, cirri are inactive at the cell's resting potential.
Depolarization of the membrane induces a power stroke directed to the cell
anterior; hyperpolarization initiates a power stroke toward the cell
posterior (Fig. 4). For an extensive account of the control of motor
organelles and behaviour of _Stylonychia_ see Machemer & Deitmer (1987).

In voltage clamp experiments, the cell membrane was held at its
resting potential of -50 mV and depolarized in steps of various amplitudes
and of 30 ms in duration (Fig. 2). Current I, which can be activated
already by small membrane depolarizations of 2-4 mV, rapidly rises and
decays in both Ca and Ba solution. Current II, however, which is
activated at potentials more positive than -40 mV, decays fast in Ca
solution, but very slowly in Ba solution (De Peyer & Deitmer, 1980;
Ballanyi & Deitmer, 1984). At membrane potentials between -40 mV and -30
mV _both_ current I and current II are activated, and, due to their
different kinetics in Ba solution, they can be clearly identified.
Current I carried by Ba ions apparently inactivates completely.
Furthermore, in Ba solution no net outward current is observed up to
positive potentials (+10 mV), presumably because the outward current
apparent at potentials of -25 mV and more positive in Ca solution, is
largely Ca-dependent (De Peyer & Deitmer, 1980). Fig. 3 shows inward
current-voltage relationships with either Ca or Ba as charge carrier.
There are two maxima in both current-voltage relationships, in Ca solution
at -45 mV and -17 mv, and in Ba solution at -44 mV and -25 mV; they
represent the maximum current through channel I and channel II,
respectively.

In an attempt to isolate the inward current from other current

components, especially K outward currents, drugs known to block K
channels, such as Cs, tetraethylammonium and 4-aminopyridine were used
(see Deitmer, 1984). The outward currents were considerably reduced, but
not completely inhibited by these substances in Ca solution. A fast,
transient outward current apparent in some cells could be abolished. The
inward currents remained virtually unaltered except for a somewhat slower
decay due to the reduction of outward current, which normally produces an
apparent, fast inward current decay. The peak inward currents, and hence
the inward current-voltage relationships, were not changed and still
showed the same amplitudes of the two distinct maxima.

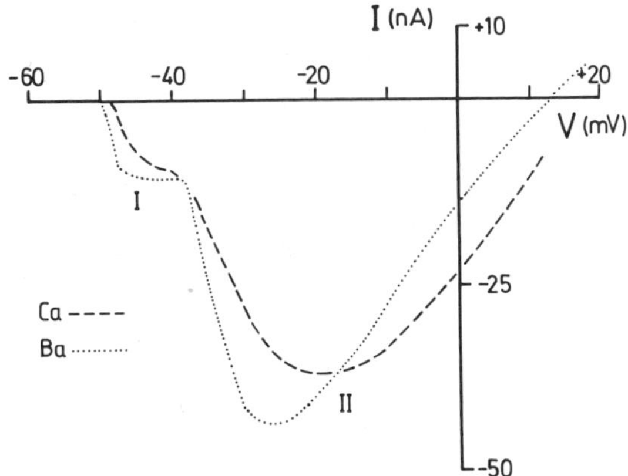

Fig. 3.   Voltage relationships of inward currents obtained with Ca
          (broken line) and Ba (dotted line) as charge carrier. Note
          the inflection and the two peaks (I and II) in the
          current-voltage relationships, indicating currents through
          channel I and channel II respectively. The holding
          potential was -50 mV.

LOCALIZATION OF THE TWO CALCIUM CHANNELS

Deciliation experiments on the holotrichous ciliate _Paramecium_ had
shown that the voltage-dependent Ca channels reside in the membrane of the
cilia, but not in the cell body membrane (Ogura & Takahashi, 1976; Dunlap,
1977; Machemer & Ogura, 1979). From the different types of compound cilia
forming functionally different motor organelles in _Stylonychia_, the
possibility of an inhomogenous distribution of Ca channels in the membrane
was tested. For this, cells were cut transversely, or parts of the cells
were removed to obtain cell fragments (Deitmer, 1984). The remaining,
resealed anterior or posterior cell parts were used for intracellular
recording. In cells without the anterior part, i.e. without their

membranellar band, the second action potential component and inward
current I were absent. When the posterior half or third of a cell was
removed to record from an intact anterior cell fragment with its
functional membranellar band, the action potential still had two peaks,
and both inward current I and current II could be observed.

In a cell which had released the anterior part including its
membranellar band, the second action potential component disappeared, and
inward current I could not be recorded. The inward current–voltage
relationship revealed only one maximum at –18 mV, which was typical for
that of inward current II (Deitmer, 1984).

This strongly suggests that the two Ca channels reside in different
parts of the cell membrane, one in the membrane of the membranelles, and
the other in the membrane of the cirri (Fig. 4). Recording from cells
which were in the process of encystment, during which period the cells
resorb all their ciliary organelles, confirmed this interpretation. In
these rounded cells with almost all membranelles and cirri resorbed,
current II was reduced to less than 10%, while current I was completely
absent (Deitmer, 1987). In addition, young cells of less than 30 min
after cell division, have a relatively larger amount of current I than
current II, suggesting that the inward currents are coupled to the
morphogenetic development of the ciliary organelles. Shortly after cell
division, when the formation of new membranellar bands is nearly completed
in the two daughter cells, current I amounts to 72% of that in 'adult'
cells. In contrast, current II only amounts to 44% of that found in
'adult' cells, which corresponds to the rudimentary state of development
of cirri at that stage (Deitmer et al., 1986).

It is therefore tempting to speculate that inward current I,
restricted to the membranelles, is involved in controlling membranellar
activity, while inward current II controls the beating activity of ventral
and marginal cirri. The two Ca channels might thus provide the means for
separate gears for independent motor control of membranelles and cirri.

PROPERTIES OF THE TWO CALCIUM CHANNELS

There is a variety of characteristics by which channel I and
channel II can be distinguished (Table 1). A striking difference is that
channel I, but not channel II, can be inhibited by low concentrations of
the plant lectin concanavalin A (Con A, 0.2–0.5 µg/ml, equivalent to 2–5
nM, for full inhibition within minutes). Incubation of cells in
fluorescein–labeled Con A produced a strong fluorescence of the cirri and
of the somatic membrane (Ivens & Deitmer, 1986). This, again, is in line
with the finding that channel I is localized on the membranelles.

The inhibitory effect of Con A on current I could be prevented by
10–30 mM α–methyl D–mannoside, indicating that the Con A effect was
mediated by binding to specific sugar residues on the excitable membrane.
The lectin wheat germ agglutinin (20 ʋg/ml) was ineffective. The
succinylated dimeric derivative of Con A did not inhibit current I. The
selective inhibition of only Ca channel I indicates that this channel
might be a glycoprotein with Con A–specific carbohydrate residues. With
this tool the two inward currents can now be separated easily (Ivens &
Deitmer, 1986; Deitmer, 1986b).

The inactivation of the two channels occurs by two different
mechanisms. Inward current I decays with a similar time course when
either Ca, Sr or Ba ions act as charge carrier. This is in contrast to
inward current II which decays rapidly in Ca– and in Sr–solution, even in

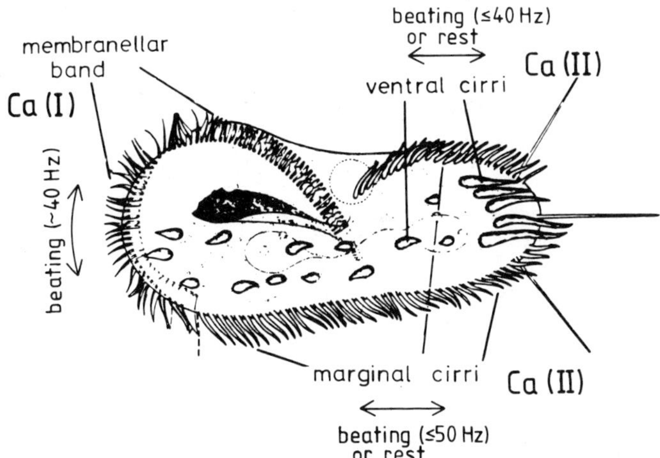

Fig. 4.   Ventral view of <u>Stylonychia</u> (after Machemer), indicating the ciliary organelles and their maxium beating frequencies, and the presumed localization of Ca channel I in the membranellar band and Ca channel II in the ventral and marginal cirri.

Table 1.   Properties of the two calcium channels

Property	Ca channel I	Ca channel II
selectivity	Ca, Sr, Ba	Ca, Sr, Ba
activation kinetics $V_{0.5}$	fast, transient -47 mV ($V_h = -50$ mV) (varies with V_h)	fast, (longer component) -33 mV
inactivation mechanism kinetics	voltage-dependent 20–60 ms	Ca-dependent 44 ms, 218 ms
channel blockers	Co, Mn Cd ($K_{0.5}{\sim}10^{-4}$ M) Con A (5 nM)	Co, Mn Cd ($K_{0.5}{\sim}10^{-5}$ M)
peak current, nA (% total)	5 – 10 nA (20%)	25 – 40 nA (80%)
stimulates $I_K(Ca)$	no	yes
localization	membranelles	cirri

the presence of K channel blockers, but very slowly and incompletely in
Ba-solution (see Fig. 2). The amplitude of inward current II when carried
by Ba ions is unaltered, if activated by one long, or by multiple short
voltage steps. In Ca-free Ba-solution a large inward current II is
maintained over several seconds. This rules out inactivation of current
II, during short pulses, due to extracellular depeletion of the charge
carrying ion (Deitmer, 1986b).

Experiments using double pulses have suggested that channel I
inactivates by a voltage-dependent mechanism, and channel II by a
predominantly Ca-dependent mechanism (Deitmer, 1984, 1986b). Accordingly,
injection of EGTA into the cells did not affect inactivation of channel I,
but reduced inactivation of channel II. On the other hand, the larger Ca
current II, the greater its inactivation, as observed during a subsequent
second pulse. Although both voltage-dependent and Ca-dependent
inactivation mechanisms have been described in a variety of preparations
(see Eckert & Chad, 1984), this appears the first example for the
occurrence of both mechanisms of Ca channel inactivation in <u>one</u> cell.

The removal of inactivation would therefore also be expected to be
different for the two types of Ca channels. Indeed, the removal of
inactivation of channel I and channel II displayed quite different time
courses. Ca current I reappears within 5-10 ms after inactivation by a
brief depolarizing pulse of 5-10 mV. It takes, however, a 20 to 60 ms
interval of repolarization after a pulse to activate maximum current I by
a second pulse. Ca current II recovers with a double exponential time
course after inactivation. With two 10 ms pulses from -50 mV to -20 mV,
separated by an increasing interval, the mean time constants for the
removal of current II inactivation are 44 ±15 ms and 218 ±60 ms
(±S.D., n=4-5). The complete removal of inactivation was established
only after a few seconds. When the external Ca concentration was reduced
from 1 mM to 0.1 mM, the time course of removal of inactivation became
faster, the time constants being 32 ±2 ms and 183 ±45 ms (n=3). A
faster time course of removal of inactivation was also observed, when the
external solution contained 0.5 mM $CoCl_2$. The amplitude of Ca current
II decreased by 30-80% after reducing the external Ca concentration or
after addition of Co. Thus, a smaller Ca current II produces a smaller
increase in intracellular free Ca, and hence less inactivation and a
faster recovery from inactivation.

An interesting feature of the two Ca channels is that activation of
channel I is never accompanied or followed by an outward current, while Ca
current through channel II activates a substantial K outward current.
This K current appears to be induced directly by Ca current II, since
reduction of this Ca current, e.g. by lowering the external Ca
concentration, by predepolarization to inactivate Ca current II, by
addition of inorganic Ca antagonists such as Co or Cd, or by injection of
the Ca chelator EGTA, decreases the outward current. This may be related
to the fast repolarization occurring after the first, large and graded
action potential peak, and the relatively slow repolarization following
the second, all-or-none action potential component.

DEPENDENCE ON MEMBRANE POTENTIAL AND CALCIUM

The voltage-dependence of channel I displays an unusual behavior; it
shifts with the holding potential along the voltage axis (Deitmer,
1986b). When the holding potential was varied from -50 mV to values
between -45 mV and -65 mV, the activation of current I was shifted in the
same direction by nearly 8 mV per 10 mV change in holding potential. The
inward current II remained unaffected in its time course by this variation
in holding potential (Fig. 5).

Increasing the external Ca concentration resulted in a positive shift
of the resting (and holding) potential and current-voltage relationships
(Deitmer, 1986b). The holding potential was adjusted to the resting
potential, which was -60 mV in 0.1 mM Ca, -50 mV in 1 mM Ca, and -40 mV in
5 mM Ca. The shifts of Ca inward and K outward currents were similar and
amounted to 19-22 mV, when the external Ca concentration was increased
from 0.1 to 5.0 mM. This shift could be explained by a change in negative
surface charges at the cell membrane (McLauthlin et al., 1971; Schauf,
1975).

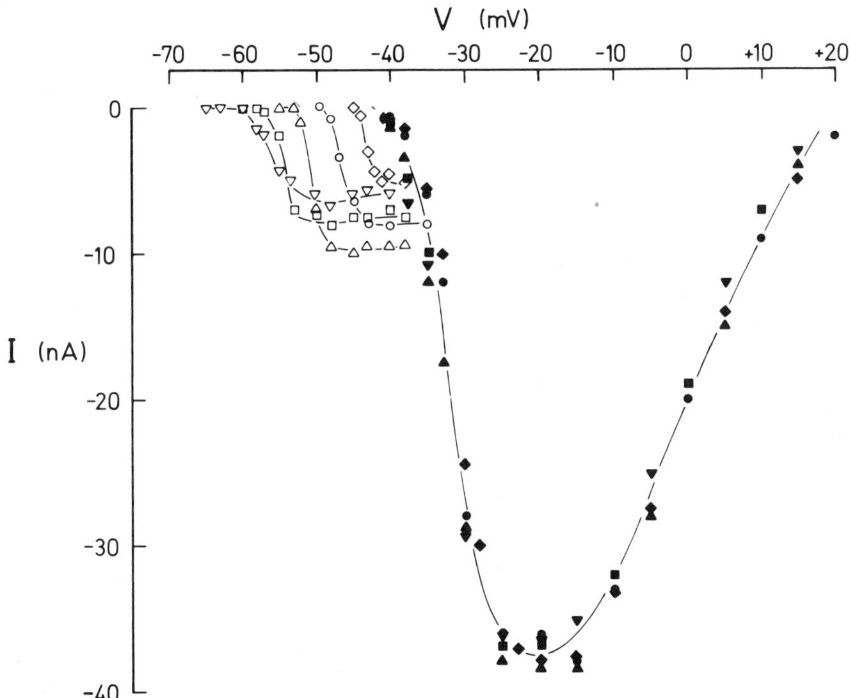

Fig. 5.   Inward current-voltage relationships as obtained from the
           peak of current I (open symbols) and the peak of current II
           (filled symbols) in Ba solution at five different holding
           potentials between -45 mV and -65 mV (different symbols).
           From Deitmer (1986b).

The amplitude of the two inward currents also varied with the
external Ca concentration, as would be expected for Ca currents. The
relative increase of current I was 80%, and that of current II 35%, when
raising the Ca concentration from 0.1 to 5.0 mM. This may suggest a
different binding affinity (or saturation of binding sites) for Ca ions to
channel I and channel II. Similar observations were made when the charge
carrying ion was Ba, and the Ba concentration was varied. The voltage
shift of current I with different holding potentials occurred similarly
when either Ca or Ba acted as charge carrier, and it could not only be
explained with the removal of voltage-dependent

inactivation. Amplitude and rate of rise were maximal with a holding
potential between -50 mV and -55 mV, and similar or even smaller for more
negative holding potentials. A shift of outward currents with varying
holding potentials was never observed, and the presence of K channel
blockers did not affect the shift of current I. In solutions containing
higher divalent cation concentrations (5-10 mM), the holding
potential-dependent shift of current I appeared to be reduced. The
mechanism responsible for the voltage shift of current I is yet unknown,
but a change in surface charges by varying the membrane potential has been
hypothesized to occur due to phospholipid flip-flop in the excitable
membrane (McLaughlin & Harary, 1974). These flip-flops could move and
reorient charged sites according to the actual potential and thus rapidly
change the surface potential. Indeed, the voltage shift of current I
could be achieved by a 10-30 ms prepulse.

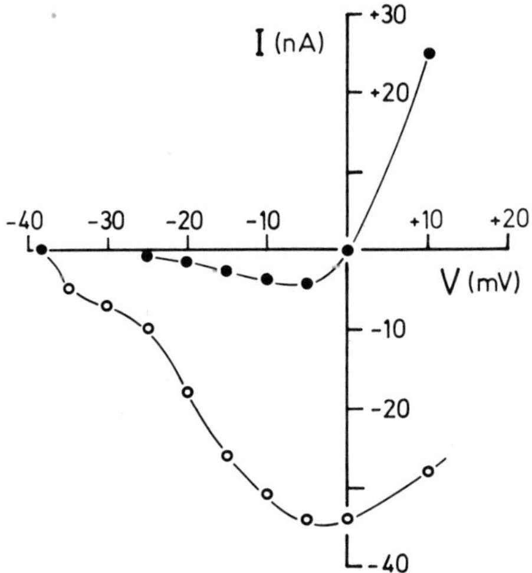

Fig. 6. Current-voltage relationship in Ca solution of a cell with
intracellular EGTA; peak inward currents (open circles) and
current 30 ms after pulse onset (filled circles) are shown.
The holding potential was -40 mV.

Lowering the intracellular ionized Ca level by EGTA (0.5 M EGTA in
the current-injecting microelectrode) had a variety of effects on the
cells' electrical excitability. The resting potential depolarized to
approximately -40 mV. The spontaneous action potentials were increased in
amplitude to more than 40 mV to become overshooting, and they were
prolonged 4-6 fold to approximately 1 s.

The inward current-voltage relationships (Fig. 6) shifted by 12-14 mV
to more positive potentials as compared to the control (without
intracellular EGTA). The amplitudes of both types of inward current were,

however, not affected by intracellular EGTA. While current I remained
unaltered also in its fast, transient time course, current II displayed a
component which decayed only very slowly over seconds. In the
current-voltage relationship taken 30 ms after the onset of the command
pulse, this component appears as a negative resistance between -25 mV and
0 mV, with a peak of 5 nA at -5 mV. This coincides with the peak of
current II, which was at -3 mV.

The apparent shift of the outward current was even larger after EGTA
injection; however, this might be partially produced by reduction of the
Ca-dependent component of the outward current. At negative potentials
there was no net outward current (Fig. 6). This, and the non-inactivating
component of inward current II presumably produce the prolongation of the
action potential, which is also observed in Ca-free Ba solution (Ballanyi
& Deitmer, 1984).

DISCUSSION

The different properties of the two Ca channels in the ciliate
Stylonychia suggest a functional significance of the two separate Ca
influx mechanisms. The different activation voltage of the Ca channels
implies that, depending on the resting (holding) potential and the step
depolarization, the two populations of Ca channels can be activated
independently from each other. This and the different amplitudes and
kinetics of the two currents result in different Ca flow across the cell
membrane, and hence different intracellular Ca transients. Moreover,
these currents occur at different cell sites, current I at the
membranelles and current II at the cirri, and thus may not even interact
(Deitmer, 1983).

The low-threshold and steep activation kinetics of Ca channel I
appears to favor a role of channel I in the generation of spontaneous
action potentials. Inhibition of channel I by concanavalin A
significantly reduces the frequency of spontaneous action potentials.
Since each action potential triggers a back-and-forth movement of the
cells, channel I has a direct influence on cellular behavior. The unique
voltage-dependence of channel I enables the cells to maintain their
electrical excitability by keeping activation of current I close to the
resting potential, even if this varies by tens of millivolts. This may
well happen when the cells' natural habitat, fresh-water ponds, change
their ionic composition during rainfall or drought. The
voltage-dependence of channel I may thus be a mechanism which allows the
cells to adjust to quite a wide range of ionic concentrations without a
change in their electrical excitability.

The low-threshold Ca channel I has also been suggested to play a role
in modulating spike pattern in neurons (Llinas & Yarom, 1981; Carbone &
Lux, 1984; Bossu et al., 1985). Channel I may also be involved in
generating rhythmic activity in heart cells (Bean, 1985; Nilius et al.,
1985).

An important factor to understand channel function may be their
localization in the cell membrane. The presence of Ca channels in
different membrane areas has actually been known for some time in nerve
cells. Neurons may produce Ca action potentials in their cell bodies and
release neurotransmitter at their terminals due to local Ca influx through
voltage-gated Ca channels (Katz, 1969). The Ca influx into these neurons
is separated in time and space by the conduction of the nerve impulse from
their somata to the presynaptic terminals. Recent recordings with
intracellular Ca sensors have increased the evidence that

voltage-dependent Ca influx and intracellular Ca transients are localized
(Connor, Smith, this volume). In the presynaptic terminal region of
photoreceptor axons in the giant barnacle, Ca entry was monitored only in
restricted areas of less than 50 μm in length (Stockbridge & Ross,
1984). In crab stomatogastric ganglion cells membrane potential
oscillations produced intracellular Ca oscillations in the neuropil region
but not in the soma region (Graubard & Ross, 1984). In crab
stomatogastric ganglion cells, membrane potential oscillations produced
intracellular Ca oscillations in the neuropil region but not in the soma
region (Graubard & Ross, 1985). The site of the Ca entry during
electrical activity may also be restricted to neuronal soma, dendrites,
neurites and/or growth cones (Anglister et al., 1982; Hirst & McLachlan,
1986; Bolsover & Spector, 1986).

In the ciliate <u>Paramecium</u>, where the voltage-dependent Ca channels
are localized on the <u>ciliary</u> membrane only (Ogura & Takahashi, 1976;
Dunlap, 1977), the current flows through the ciliary membrane to produce a
Ca increase within the intraciliary space which activates the axoneme.
The ciliary beating activity (towards the cell anterior), and hence
backward swimming of the cell, is directly related to the Ca current
during the action potential. For the more advanced hypotrichous ciliate
<u>Stylonychia</u> the functionally different ciliary organelles employ two
distinct types of Ca channels presumably to achieve different Ca
transients in the two cell compartments, and hence independent motor
control of these organelles.

In most cells little is yet known about the spatial distribution and
functional differences of multiple types of Ca channels, although this
information may provide important clues to the whereabouts of Ca influx,
intracellular Ca transients and Ca-dependent processes. The evidence for
regional Ca influx and intracellular Ca transients indicates that the
localization of different types of Ca channels in specific membrane areas
may well be a more common feature of other cells, such as nerve and muscle
cells.

Hence it may also help us to understand the Ca-dependent control of
cellular activities if we know more about the variability of Ca influx.
Multiple Ca channels may be activated independently - e.g. due to the
different voltage and kinetics of their activation and inactivation - and
they may be distributed inhomogeneously in the cell membrane by being
restricted to certain membrane areas. Due to their distinctive properties
these different Ca channel populations may provide the means to regulate
Ca influx more effectively. The initiation of Ca-induced processes may
become faster and more easily controlled, and the energy expenditure for
Ca extrusion could be reduced if the Ca influx only occurred only at sites
where Ca ions are 'needed'. This could be in cell compartments such as
the presynaptic terminal, in a growth cone or in a cilium.

<u>Acknowledgements</u>. The receipt of a Heisenberg-Fellowship by the Deutsche
Forschungsgemeinschaft is gratefully acknowledged. The experimental work
was carried out at the Ruhr-Universität Bochum with financial support
through the SFB 114, TP A5 of the D.F.G.

REFERENCES

Anglister, L., Farber, I.C., and Grinvald, A., 1982, Localization of
    voltage-sensitive calcium channels along developing neurites: Their
    possible role in regulating neurite elongation, <u>Dev. Biol.</u>, 94:351.
Ballanyi, K., and Deitmer, J.W., 1984, Concentration-dependent effects
    of Ba on action potential and membrane currents in the ciliate

Stylonychia, <u>Comp. Biochem. Physiol.</u>, 78A:575.

Bean, B.P., 1985, Two kinds of calcium channels in canine atrial cells. Differences in kinetics, selectivity, and pharmacology, <u>J. Gen. Physiol.</u>, 86:1.

Bolsover, S.R., and Spector, I., 1986, Measurements of calcium transients in the soma, neurite and growth cone of single cultured neurones, <u>J. Neurosci.</u>, 6:1934.

Bossu, J.L., Feltz, A., and Thomann, J.M. 1985, Depolarization elicits two distinct calcium currents in vertebrate sensory neurones, <u>Pflügers Arch.</u>, 403:360.

Carbone, E., and Lux, H.D., 1984, A low voltage-activated fully inactivating Ca channel in vertebrate sensory neurones, <u>Nature</u>, 310:501.

Deitmer, J.W., 1983, Ca channels in the membrane of the hypotrich ciliate Stylonychia, <u>in</u>: "The Physiology of Excitable Cells", A.D. Grinnell, and W.J. Moody, eds., A.R. Liss. Inc., New York.

Deitmer, J.W., 1984, Evidence for two voltage-dependent calcium currents in the membrane of the ciliate Stylonychia, <u>J. Physiol.</u>, 355:137.

Deitmer, J.W., 1986a, Properties of two voltage-dependent calcium channels in a ciliate, <u>in</u>: "Membrane Control of Cellular Activity", H.C. Lüttgau, ed., <u>Progress in Zoology</u>, 33:111.

Deitmer, J.W., 1986b, Voltage dependence of two inward currents carried by calcium and barium in the ciliate <u>Stylonychia</u> <u>mytilus</u>, <u>J. Physiol.</u>, 380:551.

Deitmer, J.W., 1987, Loss of electrical excitability during encystment of the hypotrichous ciliate <u>Stylonychia</u> <u>mytilus</u>, <u>Naturwissenschaften</u>, (in press).

Deitmer, J.W., Ivens, I., and Pernberg, J., 1986, Changes in voltage-dependent calcium currents during the cell cycle of the ciliate <u>Stylonychia</u>, <u>J. comp. Physiol.</u>, 154:113.

De Peyer, J.E., and Deitmer, J.W., 1980, Divalent cations as charge carriers during two functionally different membrane currents in the cyliate <u>Stylonychia</u>, <u>J. exp. Biol.</u>, 88:73.

De Peyer, J.E., and Machemer, H., 1977, Membrane excitability in <u>Stylonychia</u>: properties of the two-peak regenerative Ca-response, <u>J. comp. Physiol.</u>, 121:15.

Dunlap, K., 1977, Localization of calcium channels in <u>Paramecium caudatum</u>, <u>J. Physiol.</u>, 171:119.

Eckert, R., and Chad, J.E., 1984, Inactivation of Ca channels. <u>Progr. Biophys. mol. Biol.</u>, 44:214.

Graubard, K., and Ross, W.N., 1985, Regional distribution of calcium influx into bursting neurons detected with arsenazo III, <u>Proc. Natl. Acad. Sci. USA</u>, 82:4824.

Hagiwara, S., Ozawa, S., and Sand, O., 1975, Voltage clamp analysis of two inward current mechanisms in the egg cell membrane of a starfish, <u>J. Gen. Physiol.</u>, 65:617.

Hirst, G.D.S., and McLachlan, E.M., 1986, Development of dendritic calcium currents in ganglion cells of the rat lower lumbar sympathetic chain. <u>J. Physiol.</u>, 377:349.

Ivens, I., and Deitmer, J.W., 1986, Inhibition of a voltage-dependent Ca current by concanavalin A, <u>Pflügers Arch.</u>, 406:212.

Katz, B., 1969, The Release of Neurotransmitter Substances, Thomas, Springfield, Illinois.

Llinas, R., and Yarom, Y., 1981, Electrophysiology of mammalian interior olivary neurones in vitro. Different types of voltage-dependent ionic conductances, <u>J. Physiol.</u>, 315:549.

Machemer, H., and Deitmer, J.W., 1987, From structure to behaviour: <u>Stylonychia</u> as a model system for cellular physiology, <u>Progr. Protistol.</u>, 2: (in press).

Machemer, H., and Ogura, A., 1979, Ionic conductances of membranes in ciliated and deciliated Paramecium. <u>J. Physiol.</u>, 296:49.

McClesky, E.M., Fox, A.P., Feldman, D., and Tsien, R.W., 1986, Different types of calcium channels, <u>J. exp. Biol.</u>, 124:177.

McLaughlin, S.G.A., Szabo, G., and Eisenman, G., 1971, Divalent ions and surface potential of charged phospholipid membranes, <u>J. Gen. Physiol.</u>, 58:667.

McLaughlin, S.G.A., and Haray, H., 1974, Phospholipid flip-flop and the distribution of surface charges in excitable membranes, <u>Biophys. J.</u>, 14:200.

Nilius, B., Hess, P., Lansman, J.B., and Tsien, R.W., 1985, A novel type of cardiac calcium channel in ventricular cells, <u>Nature</u>, 316:443.

Ogura, A., and Takahashi, K., 1976, Artificial decilication causes loss of calcium-dependent responses in <u>Paramecium</u>, <u>Nature</u>, 264:170.

Schauf, C.L., 1975, The interactions of calcium with <u>Myxicola</u> giant axons and a description in terms of a simple surface charge model. <u>J. Physiol.</u>, 248:613.

Stockbridge, N., and Ross, W.N., 1984, Localized Ca^{2+} and calcium-activated potassium conductances in terminals of a barnacle photoreceptor, <u>Nature</u>, 309:266.

CALCIUM CHANNELS INCORPORATED INTO PLANAR LIPID BILAYERS: PHENOMENOLOGY,
PHARMACOLOGY, AND PHYLOGENY

Barbara E. Ehrlich° and Michael Forte[*]

°Division of Cardiology
University of Connecticut Health Center
Farmington, Connecticut 06032

[*]Vollum Institute for Advanced Biomedical Research
Oregon Health Sciences University
Portland, Oregon 97201

INTRODUCTION

Voltage-dependent calcium (Ca) channels are found in virtually all
eukaryotic cells (Hille, 1984). However, the characteristics of Ca
channels vary among cell types (Hagiwara and Byerly, 1981) and even in the
same cell (Hagiwara et al., 1975; Carbone and Lux, 1984; Armstrong and
Matteson, 1985; Nowycky et al., 1985; Fedulova et al., 1985). Did all
these channels evolve from the same precursor or was the Ca channel
invented several times during phylogeny? To answer this question one
needs to compare Ca channels from the most primitive organism known to have
Ca currents with Ca channels from other "higher" organisms.

The work of Naitoh and Eckert clearly demonstrated the existence of Ca
currents in Paramecium first using current clamp (Naitoh et al., 1972), and
later, using voltage clamp techniques (Naitoh, 1974). To date
voltage-dependent Ca currents have not been found in organisms more
primitive than Paramecium (Saimi and Kung, 1987; Martinac et al., in this
volume). Since Paramecium are large (100 um), it was possible to complete
extensive electrophysiological studies on these cells (Eckert and Brehm,
1979; Kung and Saimi, 1982). Many characteristics of the Ca currents from
Paramecium such as the voltage-dependence, ion selectivity, and
Ca-dependent inactivation are similar to those found in a variety of other
organisms including mammalian heart and nerve (Hagiwara and Byerly, 1981).
Other characteristics such as the pharmacology differ (Hennessey and Kung,
1984).

A striking feature of Paramecium is that all of the Ca channels are on
the cilia, despite the continuity of the ciliary and plasma membrane. The
localization of the Ca currents was elegantly demonstrated by showing that
deciliated cells have no Ca currents and that the return of the Ca current
is proportional to the length of the cilia that has regrown (Dunlap, 1977;
Ogura and Takahashi, 1977).

The localization of the Ca channels to the cilia presents one disadvantage and one advantage to the study of <u>Paramecium</u> Ca channels. The disadvantage is that measurement of single channel currents in intact cells is virtually impossible since the cilia are 200 nm in diameter, too small to patch clamp. However, the localization makes <u>Paramecium</u> a good starting material for reconstitution. One can grow large quantities of a single cell type and it is possible to prepare pure ciliary membrane vesicles. By using ciliary membranes one gets approximately a 100-fold purification (cilia account for ~1% of the total membrane area). In addition, cilia are largely devoid of membranous intracellular organelles, so that intracellular membrane channels do not contaminate the final preparation of ciliary membrane vesicles. Once the Ca channels have been inserted into the bilayer it is possible to measure single channel currents.

This article describes the behavior of <u>Paramecium</u> Ca channels after they have been incorporated into planar lipid bilayers. In particular, it will focus on the pharmacology of these channels since this characteristic can be used to show that the correct channel has been incorporated into the bilayer, to investigate the mechanism of channel-drug interactions, and to compare this channel with other Ca channels. Eventually the pharmacological agents described here may also be useful in the localization and biochemical analysis of this channel.

METHODS

Behavioral assay

The initial screening of pharmacological agents was done using a behavioral assay. This rapid assay generates dose-response curves that correlate well with those obtained with direct measurements of Ca currents in intact <u>Paramecium</u> (Hennessey and Kung, 1984). In this test the duration of backward swimming is measured in the presence of varying concentrations of drug. Backward swimming is induced by depolarizing the cells with potassium. The depolarization opens Ca channels which subsequently increases intracellular Ca (Ca_i). This increase in Ca_i triggers the cilia to beat in the opposite direction and the <u>Paramecium</u> swims backwards. The duration of backward swimming is measured (control values are approximately 55 s). Addition of drugs that decrease the depolarization-induced change in Ca_i will decrease the duration of backward swimming.

Ciliary membrane vesicles

Membrane vesicles were prepared as described previously (Adoutte et al., 1980). After isolation vesicles were resuspended in 10 mM KCl, 10 mM Tris, pH 7.4 at a protein concentration of 3 mg/ml and stored at -80°C until use.

Bilayers and single channel recording

Vesicles were incorporated into preformed planar lipid bilayers formed at the tip of patch-style pipettes (Coronado and Latorre, 1983). To make these bilayers a pure lipid monolayer was layered on top of a lucite well containing 0.3 ml of a buffered salt solution. By passing the pipette twice through the monolayer, a bilayer was formed at the pipette tip. In all bilayer experiments described, the composition of the bath solution was 50 mM $BaCl_2$, 0.5 mM Tris-EGTA, pH 7.2 and the composition of the pipette solution was 50 mM $MgCl_2$, 0.5 mM $BaCl_2$, and 0.5 mM Tris-EGTA, pH 7.2. Membranes were made of phosphatidylethanolamine and phosphatidylserine (1:1 by weight; Avanti Polar Lipid, Atlanta, GA).

After the membrane was formed, vesicles were added to the bath while the solution was stirred. Stirring was maintained until vesicle incorporation was detected. Channel insertion and subsequent experiments were monitored under voltage clamp conditions. The pharmacological composition of the bath was changed by moving the pipette to a new bath. Several transfers could be executed in each experiment without breaking the bilayer. Data were stored on chart and tape recorders and were transferred to a computer for analysis.

RESULTS AND DISCUSSION

Characteristics of the Bilayer-Incorporated Channel

Ca channels have many identifying characteristics which can be used to ensure that the channel in the bilayer really is the Ca channel of interest. The _Paramecium_ Ca channel has additional characteristics that can be compared since there are mutant strains that are known to lack Ca currents. In all cases studied to date the bilayer currents match with expectations from the _in situ_ currents (Ehrlich et al., 1984a). The bilayer-incorporated channel is permeable to Ca, Sr, and Ba, but not to anions, K and Mg. The channel is voltage-dependent; a 10 mV change in the membrane potential produces an e-fold change in the open channel probability. The channel is more likely to be open at positive voltages than negative voltages. These results are similar to those measured in the intact cell. The channel has a conductance of 1.5 - 2 pS, regardless of the salt concentration (1 mM - 100 mM). Again this is the expected result since _in situ_ the Ca current saturates at 1 mM. With the mutant strains that lack Ca currents, it is difficult to convince oneself that no activity really means no channels rather than no incorporation. However, preliminary results with "leaky" mutants demonstrated the appropriate decrease in activity. Taken together, these results strongly suggest that the bilayer-incorporated channel really is the Ca channel found in _Paramecium_.

The pharmacology of this channel is unlike vertebrate Ca channels. The only class of compounds that have been shown to block the _Paramecium_ Ca channel is the sulphonated naphthalenes (Hennessey and Kung, 1984). The classical Ca channel blockers, verapamil, diltiazem and the dihydropyridines are ineffective. The structures of two sulphonated naphthalenes are shown in Figure 1. These compounds were originally designed to be calmodulin antagonists (Hidaka et al., 1981). To get complete block of the _Paramecium_ Ca channel one needs 100 uM W-7. W(12)Br is approximately 100 times more effective than W-7.

W-7 $\quad$ $Cl-$ naphthalene $-SO_2NH(CH_2)_6NH_2$

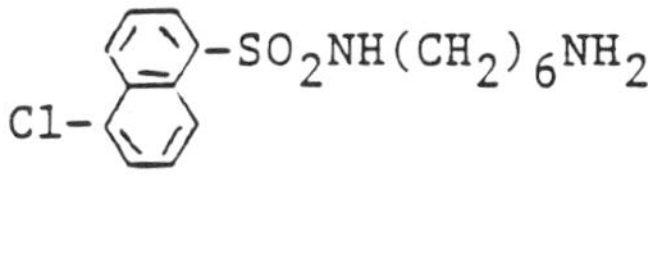

W(12)Br $\quad$ $Br-$ naphthalene $-SO_2NH(CH_2)_{12}NH_2$

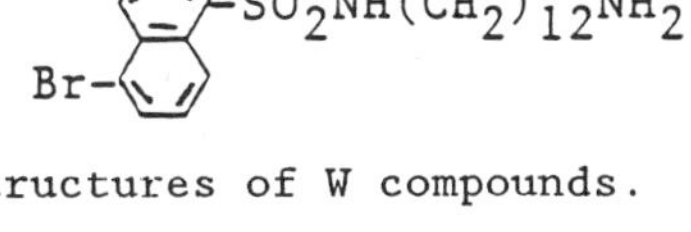

Figure 1. Structures of W compounds.

Figure 2 compares the effectiveness of W-7 on the Ca currents of the
Paramecium. The lower traces of Figure 2 are taken from the work of
Hennessey and Kung (1984). 150 uM W-7 reversibly blocks the Ca current in
intact cells. The bilayer currents (Figure 2, top traces) respond the same
way. In this series of recordings downward deflections are channel
openings. For the bilayer currents the transmembrane potential was held at
+100 mV, where the bath is ground. From the voltage-dependence of the
channel openings, and using the physiological convention that the outside
of the cell is ground, the intracellular face of the channel must be toward
the bath. Within one minute of transferring the pipette to a solution
containing 100 uM W-7, the middle trace was recorded. Two minutes after
returning to the initial bath, the final trace was obtained. Sometimes the
membrane could be transferred several times, and with each exposure to W-7
the channels were inhibited.

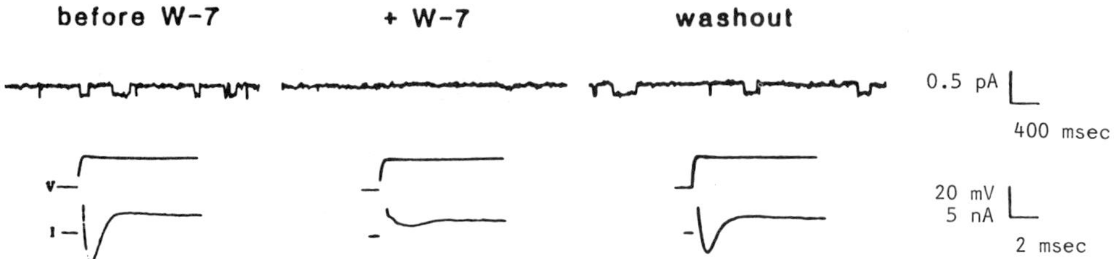

Figure 2.   The effect of W-7 on Paramecium Ca currents. The top traces
             show the current generated by a single Ca channel
             incorporated into a bilayer. The middle traces show the
             voltage and the bottom traces are the current from a
             voltage-clamped intact Paramecium (Hennessey and Kung,
             1984). Addition of W-7 blocked the Ca current in both the
             bilayer (100 uM) and the intact cell (150 uM). Currents
             return after washout of W-7.

Figure 3 shows a dose response curve for W(12)Br when it is tested on
the avoiding behavior of intact Paramecium. In this test the duration of
backward swimming is measured in the presence of varying concentrations of
W(12)Br. Backward swimming is induced by depolarizing the cells with
potassium. 100% response is the duration of backward swimming in the
presence of no drug. At 1 uM W(12)Br the behavior is completely abolished,
and the apparent K_d is approximately 0.1 uM. A similar curve is
generated when the drug is tested on Ca channels that have been
incorporated into planar lipid bilayers (Ehrlich et al., 1986). As in the
behavioral assay, 1 uM W(12)Br completely abolishes channel activity in the
bilayer and the apparent K_d is approximately 0.1 uM. As with W-7, the
effect is reversible, but the washout takes longer (4-10 min).

Drug-Channel Interactions

Preliminary results (Ehrlich et al., 1986) suggest that the drugs
interact with the channel in at least two ways and that the interactions
are independent of calmodulin. It appears that W(12)Br exerts its effect
by decreasing the mean open time of the channel and by increasing the
probability of very long silent periods. To speculate, we imagine that the
halogenated ring end of the drug is embedded in the membrane and that the
long alkyl chain intercalates between the helices of the channel protein.
The location of the alkyl chain would allow the primary amine end, which is

presumably charged at neutral pH, to intermittently enter the channel
producing "flickery" block. The long closings may result from interactions
between the halogenated ring of the drug and the gating part of the
channel.

     To test the possibility that the W compounds were acting in a
nonspecific manner in the bilayer, we tested the effectiveness of local
anesthetics on these channels. Local anesthetics were chosen because the W
compounds and the local anesthetics share some structural features. After
determining that procaine had no effect on Ca currents in the bilayer, we
discovered that procaine, lidocaine, and tetracaine had been tried more
than a decade ago on the behavioral response (Browning and Nelson, 1976),
and on the Ca currents in intact _Paramecium_ (Freidman and Eckert, 1973)
with the same negative result. This result supports the hypothesis that
the blockade by the W compounds involves a specific interaction between the
drug and some part of the channel.

     We do not think that the W compounds are inhibiting the channel
through an effect on calmodulin. The evidence supporting this hypothesis
includes 1) the W compounds are the only calmodulin antagonists that
inhibit the channel (many other classes of antagonists have been tried),
and 2) the agents work in the bilayer where calmodulin may be absent or, at
least, very dilute. It should be noted, however, that W(12)Br is a better
calmodulin inhibitor than W-7 and that the magnitude of the improvement is
similar to the improvement in channel blocking.

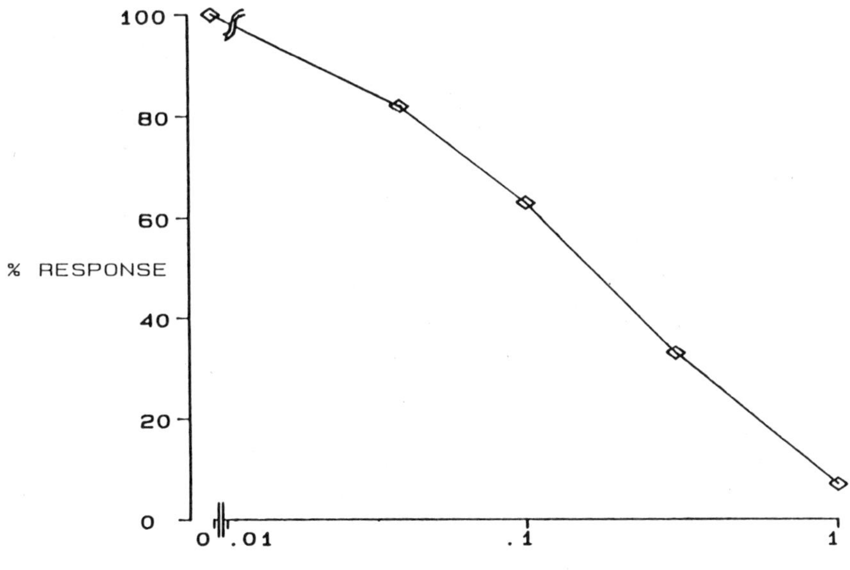

Figure 3.   The effect of W(12)Br on the duration of backward swimming
              by intact _Paramecium_. In the absence of drug, cells swim
              backwards upon depolarization with potassium for 55 s.
              1 uM W(12)Br abolishes the response. This test correlates
              well with direct measurements of Ca currents.

Comparative Pharmacology of the Ca Channel

The most primitive organism known to have Ca channels is <u>Paramecium</u>.
Is the <u>Paramecium</u> channel like mammalian Ca channels?   Since the
<u>Paramecium</u> channel is not sensitive to the dihydropyridines, it clearly
cannot be like the "L" channel found in a variety of cells (Carbone and
Lux, 1984; Armstrong and Matteson, 1985; Nowycky et al., 1985; Fedulova et
al., 1985; Fox et al., in this volume). What about the other types of Ca
channels?   Table 1 lists one or two examples from phyla that have been
shown to have Ca channels and the known pharmacological agents which block
these channels. Note that prokaryotes do not have Ca channels (Ehrlich et
al., 1984b; Saimi and Kung, 1987; Martinac et al., in this volume) and
that Ca channels have not been found in yeast (Saimi and Kung, 1987;
Martinac et al., in this volume).

Moving up the phylogenetic tree, Ca channels of protozoa, sponges,
coelenterates, and ctenophores are inhibited by the W compounds, but they
are unaffected by the dihydropyridines. The pharmacological sensitivity of
the protozoa and the coelenterate were tested electrophysiologically
(Hennessey and Kung, 1984; Takeda, Brehm, and Dunlap, personal
communication), while behavioral assays were used to test the sponge and
the ctenophore (Duhnam et al., 1983; Tamm, personal communication).

The next two phyla are known to have Ca channels (Weisblat et al.,
1976; Hagiwara et al., 1975), but the pharmacological sensitivity of
representative examples is still unknown.

The molluscs, annelids, and arthropods are sensitive to neither the
dihydropyridines nor the W compounds. However, at least one species of
arthropod (<u>Drosophila</u> <u>melanogaster</u>; Greenberg et al., 1987) shows specific
binding of verapamil. Actually, only dihydropyridines have been tested on
annelid nerve, but it appears very unlikely that this organism will be
sensitive to W-7. Electrophysiological techniques were used to test the
mollusc (Chow, personal communication; Swandulla, personal communication)
and annelid (Kleinhaus, personal communication). Behavioral and binding
assays were used to test the arthropod (Greenberg et al., 1987).

Vertebrate cells have two, and sometimes three, types of Ca currents.
None of these currents responded to W-7. Only the "L type" current is
sensitive to the dihydropyridines. Both heart and nerve cells from <u>Rana</u>
were tested and none of the Ca currents were altered by W-7 (Bean, personal
communication). The <u>Rattus</u> cells tested were cultured GH3 cells (a cell
line derived from pituitary cells). Neither current in this cell line was
altered by W-7 (Cota, personal communication).

In summary, Ca channels can be grouped by their sensitivity to
pharmacological agents. Only the vertebrates are sensitive to the
dihydropyridines, and among the vertebrate Ca channels, only the "L type"
is sensitive. Only the lower invertebrates are sensitive to W-7. The
higher invertebrates are sensitive to neither the dihydropyridines nor the
W compounds. What happened to the Ca channels? How different are they?
Are the different channels comprised of completely different proteins? Or
have there been small modifications in the channel which modify the drug
sensitivity in a manner akin to the lack of tetrodotoxin sensitivity in the
sodium channel of puffer fish?

ACKNOWLEDGMENTS

This work was supported in part by NSF grant DCB83-09110.   B.E.E. is a
PEW Scholar in the Biomedical Sciences.

Table 1. Comparative Pharmacology of the Calcium Channel

PHYLUM	GENUS	DRUG SENSITIVITY	
		W-7	DHP[*]
Archaebacteria	Halobacterium[a]	[calcium	]
Eubacteria	Escherichia[b]	channels	
Fungi	Saccharomyces[b]	not present]	
Protozoa	Paramecium[c]	yes	no
Sponges	Microciona[d]	yes	no
Coelenterates	Obelia[e]	yes	no
Ctenophores	Beroë[f]	yes	no
Nematodes	none tested	?	?
Echinoderms	none tested	?	?
Molluscs	Loligo[g]	no	no
	Helix[h]	no	no
Annelids	Hirudo[i]	?	no
Arthropods	Drosophila[j]	no	no
Chordates	Rana[k] - "L type"	no	yes
	- "T type"	no	no
	Rattus[l] - "L type"	no	yes
	- "T type"	no	no

[*]DHP, dihydropyridine
a. Ehrlich et al., 1986
b. Saimi and Kung, 1987; Martinac et al., in this volume
c. Hennessey and Kung, 1984
d. Dunham et al., 1983
e. K. Takeda, P. Brehm, and K. Dunlap, personal communication
f. S. Tamm, personal communication
g. R. Chow, personal communication
h. D. Swandulla, personal communication
i. A. Kleinhaus, personal communication
j. Greenberg, Ehrlich, and Hall, submitted for publication
k. B. Bean, personal communication
l. G. Cota, personal communication

REFERENCES

Adoutte, A., Ramanathan, R., Lewis, R.M., Duto, R.R., Ling, K., Kung, C., Nelson, D.L. (1980) Biochemical studies of the excitable membrane of *Paramecium tetraurelia*. Journal of Cell Biology 84: 717-738.

Armstrong, C.M., Matteson, D.R. (1985) Two distinct population of calcium channels in a clonal line of pituitary cells. Science 227: 65-67.

Browning, J.L. Nelson, D.L. (1976) Amphipathic amines affect membrane excitability in *Paramecium*: Role for bilayer couple. Proceedings of the National Academy of Sciences (USA) 73: 452-456.

Carbone, E., Lux, H.D. (1984) A low voltage-activated calcium conductance in embryonic chick sensory neurons. Biophysical Journal 46: 413-418.

Coronado, R., Latorre, R. (1983) Phospholipid bilayers made from monolayers on patch-clamp pipettes. Biophysical Journal 43: 231-236.

Dunham, P., Anderson, C., Rich, A., Weissman, G. (1983) Stimulus-response coupling in sponge cell aggregation: Evidence for calcium as an intracellular messenger. Proceedings of the National Academy of Sciences 80: 4756-4760.

Dunlap, K. (1977) Localization of calcium channels in *Paramecium caudatum*. Journal of Physiology (London) 271: 119-133.

Eckert, R., Brehm, P. (1979) Ionic mechanisms of excitation in *Paramecium*. Annual Reviews in Biophysics and Bioengineering 8: 353-383.

Ehrlich, B., Cohen, A., Forte, M. (1986) Calcium currents in paramecium are blocked by one class of calmodulin antagonists. Biological Bulletin 171: 492-493.

Ehrlich, B., Finkelstein, A., Forte, M., Kung, C. (1984a) Voltage-dependent calcium channels from *Paramecium* cilia incorporated into planar lipid bilayers. Science 225: 427-428.

Ehrlich, B.E., Schen, C.R., Spudich, J.L. (1984b) Bacterial rhodopsins monitored with fluorescent dyes in vesicles and in vivo. Journal of Membrane Biology 82: 82-94.

Fedulova, S.A., Kostyuk, P.G., Veselovsky, N.S. (1985) Two types of calcium channels in the somatic membrane of newborn rat dorsal root ganglion neurons. Journal of Physiology (London) 359: 431-446.

Fenwick, E.M., Marty, A., Neher, E. (1982) Sodium and calcium channels in bovine chromaffin cells. Journal of Physiology (London) 331: 599-635.

Friedman, K. Eckert, R. (1973) Ionic and pharmacological modification of input resistance and excitability in *Paramecium*. Comparative Biochemistry and Physiology 45A: 101-114.

Greenberg, R., Ehrlich, B.E., Hall, L.M. (1987) Membrane extracts from *Drosophila melanogaster* heads contain binding sites for [^{3}H]Verapamil, a calcium channel blocker. Submitted for publication.

Hagiwara, S., Byerly, L. (1981) Calcium channel. Annual Reviews in Neuroscience 4: 69-125.

Hagiwara, S., Ozawa, S., Sand, O. (1975) Voltage clamp analysis of two inward current mechanisms in the egg cell membrane of a starfish. Journal of General Physiology 65: 617-644.

Hennessey, T.M., Kung, C. (1984) An anticalmodulin drug, W-7, inhibits the voltage-dependent calcium current in *Paramecium caudatum*. Journal of Experimental Biology 110: 169-181.

Hidaka, H., Asano, M., Tanaka, T. (1981) Activity-structure relationship of calmodulin antagonists: naphthalenesulfonamide derivatives. Molecular Pharmacology 20: 571-578.

Hille, B. "Ionic Channels of Excitable Membranes," Sinauer Associates Inc., Massachusetts (1984).

Kung, C., Saimi, Y. (1982) The physiology of taxes in *Paramecium*. Annual Review of Physiology 44: 519-534.

Naitoh, Y. (1974) Bioelectric basis of behavior in protozoa. American Zoologist 14: 883-893.

Naitoh, Y., Eckert, R., Freidman, K. (1972) A regenerative calcium
response in _Paramecium_. Journal of Experimental Biology 56: 667-681.
Nowycky, M., Fox, A., Tsien, R.W. (1985) Three types of neuronal calcium
channel with different calcium agonist selectivity. Nature
316: 440-443.
Oertel, D., Schein, S.J., Kung, C. (1977) Separation of membrane currents
using a _Paramecium_ mutant. Nature 268: 120-124.
Ogura, S., Takahashi, T. (1977) Artificial deciliation causes loss of
calcium-dependent responses in _Paramecium_. Nature 264: 170-172.
Saimi, Y. Kung, C. (1987) Ion channels of _Paramecium_, Yeast and
Escherichia coli. in: "Current Topics in Membrane and Transport,"
Volume 25. Molecular Biology of Ion Channels. W.S. Agnew, ed.,
Academic Press, New York.
Weisblat, D.A., Byerly, L., Russell, R.L. (1976) Ionic mechanism of
electrical activity in somatic muscle of the nematode _Ascaris
lumbricoides_. Journal of Comparative Physiology 111: 93-113.

MODULATION OF IONIC SELECTIVITY OF Ca CHANNELS IN

THE NEURONAL MEMBRANE BY Ca IONS

Sergei L. Mironov

A. A. Bogomoletz Institute of Physiology, Kiev
252601, USSR

INTRODUCTION

Calcium channels play an important role in the behavior of nerve
cells, establishing the pathways of Ca entry. For control and modulation
of cell function it is pertinent to know the selectivity properties of Ca
channels and their possible modification by different agents.

Recently in our laboratory (Kostyuk, Mironov & Shuba, 1983) and
others (Almers & McCleskey, 1984; Hess & Tsien, 1984; Fukushima &
Hagiwara, 1985; Byerly, Chase & Stimmers, 1985) it was found, that Ca
channel permeability can be effectively modulated by Ca and other divalent
cations. These results and other relevant data will be discussed here,
using recently developed molecular models, which suggest definite
structures of ion-transporting pathways in Ca channels.

Ca^{2+} MODULATION OF CALCIUM CHANNEL PERMEABILITY

Kostyuk and Krishtal (1977) were the first to show that EDTA addition
to Ca-free external solution leads to the modification of Ca channels,
which become permeable to monovalent cations. In a detailed study of this
effect (Kostyuk, Mironov & Shuba, 1983) we found a similar action of other
Ca-chelating agents. Raising external Ca concentration into the
micromolar ranges we observed a decrease of Na current through these
modified Ca channels. This block did not depend on membrane potential in
a wide range. This may indicate that a corresponding Ca-binding site is
located at the external surface of the membrane.

Similar effects were observed for other alkaline earth metal
cations. However, they blocked Na current through Ca channels with less
potency. Corresponding dissociation constants were 0.2 µM (Ca),
3.5 µM (Sr), 14 µM (Ba) and 60 µM (Mg). This blocking effect gave
place to the development of a corresponding divalent ion current (except
Mg) in the millimolar range of their concentrations.

All these data suggest that the structure of the observed high
affinity Ca-binding site of the channel may be similar to that of sites,
found in such proteins as troponin C, calmodulin, parvalbumin etc. (Levine
& Williams, 1982). These Ca-binding sites usually contain several

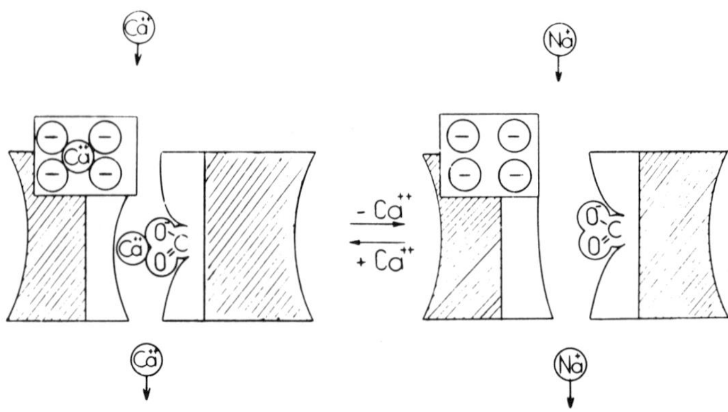

Fig. 1. Suggested structure and location of Ca binding sites in the
calcium channel and their possible function in
ion-transporting mechanisms.

carboxylic groups, establishing an octahedral coordination of Ca ion. In
studies of different Ca-binding proteins it was shown that Ca binding to
these sites is accompanied by a movement of alpha-helices, which
eventually induces a conformational rearrangement of the whole protein.

These and earlier data (Kostyuk, Moronov & Doroshenko, 1982),
obtained in a study of the relative permeability and blocking action of
different divalent cations, led us to a model for the possible
organization of ion-transporting pathways in the calcium channel, shown in
Fig. 1. According to it, channel permeability for different divalent
cations is mainly determined by so called inner selectivity filter or
Hagiwara's binding site (Hagiwara & Takahashi, 1967). Judging from
corresponding dissociation constants, we concluded that this site should
contain only one carboxylic group, thus having a relatively low affinity
to divalent cations. The stronger a given divalent cation binds to this
site, the higher is its blocking potency and the lower is its
permeability. This scheme can quantitatively explain the ability of the
channel to pass alkaline earth metal cations and the blocking effect of
transition metal cations. Thus, the presence of a single carboxylic group
inside the channel forms a necessary and sufficient condition for its
selectivity among divalent cations.

Ca channels must also possess a high-affinity Ca-binding site. If it
is placed on the ionic pathway through the channel, this lowers the single
channel conductance at least two orders of magnitude below the
experimentally observed value. Therefore we moved it away from the ionic
pore onto the external membrane surface. Such a scheme leads to a quite
natural suggestion that Ca binding in this site in normal physiological
conditions is responsible for high channel selectivity for divalent
cations. Ca exit from this site, as in other Ca-binding proteins, should
induce some conformational transition of the channel to a state, where it
can pass other, e.g. monovalent cations.

It is interesting that according to this model the second
conformational state of the channel may exist even if the Ca content in
the external solution is considerably higher than the dissociation

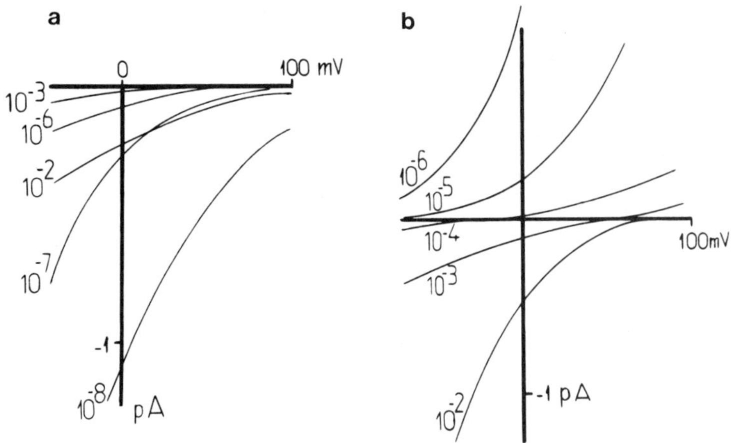

Fig. 2.   I-V curves for the model of a Ca channel with different
conformational states, calculated on the basis of the
kinetic scheme, represented by Eq. (1), at various external
Ca concentrations, indicated near each curve.   Na
concentration in the external (a) or the internal (b)
solutions was 30 mM.   Model parameters and main equations
were given elsewhere (Kostyuk & Mironov, 1986).

constant for Ca-binding to the high-affinity site. We modelled (Kostyuk &
Mironov, 1986) the blocking effect of Ca ions on the inward and the
outward sodium currents through the calcium channel, translating the
scheme, shown in Fig. 1, into the formal kinetic model

$$(Ca_1Ca_h) \leftrightarrow (O_1Ca_h) \leftrightarrow (O_1O_h) \leftrightarrow (Na_1O_h) \qquad (1)$$

where notations Ca, O and Na stand for the occupancy mode of low (index 1)
and high-affinity (index h) channel binding sites.

The behaviour of I-V curves for the Ca channel is mainly determined
by the corresponding dissociation constants, known from experiment
(Kostyuk, Mironov & Doroshenko, 1982; Kostyuk, Mironov & Shuba, 1983).
Fig. 2a shows that this model describes well the block of the inward Na
current due to an increase of external Ca. This is followed by the
development of Ca current in the millimolar range of Ca concentration.

Ca block of the outward Na current is accompanied by an increase in
the inward Ca current (Fig. 2b). Superposition of these currents gives a
clear reversal potential. Explanation of this effect is as follows: a
decrease in the driving force for extracellular Ca ions lowers the
probability of a Ca-conduction state. Thus, due to this shift in the
equilibrium occupancy of channel states, the probability of the channel
occupying the Na-conducting state increases. This effect converts the
channel at positive potentials into a state, where it can pass monovalent
cations.

A reversal potential for Ca channels was observed by Lee and Tsien
(1984) for heart cells and by Fukushima and Hagiwara (1985) for mouse
myeloma cells. In these studies Ca currents in the whole range of

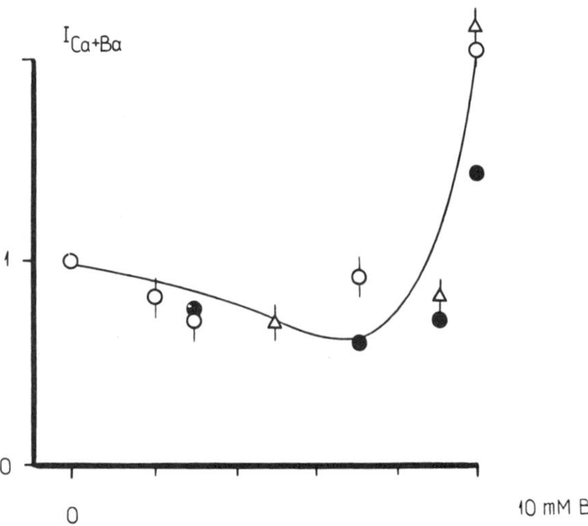

Fig. 3. Prediction of an anomalous mole fraction effect of divalent
ion current by a model Ca channel with different
conformational states. Model parameters and main equations
were given elsewhere (Kostyuk & Mironov, 1986). Calculated
curve corresponds to a zero membrane potential. Points
represent experimental data obtained by Hess and Tsien
(1984) (●), Almers and McCleskey (1984) (Δ), Byerly et
al. (1985) (o).

potentials were successfully isolated from other interfering ionic
currents. Unfortunately, neither we, nor Byerly, Chase and Stimers (1985)
could observe the same effect for neuronal membranes. Probably, this is
due to a large potassium conductance in nerve cells, a complete block of
which could not be achieved using traditional pharmacological agents.

Our model describes also an anomalous mole fraction effect, observed
recently in divalent ion mixtures (Almers & McClesky, 1984; Hess and
Tsien, 1984; Byerly, Chase & Stimers, 1985). This effect (Fig. 3) reveals
a non-monotonic dependence of total divalent ion current on Ca or Ba
concentration, when the total concentration of divalent ions is held
constant.

Such behavior of divalent ion current was used as evidence for a
multi-ion mode of Ca conductance. Both Almers and McCleskey (1984) and
Hess and Tsien (1984) developed models of a Ca channel, which contain two
high-affinity Ca-binding sites inside the channel. These sites interact
with negative cooperativity due to the Coulomb repulsion between ions.

It was pointed above that any model of a Ca channel possessing one or
several high-affinity Ca-binding sites in an ion-transporting pathway
should have an extremely low single channel conductance. Therefore, to
obtain a high channel conductance, these two models must necessarily have
additional assumptions. Thus, Almers and McCleskey suggested that the
Coulomb repulsion affects only the exit rate of the second ion from the
channel interior, whereas it has no effect on the entrance rate. The
model of Hess and Tsien is free of this assumption and they had to
increase the channel conductance by lowering the height of the terminal
potential barriers. However, in this case the entrance rate becomes much

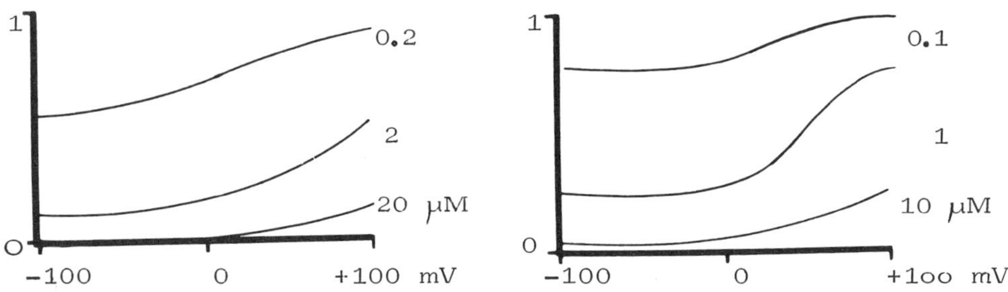

Fig. 4. Voltage dependence of Ca block of Na current through the Ca channels. Curves are calculated according to conformational (left) and multi-ion models (right). External and internal Na concentrations were 140 mM. Ca concentration in the external solution is indicated near each curve.

higher than the so called diffusion limit, which usually restricts the ion flow through the channel.

An intrinsic property of multi-ion models is the considerable difference between dissociation constants for binding the first and the second Ca ions within the channel. Therefore this effect should also be revealed for different Ca-binding proteins which usually possess several high-affinity Ca-binding sites. However, all these sites in the same protein have apparently the same dissociation constants for Ca ions and their values do not depend on the occupancy of neighbour Ca-binding sites.

Both multi-ion models successfully describe the dependence of Na and Ca currents through the Ca channel on the external Ca. These models predicted also an anomalous mole fraction behaviour, which later was observed experimentally. Finally, these theoretical considerations help to understand qualitatively the physical basis of Ca current reversal.

PREDICTING ABILITY OF CALCIUM CHANNEL MODELS

The above discussion shows that from a mathematical point of view both conformational and multi-ion models of Ca channel can describe all available experimental data equally well. Therefore, at present it seems reasonable not to dispute the physico-chemical concepts on which they are based, but instead to study in detail the predictions of these models and to search for new ways of testing them experimentally.

The simplest way to modulate the channel selectivity is to change the ionic composition of the medium. Fig. 5 presents a detailed analysis of anomalous mole fraction effects for both models. It shows the dependence of the relative amplitude of mixed divalent ion current on external Ba for different membrane potentials and total concentration of divalent cations. It is seen that an amomalous mole fraction effect is more pronounced at positive membrane potentials and for relatively small concentrations of divalent cations. However, both conformational and

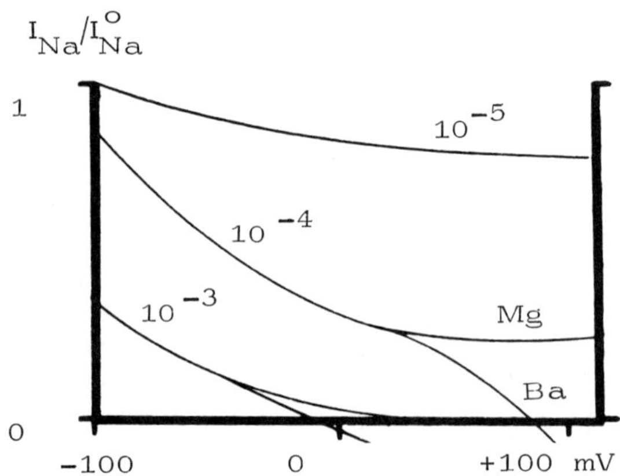

Fig. 6. Block of Na current through Ca channels by permeant (Ba) and
impermeant (Mg) cations. Curves were calculated according
to multi-ion model of Almers and mcCleskey (1984). Na
(30 mM) was present only in the external solution. External
concentration of divalent cations is indicated near each
curve.

modification of the calcium channel. The induction of the Na current was
observed only about 20 minutes later, after a train of depolarizing
pulses. It seems that this considerable delay is due to a removal of
residual Ca ions from the external solution, leaking into the cytoplasm
via Ca channels.

The most promising way to show clearly the asymmetry of Ca block of
monovalent ion current through Ca channels is to follow ideas, developed
in recent years by Prof. R. Eckert and his colleagues in studies of
mechanisms of Ca-mediated inactivation and the possible role of different
Ca-binding proteins in the maintenance of the functional state of Ca
channels. Using these experimental approaches, one can successfully
isolate possible direct interactions of Ca ions with Ca channels from
their interfering metabolic action (Chad & Eckert, 1985).

It should also be mentioned that the conformational flexibility of
Ca-binding proteins can be accounted for in the following extension of our
model. Let us suppose, that the Ca channel after its opening undergoes a
conformational transition, induced by binding of the penetrating cation in
the high-affinity site. The open state, thus obtained, may be permeable
only to this type of ion and persist for a long time. Thus, this model
unites the ion-transporting and part of the gating mechanisms.

Such an extended model explains the direct correlation between the
channel open time and the binding properties of penetrating ions, as well
as the absence of such effects for the channel closed time. These effects
were recently observed in our (Savchenko & Shuba, manuscript in
preparation) and other laboratories (Nelson et al., 1984;
Chesnoy-Marchais, 1985).

This model also explains some anomalous properties of Ca channels.
Doroshenko and Martynyuk (1985) in our laboratory observed an abrupt

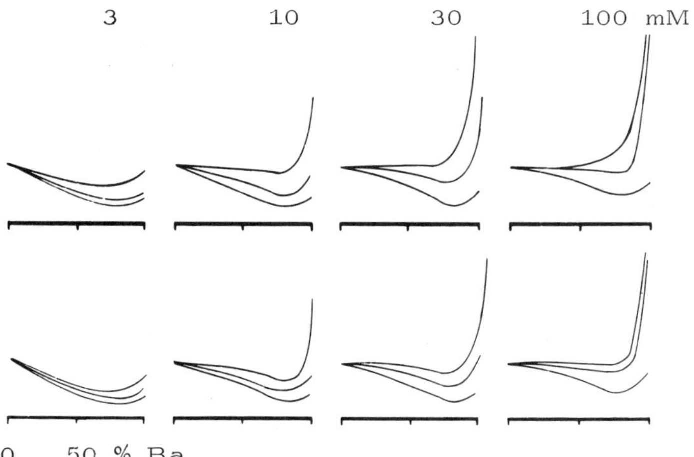

Fig. 5.   Prediction of an anomalous mole fraction effect for
conformational (bottom) and multi-ion (top) models. Total
concentration of divalent cations are shown in the uppermost
line. Upper, middle and lower curves correspond to membrane
potentials, equal to -80, 0 and +80 mV, respectively.

multi-ion models possess qualitatively similar properties and it is
unlikely that one can make a proper choice in this way.

It was also suggested that differences between two models may arise
from the voltage dependence of divalent ion block of the monovalent ion
currents through the calcium channels. However, both conformational and
multi-ion models (Fig. 4) predict almost the same weak dependence of block
on membrane voltage.

In general, the effect of internal divalent cations on the monovalent
ion current seems the most direct way to discriminate between the two
models. It is evident, that in multi-ion models the block of sodium
current by divalent cations should be the same, independent of the side of
their application. Although free Mg concentration inside the cell reaches
several mM, the presence of Mg in the internal solution did not prevent
the development of Na current through the Ca channels, as is expected from
a multi-ion model (Fig. 6). Millimolar intracellular concentrations of Ca
or Ba, in addition to blocking the monovalent ion current should also lead
to the appearance of the corresponding outward current.

In our experiments we have no direct evidence for such an effect
(Kostyuk, Mironov & Shuba, 1983). Ca ions, being applied intracellularly,
did block the inward Na current, but the development of this effect was
very retarded, compared to the time usually needed for a complete change
of intracellular solution during cell perfusion. Therefore, we can
tentatively conclude that this effect is not mediated by direct
interaction of Ca ions with the channels, but is probably connected with
some metabolical processes, which ultimately control the functional state
of Ca channels. For a comparison it should be noted, that according to
the multi-ion model, Ca block of the Na current should be almost
instantaneous on this time scale.

In addition, in a Ca-free external solution intracellular perfusion
of neurons by an EDTA-containing solution did not cause an immediate

decrease in the Ca current after replacing all external Ca with Ba. This
effect was induced only after the opening of Ca channels. It, possibly,
reflected some unknown mechanism of the functioning of the open Ca
channel, which may be related to the mechanism proposed here. McDonald et
al. (1986) also observed a dependence of an anomalous mole fraction effect
on the type of ion (Ca or Ba), with progressive replacement in the
external solution. Such hysteresis may also indicate some long term
changes (probably, conformational) in channel function, which depend on
the type of penetrating cation.

This hypothesis at present seems to be acceptable as a working
mechanism only for one type of calcium channel, namely Ca channels in
sarcoplasmic reticulum, activated by intracellular Ca ions (Martonosi,
1984). Whether this is true for other Ca channels, further experiments
will show, but we hope that this mechanism is worth detailed study and
analysis.

SUMMARY

Data on the modulation by divalent cations of ionic selectivity of Ca
channels are presented. They are described using a model, which assumes
the existence of several conformational states of the channel, possessing
different permeability properties. This model also suggests a coupling
between ion-transporting and gating mechanisms in Ca channels. Data,
supporting this hypothesis, are considered. The predicting ability of
this conformational model is compared with multi-ion models of Ca channels.

REFERENCES

Almers, W., McCleskey, E., 1984, Non-selective conductance in Ca
        channels of frog muscle: Ca selectivity in a single-file pore,
        J. Physiol., 353:585.
Byerly, L., Chase, P., Stimers, J., 1985, Permeation and interaction
        of divalent cations in Ca channels of snail neurons, J. Gen.
        Physiol., 85:491.
Chad, J., Eckert, R., 1985, Ca current inactivation is slowed in dialysed
        neurons by the substitution of ATP-γ-S for internal ATP, J. Gen.
        Physiol., 86:27a.
Chesnoy-Marchais, D., 1985, Kinetic properties and selectivity of Ca
        permeable single channels in Aplysia neurones, J. Physiol., 367:457.
Fukushima, Y., Hagiwara, S., 1985, Currents carried by monovalent
        cations through Ca channels in mouse neoplastic B lymphocytes,
        J. Gen. Physiol., 358:255.
Hagiwara, S., Takahashi, S., Surface density of Ca ions and Ca spikes
        in the barnacle muscle fiber membrane, 1967, J. Gen. Physiol., 50:583.
Hess, P., Tsien, R., 1984, Mechanism of ion permeation through Ca
        channel, Nature, 309:453.
Kostyuk, P., Mironov, S., Doroshenko, P., 1982, Energy profile of Ca
        channel in mollusc neuronal membrane, J. Membrane Biol., 70:181.
Kostyuk, P., Mironov, S., Shuba, Ya., 1983, Two ion-selecting filters
        in the calcium channel of mollusc neuronal membrane, J. Membrane
        Biol., 76:83.
Kostyuk, P., Mironov, S., 1986, Some predictions of the model of Ca
        channel with different conformational states, 1986, J. Gen. Physiol.
        Biophys., 6:681.
Lee, K., Tsien., R., 1984, High-selectivity of Ca channels in single
        dialysed heart cells of the guinea-pig, J. Physiol., 354:253.

Levine, B., Williams, R., 1982, Ca binding to proteins and other large biological anion centres, in: <u>Ca and cell function</u> W. Cheung, ed., Wiley, New York.

Martonosi, A., 1984, Mechanisms of Ca release from sarcoplasmic reticulum of skeletal muscle, <u>Physiol. Rev.</u>, 64:1240.

McDonald, T., Cavalie, A., Trautwein, W., Pelzer, D., 1986, Voltage-dependent properties of marcoscopic and elementary Ca channel currents in guinea pig ventricular myocytes, <u>Pflugers Arch.</u>, 406:448.

Nelson, M., French, R., Krueger, B., 1984, Voltage-dependent Ca channels from rat brain incorporated into planar lipid bilayers, <u>Nature</u>, 308:77.

PROTON-INDUCED TRANSFORMATION OF Ca^{2+} CHANNEL

IN DORSAL ROOT GANGLION NEURONS

Martin Morad and Hans Dieter Lux[*]

University of Pennsylvania
Department of Physiology
Philadelphia, PA, USA

*MPI for Psychiatry
8033 Planegg-Martinsried, FRG

> "O you youths, Western youths,
> So impatient, full of action, full
>   of manly pride and friendship,
> Plain I see you Western youths, see
>   you tramping with the foremost,
> Pioneers! O Pioneers!"
>   — Walt Whitman

INTRODUCTION

Ca^{2+} channels are widely distributed in peripheral and central neurons.
The voltage and time dependence of this channel and its pharmacological
sensitivity have been extensively studied in both neuronal and muscular
tissues. Based on such studies, in fact, more than one type of Ca^{2+} channel
has been identified (Carbone & Lux, 1984; Bean, 1985; Nilius et al., 1985;
Mitra & Morad, 1986; Armstrong & Matteson, 1985). Although efforts to
isolate and molecularly identify Ca^{2+} channels are well underway (Borsotto
et al., 1985; Ferry et al., 1985; Curtis & Catterall, 1984; Nakayama et al.,
1987), their characterization in neuronal or muscular cells has been
primarily based on their electrophysiological properties, gating, and ionic
selectivity.

In neuronal cells Ca^{2+} channel is thought to contribute to pacemaking,
neurohumoral secretion, and transmission of the electrical signal. The wide
distribution and high density of Ca^{2+} channels in the somatic membranes of
the central and peripheral neurons, where little presynaptic or
neurosecretory action occurs, is puzzling. It is perhaps relevant to ask
whether Ca^{2+} channels of the cell body do in fact transport Ca^{2+} under
physiological conditions? At first, this appears to be too unlikely since
Ca^{2+} channels are known to be highly selective for Ca^{2+} in the presence of
normal Na^+ concentrations. Recent studies, in fact, show that Ca^{2+} channels
transport Na^+ under voltage-gated conditions only when the $(Ca^{2+})_o$ is
reduced to micromolar levels (Hess, Lansman & Tsien, 1986; Hess & Tsien,
1984; Almers & McCleskey, 1984; Carbone & Lux, 1987). Half maximum blockade
of Na^+ current through the Ca^{2+} channel occured around 0.1 to 10 μM. Though

the Na^+ transporting property of the Ca^{2+} channel is of considerable
biophysical interest, no physiological or pathological role for the Na^+
selectivity of the Ca^{2+} channel has been proposed, primarily because $(Ca^{2+})_o$
never drop to such low levels under physiological or pathological
conditions.

In this report we shall describe a condition under which Ca^{2+} channels
lose transiently their Ca^{2+} specificity and voltage-gating and transport Na^+
to a high degree of selectivity over Ca^{2+} or other monovalents. We will
show that _small_ but rapid increases in the extracellular proton
concentrations may activate large Na^+ currents through the Ca^{2+} channel
which inactivate slowly within 1-2 seconds. The increase in the
extracellular proton concentration appears to modify the Ca^{2+} channel from a
Ca^{2+} to a Na^+ transporting state. Since changes in the extracellular pH of
neuronal cells are known to occur under a number of physiological and
pathological conditions (Lehmenkühler et al., 1981; Kraig et al., 1983,
Urbanics et al., 1978), the Na^+ transporting properties of Ca^{2+} channel may
be biologically significant.

METHODS

Two- to three-day-old chick cultured DRG neurons were studied using the
patch clamp technique in either whole cell or the isolated membrane patch
configuration (Hamill et al., 1981). Cells were dialized with solutions of
high proton and Ca^{2+} buffering capacity. Generally 11mM EGTA, 20 mM Hepes,
5mM $Mg^{2+}ATP$ and 0.2 mM cAMP were included in the internal pipette solution.
In some experiments 25 mM Bapta (a pH-insensitive Ca^{2+}-chelator) and 90 mM
Hepes were also used (See: Neher, 1986; Byerly & Moody, 1985). The
extracellular solutions were modifications of standard tyrodes, containing
100-140 mM Na^+, zero K^+, 10-20 mM TEA, and 1-5 mM Ca^{2+} (see Konnerth, Lux &
Morad, 1987). Na^+ channels were routinely blocked by 5-10 μM TTX.
Extracellular solutions were changed by a multi-barrelled perfusion pipette
capable of exchanging the solutions around the experimental cells within 10-
50 ms (Konnerth, Lux & Morad, 1987). In isolated membrane patches the speed
of perfusion system could be optimized to about 1.0 ms (Davies, Lux & Morad,
1987).

RESULTS

Figure 1A (heavy trace) shows the time course of activation of an
inward current which was activated when proton concentrations were step-
increased from pH 7.9 to 6.7. The current could be activated in whole cell
or in an outside-out patch (Fig. 1B), but could not be elicited in the
inside out patches. The activation was rapid, but the relaxation of current
back to the base line occured slowly with a time constant of 200-300 ms.
The threshold for activation of the current was about pH 7.0 with a maximum
around pH 6.0. Elevation of $(Ca^{2+})_o$ shifted the activating proton
concentrations to higher values. The current could be rapidly deactivated
by step decreases of $(H^+)_o$ (Fig 1, light trace and Fig. 1B). The
deactivation half times in the outside-out patches were about 2.0 ms. The
activation half time was proton dependent, ranging from 1-20 ms for
activating proton concentration from pH 6.0 to 7.2 (Davies, Lux & Morad,
1987). Neither the activation nor the relaxation phase of the current were
significantly voltage-dependent.

Small proton concentrations which by themselves failed to activate the
current, markedly reduced the current accompanying larger pulses. This
property was associated with an independent inactivation process and was
responsible for the slow relaxation of the current in the presence of

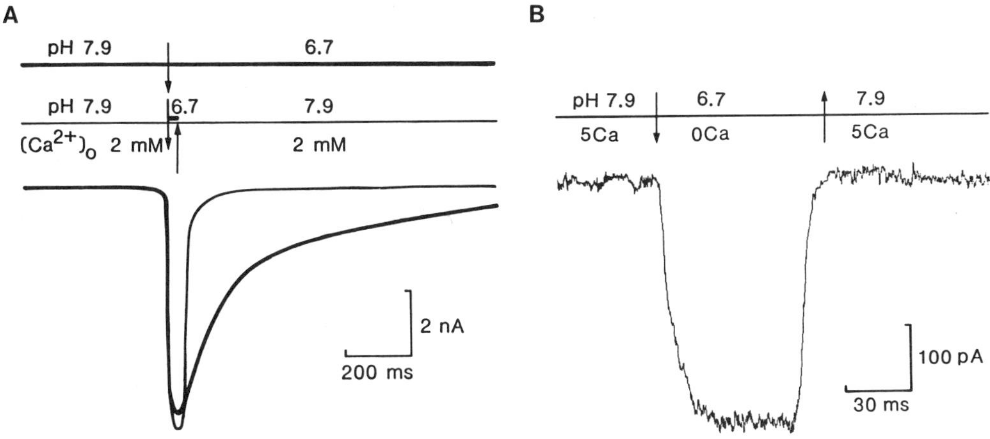

Figure 1.   Comparison of $I_{Na(H)}$ in whole cell, and isolated outside−out patch, following a rapid change in pH from 7.9 to 6.7. A, whole cell, heavy trace shows the time course of activation and inactivation of $I_{Na(H)}$ during the continued presence of pH 6.7. The light trace shows the time course of deactivation induced by step reduction of $(H^+)_o$ to pH 7.9 as indicated on the top of each trace. The holding potential in each case was −80 mV, and the external solution contained 5 mM Ca. B, the time course of activation and deactivation of $I_{Na(H)}$ in an outside−out patch, upon step elevation and step conduction of $(H^+)_o$ as indicated (from, Davies, Lux & Morad, 1987).

maintained high proton concentrations. Thus this current appears to activate, deactivate and inactivate in response to step changes of proton concentration, in a manner similar to the voltage-gated channels' response to voltage. In other words, the channel appears to be proton gated.

Proton-induced Current Is Carried by Na^+

Two types of evidence suggested that the current recorded in response to step changes in proton concentration was carried primarily by Na^+. Fig. 2A shows that the replacement of Na^+, by choline or Tris, markedly reduced the proton-induced current in whole cell or outside out patches. Examination of the voltage-dependence of the proton-induced current also showed that this current reversed at the Na^+ equilibrium potential (Fig. 2B). The reversal potential changed as predicted from the equilibrium potential for Na^+ when either the extra or intra-cellular concentrations of Na^+ were altered (Konnerth, Lux & Morad, 1987). Thus we labelled the proton-induced current as Na^+ current and abbreviated it as $I_{Na(H)}$.

$I_{Na(H)}$ Is Blocked by Ca^{2+} Channel Blockers

The finding that $I_{Na(H)}$ is carried by Na^+, but was not blocked by high concentrations of TTX (1–10 μM), suggested that $I_{Na(H)}$ does not flow through the fast Na^+ channel previously identified in these cells (Carbone & Lux, 1986). Figure 3 shows that addition of Cd^{2+} strongly suppressed $I_{Na(H)}$. Washout of Cd^{2+} relieved the block. Ni^{2+}, Cd^{2+}, verapamil and diltiazem all blocked $I_{Na(H)}$. Generally similar concentrations of both organic but somewhat higher concentrations of inorganic Ca^{2+} channel blockers were

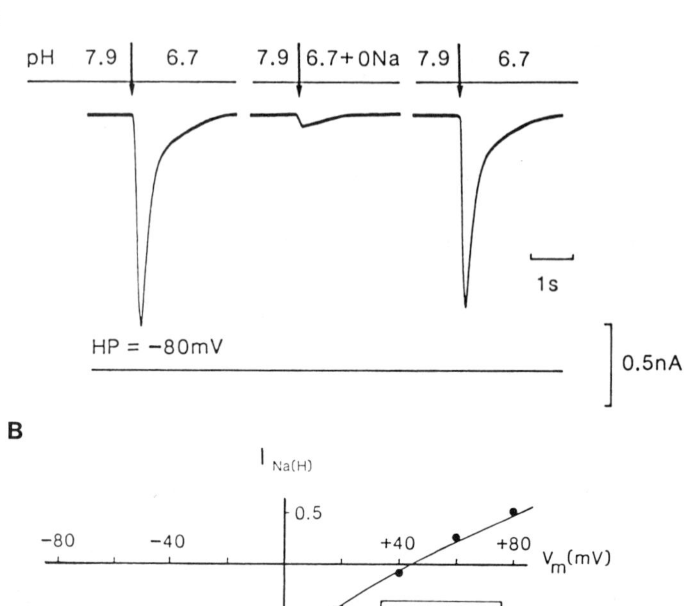

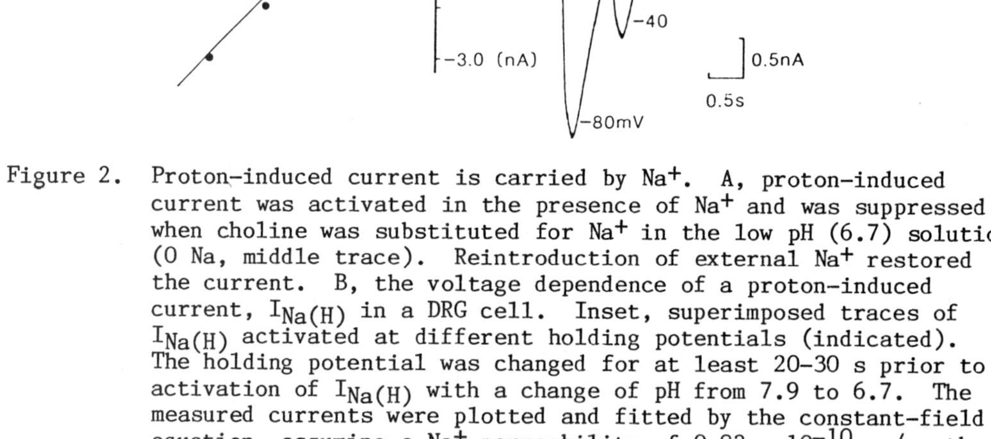

Figure 2. Proton-induced current is carried by Na^+. A, proton-induced current was activated in the presence of Na^+ and was suppressed when choline was substituted for Na^+ in the low pH (6.7) solution (0 Na, middle trace). Reintroduction of external Na^+ restored the current. B, the voltage dependence of a proton-induced current, $I_{Na(H)}$ in a DRG cell. Inset, superimposed traces of $I_{Na(H)}$ activated at different holding potentials (indicated). The holding potential was changed for at least 20–30 s prior to activation of $I_{Na(H)}$ with a change of pH from 7.9 to 6.7. The measured currents were plotted and fitted by the constant-field equation, assuming a Na^+ permeability of 0.83×10^{-10} cm/s; they reversed direction at potentials positive to +40 mV. The calculated E_{Na} was +45 mV. $[Na^+]_i$ = 20 mM; $[Na^+]_o$ = 120 mM; RT/F = 58 mV (modified from, Konnerth, Lux & Morad, 1987).

necessary to block $I_{Na(H)}$. Considering the pharmacological specificity of the organic and inorganic Ca^{2+} channel blockers for the voltage-gated Ca^{2+} channel, we concluded that $I_{Na(H)}$ must flow through either the Ca^{2+} channel or that a new channel, not described previously and specifically inhibited by Ca^{2+}-antagonists, carried $I_{Na(H)}$. It should be remembered, however, that if $I_{Na(H)}$ were to flow through the Ca^{2+} channel, the channel was not in its usual voltage-gated state, rather in a novel voltage-insensitive "proton-gated" state. This possibility, therefore, required that Ca^{2+} channels should exist in two different conformational states; a voltage-gated Ca^{2+} permeable state, and a proton-gated Na^+ selective state.

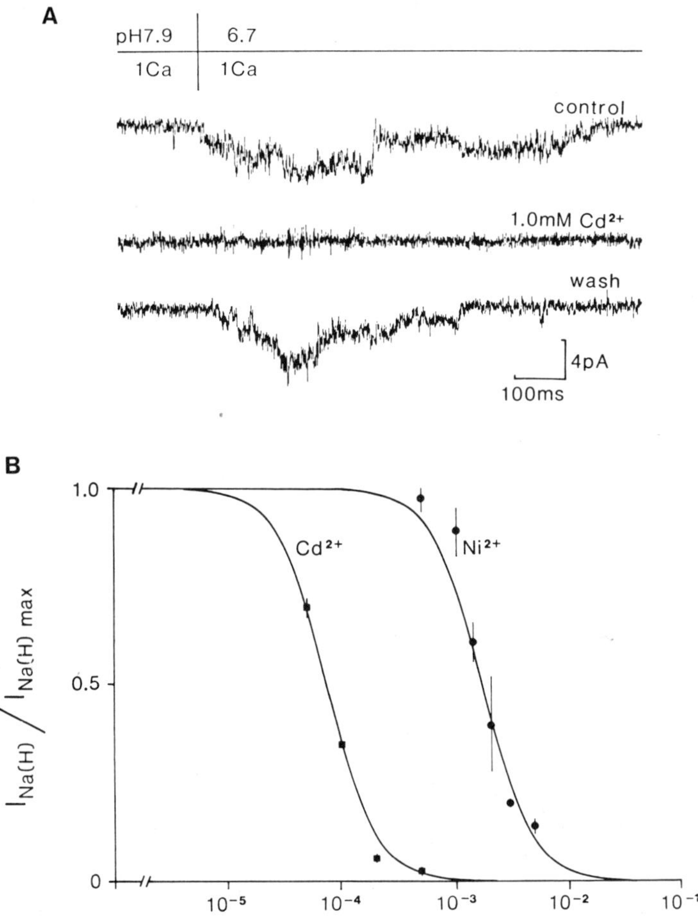

Figure 3. Block of $I_{Na(H)}$ by Cd^{2+} and Ni^{2+}. A, 1 mM Cd^{2+} totally abolished $I_{Na(H)}$ in an outside-out patch (Hp −80mV). The unitary events recovered upon removal of Cd^{2+}. B, concentration-response relationship for the blockade of $I_{Na(H)}$ in whole cells by Cd^{2+} and Ni^{2+}. The curves drawn according to 2:1 stoichiometry. Two ions appear to be required to block $I_{Na(H)}$. The vertical lines indicate the standard error. $(Ca^{2+})_o = 5$ mM; Hp = −80mV; $pH_i = 7.3$. Six cells are used in each measurement (modified from Davies, Lux & Morad, 1987).

<u>Evidence that $I_{Na(H)}$ Flows Through a Transformed State of the Ca^{2+} Channel</u>

Though the specificity of Ca^{2+} channel blockers in suppressing $I_{Na(H)}$ provides suggestive evidence that Ca^{2+} channel is the conduit for $I_{Na(H)}$, more direct evidence that $I_{Na(H)}$ flows through a proton-modified state of the Ca^{2+} channel was obtained when an attempt was made to measure simultaneously $I_{Na(H)}$ and I_{Ca}. The ionic independence hypothesis (Hodgkin & Huxley, 1952) not only states that ionic currents flow through independent channels, individually separable by specific channel blockers or removal of permeant ion, but also requires that different channels operate simultaneously to carry the specific permeant species. Thus, by Kirchhoff's law Na^+ and Ca^{2+} current, in response to a depolarizing voltage pulse, flow with their independent time courses which may be separated temporally and pharmacologically. Figure 4 illustrates an experiment in which we attempted

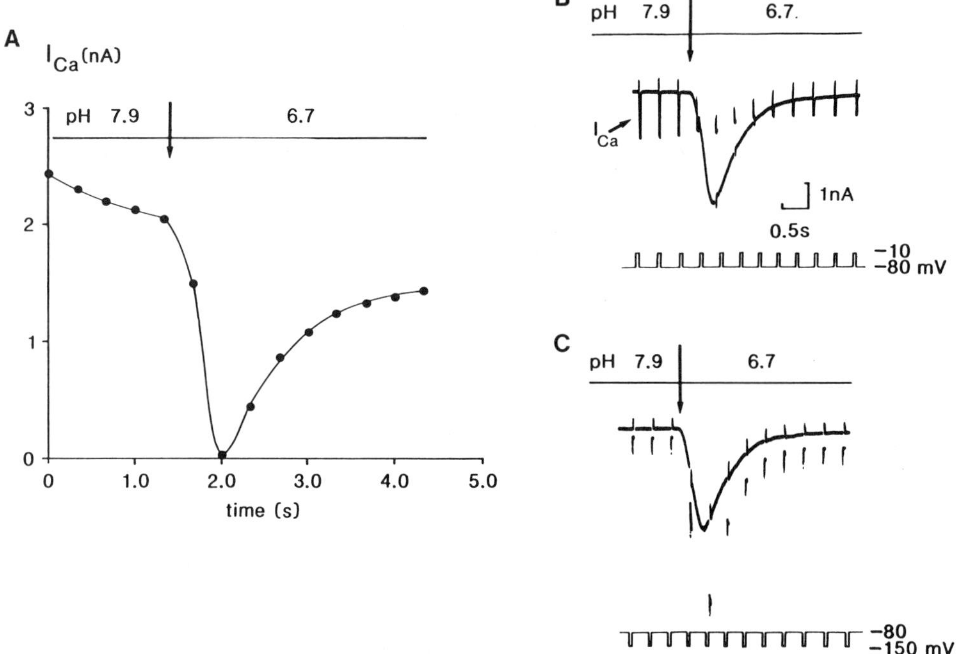

Figure 4. Transient transformation of Ca^{2+} channel from a voltage-gated to a non-gated state. B shows inward I_{Ca}s activated at a frequency of 3 Hz by short, 30 ms depolarizing pulses before and during the time course of activation of $I_{Na(H)}$. Prior to application of the pH step, I_{Ca} was corrected by subtracting the "leak" current measured during hyperpolarizing pulses (C). To determine the amplitude of I_{Ca} during the step in pH, $I_{Na(H)}$ was subtracted from the total current during the pulses. A linear I-V relation of $I_{Na(H)}$ was assumed in line with earlier observations (Fig. 2B). I_{Ca} thus determined was plotted prior to and during the application of the pH step (A). At the arrow (panel A, when proton concentration is increased) I_{Ca} rapidly decreased as $I_{Na(H)}$ activated. Slow relaxation of $I_{Na(H)}$ occurred simultaneously with the recovery of the voltage-gated I_{Ca}. The decrease in magnitude of I_{Ca} in solutions buffered at pH 6.7 represents the depressive effect of low-pH solutions on the voltage-gated Ca^{2+} channel. $[Ca^{2+}]_o$ = 5 mM; [TTX] = 2 μM; room temperature; pH_i = 7.3; $[Ca^{2+}]_i$ = 10^{-9}M (from Konnerth, Lux & Morad, 1987).

to measure $I_{Na(H)}$ and I_{Ca} simultaneously. Since $I_{Na(H)}$ lasted about 1-2 seconds and could be activated at resting potentials by step-increases in $(H^+)_o$, it allowed the measurement of the voltage-gated I_{Ca}, during the time course of $I_{Na(H)}$. Thus by clamping the membrane to zero mV for 30-50 ms repetitively (3-5 Hz), we attempted to activate I_{Ca} during the time course of $I_{Na(H)}$. Figure 4 shows that activation of $I_{Na(H)}$ by step increases of protons to pH 6.7 resulted in marked decrease and complete disappearance of the voltage-gated I_{Ca}. Ca^{2+} current then recovered slowly as $I_{Na(H)}$ inactivated despite the presence of high proton concentrations. Thus it appears that I_{Ca} and $I_{Na(H)}$ are mutually exclusive, supporting the idea that both currents are carrried by the same channel. In the voltage-gated state the channel conducts Ca^{2+}, while in the proton-gated state the channel carries Na^+. This finding thus implies that sudden elevation of proton concentrations at the external "mouth" of the Ca^{2+} channel leads to its transformation from a Ca^{2+} permeable to a Na^+ permeable state.

The possible transient transformation of Ca^{2+} channel, from Ca^{2+} to a Na^+ carrying state, in response to increases of extracellular proton concentration, though quite novel in concept, is not particularly surprising. It has been known for some time, for instance, that Ca^{2+} channel is permeable to Na^+. In fact as pointed out above, at $(Ca^{2+})_o$ below 1-10 μM, Ca^{2+} channel transports Na^+ very effectively in neuronal and muscular tissues. Such a Na^+ permeable state is not very dissimilar from that caused by step elevation of protons. Comparison of single channel conductance measured at the two states, suggests that the proton-transformed single channel conductance of 20-28 ps (Morad & Lux, 1987) is quite similar to that of the voltage-gated Na^+ transporting Ca^{2+} channel. Thus it appears that either elevation of protons or removal of Ca^{2+} leads to increased Na^+ permeability of Ca^{2+} channel, suggesting the possibility that a similar mechanism is responsible for both events.

Ca^{2+} and H^+ Interacts at a Site Associated with the Ca^{2+} Channel to Induce or Prevent Transformation of Ca^{2+} Channel

We tested this assertion by activating $I_{Na(H)}$ in the presence of different $(Ca^{2+})_o$. Fig. 5C shows that as $(Ca^{2+})_o$ were increased the magnitude of $I_{Na(H)}$ decreased, such that at concentrations above 25 mM little or no $I_{Na(H)}$ could be activated. Simultaneous step elevation of $(H^+)_o$ and $(Ca^{2+})_o$ to 40 mM not only blocked the activation of $I_{Na(H)}$, but also increased the voltage-gated I_{Ca} (Konnerth, Lux and Morad, 1987), suggesting that high concentrations of Ca^{2+} prevented transformation of the channel, as if displacement of Ca^{2+} by H^+, from a site closely associated with the Ca^{2+} channel was necessary to transform the channel gating and permeability. Consistent with this assertion step reduction of $(Ca^{2+})_o$ (from 10^{-3} to 10^{-8}M) in solutions buffered at pH 7.3 caused marked activation of $I_{Na(H)}$ (Fig. 5A, inset; and Hablitz et al., 1986). $I_{Na(H)}$ failed to activate by such reduction of $(Ca^{2+})_o$ when proton concentration were buffered at pH 7.9 (Fig. 5B). These results imply that Na^+ current activated by step removal of Ca^{2+} is also mediated by the binding of protons to a site closely associated with the Ca^{2+} channel. Thus it appears that Ca^{2+} and H+ interact in transformation of Ca^{2+} channel and activation of $I_{Na(H)}$.

Summary, Conclusions and Implications

Our results show that a large slowly inactivating (1-2 s) Na^+ current may be activated in whole cells or isolated membrane of chick dorsal root ganglion cells in response to a small or moderate step elevation of the extracellular proton concentrations. The amplitude of the current was enhanced by elevation of $(H^+)_o$, decrease of $(Ca^{2+})_o$, reduction of $(Na^+)_o$ and addition of organic or inorganic Ca^{2+} channel blockers. Binding of H^+ to a

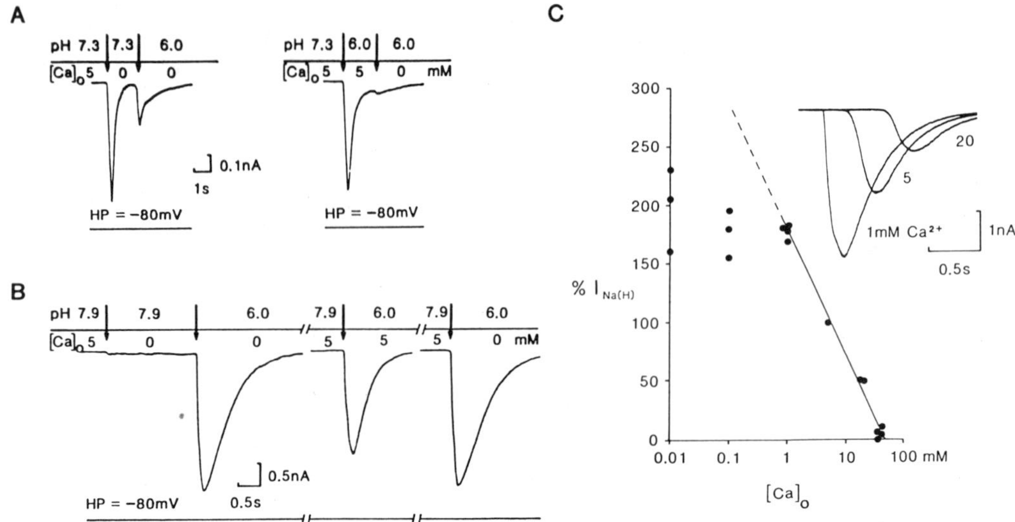

Figure 5. Ca^{2+} and H^+ interacts at a site associated with the Ca^{2+} channel to induce or prevent transformation of Ca^{2+} channel. A, the time course of activation of zero Ca^{2+}-induced Na^+ current followed by a step increase in $[H^+]_o$. If the current was activated first by application of proton, the zero Ca^{2+} solution (Ca^{2+} omitted plus 1 mM EGTA) failed to activate any inward current. B, zero-Ca^{2+} induced $I_{Na(H)}$ failed to activate when $[H^+]_o$ was reduced to pH 7.9. Comparison of the simultaneous application of zero Ca^{2+} (no Ca^{2+} added and 1 mM EGTA) and the proton step with that where the cell was incubated first, for a short time, in a zero Ca^{2+} solution followed then by elevation of $[H^+]_o$ showed no difference in the time course or the magnitude of $I_{Na(H)}$. Room temperature; $[Na^+]_i$ = 20 mM; [TTX] = 2 μM; $(pH)_i$ = 7.3. C, the time course of $I_{Na(H)}$ at 1.0, 5.0 and 20 mM $[Ca^{2+}]_o$ in response to a step change in pH from pH 7.9 to 6.7. $I_{Na(H)}$ is completely blocked by 35–40 mM $[Ca^{2+}]_o$. The graph shows considerable scatter at concentrations below 1 mM. The results from five cells were normalized with respect to $I_{Na(H)}$ measured at 5 mM Ca^{2+}. At least three different Ca^{2+} concentrations were used in each cell. [TTX] = 2 μM; room temperature (modified from, Konnerth, Lux & Morad, 1987).

divalent specific site of the Ca^{2+} channel induced loss of voltage gating and transformation of Ca^{2+} channel from a Ca^{2+} specific to a Na^+ permeable channel. $I_{Na(H)}$ and I_{Ca} appear to be mutually exclusive such that activation of $I_{Na(H)}$, completely suppresses I_{Ca}, providing strong evidence that Ca^{2+} channel is the conduit to both currents. Thus it was postulated that Ca^{2+} channel may exist in two mutually exclusive states, a voltage-gated Ca^{2+} permeable state and a proton-gated Na^+ transporting state. The transition from the voltage- to the proton-gated state occured by rapid elevation of $(H^+)_o$. The proton-gated state appears to be unstable and relaxes or inactivates within 1–2 s. The inactivation of $I_{Na(H)}$ was accompanied by the recovery of the voltage-gated state of Ca^{2+} channel. It may appear surprising that inactivation of $I_{Na(H)}$ is accompanied by recovery of voltage-gated state. It should be remembered, however, that energy is always applied with a voltage pulse to activate I_{Ca} during $I_{Na(H)}$. The inactivated state thus may be equivalent to the voltage-gated state, i.e. the transition between the proton-inactivated state and the voltage-gated state may be very rapid.

It is intriguing to speculate whether such a transformation of Ca^{2+} channel may occur under some physiological or pathological conditions in central or peripheral neurons. We have recently shown that a variety of cultured central or peripheral neurons, which have the voltage-gated Ca^{2+} current also have the $I_{Na(H)}$ system (Davies, Lux & Morad, 1987). It has been also shown that large and rapid increases in $(H^+)_o$ occur in central nervous system, on repetitive stimulation which are often accompanied by significant depletion of $(Ca^{2+})_o$ (Kraig, Ferreira-Filho & Nicholson, 1983). Though the measured changes in $(H^+)_o$ appear moderate and slow in such studies, it should be remembered that the ion-selective microelectrodes do not measure accurately and rapidly the magnitude and the time course of variations of the ionic concentrations in the paracellular space. This is in part due to the geometry of the extracellular space and the close (10-20 nm) proximity of cellular membranes to each other. Thus larger ionic concentration spikes in such spaces would remain undetected by the extracellularly located electrodes, which necessarily create their own extracellular pools with diameters of 100s of μm. It is, therefore, likely that under some conditions a train of impulses could induce accumulation of $(H^+)_o$ which would in turn activate $I_{Na(H)}$ causing large and slowly decaying depolarization waves. It has already been postulated that such depolarization waves may induce or occur in epileptic seizures or in spreading depression (Lehmenkühler et al., 1981; Urbanics et al, 1978). One may speculate that such long lasting depolarizations when mediated by influx of Na^+ rather than Ca^{2+} are likely to be less damaging to the cell. Independent of possible physiological role of the Na^+ carrying Ca^{2+} channels in neuronal cells, our results clearly show that Ca^{2+} channel can transform their permeability and gating transiently in response to increases in $(H^+)_o$. So far it appears that transformation is a unique property of neuronal Ca^{2+} channels.

A PERSONAL NOTE:

Roger visited us in April of 1986 in Munich when we were carrying out these experiments. I would not tell him what I was doing. It was a game we often played. I did promise, however, that he would be the first to know! Roger, old friend, you cheated me out of the pleasure.
     -Martin

REFERENCES

Almers, W., McCleskey, E.W. and Palade, P.T., 1984, A non-selectivecation conductance in frog muscle membrane blocked by micromolar external calcium ions, _J. Physiol._, 353:565-583.

Armstrong, C.M. and Matteson, D.R., 1985, Two distinct populations of calcium channels in a clonal line of pituitary cells, _Science_, 227:65-67.

Bean, B.P., 1985, Two kinds of calcium channels in canine atrial cells. Differences in kinetics, selectivity and pharmacology. _J. Gen. Physiol._, 86:1-30.

Borsotto, M., Barhanin, J., Fosset, M. and Lazdunski, M., 1985, The 1,H-dihydropyridine receptor associated with the skeletal muscle voltage-dependent Ca^{2+} channel. _J. Biological Chem._, 260:14.

Byerly, L. and Moody, W.J., 1986, Membrane currents of internally perfused neurones of the snail, Lymoraea Stagnalis, at low intracellular pH, _J. Physiol._, 376:477-491.

Carbone, E. and Lux, H.D., 1984a, A low-voltage activated calcium conductance in embryonic chick sensory neurons, _Biophys. J._, 46:413-418.

Carbone, E. and Lux, H.D., 1984b, A low-voltage activated, fully inactivating calcium channel in vertebrate sensory neurones, <u>Nature</u>, 310:501-502.

Carbone, E. and Lux, H.D., 1986, Sodium channels in cultured chick dorsal root ganglion neurons, <u>Eur. Biophys. J.</u>, 13:259-271.

Carbone, E. and Lux, H.D., 1987a, Kinetics and selectivity of a low-voltage-activated calcium current in chick and rat sensory neurones, <u>J. Physiol.</u>, 386:547-570.

Carbone, E. and Lux, H.D., 1987b, External Ca^{2+} ions block unitary Na^+ currents through Ca^{2+} channels of cultured chick sensory neurones by favouring prolonged closures, <u>J. Physiol.</u>, 382:124P.

Curtis, B.M. and Catterall, W.A., 1984, Purification of the calcium antagonist receptor of the voltage-sensitive calcium channel from skeletal muscle transverse tubules, <u>Biochem.</u>, 23:2113-2118.

Davies, N.W., Lux, H.D. and Morad, M., <u>J. Physiol.</u>, in Press.

Ferry, D.R., Kampf, K., Cyroll, A. and Glossmann, H., 1985, Subunit composition of skeletal muscle transverse tubule calcium channels evaluated with the 1,4-dihyrdropyridine photoaffinity probe, [^{3}H] azidopine. <u>EMBO J.</u>, 4:1933-1940.

Hablitz, J.J., Heinemann, U. and Lux, H.D., 1986, Step reductions in extracellular Ca^{2+} activates a transient inward current in chick dorsal root ganglion cells, <u>Biophys. J.</u>, 50: 753-757.

Hamill, O.P., Marty, A., Neher, E., Sakmann, B. and Sigworth, F.J., 1981, Improved patch-clamp techniques for high-resolution current recording from cells and cell-free membrane patches, <u>Pflugers Arch.</u>, 391:85-100.

Hess, P., Lansman, J.B. and Tsien, R.W., 1986, Calcium channel selectivity for divalent and monovalent cations. Voltage and concentration dependence of single channel current in ventricular heart cells. <u>J. Gen. Physiol.</u>, 88:293-319.

Hodgkin, A.L. and Huxley, A.F., 1952, A quantitative description of membrane current and its application to conduction and excitation in nerve. <u>J. Physiol.</u>, 117:500-544.

Konnerth, A., Lux, H.D. and Morad, M., 1987, Proton-induced transformation of Ca^{2+} channels in chick dorsal root ganglion cells, <u>J. Physiol.</u>, 386:603-633.

Kraig, R.P., Ferreira-Filhs, C.-R., and Nicholson, C., 1983, Alkaline and acid transients in cerebellar microenvironment. <u>J. Neurophysiol.</u>, 49:831-850.

Lehmenküler, A., Zidek, W., Staschen, M. and Caspers, H., 1981, Cortical pH and pCa in relation to DC potential shifts during spreading depression and asphyxiation, in: "Ion-selective Microelectrodes and Their Use in Excitable Tissues," E. Sykova, ed., Plenum Press, New York.

Mitra, R. and Morad, M., 1986, Two types of calcium channels in guinea pig ventricular myocytes, <u>PNAS</u>, 93:5340-5344.

Morad, M. and Lux, H.D., 1987, Single unit analysis of the proton-induced transformation of Ca^{2+} channel in sensory neuron, <u>Biophys. J.</u>, 51:32a.

Nakayama, N., Kirley, T.L., Vaghy, P.L., McKenna, E. and Schwartz, A., 1987, Purification of a putative Ca^{2+} channel protein from rabbit skeletal muscle, <u>J. Biological Chem.</u>, 262:6572-6576.

Neher, E., 1986, Concentration profiles of intracellular calcium in the presence of a diffusible chelator, in: "Calcium Electrogenesis and Neuronal Functioning," ed. U. Heinemann, M. Klee, E. Neher, and W. Singer, Springer-Verlag, Berlin.

Nilius, B., Hess, P., Lansman, J.B. and Tsien, R.W., 1985, A novel type of cardiac calcium channel in ventricular cells, <u>Nature</u>, 316: 443-446.

Urbanics, R., Leniger-Follert, E. and Lubbers, D.W., 1978, Time course of changes of extracellular H^+ and K^+ activities during and after direct electrical stimulation of the brain cortex, <u>Pflugers Arch.</u>, 378:47-53.

PHYSIOLOGY OF MULTIPLE CALCIUM CHANNELS

A.P. Fox, L.D. Hirning, D.V. Madison[1], E.W. McCleskey[2],
R.J. Miller, M.C. Nowycky[3], and R.W. Tsien[1]

Dept. of Pharmacological and Physiological Sciences
University of Chicago
[1]Dept. of Physiology, Yale University School of Medicine
[2]Dept. of Cell Biology and Physiology, Washington University
[3]Dept. of Anatomy, Medical College of Pennsylvania
(R.J.M. is a Guggenheim fellow.)

Introduction

Calcium ions entering cells via voltage-dependent calcium channels
play a fundamental role in many cellular processes. For instance, calcium
is involved in both synaptic transmission and hormone release in neuronal
and secretory cells. In many excitable cells, calcium ions act as a
charge carrier. Calcium ions activate the contractile machinery of both
heart and skeletal muscle. In heart the slow inward Ca current is
critically involved in maintaining the normal rhythmicity needed to keep
the heart beating. As can be seen from the few examples listed above, Ca
channels play a dual role in many excitable cells. They are capable of
responding to and influencing a cell's electrical properties; in many
cells the Ca channels alone, or in combination with sodium channels,
generate the cellular action potentials. The Ca channels' second function
is that of a transducer. By opening or closing in response to the cell
membrane potential they help regulate on a millisecond time scale the
influx of calcium ions, one of the best characterized intracellular
second messengers, into the cell. Elevated intracellular levels of
calcium will activate other ion channels, such as Ca-dependent K channels
or Ca-dependent non-selective channels, leading to complex electrical
behaviors. It is also possible that calcium entering the cell through Ca
channels may be involved in modulating a variety of cellular processes
including Ca-dependent enzyme systems such as Ca/calmodulin-dependent
protein kinase, or protein kinase C. (Hagiwara and Byerly (1981, 1983),
Pallotta, Magleby and Barrett (1982), Yellen (1982), Carafoli (1983),
Reuter (1983), Hille (1984), Nishizuka (1984, 1986), Carafoli and
Penniston (1985), Tsien (1986)).
One of the most interesting features of calcium channels is that
they are highly plastic. Neurotransmitters, hormones and drugs modulate
the gating of voltage-dependent calcium channels. (Dunlap and Fischbach
(1978, 1981), Reuter (1983), Canfield and Dunlap (1984), Tsien (1986)).
To complicate the story even further, several different types of calcium
channels have recently been discovered (Hagiwara, Ozawa and Sand (1975),
Llinas and Sugimori (1980), Fox and Krasne (1981, 1984), Carbone and Lux
(1984a,b), Deitmer (1984), Armstrong and Matteson (1985), Bean (1985),
Bossu, Feltz and Thomann (1985), Fedulova, Kostyuk and Veselovsky (1985),
Nilius, Hess, Lansman and Tsien (1985), Nowycky, Fox and Tsien (1985b),

Reuter (1985), Mitra and Morad (1986), Narahashi, Tsunoo and Yoshii
(1987)). At present many questions still remain as to the differences in
biophysical and pharmacological properties between the different types of
channels, as well as the cellular processes they subserve. In addition,
several different types of non-selective ligand gated channels, like the
NMDA or ATP receptors, have recently been shown to be quite permeable to
Ca ions, suggesting that they may have a significant role in the
regulation of calcium fluxes through the plasma membrane (Ascher and
Nowak (1986), MacDermott, Mayer, Westbrook, Smith, and Barker (1986),
Benham and Tsien, in press)). In this paper we review some of our work on
multiple calcium channels and a possible physiological role for one kind
of Ca channel.

Biophysical and pharmacological properties of three types of Ca channels

Figure 1 shows evidence for three distinct components of Ca channel
current in whole-cell recordings in DRG neurons, obtained under ionic
conditions that minimize contamination by other currents. The components
were distinguished kinetically by applying depolarizing test pulses to

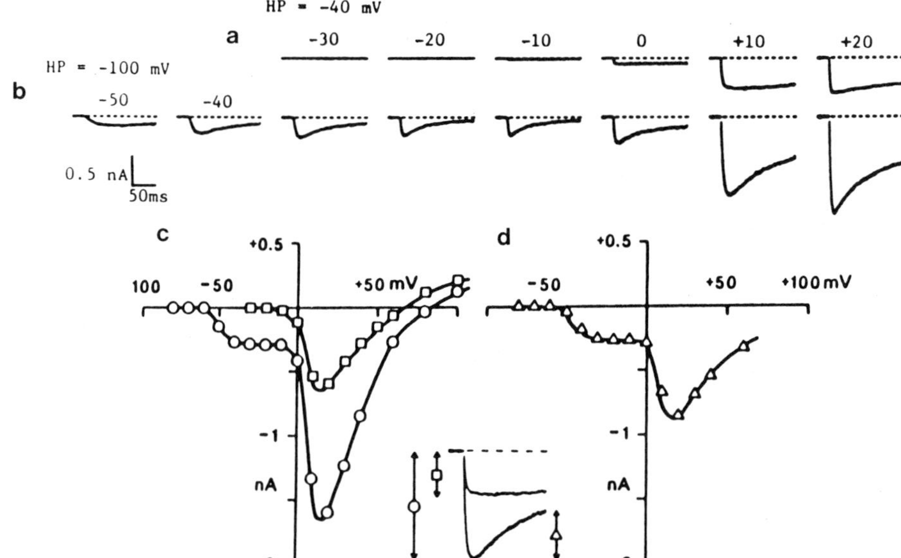

Fig. 1. **Three components of Ca channel current in whole-cell recordings
from a chick DRG neuron with 10 mM extracellular Ca. a,b,
currents evoked from HP = -40 mV (a) and HP = -100 mV (b). Test
potential indicated above. Linear leak and capacitance currents
have been subtracted with 1 ms blanked at both the on and off
of the depolarizing pulse. c, plot of peak current versus test
potential from HP = -40 mV (squares) and HP = -100 mV
(circles). d, plot of decaying current versus test potential
from HP = -100 mV. The current relaxation is the late current
subtracted from the peak current. Inset repeats records
obtained at +10 mV and shows how peak and relaxing currents
were measured. The internal solution (pipette) contained in mM:
100 CsCl, 10 Cs-EGTA, 5 MgCl$_2$, 40 HEPES, 2 ATP, 0.25 cyclic-
AMP, pH 7.3. The external solution was exchanged after
formation of a gigaseal from a standard Tyrode to one
containing (mM): 10 CaCl$_2$, 135 TEACl, 10 HEPES, and 200 nM TTX,
pH 7.3.**

various levels from different holding potentials (HP). With HP = -40 mV
(fig. 1a), strong depolarizations above -10 mV are required to activate
any inward current; the peak current-voltage relationship (fig. 1c,
squares) is typical for a single current component. Because this inward
current component decays very slowly ($t_{1/2}$ of hundreds of milliseconds),
it was designated "L" for long lasting.

Weak depolarizations from HP = -100 mV evoke a different Ca channel
current (fig. 1b), seen as a prominent shoulder at negative test
potentials in the associated peak current-voltage plot (fig. 1c,
circles). The additional current becomes noticeable at -60 mV, and is
nearly constant in amplitude between -40 mV and -10 mV, consistent with a
very negative range of activation. Since this component decays relatively
rapidly, (tau~ 25 msec at -30 mV), it has been called "T" (transient). A
third component of calcium current appears with strong depolarizations
from HP = -100 mV (i.e. traces at +10 or +20 mV in fig. 1b). It is
distinguished most easily in a plot of the decaying current amplitude
versus test potential (fig. 1d, triangles). The decay amplitude shows a
clear plateau between -30 and -10 mV, as expected for component T;
however, it grows substantially with stronger test pulses, reaching a
peak at +20 mV. The extra decaying current is attributed to a third
component called "N" (neither T nor L). Like T, but unlike L, component N
contributes phasic current and requires strongly negative holding
potentials for complete removal of inactivation. Like L, but unlike T, N
currents require strong depolarizations for activation. All three
components are present in whole-cell recordings although the relative
amplitudes vary considerably between cells and with different ionic
conditions. N current was particularly prominent with 55 or 110 mM Ba as
the charge carrier; in whole-cell recordings from four cells (average
capacitance 30 pF), the maximal amplitude of the decaying N current at
+20 was 1340 ± 246 pA, roughly equal to the L current at +20 mV and about
eight times greater than the peak T current at -10 mV (peak T).

Recording from cell-attached patches on DRG cell bodies show three
types of unitary Ca channel current, readily distinguished by different
slope conductances in 110 mM Ba (fig. 2). The different channel types
show kinetic properties that correspond well with components T, N, and L
in whole-cell recordings. Fig. 2a shows unitary current activity that is
most closely associated with the T component. Inactivation of this
channel is complete positive to -40 mV, and is fully removed only at
holding potentials more negative than -80 mV. Activation becomes
detectable positive to -50 mV. With test depolarizations to -20 or -10
mV, the average current decays with a time course very similar to
component T in whole-cell recordings. The T-type calcium channel has a
tiny-slope conductance (about 8 pS in 110 Ba).

Fig. 2b illustrates the third type of unitary activity called N
because its properties account for component N in whole-cell recordings.
N-type calcium channels differ from L- and T-type in several respects.
Their slope conductance is 13 pS with 110 mM Ba, larger that T but
smaller than L-type channels. The difference in unitary current size is
highly significant; for example at -20 mV, average values (+ s.e.m.) for
T, N, and L were 0.62 ± 0.03 (5 patches), 1.22 ± .03 (10 patches), and
2.07 ± 0.09 (6 patches), respectively. At +20 mV, the test potential
illustrated in fig. 2b and 2c, the difference between average values for
N and L (0.824 and 1.12 pA) is smaller in absolute terms but is still
appreciable.

N-type calcium channels can also be distinguished from L-type
calcium channels by their inactivation properties. The time-dependence of
inactivation of N-type Ca appears as a clustering of openings near the
beginning of the depolarization and an absence of openings near the end
(fig. 2b); the averaged current record (not shown) decays almost
completely by the end of a 136 msec test pulse. The steady voltage-
dependence of inactivation was studied by varying the holding potential.

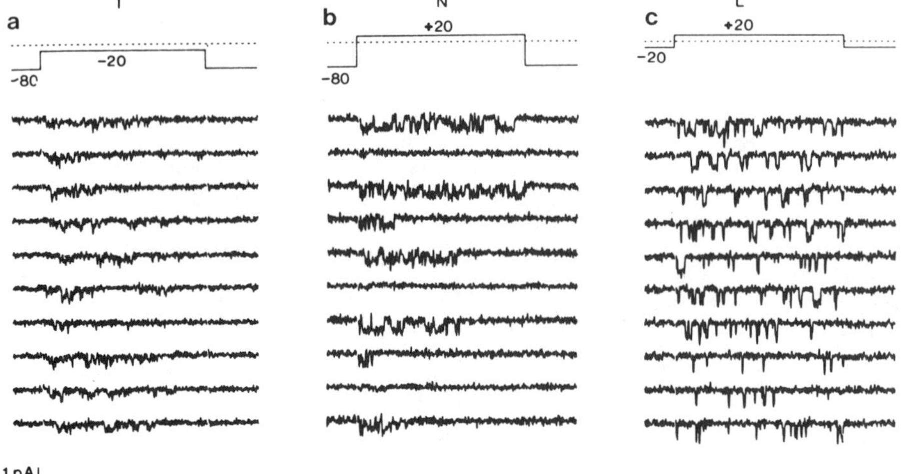

Fig. 2.   Three types of unitary Ca channel activity seen in cell-attached patch recordings with Ba as the charge carrier. A, T-type channel activity, cell N85A. B, N-type channel, cell N75H. C, L-type channel, cell N67J. Patch pipettes contained (in mM): 110 BaCl$_2$, 10 HEPES, and 200 nM TTX, pH 7.4. To put patch membrane potential on an absolute scale, cell resting potential was zeroed with an external solution containing (in mM): 140 K-aspartate, 10 K-EGTA, 10 HEPES, 1 MgCl$_2$, pH 7.4. Current signals were filtered at 1 KHz and sampled at 5 KHz. Within each panel traces are consecutive. Values of single-channel amplitude and slope conductance given in text are were partially obtained from amplitude histograms. The unitary current amplitude of the T-type channel at −20 mV was given by the average of amplitudes from 3 different patches exhibiting long well-resolved openings. We have seen all three channel types in isolation and in all possible combinations. The number of channels under the patch pipette varied considerably from one channel to hundreds of channels.

Unitary N-type activity was completely abolished by holding the patch at −20 mV, a potential at which L-type calcium channels remain largely available for opening (see fig. 2c). N- and T-type calcium channels are fairly similar in their inactivation properties but very different in the voltage-dependence of activation. With isotonic Ba in the pipette, significant activation of N-type channels requires depolarizations to −20 mV, while opening of T-type channels first becomes detectable beyond −50 mV.

In addition to differences in kinetics, unitary conductance and sensitivity to Cd, the three channel types differ in their responsiveness to Bay K 8644. This dihydropyridine (DHP) calcium channel agonist elevates calcium influx by promoting a pattern of calcium channel gating with very long openings in neurones and heart cells (Hess, Lansman and Tsien (1984), Kokubun and Reuter (1984), Nowycky (1985a)). As fig. 3 (bottom traces) illustrates, Bay K 8644 strongly enhances averaged L-type calcium currents in DRG neurons. In contrast, Bay K 8644 produced no change in the current carried by T channels (6 patches), or N-type channels (4 patches). These results are consistent with those found in

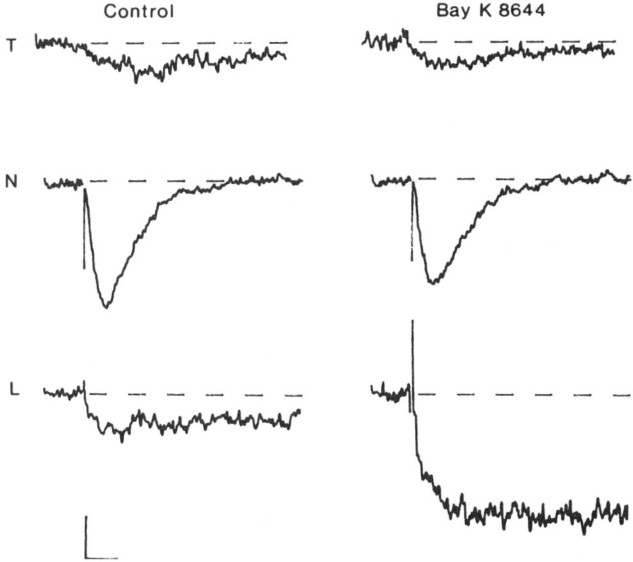

Fig. 3. Single channel current averages before and after the addition of Bay K 8644. All recording conditions were identical to those described in Fig. 2 legend. Averaged currents for the three channel types before (left) and after exposure to 5 uM Bay K 8644. (right). Top T-type activity, voltage step from −80 to −30 mV. Patch 86D. Middle: N-type activity, voltage step from −80 to +10 mV. Patch 84K. Bottom L-type activity, voltage steps from −40 to +10 mV. Patch 89F. Vertical calibration bar gives 1 pA for middle panel and 0.25 pA for top and bottom panels. Horizontal calibration bar gives 20 ms for top and bottom panels and 10 ms for middle panel.

heart for T and L-type channels (Nilius et. al. 1985). The existence of DHP-resistant channels in neurons might help explain the partial or complete lack of effect of DHPs on depolarization-induced Ca entry or transmitter release in many systems.

In the past, toxins specific for ACh receptor and sodium channels have been invaluable aids both for electrophysiological and biochemical studies (Hille (1984)). Recently, we have attempted to characterize the effects of the Ca channel toxin w−Conus Geographus V1A (w−CgTx) (Kerr and Yoshikami (1984), Olivera, Gray, Zeikus, McIntosh, Varga, Rivier, De Santos and Cruz (1985)) on the calcium channels in a variety of preparations (McCleskey, Fox, Feldman, Cruz, Olivera, Tsien, and Yoshikami (1987)).

Fig. 4 shows the different voltage-clamp protocols that were used to evoke the various Ca current components in a DRG neuron. Control recordings are shown superimposed on traces following a brief application of toxin (marked *). The T component was not affected by w−CgTx (fig. 4a) except for a transient reduction during exposure to toxin (data not shown). In contrast, both L and N components were rapidly blocked by w−CgTx and remained so even after the free toxin either diffused or was washed away (fig. 4b and 4c). These results are consistent with the idea that w−CgTx produces persistent block of N and L but not T currents in DRG neurons. Interestingly, the toxin affects vertebrate neurons, but not muscle cells. There is a persistent block of L and N currents in chick DRG, rat hippocampal neurons, and rat sympathetic neurons. In contrast the w−CgTx was ineffective at inhibiting L currents found in guinea-pig

ventricle, frog atrial or ventricle cells, chick myotubes or smooth
muscle cells from the rabbit ear artery (McCleskey, Fox, Feldman,Cruz,
Olivera, Tsien, and Yoshikami (1987)).

Table 1 summarizes some of the whole-cell and single-channel
properties of the three types of calcium channels found in chick DRG
neurons. In addition to difference in their voltage-dependencies, their
sensitivities to DHP agonists and antagonist, and the differential
effects of w-CgTx, the three channel types vary in their responses to
inorganic calcium channel blockers. For instance, T channels are
relatively insensitive to the inorganic blocker cadmium, while N and L
are very sensitive. For example, in six cells, 50 uM cadmium almost
completely abolished L and N current but left 55 + 4% of T. Interestingly
100 uM Ni had just the opposite effect as it blocked >90% of T-type
currents but showed little effect on the N or L components. These results
show that while the open channel properties of the three channel types
may be similar, they will not be identical (Fox, Nowycky, and Tsien, in
press).

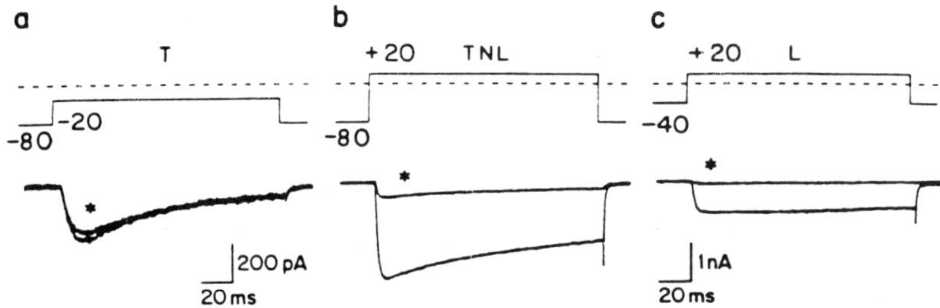

Fig. 4.   **w-CgTx VIA distinguishes different types of Ca channel types in
DRG neurons. Whole-cell voltage-clamp recordings done under
identical conditions to those described in Fig. 1 legend.
Traces marked with * indicate data taken 3-5 minutes after a
brief application of the toxin. a-c, voltage protocols and
current traces showing T current (a) and T,N, and L current
currents (b) and L current (c). Note that panel a has different
vertical scale bar than b and c. Toxin pipette contained: 10 uM
w-CgTx VIA plus 1 mg/ml cytochrome c in external solution. Cell
D01D.**

N-type Ca channels are involved in neurotransmitter release

Finding multiple types of calcium channels with disparate
biophysical and pharmacological properties immediately brings up the
question of which cellular functions are subserved by the various
channels? T channels might contribute to threshold behavior or rhythmic
activity (Llinas & Sugimori (1980), Llinas & Yarom (1981), Fox, Nowycky &

Table 1. Electrical and pharmacological properties of the three types of calcium channels found in chick DRG neurons.

	T	N	L
ACTIVATION RANGE (for 10 Ca)	positive to −70 mV	positive to −30 mV	positive to −10 mV
INACTIVATION RANGE (for 10 Ca)	−100 to −60 mV	−120 to −30 mV	−60 to −10 mV
INACTIVATION RATE (0 mV, 10 Ca or 10 Ba)	rapid (tau~20−50 ms)	moderate (tau~50−80 ms)	very slow (tau > 500 ms)
SINGLE−CHANNEL CONDUCTANCE (110 Ba)	8 pS	13 pS	25 pS
SINGLE−CHANNEL KINETICS	late opening, brief burst, inactivation.	long burst, inactivation.	hardly any inactivation.
RELATIVE CONDUCTANCES	Ba = Ca	Ba > Ca	Ba > Ca
Cd BLOCK	resistant	sensitive	sensitive
Ni BLOCK	sensitive	resistant	resistant
w−CgTx VIA BLOCK	weak,reversible	persistent	persistent
DIHYDROPYRIDINE SENSITIVITY	no	no	yes

Tsien, in press). Roles for L and N-type calcium channels have not yet been firmly established. The existence of multiple types of calcium channels in neurons raises new questions as to the identity of the channel mediating neurotransmitter release. To address this question we studied norepinephrine (NE) release from rat sympathetic neurons (superior cervical ganglion) in an effort to relate release with the known pharmacology of the different calcium channels. Patch-clamp studies revealed two types of Ca channels that had similar biophysical and pharmacological properties as compared to DRG L and N channels with one interesting exception, the rate of current decay for the sympathetic L- and N-type channels were about an order of magnitude slower than for those of DRG neurons. The sympathetic neurons had little or no T currents. As in DRG neurons the L-type calcium channels were sensitive to DHPs while N channels were not. 10 uM Cd almost completely abolished the currents through both types of channels. w−CgTx persistently blocked both types of channels. To study neurotransmitter release, cells were loaded with radiolabelled ^{3}H-norepinephrine (^{3}H-NE) by pre-incubating them for one hour; the ^{3}H-NE was then carefully washed out of the bath. Neurotransmitter release was triggered by exposing the cells to high K solutions for varying periods of time and then collecting the released ^{3}H-NE for analysis. The release process was shown to be Ca-dependent; removing Ca from the bathing medium (data not shown) or adding the Ca channel blockers Cd to the bath suppressed ^{3}H-NE release (see Fig. 5). In addition, Fig. 5 also shows that transmitter release was

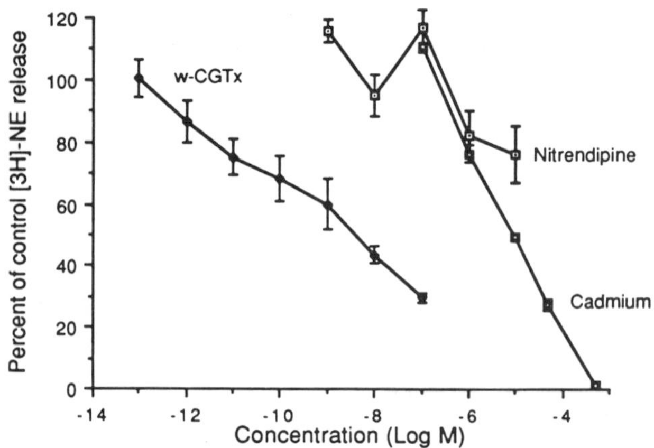

Fig. 5. The pharmacology of evoked ^{3}H–norepinephrine (^{3}H–NE) release from rat sympathetic neurons. Cultured rat superior cervical ganglion neurons were pre-incubated in ^{3}H–NE for one hour; aliquots of the bath solution were removed after 5 minutes of exposure to a depolarizing 70 mM K solution. Percent ^{3}H–NE release, normalized by transmitter release with no drugs present, plotted as function of drug concentration. Open squares show response to nitrendipine, filled diamonds to w–CgTx VIA, and closed squares to Cd.

unaffected by the DHP blocker nitrendipine at concentrations up to 10 uM. Conversely, the calcium channel toxin w–CgTx reduced release even at concentrations as low as 10 picomole. These results suggest that it is the DHP insensitive but w–CgTx sensitive N–type of Ca channel that is responsible for ^{3}H–NE release in rat sympathetic neurons. In contrast to these results, substance P release from rat DRG neurons showed a clear DHP sensitivity suggesting that L–type calcium channels were involved in in the release process (Perney, Hirning, Leeman and Miller (1986)).

To augment the data obtained from the rat sympathetic neurons, an additional series of experiments exploring the question of which type of Ca channel(s) participates in neurotransmitter release, was done on cultured neurons isolated from the CA_3 region of the rat hippocampus. Current records were obtained using the different voltage–clamp protocols best suited for isolating the various Ca current components (data not shown), before and after the application of 10 uM 2–chloroadenosine (2–CA). Neither the T nor the L components were affected by the 2–CA. In contrast, the N component, seen as an additional component of decaying current from negative holding potentials and using strong test depolarizations, was rapidly but reversibly blocked by the 2–CA. Adenosine has been shown to block excitatory synaptic transmission between the Schaffer Collateral/ Commissural pathway and CA1 (Sch–CA1) neurons (Proctor and Dunwiddie 1983). These results are consistent with the idea that N–type calcium channels mediate excitatory transmission in the Sch–CA1 pathway, as they do for NE release in rat sympathetic neurons. Inconsistent with our results are experiments done by Brown et al. (1985) showing that synaptic transmission in the Sch–CA1 pathway was blocked by suitably low concentrations of nimodipine indicating that the dihydropyridine sensitive L–type calcium channels were involved. More experimental data are needed to resolve this and other important questions pertaining to the cellular functions mediated by the different calcium channels. Determination of the cellular distribution of the channel types and their sensitivity to neuromodulators will also help in the elucidation of their physiological roles.

References

Armstrong, C.M., & Matteson, D.R., (1985). Two distinct populations of calcium channels in a clonal line of pituitary cells. Science 227: 65-67.

Ascher, P., & Nowak, L., (1986). Calcium permeability of the channels activated by N-methyl-D-aspartate (NMDA) in mouse central neurons. J. Physiol. 377, 34p.

Bean, B.P., (1985). Two types of calcium channels in canine atrial cells. Differences in kinetics, selectivity and pharmacology. J. Gen. Physiol. 86: 1-30.

Benham, C.D., and Tsien, R.W., (1987). Receptor-operated, Ca permeable channels activated by ATP in arterial smooth muscle cell. Nature, in press.

Bossu, J.L., Feltz, A. and Thomann, J.M., (1985). Depolarization elicits two distinct calcium currents in vertebrate sensory neurons. Pfluegers Archiv. 403: 360-368.

Brown D.A., Docherty, R.J., Gahwiler, B., and Halliwell, J.V., (1985). Calcium currents in mammalian central neurons. In: Cardiovascular effects of dihydropyridine-type calcium antagonists and agonists. Eds: A. Fleckenstein, C. Van Breemen, R. Grob, and F. Hoffmeister. Bayer-Symposium IX.

Canfield, D.R., & Dunlap, K., (1984). Pharmacological characterization of amine receptors on embryonic chick sensory neurones, Br. J. Pharmac.

Carafoli, E., & Penniston, T.J., (1985). The calcium signal. Sci. Amer. 253, No. 5, 70-78.

Carbone, E., & Lux, H.D., (1984a). A low voltage-activated calcium conductance in embryonic chick sensory neurons. Biophys. J. 46: 413-418.

Carbone, E., & Lux, H.D., (1984b). A low voltage-activating, fully inactivating Ca channel in vertebrate sensory neurons. Nature 310: 501-502.

Cavalie, A., Ochi., R., Pelzer, D., & Trautwein, W., (1983). Elementary currents through Ca^{2+} channels in guinea pig myocytes. Pflugers Archiv. 398: 284-297.

Cohen, C.J., & McCarthy, R.T., (1985). Differential aspects of dihydropyridines on two populations of Ca channels in anterior pituitary cells. Biophys. J. 47: 513a.

Deitmer, J.W., (1984). Evidence for two voltage-dependent calcium currents in the membrane of the ciliate _Styloncia_. J. Physiol. 355: 137-159.

DeRiemer, S.A., Strong, J.A., Albert, K.A., Greengard, P., Kaczmarek, L.K., (1985). Enhancement of calcium current in _Aplysia_ neurones by phorbol ester and protein kinase C. Nature 313: 313-316.

Dunlap, K., & Fischbach, G.D., (1978). Neurotransmitters decrease the calcium component of sensory neurone action potentials. Nature 276, 837-839.

Dunlap, K., & Fischbach, G.D., (1981). Neurotransmitters decrease the calcium conductance activated by depolarization of embryonic chick sensory neurones. J. Physiol. 317, 519-535.

Eckert R., & Chad, J.D. (1984). Inactivation of calcium channels. Progress in Biophysics and Molecular Biology. 44: 215-267.

Fabiato, A., (1983). Calcium-induced release of calcium from the cardiac sarcoplasmic reticulum, Brief Review, American Physiological Society. C1-C14.

Fedulova, S.A., Kostyuk, P.K., & Veselovsky, N.S., (1985). Two types of calcium channels in the somatic membrane of new-born rat dorsal root ganglion neurones. J. Physiol. 359: 431-446.

Feldman, D.H., & Yoshikami D. (1985). A peptide toxin from the marine mollusc Conus Geographicus blocks voltage-gated calcium channels. Soc. Neurosci. Abs. :517.

Fox, A.P., & Krasne, S. (1981). Two calcium currents in egg cell. Biophys. J. 33:145a.

Fox, A.P., & Krasne, S. (1984). Two calcium currents in Neanthes Arenaceodentata egg cell membranes. J. Physiol. J. Physiol. 356: 491-505.

Fox, A.P., Nowycky, M.C., & Tsien, R.W., (1987). Kinetic and pharmacological properties distinguishing three types of calcium currents in chick sensory neurons. J. Physiol. in press.

Fox, A.P., Nowycky, M.C., & Tsien, R.W., (1987). Single channel recordings of three types of calcium channels in chick sensory neurons. J. Physiol. in press.

Hagiwara, S., & Byerly, L., (1981). Calcium channel. Ann. Rev. Neurosci. 4: 69-125.

Hagiwara, S., & Byerly, L., (1983). The calcium channel. TINS, 189-193.

Hagiwara, S., Ozawa, S., & Sand, O., (1975). Voltage-clamp analysis of two inward currents mechanisms in the egg cell membrane of a starfish. J. Gen. Physiol. 65: 617-644.

Hess, P., Lansman, J.B., & Tsien, R.W., (1984). Different modes of Ca channel gating behavior favored by dihydropyridine Ca agonists and antagonists. Nature 311: 538-544.

Hille, B., (1984). Ionic channels in excitable membranes. Sinauer Associates, Sunderland, Mass., pp 426.

Kerr, L.M., & Yoshikami, D., (1984). A venom peptide with a novel presynaptic blocking action. Nature 308: 282-284.

Kokubun, S., & Reuter, H., (1984). Dihydropyridine derivatives prolong the open state of Ca channels in cultured cardiac cells. Proc. Natl. Acad. Sci. 81: 4824-4827.

Llinas, R., & Sugimori, M., (1980). Electrophysiological properties of in vitro Purkinje cell somata in mammalian cerebellar slices. J. Physiol. 305: 171-195.

Llinas, R., & Yarom, Y. (1981). Electrophysiology of mammalian inferior olivary neurones *in vitro*. Different types of voltage-dependent ionic conductances. J. Physiol. 315: 549-567.

MacDermott, A.B., Mayer, M.L., Westbrook, G.L., Smith, S.J., and Barker, J.L., (1986). NMDA-receptor activation increases cytoplasmic calcium concentration in cultured spinal cord neurones. Nature 321, 519-522.

McCleskey, E.W., Fox, A.P., Feldman, D., Cruz, L.J., Olivera, B.M., Tsien, R.W., & Yoshikami, D., (1987). Calcium channel blockade by a peptide from Conus: Specificity and mechanism. Proc. Natl. Acad. Sci. in press.

Mitra, R. & Morad, M. (1986). Two types of calcium channels in guinea pig ventricular myocytes. Proc. Natl. Acad. Sci. U.S.A. 83, 5340-5344.

Narahashi, T., Tsunoo, A., & Yoshii, M., (1987). Characterization of two types of calcium channels in mouse neuroblastoma cells. J. Physiol. in press.

Nilius, B., Hess., Lansman, J.B., & Tsien, R.W., (1985). A novel type of cardiac calcium channel in ventricular cells. Nature 316:443-446.

Nishizuka, Y., (1984). The role of protein kinase C in cell surface signal transduction and tumour promotion. Nature 308, 693-698.

Nishizuka, Y., (1986). Studies and perspectives of protein kinase C. Science 233, 305-312.

Nowycky, M.C., Fox, A.P., & Tsien, R.W., (1985a). Long-opening mode of gating of neuronal calcium channels and its promotion by the dihydropyridine calcium agonist Bay K 8644. Proc. Natl. Acad. Sci. 82: 2178-2182.

Nowycky, M.C., Fox, A.P., & Tsien, R.W., (1985b). Three types of calcium channel with different calcium agonist sensitivity. Nature 316: 440-443.

Olivera, B.M., Gray, W.r., Zeikus, R., McIntosh, J.M., Varga, J., Rivier, J., De Santos, V., & Cruz, L.J., (1985). Peptide neurotoxins from fish-hunting cone snails. Science 230:1338-1343.

Pallotta, B.S., Magleby, K.L., & Barrett, J.N., (1981). Single channel recordings of Ca2+-activated K+ currents in rat muscle cell culture. Nature 293, 471-474.

Perney, T.M., Hirning, L.D., Leeman, S.E., & Miller, R.J., (1986). Multiple calcium channels mediate neurotransmitter release from peripheral neurones. Proc. Natl. Acad. Sci. 83: 6656-6659.

Reuter, H. (1983). Calcium channel modulation by neurotransmitters, enzymes and drugs. Nature 301: 569-574.

Reuter, H. (1985). A variety of calcium channels. Nature 316: 391.

Strong, J.A., Fox, A.P., Tsien, R.W., & Kaczmarek, L.K., (1987). Stimulation of protein kinase C recruits covert calcium channels in Aplysia bag cell neurons. Nature in press.

Tsien, R.W., (1986). Calcium channels in heart cells and neurons. In: Neuromodulation, L.K. Kaczmarek & I.B. Levitan eds., Oxford University Press.

Yellen, G., (1982). Single Ca2+-activated nonselective cation channels in
neuroblastoma. Nature 296, 357-359.

EXPRESSION OF PRESYNAPTIC CALCIUM CHANNELS IN <u>XENOPUS</u> OOCYTES

Joy A. Umbach and Cameron B. Gundersen

Department of Pharmacology and Jerry Lewis Neuromuscular
Research Center, UCLA School of Medicine
Los Angeles, CA 90024

INTRODUCTION

The release of neurotransmitter from nerve terminals is triggered by
the influx of calcium ions through voltage-activated channels[1-3]. In
excitable tissues there appear to be a number of different types of Ca
channels, based on physiological parameters (single channel conductance,
ion selectivity, voltage-dependent activation and inactivation properties)
and sensitivity to pharmacological agents (especially dihydropyridines,
w-conotoxin (wCgTx) and Cd ions). The type of Ca channel underlying
synaptic transmission has not been established, although recent studies of
mammalian neurons suggest that N-type channels are involved[4,22].

The direct measurement of Ca currents in nerve terminals has largely
been confined to the squid giant synapse because of its large size and
favorable anatomy[5-7]. Recently, some advances in the study of
presynaptic Ca channels have been made with extracellular recordings of
currents at vertebrate motor-nerve terminals[8,9], using optical dyes[10]
and by incorporating synaptosomal Ca channels into lipid bilayers[11]. In
this work we use an alternative approach involving mRNA-injected <u>Xenopus</u>
oocytes.

In the early 1970's John Gurdon and his colleagues demonstrated that
oocytes from the South African clawed frog, <u>Xenopus</u> <u>laevis</u>, when injected
with reticulocyte mRNA, efficiently translated that RNA to yield
globin[12]. A decade later, Katumi Sumikawa, Eric Barnard, Ricardo Miledi
and their coworkers[13,14] introduced this translation system to
neurobiologists by demonstrating the expression of functional nicotinic
ACh receptors in oocytes injected with <u>Torpedo</u> electroplax mRNA.
Subsequently, numerous laboratories have used the <u>Xenopus</u> oocyte system to
induce expression of receptors and channels and ultimately to correlate
structure with function, as exemplified in the investigations of Numa,
Sakmann and coworkers on the ACh receptor[15].

We sought to induce <u>Xenopus</u> oocytes to express a calcium channel
bearing the hallmarks of a nerve terminal calcium channel by injecting
mRNA from a source rich in presynaptic Ca channels. Electric fish possess
densely innervated electrocytes which comprise the electric organs. The
electroplax tissue has been widely employed by biochemists to isolate
proteins and organelles involved in synaptic transmission. We anticipated

that the cell bodies of the neurons which innervate this tissue would be a
good source of mRNA coding for presynaptic proteins. Recently, we
reported that oocytes injected with mRNA from the electric lobes of
Torpedo californica expressed a unique Ca channel with properties unlike
those previously observed in oocytes[16]. This paper considers in more
detail the expression and characteristics of this nerve-ending-like
calcium channel and points toward the oocyte system as a strategy for
isolating cDNA clones coding for a presynaptic calcium channel.

METHODS

    This investigation used standard techniques for isolating
poly(A)$^+$RNA and for injecting and culturing oocytes[17,18]. The source
of mRNA, experimental protocol and solution composition have been
described in a previous paper[16]. Briefly, chloroform-phenol was used to
extract RNA from the electric lobes of Torpedo californica. Poly(A)$^+$RNA
was selected by oligo-dT-cellulose chromatography. Oocytes were injected
with 40-60ng of poly(A)$^+$RNA in 40-60nl of water. Oocytes were cultured
in Barths medium plus gentamicin at 18°C. To facilitate impalement with
microelectrodes, the enveloping cellular layers of the oocyte were
manually removed following 1 hr collagenase treatment (1 mg-ml^{-1}).
Oocytes were voltage clamped with a conventional two electrode voltage
clamp. To maximize currents through Ca channels in oocytes and to
minimize activation of the endogeneous Ca-activated chloride current,
oocytes were bathed in a Ba (40mM)-methanesulfonate solution described by
Dascal et al.[19] to which 1 mM tetraethylammonium bromide was added to
block K currents. Oocytes injected with Torpedo electric lobe in mRNA are
referred to as injected oocytes, while uninjected oocytes from the same
frog are designated control oocytes. All experiments were conducted at
20-22°C.

RESULTS

    Uninjected oocytes exhibit voltage-activated channels which are
permeable to Ba and blocked by Cd ions (Fig. 1C). As reported
earlier[19], this Ba current is insensitive to the removal of Na ions or
application of 1μM TTX. If oocytes from the same frog are injected with
Torpedo electric lobe mRNA, the inward currents are much larger
(Fig. 1A). As in control oocytes, these currents are blocked by Cd
(Fig. 1B) and are unaffected by replacement of Na with choline or by 1μM
TTX. However, the Ba current of injected oocytes is only partially
inactivated during prolonged depolarizing pulses that completely
inactivate the endogeneous Ba current. Current-voltage relations of the
Ba current of injected oocytes reveal that maximum current is elicited at
more positive voltages than the Ba current of control oocytes (Fig. 1D,
3B).

    With the mRNA preparation we have used, it requires 6-7 days after
mRNA injection for large-amplitude Ba currents to develop (Fig. 2). (It
should be noted that oocytes are manually defolliculated the third day
after mRNA injection. This disturbance of the oocyte may contribute to
the delay in the appearance of channel proteins.) The ideal period to
screen oocytes for the expression of these Ba currents is between 8-14
days. However, even in oocytes that remain viable after 4 weeks, Ba
currents persist in injected oocytes that are larger than the endogeneous
Ba current of uninjected cells.

    We have demonstrated[16] that wCgTx allows one to distinguish
components of the Ba current of injected oocytes (Fig. 3). At

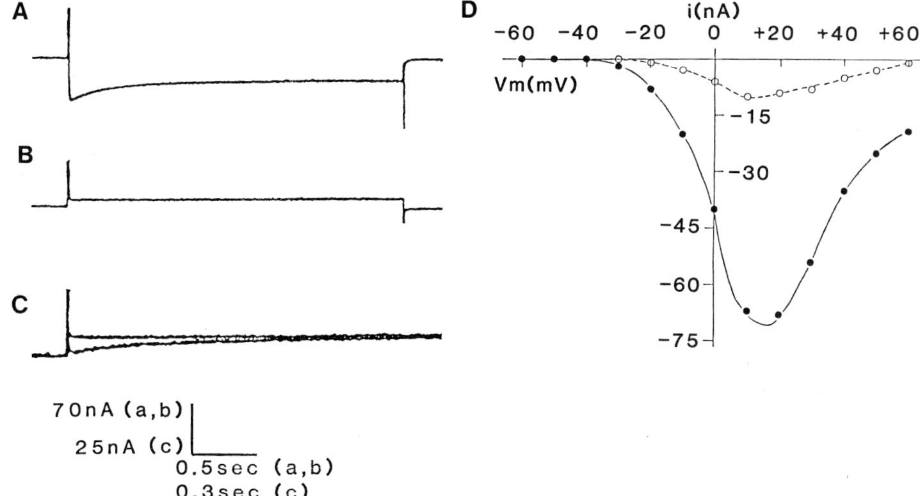

Fig. 1. Ba currents of injected and control oocytes. A) Current recorded
with a +70mV step from V_{hold}=-60mV for an oocyte in Ba Ringer 9
days after mRNA injection. B) Same oocyte from (A) in Ba Ringer
+2mM $Cd(OH)_2$. C) Superimposed current records during the same
test protocol (V_{hold} = -60mV, test pulse to +10mV) in a control
oocyte bathed in Ba Ringer (lower trace) and Ba Ringer +2mM
$Cd(OH)_2$ (upper trace). D) Current-voltage relations for peak
i_{Ba} of an injected oocyte (●) and a control oocyte (o). i_{Ba}
was measured as the peak inward current (500 msec voltage step
from V_{hold} = -60mV in Ba Ringer) less the residual current
(elicited with the same voltage step) in Ba Ringer +2mM
$CD(OH)_2$. (from 16).

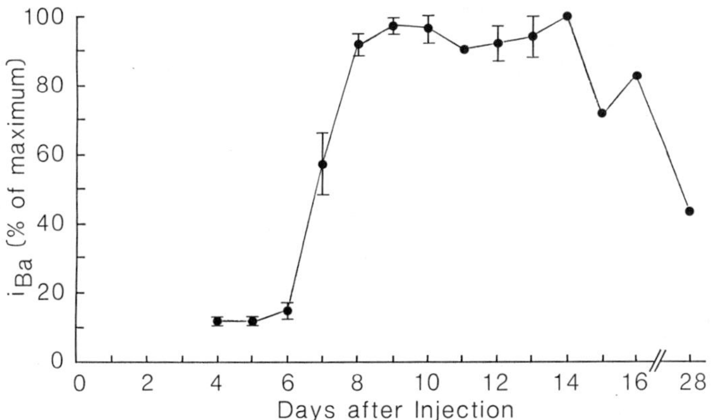

Fig. 2. Time course of development of additional Ba current after
injection of _Torpedo_ electric lobe mRNA. Normalized peak Ba
current (mean ± S.E.) elicited in response to a +70mV step from
V_{hold} = -60mV is provided for oocytes from 4 different frogs.
Single points (no error bars) are measurements from individual
oocytes of the same donor. The Ba current amplitude 4 and 5 days
after mRNA injection was indistinguishable from uninjected
oocytes. The Ba current of uninjected oocytes does not increase
over the time period examined.

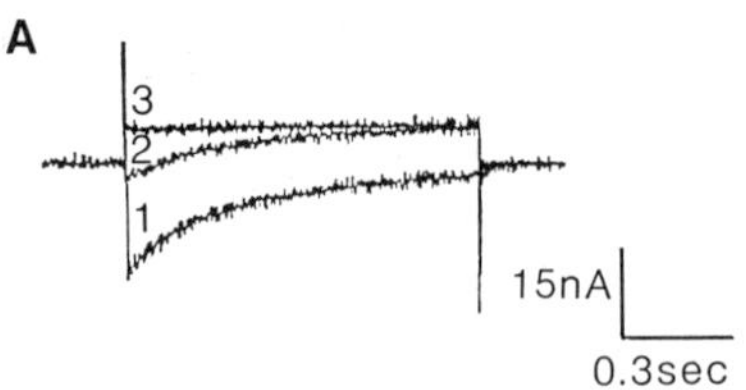
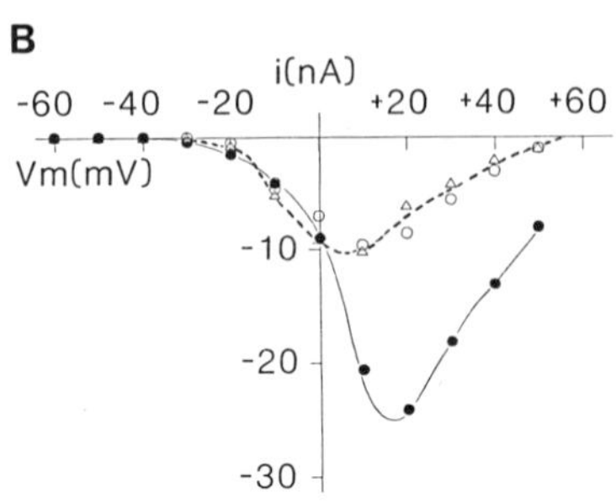

Fig. 3. Action of wCgTx on Ba currents. A) Ba current of an oocyte (8
        days after mRNA injection) evoked by a +70 mV step from V_{hold} =
        -60 mV; 1. in Ba Ringer; 2. after 30 min. in Ba Ringer plus
        wCgTx (10 μM); 3. in Ba Ringer plus 2 mM $CdCl_2$. B) Current
        voltage relations for residual-current-corrected (i.e., Cd
        subtracted, see Fig. 1 legend) Ba currents. Δ,
        wCgTx-insensitive i_{Ba} (currents that remain after 30 min in Ba
        Ringer plus 10 μM wCgTx) of an injected oocyte; ●,
        wCgTx-sensitive i_{Ba} from the same oocyte. Values for
        wCgTx-sensitive currents were obtained by subtracting the
        wCgTx-insensitive current (Δ) from the total i_{Ba} measured in
        Ba Ringer. o, i_{Ba} measured in a control oocyte bathed in Ba
        Ringer (note; exposure of this oocyte to wCgTx for 60 min. had no
        effect on the amplitude or shape of this current), (from 16).

10μM, wCgTx completely eliminates a sizeable fraction of the Ba current
of injected oocytes. However, the endogeneous Ba current of control or
injected oocytes is insensitive to wCgTx at 10-20μM with exposure times
of 30 min to 2 hrs. The wCgTx (10μM) mediated inhibition of the Ba
current of injected cells is complete within 30-35 min and is half maximal
at 10 min (Fig. 4). This effect is not reversed by 1 hour of washing in
wCgTx-free Ba Ringer.

     The degree to which the Ba current of injected oocytes is blocked by
wCgTx depends on the amount of toxin present (Fig. 5). More than 50% of
the wCgTx-sensitive Ba current of injected oocytes is blocked at 1μM,
while no further inhibition is seen at concentrations above 10μM. The
wCgTx-insensitive current of an injected oocyte has the same IV relation
as an uninjected oocyte from the same donor. The maximum Ba current
elicited in an injected oocyte in Ba Ringer occurs at +20mV, whereas the
wCgTx-insensitive component of this current peaks at +10mV (Fig. 5B).

     The Ba current observed in oocytes after the injection of Torpedo
electric lobe mRNA and the endogenous Ba current of uninjected oocytes
also exhibit different sensitivity to Cd ions[16]. The dose of Cd needed
to inhibit 50% of the wCgTx-sensitive Ba current of injected oocytes is
approximately 0.1μM, whereas 50% inhibition of the endogeneous Ba current

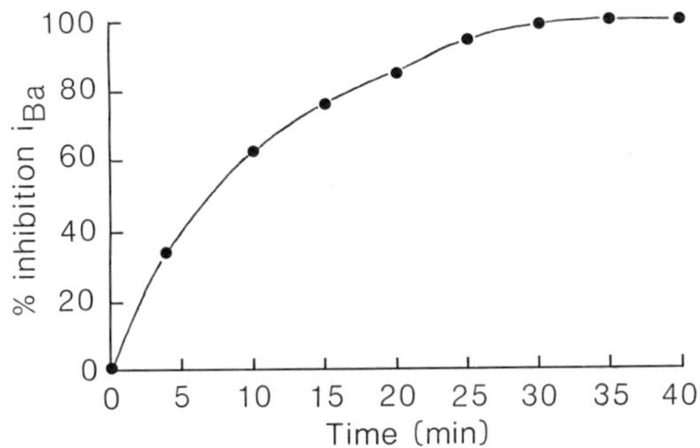

Fig. 4. Time course of wCgTx action. A mRNA-injected oocyte is exposed to
10 μM wCgTx at time 0. At subsequent times the reduction of
peak Ba current (elicited by a +70 mV step from V_{hold} = −60mV)
is normalized to the maximum Ba current blocked by wCgTx.

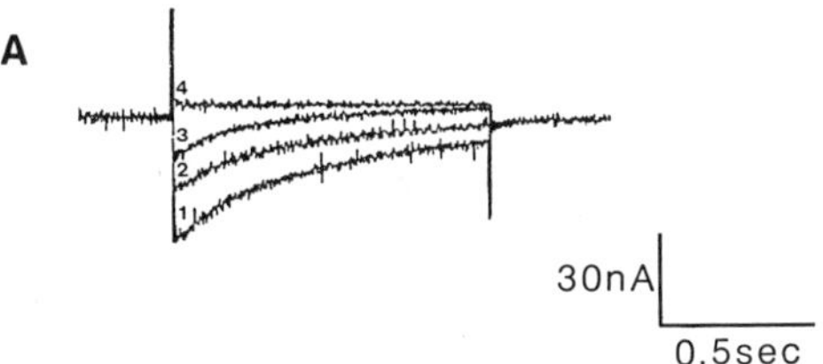

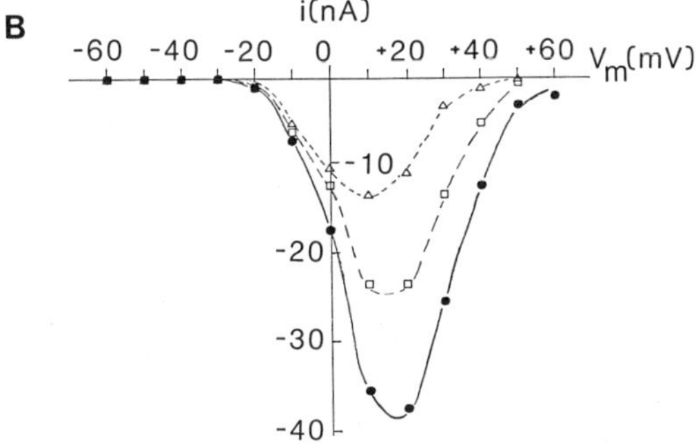

Fig. 5. Action of 1 μM vs. 10 μM wCgTx. A. Current responses of an
injected oocyte to a +70 mV step from V_{hold} = −60 mV in: 1) Ba
Ringer; 2) Ba Ringer plus 1 μM wCgTx (30 minutes); 3) Ba
Ringer plus 10 μM wCgTx (30 minutes); 4) Ba Ringer plus 2 μM
Cd(OH)$_2$. B. IV plot of total peak Ba current of the injected
oocyte in Ba Ringer (●); Ba current remaining after 30 minute
treatment with 1 μM wCgTx (□) and i_{Ba} remaining after
exposure to 10 μM wCgTx (30 minutes)(Δ).

occurs at 10–20μM. Also, we found[16] that 1μM Cd completely blocks
the wCgTx–sensitive Ba current that appears after mRNA injection while
leaving the Ba current of control oocytes unaffected.

Consistent with the interpretation that 1μM Cd and 10μM wCgTx
both act on that component of the Ba current which develops after mRNA
injection is the observation that both agents block the same amount of
current in a single, injected oocyte. As can be seen in Fig. 6, the
large, sustained inward current of an injected oocyte in Ba Ringer is
blocked substantially by 1μM Cd. The IV relation of the residual (1μM
Cd–insensitive) Ba current is similar to that of uninjected oocytes (data
not shown). The Cd block is reversible and upon washing, the Ba current
recovers to its initial level. Subsequent exposure to wCgTx (10μM for
30 min) reduces the current amplitude to the same level found in 1μM Cd
(Fig. 6). This remaining current is not affected further by 1μM Cd;
rather, it exhibits the same IV relation and Cd sensitivity (i.e. IC_{50}
at 10μM) as the endogeneous Ba current of a control oocyte.

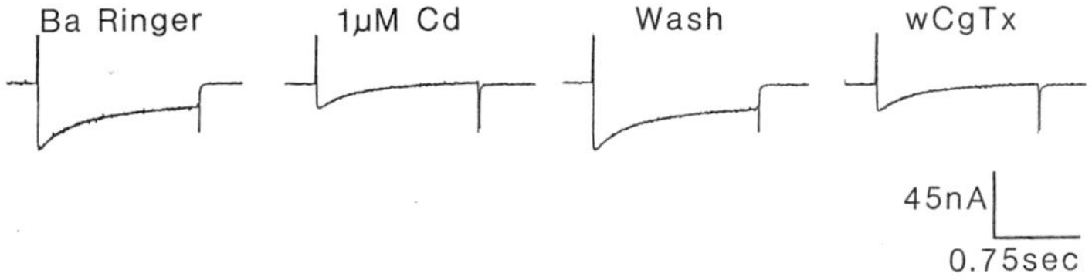

Fig. 6. Inhibition of the barium current of injected oocytes by wCgTx or
cadmium. wCgTx (10 μM) and cadmium (1 μM) eliminate
essentially the same proportion of the barium current of an
injected oocyte. Currents elicited by +70 mV pulses from
$V_{hold} = -60$ mV.

We previously reported that the curve describing the steady–state
inactivation of the wCgTx–sensitive component of the Ba current of
injected oocytes is shifted to more positive values (with 50% inactivation
at approximately–18mV) than the curves for the wCgTx–insensitive component
of the Ba current of injected oocytes or the endogeneous Ba current of
control oocytes (50% inactivation at approximately – 48 mV)[16]. The same
distinction is seen if one compares the steady–state inactivation of those
components of the Ba current of injected oocytes that are and are not
blocked by 1μM Cd (Fig. 7). Again, the Ba current that is resistant to
1μM Cd exhibits steady–state inactivation properties like controls
(Fig. 7). Neither 1μM Cd nor 10μM wCgTx affects Ba–current
inactivation of control oocytes.

Two–pulse experiments have shown that the induced Ba current of
injected oocytes is less prone to inactivation by a short term (500 msec)
prepulse of variable amplitude than the Ba current of uninjected
oocytes[16]. By systematically varying the duration of a fixed amplitude
prepulse, it was also demonstrated that complete inactivation of the Ba
current expressed after mRNA injection occurs much more slowly than the
native Ba current of control cells[16]. Accentuating the differences
between induced and endogeneous Ba currents, variation of interpulse

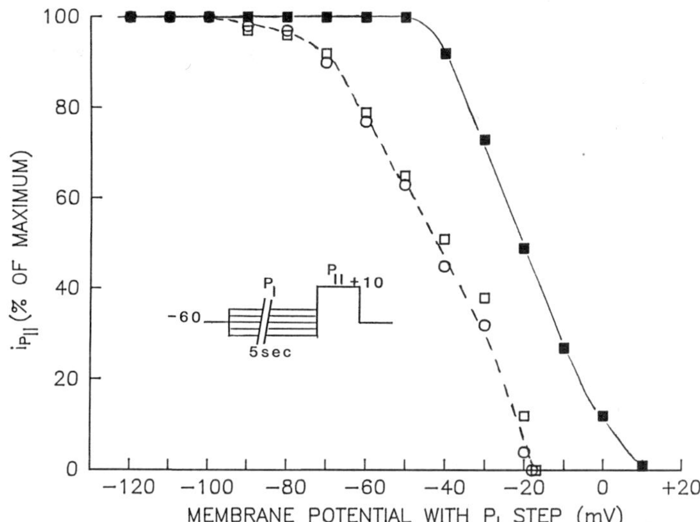

Fig. 7. Effects of 1 µM Cd on steady-state inactivation. Protocol is
given in inset. Peak inward current during P_{II} is normalized to
the maximum current observed and given as a function of membrane
potential during P_I. ■, 1 µM Cd-sensitive Ba current of an
injected oocyte; □, 1 µM Cd-insensitive Ba current of the same
injected oocyte; o, control oocyte in Ba Ringer (the steady state
inactivation of the control oocyte was unaltered by 1 µM Cd).
Lines are best fit by eye.

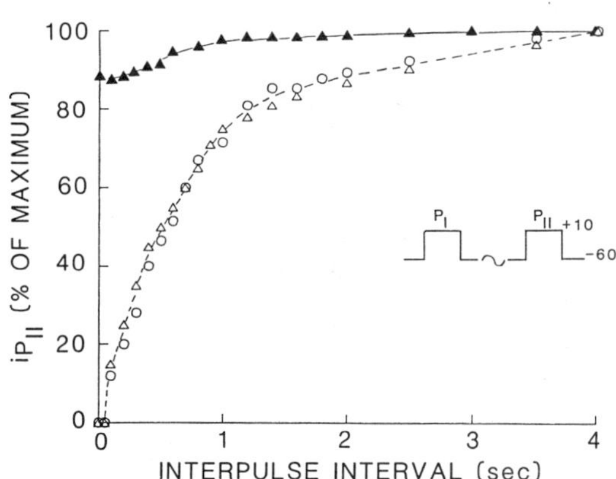

Fig. 8. Time course of Ba current recovery from inactivation by a
prepulse. Protocol is given in inset. Test pulses are 500 msec
in duration. ▲, wCgTx-sensitive component of Ba current of an
injected oocyte. Δ, wCgTx-insensitive component of Ba current
from the same oocyte. o, control oocyte.

interval reveals that the wCgTx-sensitive Ba current of injected oocytes recovers from inactivation more rapidly than the native Ba current of control oocytes (Fig. 8).

In an effort to find other compounds that would selectively affect either the endogeneous or induced Ba currents, we tested a number of dihydropyridines[16]. Neither nisoldipine nor nitrendipine (1μM) affected the Ba current of injected or control oocytes (see Fig. 9). As a positive control to ensure that oocytes were capable of translating mRNA to produce a Ca channel with a dihydropyridine binding site, 1μM nisoldipine was found to reduce the Ba current of oocytes injected with neonatal rat heart mRNA (Fig. 9), confirming the work of Dascal and coworkers[19].

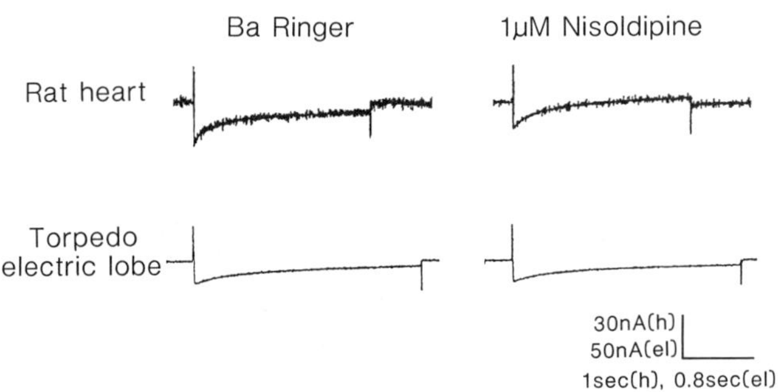

Fig. 9. Effect of nisoldipine on Ba currents of mRNA injected oocytes. Current traces are responses to a +70 mV step from V_{hold} = -60 mV in Ba Ringer and Ba Ringer + 1 μM nisoldipine. mRNA was prepared from the hearts of 14d old Sprague-Dawley rats using the same protocol as for _Torpedo_ mRNA. Oocytes are from the same frog.

A pressing question that concerns most uses of the oocyte as an expression vehicle is whether or not the protein under study is a direct translation product of the injected mRNA. We reported previously that when oocytes were cultured continuously with actinomycin D at concentrations that block 98-99% of oocyte RNA synthesis, we saw no suppression of the Ba current of injected oocytes[16]. Figs. 10 A and B show Ba currents of injected and control oocytes that were cultured in actinomycin D. Data are also presented (Fig. 10C) for an injected oocyte that was not exposed to actinomycin D. Regardless of whether or not the injected oocytes were exposed to actinomycin D, the resultant Ba currents were very similar and were distinctly larger than the endogeneous Ba current. A similar degree of sensitivity to wCgTx was observed for the Ba current of injected oocytes cultured with or without actinomycin D. Therefore, the expression of a unique Ba current in injected oocytes is not a consequence of endogeneous synthesis of RNA by the oocyte.

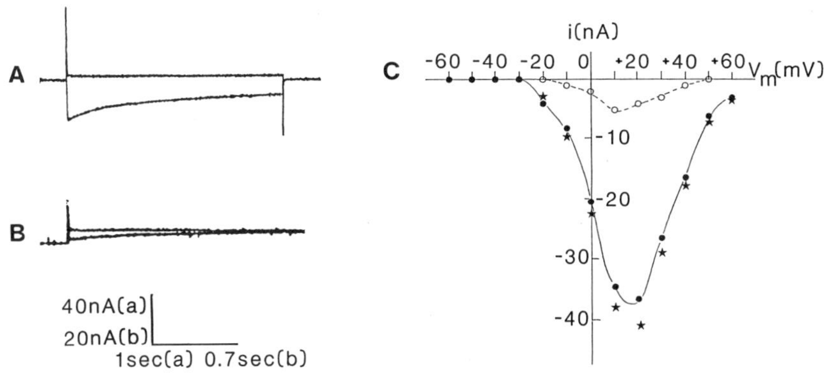

Fig. 10. Action of actinomycin D on Ba currents appearing after mRNA injection. Ba currents evoked by a +70 mV step from V_{hold} = -60mV in an injected oocyte (A) and control oocyte (B), both incubated for 13 days in 50 mg-ml^{-1} actinomycin D (For A and B, lower trace in Ba Ringer, upper trace Ba Ringer +2 μM $Cd(OH)_2$). C. I-V relations for control oocyte in actinomycin D,O; injected oocyte in actinomycin D,●; and an injected oocyte incubated in the absence of actinomycin D,*. (Control oocytes that were not exposed to actinomycin D had slightly larger currents.)

DISCUSSION

The data presented in this report and our previous communication[16] indicate that mRNA from _Torpedo_ electric lobe induces _Xenopus_ oocytes to express a unique Ca channel. This interpretation relies heavily on our ability to distinguish between the endogeneous Ca channels of the oocyte and the new population of channels that appears after injection of mRNA. Without detectably affecting the intrinsic Ba current of control oocytes, both 1μM Cd and μM quantities of wCgTx reduce the amplitude of the Ba current of injected oocytes to the control level. These same agents produce a shift in the IV characteristics and inactivation parameters of the Ba currents of injected oocytes such that they resemble controls. Moreover, the induced Ba currents are 100-fold more sensitive to inhibition by Cd than the endogeneous Ba currents. As a final distinction, the Ba currents of control oocytes inactivate more rapidly and recover more slowly with a fixed prepulse voltage step than those currents of injected oocytes.

The Ca channel that appears in oocytes after injection with mRNA from _Torpedo_ electric lobe has properties that are similar to those reported for Ca channels controlling the release of transmitter in other systems. Studies by several groups[4,20-22] indicate that presynaptic Ca channels are particularly sensitive to inhibition by W-conotoxins. For instance, the K-evoked release of transmitter (measured as ATP output) from electric fish synaptosomes is reduced by 50% at approximately 0.5μM wCgTx with a half time of 15 minutes[21]. Inhibition of release approaches saturation at 5μM wCgTx[21]. The dose and time course of wCgTx inhibition of the novel Ba current of injected oocytes is within the same range. In both electric fish synaptosomes and rat brain synaptosomes, dihydropyridine antagonists have been reported to be ineffective in blocking the voltage-activated Ca channels[21,4]. Dihydropyridines do not affect the Ba currents of oocytes injected with _Torpedo_ mRNA. Moreover, presynaptic

Ca channels appear to be quite sensitive to blockade by Cd ions[4]. Depolarization induced $^{45}Ca^{2+}$ influx into rat brain synaptosomes is inhibited by cadmium with an IC_{50} of approximately 1μM (Ref. 4). Recently, it was observed that the depolarization-induced release of ATP from electric fish synaptosomes was blocked by Cd with an IC_{50} of about 3μM (G. Miljanich, personal communication). The Cd sensitivity of the Ba current that appears after mRNA injection is in the submicromolar range. All of these similarities support the hypothesis that mRNA from _Torpedo_ electric lobe induces the expression of presynaptic Ca channels in oocytes.

Ideally, it would be desirable to establish with greater certainty the similarity between the Ca channels of the electric fish nerve ending and the Ca channel that we have "transplanted" to the oocyte. Presently, experiments are underway to obtain single channel records from both the _Torpedo_ nerve ending and injected oocytes. Enrichment of our _Torpedo_ electric lobe mRNA preparation (by size fractionation or cDNA cloning) in transcripts coding for this Ca channel protein should elevate the density of channels expressed in the oocyte to a level suitable for patch clamping. Fusion of _Torpedo_ synaptosomes[23,24] or the double-dip patch method[25] with a synaptosome preparation should permit single channel recording from the presynaptic Ca channels produced by the electromotor neurons. If the oocyte and nerve ending Ca channels prove to be dissimilar, we will attampt to record from the cell bodies of the electromotoneurons to determine if they are the location of the Ca channels that are expressed in the oocyte.

The most straightforward explanation for the appearance of nerve-ending-like Ca channels in our mRNA primed oocytes is that these channels are formed as a consequence of the translation of the appropriate _Torpedo_ mRNAs. The actinomycin D experiments show that transcription of _Xenopus_ DNA does not play a direct role in the appearance of these channels. Moreover, in our hands only mRNA from _Torpedo_ electric lobe has produced Ca channels with high sensitivity to Cd and wCgTx[16]. Poly (A)$^+$RNA isolated from rat brain, rat heart, human brain, _Torpedo_ electric organ, _Torpedo_ mid-brain, _Paramecium_, _Xenopus_ oocytes and PC_{12} & DDT_1 clonal cell lines has not yielded similar currents in oocytes. Moreover, Leonard and colleagues[26] also fail to detect wCgTx sensitivity of Ca channels induced by rat brain mRNA. However, at this time, we cannot exclude more circuitous possibilities for the appearance of these currents, such as the unmasking of dormant channels or modification of existing channels. Indeed, this brings up the question of uniformity of calcium channel types; whether post-translational modifications can transform, for example, an L channel into an N channel. Future studies should resolve whether the multiplicity of different Ca channel types can be ascribed to genes coding for each channel type. Ultimately, the oocyte system should be amenable to use as a bioassay to select cloned cDNA for Ca channel proteins and to compare their structural features with functional characteristics.

<u>Acknowledgements</u>

This work would not have been possible without the loan of equipment from Roger. His advice, criticism, wit, humor and friendship will be deeply missed. Funds for this work were supplied by a contract from the ARO (DAA629-85-K-0113), and the American Heart Association. This work was also aided by Basil O'Connor Starter Scholar Research Award No. 5-571 from the March of Dimes Birth Defects Foundation. JAU was supported by the Laubisch Fund of UCLA and by NS 23851. CBG is a Sloan Foundation fellow in Neuroscience and is supported by NS 00827.

REFERENCES

1. B. Katz and R. Miledi, A study of synaptic transmission in the
    absence of nerve impulses, _J. Physiol. (Lond.)_ 192:407 (1967).
2. B. Katz and R. Miledi, Tetrodotoxin-resistant electrical activity in
    presynaptic terminals, _J. Physiol. (Lond.)_ 203:459 (1969).
3. B. Katz, "The Release of Neural Transmitter Substances", Charles C.
    Thomas, Springfield, Il., (1969).
4. I. J. Reynolds, J. A. Wagner, S. H. Snyder, S. A. Thayer, B. M.
    Olivera and R. J. Miller, Brain voltage-sensitive calcium channel
    subtypes differentiated by w-conotoxin fraction GVIA, _Proc. Natl._
    Acad. Sci. USA. 83:8804 (1986).
5. R. Llinas, I. Z. Steinberg and K. Walton, Presynaptic calcium current
    in squid giant synapse, _Biophys. J._ 33:289 (1981).
6. M. P. Charlton, S. J. Smith and R. S. Zucker, Role of calcium ions
    and channels in synaptic facilitation and depression of the squid
    giant synapse, _J. Physiol. (Lond.)_ 323:173 (1982).
7. G. J. Augustine, M. P. Charlton and S. J. Smith, Calcium entry into
    voltage-clamped presynaptic terminals of squid, _J. Physiol. (Lond.)_
    367:143 (1985).
8. C. B. Gundersen, B. Katz and R. Miledi, The antagonism between
    botulinum toxin and calcium in motor nerve terminals, _Proc. R Soc._
    (Lond.) B. 216:369 (1982).
9. J. L. Brigant and A. Mallart, Presynaptic currents in mouse motor
    nerve endings, _J. Physiol. (Lond.)_ 333:619 (1982).
10. B. M. Salzberg, A. L. Obaid, D. M. Senseman and H. Gainer, Optical
    recording of action potentials from vertebrate nerve terminals using
    potentiometric probes provides evidence for sodium and calcium
    components, _Nature_ 306:36 (1983)
11. M. T. Nelson, R. J. French and B. K. Krueger, Voltage-dependent
    calcium channels from brain incorporated into planar lipid bilayers,
    Nature 308:77 (1984).
12. J. B. Gurdon, C. D. Lane, H. R. Woodland and G. Marbaix, Use of frog
    eggs and oocytes for the study of mRNA and its translation in living
    cells, _Nature_ 233:177 (1971).
13. E. A. Barnard, R. Miledi, and K. Sumikawa, Translation of exogenous
    mRNA encoding for nicotinic acetylcholine receptors produces
    functional receptors in _Xenopus_ oocytes, _Proc. R. Soc. (Lond.) B_
    215:241 (1982).
14. K. Sumikawa, M. Houghton, J. Emtage, B. Richards and E. Barnard,
    Active multi-subunit ACh receptor assembled by translation of
    heterologous mRNA in _Xenopus_ oocytes, _Nature_ 292:862 (1981).
15. B. Sakmann, C. Methfessel, M. Mishina, T. Takahashi, T. Takai,
    Kursaki, M., Fukuda, K. and Numa, S, Role of acetylcholine receptor
    subunits in channel gating, _Nature_ 318:538 (1985).
16. J. A. Umbach and C. B. Gundersen, Expression of an
    w-conotoxin-sensitive calcium channel in _Xenopus_ oocytes injected
    with mRNA from _Torpedo_ electric lobe, _Proc. Natl. Acad. Sci. USA_ (in
    press) (1987).
17. C. B. Gundersen, R. Miledi and I. Parker, Slowly inactivating
    potassium channels induced in _Xenopus_ oocytes by messenger
    ribonucleic acid from _Torpedo_ brain, _J. Physiol. (Lond.)_ 353:231
    (1984).
18. C. B. Gundersen, D. J. Jenden and R. Miledi, Choline
    acetyltransferase and acetylcholine in _Xenopus_ oocytes injected with
    mRNA from the electric lobe of _Torpedo_, _Proc. Natl. Acad. Sci. USA_
    82:608 (1985).
19. N. Dascal, T. P. Snutch, H. Lubbert, N. Davidson and H. A. Lester,
    Expression and modulation of voltage gated-calcium channels after RNA
    injection in _Xenopus_ oocytes, _Science_ 231:1147 (1986).

20. L. M. Kerr and D. Yoshikami, A venom peptide with a novel presynaptic blocking action, _Nature_ 308:282 (1984).

21. R. E. Yeager, O. Yoshikami, J. Rivier, L. J. Cruz and G. P. Miljanich, Transmitter Release from presynaptic terminals of electric organ: inhibition by the calcium channel antagonist, omega _Conus_ toxin, _J. Neurosci_. (in press) (1987).

22. E. W. McCleskey, A. P. Fox, D. Feldman and R. W. Tsien, Different types of calcium channels, _J. Exp. Biol_. 124:177 (1986).

23. J. A. Umbach, C. B. Gundersen and P. F. Baker, Giant synaptosomes, _Nature_ 311:474 (1984).

24. S. A. deReimer, R. Martin, R. Rahamimoff, B. Sakmann and H. Stadler, Use of fused synaptosomes or synaptic vesicles to study ion channels involved in neurotransmission, _in_:this volume.

25. W. Hanke, C. Methfessel, U. Wilmsen and G. C. Boheim, Ion channel reconstitution into lipid bilayer membranes on glass patch pipettes, _Bioelectrochem. Bioenerget_. 12:329 (1984).

26. J. P. Leonard, J. Nargeot, T. P. Snutch, N. Davidson and H. Lester, Ca channels induced in _Xenopus_ oocytes by rat brain mRNA, _J. Neurosci_. 7:875 (1987).

SECTION 2

INTRACELLULAR CALCIUM AND CELL FUNCTION: SENSORY

TRANSDUCTION, MODULATION OF EXCITABILITY, AND NEUROSECRETION

CONTROL OF A LIGHT EMITTING PHOTOPROTEIN BY CALCIUM CHANNELS IN A HYDROZOAN COELENTERATE

Kathleen Dunlap and Paul Brehm

Department of Physiology, Tufts Medical School, Boston, MA., 02111

Bioluminescence has been studied in representative species of all the major invertebrate phyla, from single cell protozoans to echinoderms (1). Comparison of the various mechanisms underlying light emission in these different animals has revealed that it is often triggered by activation of membrane channels. Consequently, many investigators have sought a better understanding of excitation-effector coupling through electrophysiological studies on the control of cellular luminescence. Of the luminescent systems examined, none has yielded more information about membrane control of light emission than the single celled dinoflagellate *Noctiluca*. In experiments performed more than 20 years ago by Roger Eckert, it was shown that a propagating action potential in the vacuolar membrane triggered the activation of a large number of intracellular light emitting organelles (2-4). Because the action potential propagated via an intracellular membrane and not the plasma membrane, it was not possible to ascertain the ionic requirements for luminescence or the action potential. This technical limitation of the system prevented a determination of the nature of coupling between ion channel activation and light emission.

Relating activation of membrane channels to luminescence requires a thorough understanding of the biochemistry of the light emitting molecules. In many luminescent systems, particularly those involving luciferin-luciferase interactions, the link between membrane depolarization and light emission may be complicated. However, in hydrozoan coelenterates, a simple coupling between channel activation and luminescence may exist. In the luminescent cells of these animals there is a well characterized class of light emitting molecules, termed the calcium-activated photoproteins (5). The best characterized of these photoproteins, aequorin, has been shown to require only calcium for its activation, making it an excellent indicator for intracellular calcium (6). In fact, aequorin has traditionally been used as a calcium indicator following its injection into muscle, nerve, and egg cells (7). In spite of the appealing characteristics of this photoprotein, there have been no studies on membrane excitation-luminescence coupling in the cells of coelenterates.

A few years ago we became interested in the triggering mechanism of the native calcium-activated photoprotein. It was clear from early studies on hydrozoans that luminescence was obligatorily coupled to electrical activity. Work on *Obelia geniculata*, in particular, detailed the relationship between the extracellularly recorded action potential that propagates throughout the animal's epithelium and the activation of the specialized light-emitting cells (8,9). *Obelia* offered the further advantage that the calcium-activated photoprotein, obelin (10), was intimately associated with an endogenous fluorescent molecule (11). This association is important to studies of luminesence because it provides a specific fluorescent marker for the light emitting cells (ref. 12 and see below). The additional fact that the photoprotein complex was cytoplasmic in *Obelia* suggested that it would be an ideal preparation in which to investigate the electrical control of luminescence. In this chapter we review our findings to date. Some of these results have been recently published (13).

CELLULAR SOURCE OF LUMINESCENCE IN OBELIA

Obelia geniculata is a species of hydrozoan which spends much of its life cycle in the polypoid stage. A single animal is composed of many feeding polyps which communicate through a common tissue composed of only two cell layers. It is within the most central (endodermal) layer of this tissue that the light emitting cells, termed photocytes (12), are found (Fig. 1). It has not been possible to distinguish the photocytes from the surrounding non-luminescent cell types on the basis of morphology. Because the photocytes comprise only a fraction of the total cell number, we have relied on the endogenous fluorescence of the photocytes for their identification. This fluorescence is the result of an intracellular green fluorescent protein (GFP) which is found only in the photocytes (12).

The GFP serves as the light emitter both for fluorescence and for the native luminescence of the intact animal (11). However, it is the calcium-activated photoprotein (obelin) and not the GFP which is activated by a rise in intracellular free calcium (5,6). In the intact photocyte the luminescence is thought to occur in the following manner: obelin binds two calcium ions causing a self-triggered conformational change. Then, an efficient transfer of energy from obelin to the GFP occurs, followed by GFP emission of green light at 508 nm (11). In contrast to the natural activation of the GFP by obelin, GFP can be artificially activated by direct illumination with 460 nm light (blue), bypassing the activation of obelin (12). This probably results from the similarity between the activation energy provided by the blue light and that of activated obelin. Experimentally, we have used this native fluorescence of the GFP in order to identify the photocytes both in the intact animal and following dissociation of the tissue. Due to the enormous concentration of GFP/obelin complex present in photocytes, illumination by even weak light sources was adequate to identify the luminescent cells. It is important to keep in mind that using fluorescence to identify photocytes does not expend any of the available obelin.

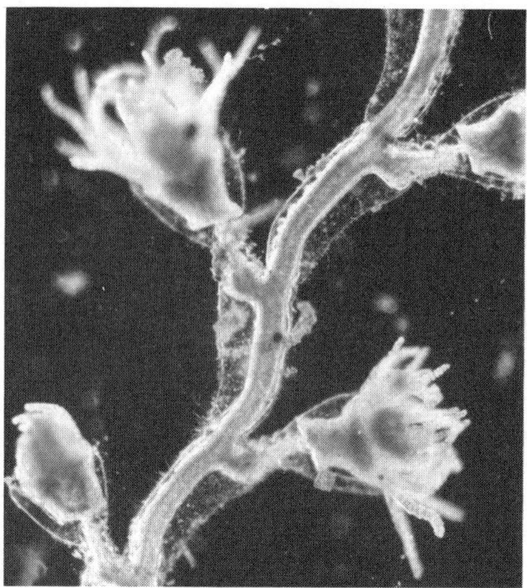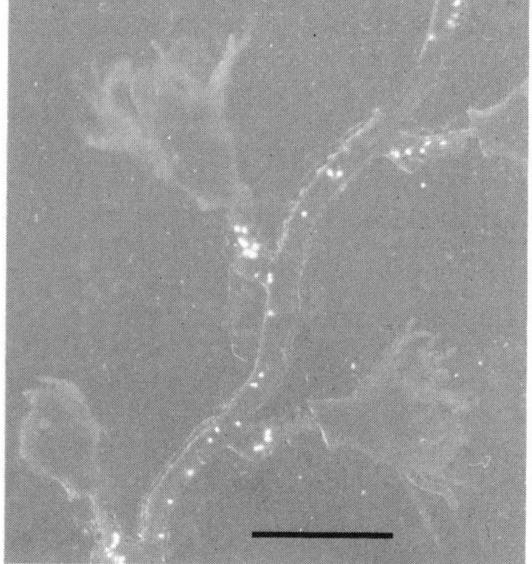

Figure 1. Brightfield (left) and fluorescence micrographs of <u>Obelia</u> <u>geniculata.</u> For the fluorescence picture, animals were exposed to 460 nm excitation light from a mercury lamp filtered through FITC. Slight backlighting with tungsten light was used to allow comparison of the fluorescence image with the brightfield one. The fluorescent cells (photocytes) contain the green fluorescent protein and the calcium-activated photoprotein obelin. Calibration bar = 0.5 mm.

In the intact animal the photocytes are scattered throughout the endoderm, with the highest density in the growing tips of the colony (figure 1). Each photocyte is distinct, with short processes extending between adjacent non-luminescent cells. Visualization of the photocytes in the animal is limited by the presence of a perisarc which covers the soft tissue between feeding polyps. This rigid exoskeleton prevents access to the photocytes by recording electrodes. Our studies of the bioelectric control of light emission, therefore, required dissociation of *Obelia* into a mixture of cell clusters and single cells prior to patch clamp recording. Some of the photocytes in large cell clusters retained the characteristic stellate appearance, but those in either small clusters or in isolation tended to round up. We restricted our electrophysiological recordings to those photocytes which were isolated or in contact with fewer than 4 or 5 non-luminescent neighbor cells.

IONIC BASIS OF THE ACTION POTENTIAL IN *OBELIA*

Bioluminescence in *Obelia* can be elicited by chemical, mechanical, or electrical stimulation (9). Associated with the light emission is an action potential which propagates through the endodermal tissue at a rate of approximately 20 cm/sec (8). A single stimulus can set off a volley of action potentials, each one causing a rapid flash of light as it passes through the region of tissue containing the photocyte (figure 2). In this way the sequential flashing of photocytes along the length of the colony reflects the actual conduction velocity of the propagating action potential. It appears that photocytes only flash once in response to a single action potential and that an action potential is necessary for light emission to occur.

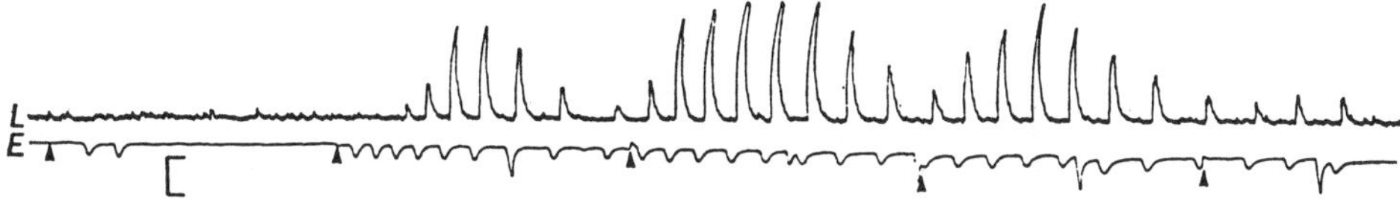

Figure 2. Luminescent flashes (L) and extracellularly-measured luminescent potentials (E) recorded from Obelia in response to a train of electrical stimuli applied at 2 Hz (arrowheads). The light responses are triggered by the luminescent potentials. Calibration bars: 1mV, 1 second. (Taken from Morin, J.G. and Cooke, I.M., 1971, J. exp. Biol., 54:707-721.)

The action potential which triggers light emission in *Obelia* was termed the "luminescent potential" by Morin (11). This potential could be easily recorded by extracellular electrodes, suggesting that a large current source was responsible for its generation. On the basis of the current amplitude, it was proposed that propagation of the luminescent potential occurred by means of electrically coupled epithelia rather than by specialized neurons. This idea is supported by the following: first, electron microscopic analysis of several species of the campanularid hydrozoans, including *Obelia* (personal observations), have failed to detect neurites in the tissue connecting the feeding polyps (14); second, in related hydrozoans, it has been directly shown that the generalized epithelial cells of the body wall can generate an action potential (15-18); finally, electron microscopic studies have shown that the cells composing the body wall of many hydrozoans are interconnected by gap junctions, which establishes electrical coupling between cells (19).

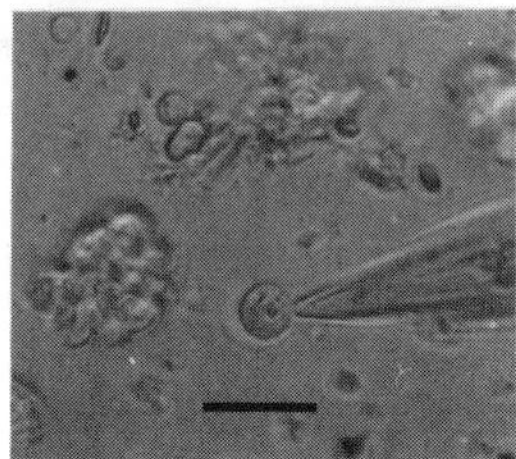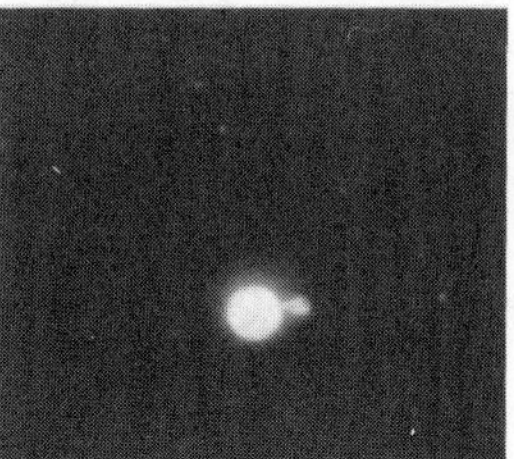

Figure 3. Left, brightfield micrograph of dissociated cells from Obelia with the patch pipette touching an isolated photocyte, as indicated by its fluorescence (center). The fluorescence picture also depicts the effect of applying negative pressure to the cell membrane during seal formation with the pipette. Right, rupturing of the plasma membrane in the patch (following pipette seal formation) to achieve the whole cell recording configuration results in an exchange of photoprotein with the pipette solution, indicating that obelin is not bound to cytoplasmic organelles. Calibration bar: 20 μm.

    We performed whole cell recordings (Fig. 3) from dissociated cells of *Obelia* (either isolated or in small cell clusters) in order to further ascertain which cell types were able to generate action potentials. Recordings of membrane potential from cells in artificial sea water indicated resting potentials on the order of -50 mV. Depolarization of non-luminescent cells indicated, that many of them are capable of generating voltage-dependent inward current and overshooting action potentials (Fig 4A). The action potential was the result of calcium channel activation and either calcium, barium, or strontium was able to carry the inward current. We found no evidence for a sodium dependent action potential in any of the cells examined. These data indicate that the luminescent potential, which was originally described by extracellular recording (8), is the result of activation of voltage-dependent calcium channels. Consistent with this conclusion is the slow conduction velocity of the luminescent potential, a characteristic commonly associated with propagation of calcium action potentials. The properties of the calcium channels in non-luminescent cells will be further detailed in the next section of this paper.

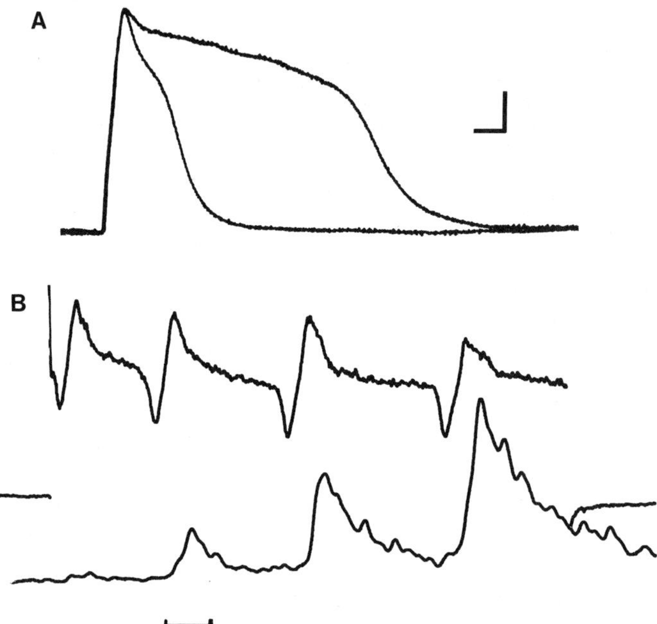

Figure 4. Calcium channel activation underlies the action potential and triggers luminescence. A. Two superimposed action potentials recorded under the whole cell configuration from a small cluster of non-photocytes in calcium and barium-containing (longer action potential) artificial sea water. The resting potential of the cell was -48 mV. The sea water contained (in mM) 40 CaCl2 (or BaCl2), 377 NaCl, 9 KCl, 40 MgCl2 and 2 NaHCO3, pH 7.7. The pipette solution contained (in mM) 140 KCl, 11 EGTA, 10 HEPES, 1 CaCl2, and 696 dextrose, pH 7.0. B. Current recording (upper trace) from a cluster of non-photocytes following a step depolarization to 40 mV from a holding potential of -60 mV in 40 mM calcium artificial sea water (as in A). Because the cell cluster was only poorly clamped, several calcium inward current spikes resulted from the depolarization. Each spike of current evoked a light response (lower trace) from a photocyte embedded in the cluster of non-luminescent cells. Light was measured with an EMI side-window photomultiplier tube attached to the camera port of the microscope. The light response shows facilitation with successive stimuli, a hallmark of calcium-activated luminescence. Calibrations: A, 10 mV, 50 msec; B, 20 msec. Light was not calibrated.

    The calcium action potential is responsible for the sequential activation of the photocytes which are embedded within the excitable non-luminescent cells. Close coupling between action currents in non-luminescent cells and luminescence from neighboring photocytes can be observed by simultaneously recording light production and calcium current activation in multicellular clusters containing a single photocyte and a few non-luminescent neighbor cells. An example shown in Figure 4B illustrates a poorly clamped current as indicated by the presence of four inward current transients following depolarization of the cell cluster. Associated with each of these sequential inward currents was a rapid flash of light from the photocyte in the cluster, with a time course similar to that observed *in vivo* (9). The luminescence was activated within 20 msec from the onset of inward current, decayed within an additional 20-100 msec (depending upon its peak amplitude), and facilitated with repetitive stimuli, properties also comparable to those in the intact animal (Fig. 2). Both *in vivo* and in dissociated cell clusters, luminescence was not observed when the extracellular calcium was removed, further supporting a dependence of luminescence on calcium which flows in during an action potential.

On the basis of these observations we conclude that the luminescent potential in *Obelia* is propagated by means of a calcium action potential. Furthermore, we support the assertion made by earlier investigators that the ability to generate an action potential is common to many of the cells in the tissue (15-18). Therefore, one important role played by the gap junctions is to mediate the propagation of the action potential between these small diameter epithelial cells. However, not all of the cells are capable of generating calcium-dependent action potentials. Among these are the photocytes themselves (see below). The absence of voltage-dependent calcium channels in the photocyte membrane leads to the question of how the requisite rise in intracellular calcium, responsible for triggering the luminescence, is achieved in the photocyte. As will be shown in a later section of this paper, the gap junctions appear to allow the diffusion of a chemical signal from neighbor cells to photocytes which is critical for the activation of luminescence.

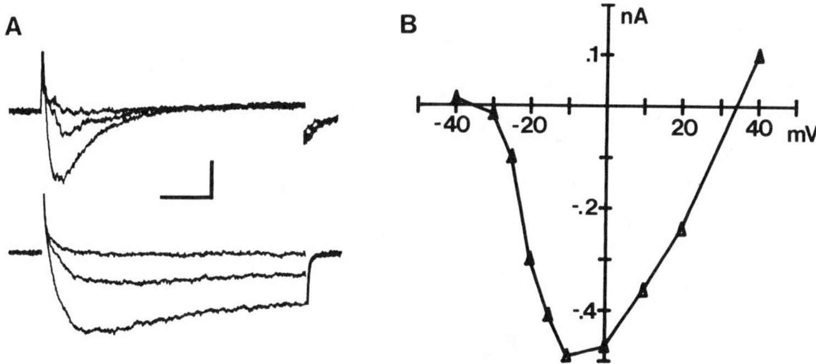

Figure 5. Calcium channel activation. A. Superimposed current traces in a non-luminescent cell following step depolarizations of 40, 45, and 50 mV from a holding potential of -60 mV in artificial sea water containing either 40 mM calcium (upper traces) or 40 mM barium. Pipette solution was the same as in Figure 4 legend. Calibration: 200 pA, 40 msec. B. Current-voltage relationship for barium current recorded from a small cluster of non-luminescent cells. Leakage subtraction not performed.

PROPERTIES OF THE VOLTAGE-DEPENDENT CALCIUM CHANNEL

We have never observed a voltage dependent calcium or barium current in recordings from photocytes. This is true for photocytes in isolation as well as for photocytes in multicellular clusters containing non-luminescent cells. By contrast, we often observed voltage-activated calcium current or barium current in recordings from non-luminescent cells (Fig. 5A). The threshold for activation of calcium current in *Obelia* was approximately -30 mV; the current-voltage relations (Fig. 5B) for the channel are similar to those described for high threshold calcium channels in other invertebrate and vertebrate preparations (20). Generally, barium was used as the permeant ion for our studies, because the inward barium current tends to be larger and shows slower inactivation than calcium current (Fig. 5A). Strontium is also capable of carrying inward current through the calcium channel, an observation which is consistent with the selectivity of calcium channels in more highly evolved animals. The calcium and barium currents are both effectively blocked by 1 mM cadmium but are not appreciably affected by the organic calcium channel blocker, nifedipine at 1μM.

The increased rate of inactivation of calcium current compared to barium current suggested that the inactivation of the calcium channel was mediated by intracellular calcium accumulation rather than by voltage (Fig.6). In order to determine whether inactivation was voltage- or ion-dependent, the inactivation of calcium and barium currents were studied by a two pulse protocol. For this purpose, a variable amplitude conditioning depolarization was elicited 20 msec before a fixed test pulse to 0 mV. The amplitude of the test pulse was chosen because it was sufficient to maximally activate inward current. Inactivation of calcium current was quantitated as the ratio of peak inward test pulse current measured with and without a prior conditioning depolarization. Inactivation of the calcium channel was observed at all potentials sufficient to activate inward current. In fact, complete inactivation of inward calcium current was observed with conditioning depolarizations to -10mV (Fig. 6), which was also the potential that maximally activated the inward current (Fig. 5). Further increases in conditioning depolarizations beyond 0 mV became increasingly less effective in causing inactivation. A reduction in inactivation with strong depolarizations was observed for both barium and calcium current although, at all potentials, the inactivation of the calcium current was greater than that for barium. This reduction in inactivation with strong conditioning depolarizations can be attributed to a decrease in divalent entry during the conditioning pulse, supporting a role for divalent ions in channel inactivation. Because the inactivation of calcium current was not completely eliminated with depolarizations to the suppression potential for calcium ions (not shown), we can not exclude the possibility that inactivation is also voltage dependent. However, the difference in inactivation measured with barium and calcium, the U-shaped voltage dependence of the inactivation measured by the two pulse protocol, the correlation between the voltage-dependence of inward current activation and inactivation, and the alterations of inward current decay in the presence of barium all argue that inactivation is dependent on the permeant ion.

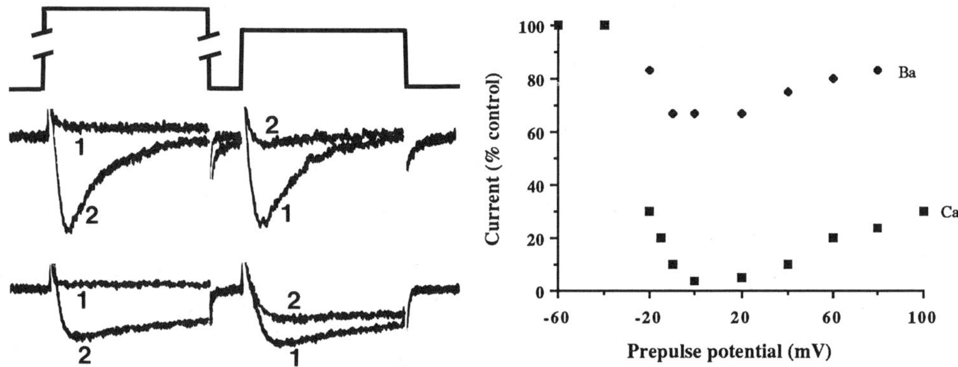

Figure 6. Calcium channel inactivation. Left, in the lower two traces are shown calcium (upper) and barium currents recorded in a non-luminescent cell during a 100 msec test depolarization to 0 mV from a holding potential of -60 mV. The voltage protocol is illustrated schematically in the top trace. The test pulse was preceded (20 msec interval) by a 100 msec variable amplitude prepulse. Two representative current traces are shown for prepulses of 20 mV ("1") or 60 mV ("2"). The corresponding test pulse currents are so labelled. The internal solution contained 140 NaCl in place of KCl to minimize outward current contamination; all other constituents were the same as in figure 4 legend. Right, inactivation curve for calcium channel current. The ratio of the current measured during the test pulse to the maximal current (measured without a prepulse) is plotted as a function of prepulse potential.

Like many other preparations the calcium current disappears or 'washes out' during whole cell recording (see chapters by Byerly & Hagiwara and Kalman & Armstrong). This washout of inward current, like inactivation, is slowed if barium is substituted for calcium as the permeant ion. In fact, relatively stable recordings of barium current can be rapidly destabilized if the cell is transiently exposed to extracellular calcium and subsequently depolarized. Removal of the calcium and return to barium solution will once again lead to a barium current, but the peak amplitude of the barium current will be permanently reduced by the prior influx of calcium. Every repeated exposure of the cell to calcium will further reduce the pure barium current resulting finally in a total, irreversible loss of barium current. These findings suggest that washout, like inactivation, is current dependent and that the influx of calcium, is more effective than barium. They further suggest the possibility that calcium channel inactivation and washout may share a common mechanism.

VOLTAGE-DEPENDENT CHANNELS OF THE PHOTOCYTE

As mentioned above, no voltage-activated inward current was observed in whole cell recordings from photocytes, whether in complete isolation or in cell clusters. The only prominent current to be recorded in these cells was a transient outward current which was activated at potentials more positive than 0 mV (Fig. 7). Measurement of tail current reversal potential indicated that it is carried by potassium, similar to the 'A-type current' described for a variety of preparations such as coelenterate egg, molluscan neuron, drosophila muscle, and mammalian cardiac muscle (22). As with these other preparations, the transient outward current in *Obelia* does not depend on calcium for its activation, and it is completely and reversibly blocked by 1mM 4-aminopyridine.

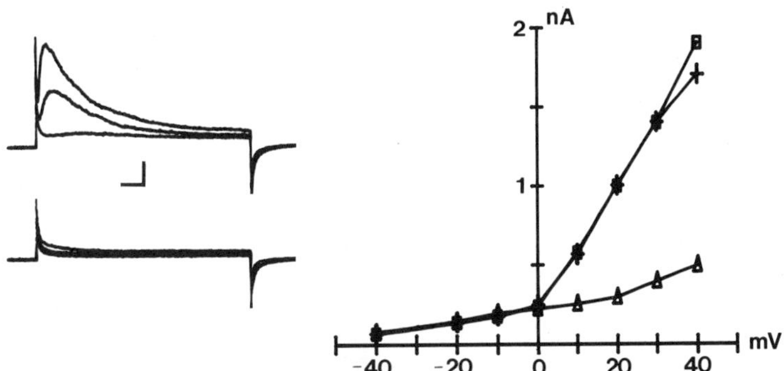

Figure 7. Photocyte current. Left, superimposed outward currents recorded from an isolated, voltage clamped photocyte following depolarizing voltage steps to 0, 20 and 40 mV from a holding potential of -60 mV before (top traces) and during application of 1 mM 4-aminopyridine (4-AP). Cal: 500 pA, 15 msec. Right, current-voltage relationship for peak outward current measured in control cells (□), in the presence of 4AP (Δ), and following recovery from 4-AP (+). The external solution was artificial sea water with 40 mM calcium and the internal solution was the same as in figure 4 legend.

The most striking feature of this outward current is the voltage-dependence of its inactivation. Steady-state inactivation curves (Fig. 8A) indicate a steep dependence of inactivation on membrane potential. No inactivation was observed at holding potentials more negative than -60mV whereas total inactivation occurred at -40mV. Half of the channels were inactivated at a mean membrane potential of -53mV. Using a two pulse protocol (with variable duration prepulses) it was found that the rate at which inactivation developed was also voltage-dependent (Fig. 8B). The half-time for development of inactivation measured 30 msec at -40mV. When less positive prepulses were applied, the inactivation developed more slowly. Recovery from inactivation exhibited a dependence on membrane potential which was the inverse to that observed for development of inactivation (Fig. 8C). For this experiment, cells were held at -20 mV to produce complete inactivation of the outward current, and hyperpolarizing pulses of variable duration preceded the test pulse. As the prepulse potential was made more negative, the current during the test pulse exhibited a faster recovery from inactivation. Therefore, as reported for more highly evolved organisms, the inactivation and recovery from inactivation of this channel in *Obelia* exhibits a strong voltage-dependence (Fig 8D).

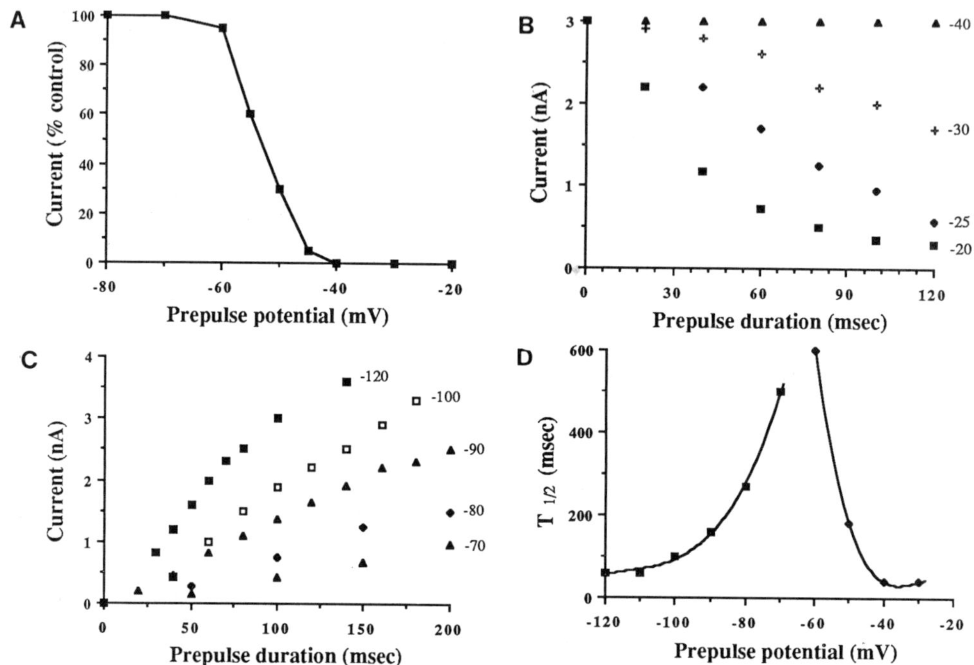

Figure 8. Inactivation properties of the A current. A. Steady-state inactivation. In a cell held at -80 mV, one second, variable amplitude step depolarizations preceded a test pulse to 40 mV. The plot shows peak current (as percentage of the maximal current measured without a prepulse) as a function of prepulse potential. Solutions were the same as in figure 7 legend. B. Development of inactivation. In a cell held at -80 mV, variable duration prepulses were followed by test depolarizations to 20 mV. The magnitude of peak current during the test pulse is plotted as a function of prepulse duration for 4 different prepulse potentials (shown next to data points). C. Recovery from inactivation. The same cell was the held at -20 mV to produce complete inactivation of the A current. Hyperpolarizing prepulses of variable duration were then applied prior to the test pulse to 20 mV to study the rate of removal of inactivation. Peak current during the test pulse is plotted as a function of prepulse duration for 5 different prepulse amplitudes. D. Half time for development (■) and removal (◆) of inactivation is plotted as a function of prepulse potential.

It is curious that the transient outward current is the only voltage-dependent current to be recorded in the photocyte. Because of the lack of inward current, the photocytes can not generate an action potential. This being the case, it is logical to question the physiological significance of the voltage-activated outward current. One possible role for the transient potassium current in the photocytes might be to counteract the depolarization which occurs in response to the calcium influx during the action potential in neighboring non-luminescent cells. As shown in the next section, the photocytes are electrically-coupled to cells which generate the action potentials and would, therefore, be expected to depolarize when an action potential invades the tissue. However, activation of potassium channels in the photocyte may effectively clamp the photocyte at a potential more negative than its neighbor cells. It is interesting in this regard that the kinetics of the transient outward current and the calcium current are similar (compare figs. 5 and 7). Thus, the transient outward current may play an important role in controlling luminescence: large depolarizations of the cells might lead either to closure of the gap junctions (23) or to a decrease in the electrical driving force between the cells, thereby interfering with the chemical triggering of luminescence between neighboring cells and photocytes. There is currently no information which would either support or refute either of these proposed roles for the transient potassium channel.

ROLE OF ION CHANNEL ACTIVATION IN TRIGGERING BIOLUMINESCENCE

The calcium action potential is intimately coupled to light emission from the photocyte. Since activation of the photoprotein requires an elevation of cytosolic calcium, it seemed likely that the calcium entering during the action potential activates the obelin. Consistent with this hypothesis is the observation that luminescence is abolished either by removal of extracellular calcium or by addition of 1 mM cadmium to the sea water (fig. 11). However, only non-luminescent cells can generate calcium entry; the hypothesis would therefore predict that photocytes would require the presence of neighboring non-photocytes to support the luminescent response. This is, in fact, the case.

The dependence of luminescence on contacts between photocytes and non-luminescent cells was shown by measuring depolarization-induced light emission. Isolated photocytes are not capable of luminescence when depolarized under voltage-clamp or by application of high external potassium. Yet, application of calcium ionophore (A23187) to the isolated cells causes prolonged luminescence, indicating that the photoprotein is functionally intact. However, if as few as one non-luminescent cell is in contact with the photocyte, light emission can be evoked in response to depolarization of the non-luminescent cell (Fig. 9).

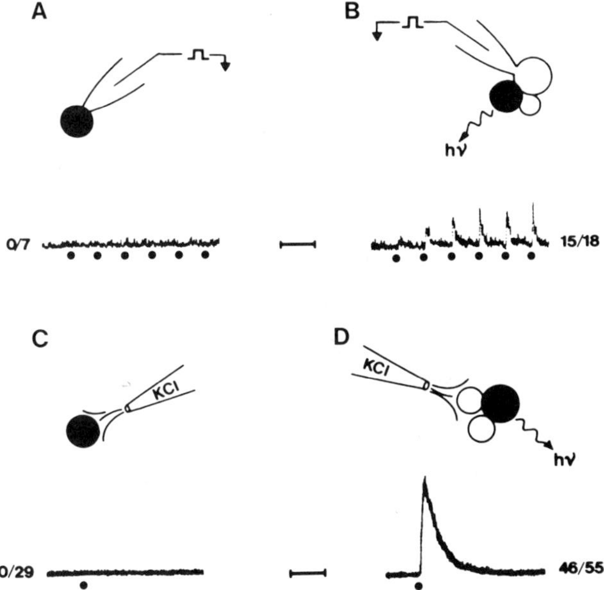

Figure 9. Luminescence response of photocytes, either isolated (A and C) or in small cluster of cells (B and D). The method of stimulation is represented schematically at the top of each panel, with the photocyte shaded; the lower traces represent the light emission from photocytes following electrical (A and B) or high KCl (C and D) depolarization. Dots, times of stimulation; numbers to the side of each trace, proportion of cells that produced light. Calibration bars: 400 msec (A and B), 2 sec (C and D); emitted light uncalibrated (same gain A-D). (Taken from Dunlap, K., Takeda, K., and Brehm, P., 1987, Nature 325:60-62.)

Is calcium influx into the neighbor cell the critical factor supporting luminescence from the photocytes in small cell clusters? A direct dependence of the luminescence on the amount of calcium entering the non-luminescent cells was shown by whole cell voltage clamp of cell clusters containing a single photocyte. Depolarization of the non-luminescent cell triggered prolonged luminescence from an adjacent photocyte. Importantly, when the level of depolarization approached the suppression potential for calcium influx into the neighbor cell, the resultant luminescence from the photocyte was virtually eliminated (Fig. 10). These data are consistent with a direct dependence of luminescence on calcium entering non-luminescent cells. We cannot, however, exclude the alternative possibility that the depolarization was having additional effects such as closure of gap junctions between photocytes and non-luminescent cells.

How does a rise in non-photocyte calcium concentration lead to luminescence from a neighboring photocyte? Lucifer yellow injection of non-luminescent cells and electrophysiological measurements indicated that cells in *Obelia* are both dye- and electrically-coupled. It seemed plausible, therefore, that calcium, or a calcium-dependent messenger might diffuse into the photocyte from its neighbor cells through the intercellular junction, thereby triggering light emission. We tested for a role of intercellular channels in the control of luminescence by blocking gap junctions either with intermediate length alcohols or by decreasing internal pH (23). As shown in figure 11, both techniques rapidly and reversibly blocked the depolarization-induced luminescence of cell clusters (without blocking calcium current). These experiments suggest an involvement of gap junctions in mediating luminescence.

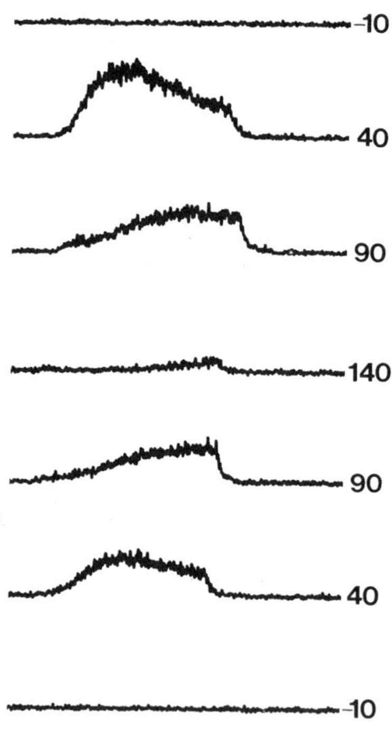

Figure 10. Suppression potential for the light response. Light emission from a photocyte was measured (as described in figure 4 legend) during a one second depolarizing voltage step delivered to one non-luminescent support cell in a small cluster surrounding the photocyte. The potentials applied are noted at the right of each light recording. Following depolarization to potentials at which calcium current was minimized (near ECa), the light response was suppressed.

What sort of chemical signal might be passing through the junctions, and how would it be related to calcium influx in the non-luminescent cell? The simplest explanation is that the chemical is calcium itself and that it enters through voltage-dependent calcium channels in the non-luminescent cell and then diffuses into the photocyte via gap junctions. This scheme is consistent with the observed dependence of luminescence on extracellular calcium, the suppression of luminescence with strong depolarization, and the facilitation of light with repetitive depolarizations. We can not exclude the possibility that the calcium which enters non-luminescent cells promotes the generation and diffusion of a chemical signal other than calcium through the gap junction. Products of phosphatidylinositol metabolism, for instance, could potentially act as chemical mediators for elevating internal calcium in the photocyte. However, luminescence cannot be evoked by the calcium ionophore A23187 in calcium-free sea water, suggesting that photocytes do not contain a sufficiently large releasable pool of internal calcium. Identifying the chemical signal that diffuses between cells will be an important goal for future research on *Obelia*.

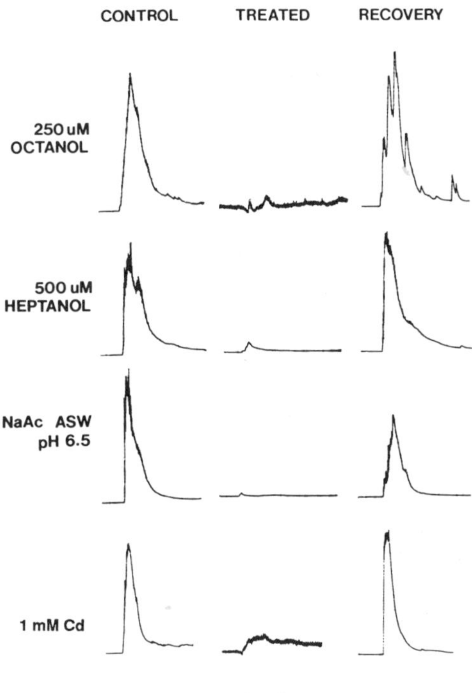

Figure 11. Block of luminescence. The effect of various agents on light responses in Obelia *evoked by high-potassium-induced depolarization of partially dissociated animals. Light responses on the left are controls. Brief treatment (1-3 minutes) with the agents indicated at the left blocked the potassium-induced luminescence as shown by the middle traces. All effects were reversible (right traces). Calibration bar = 2 sec. Light is uncalibrated but all traces are at the same gain (except center traces in cadmium and octanol are 10 times the others).*

ADAPTIVE SIGNIFICANCE FOR THE CONTROL OF BIOLUMINESCENCE BY INTERCELLULAR SIGNALLING

Although the biological significance of luminescence in *Obelia* is not understood, light emission in other organisms is known to subserve a variety of functions, including intraspecies (mating) and interspecies (predator-prey) communication (24). It seems likely that luminescence in *Obelia* too is biologically important, and, for this reason, it has been entrained to the action potential evoked by external stimuli. The absence of calcium channels

99

in the photocytes themselves may be a mechanism to preclude random luminescence that would occur with spontaneous calcium channel openings, thereby ensuring that light emission is always coupled to an appropriate external stimulus. By confining calcium entry to the neighboring cells, the small amount of calcium influx resulting from random opening of the channels might be effectively buffered and, thereby, prevented from entering the photocyte altogether. Alternatively, since proper functioning of a calcium channel depends upon its regulation by cytoplasmic calcium, the abnormally high calcium buffering in the photocyte (due to the presence of obelin) might interfere with necessary control of channel gating.

Our present hypothesis for the control of bioluminescence in *Obelia* involves the diffusion of calcium through the cytoplasm of non-luminescent cells surrounding photocytes. Previous studies have suggested that, because of multiple sites for calcium buffering in cell cytoplasm, the diffusion rate for calcium is extremely slow. In squid axon, for instance, calcium ions have been suggested to move at rates of 0.1 μm/sec (25). The rapid stimulus/effector coupling for luminescence in *Obelia* (delay of 10-20 msec) might suggest that calcium is not the chemical messenger in this system. However, several considerations must be kept in mind. 1) The site of calcium entry in the neighbor cells, and thus the path length for the calcium, is not known. Diffusion distances may be quite short. 2) The cytoplasmic buffering of intracellular calcium in the non-photocytes is also unknown. It is possible that these cells are designed for the purpose of directing calcium to neighboring photocytes. It appears clear that estimates of for calcium mobility vary from preparation to preparation. Work described in the present volume, for example, suggests that in squid presynaptic terminal and in molluscan neurons, calcium diffusion occurs more rapidly than previously supposed (see chapters by Smith, Osses, and Augustine and Tillotson and Nasi). However, further experiments are required before this assertion can be made for cells of *Obelia*.

In conclusion, our studies on *Obelia* have yielded a great deal of information about ion channel function. *Obelia* is the most primitive multicellular animal in which ion channel function has been examined. Remarkably, the transient outward current and calcium current are virtually identical to those described for many invertebrate and vertebrate preparations. Therefore, these two channel types must have been among the first to evolve. Our studies have also disclosed an unexpected role of intercellular signalling in the control of luminescence. The chemical messenger which passes through the gap junction has not been unequivocally identified, but preliminary findings point to the possibility that calcium itself is the messenger. Thus, our studies of the endogenous calcium-activated photoprotein in this coelenterate have led not only to a confirmation of the role played by gap junctions in action potential propagation but also to the conclusion that they provide biologically important pathways for the movement of chemical messengers between cells.

ACKNOWLEDGEMENT

This work was supported by a grant from the NSF.

REFERENCES

1. Herring, P.J., "Bioluminescence in Action," Academic Press, London (1978).
2. Eckert, R., Bioelectric control of bioluminescence in the dinoflagellate *Noctiluca*. I. Specific nature of triggering events, Science. 147:1140 (1965).
3. Eckert, R., Bioelectric control of bioluminescence in the dinoflagellate *Noctiluca*. II. Asynchronous flash initiation by a propagated triggering potential, Science. 147:1142 (1965).
4. Eckert, R., The flash-triggering action potential of the luminescent dinoflagellate *Noctiluca*, J. Gen. Physiol. 52:258 (1968).
5. Shimomura, O., Johnson, F.H., and Saiga, Y., Extraction, purification, and properties of aequorin, a bioluminescent protein from the luminous hydromedusan, *Aequorea*, J. Cell. Comp. Physiol. 59:223 (1962).
6. Hastings, J.W. and Morin, J.G., Kinetics of calcium-triggered luminescence of aequorin and other photoproteins, in: "Contractility of Muscle Cells and Related Processes," R. Podolsky, ed., Prentice-Hall, N.Y. (1971).

7. Campbell, A.K., Living Light: chemiluminescence in the research and clinical laboratory, Trends in Biochem. Sci. 11:104 (1986).

8. Morin, J.G. and Cooke, I., Behavioral physiology of the colonial hydroid *Obelia*. II. Stimulus-initiated electrical activity and bioluminescence, J. exp. Biol. 54:707 (1971).

9. Morin, J.G. and Cooke, I., Behavioral physiology of the colonial hydroid Obelia. III. Characteristics of the bioluminescent system, J. exp. Biol. 54:723 (1971).

10. Campbell, A.K., Extraction, partial purification and properties of obelin, the calcium-activated photoprotein from the hydroid *Obelia geniculata*, Biochem. J. 143:411 (1974).

11. Morin, J.G. and Hastings, J.W., Energy transfer in a bioluminescent system, J. Cell. Physiol. 77:313 (1971).

12. Morin, J.G. and Reynolds. G., The cellular origin of bioluminescence in the colonial hydroid, *Obelia*, Biol. Bull. 147:397 (1974).

13. Dunlap, K., Takeda, K. and Brehm, P. Activation of a calcium-dependent photoprotein by chemical signalling through gap junctions, Nature 325:60 (1987).

14. Brock, M., Strehler, B.L., and Brandes, D. Ultrastructural studies on the life cycle of a short-lived metazoan, *Campanularia flexuosa*, J. Ultrastructure Res. 21:281 (1968).

15. Josephson, R. and Macklin, M., Transepithelial potentials in *Hydra*, Science, 156:1629 (1967).

16. Mackie. G.O., Propagated spikes and secretion in a coelenterate glandular epithelium, 68:313 (1976).

17. Josephson, R.K. and Schwab, W.E., Electrical properties of an excitable epithelium, J. Gen. Physiol., 74:213 (1979).

18. Anderson, P.A.V., Epithelial conduction: its properties and functions, Prog. in Neurobiology, 15:161 (1980).

19. Mackie,G.O., Anderson, P.A.V., and Singla, C.L., Apparent absence of gap junctions in two classes of *Cnidaria*, Biol. Bull., 167:120 (1984).

20. Hagiwara, S. and Byerly, L., Calcium channel, Ann. Rev. Neurosci., 4:69 (1981).

21. Tsien, R.W., Calcium currents in heart cells and neurons, in: "Neuromodulation," L. Kaczmarek and I. Levitan, eds., Oxford Univ. Press, London (1986).

22. Rogawski, M.A., The A-current: how ubiquitous a feature of excitable cells is it?, Trends in Neurosci., 8:214 (1985).

23. Spray, D.C., White, R.L., Campos de Carvalho, A., Harris, A.L., and Bennett, M.V.L., Gating of gap junction channels, Biophys. J. 45:219 (1984).

24. Harvey, E.N., "Bioluminescence", Academic Press, N.Y. (1952).

25. Hodgkin, A.L. and Keynes, R.D., Movements of labelled calcium in squid giant axons, J. Physiol. 138:253 (1957).

CALCIUM IN PHOTORECEPTORS

Gordon L. Fain and Walter H. Schröder[*]

Jules Stein Eye Institute
UCLA School of Medicine
Los Angeles, California 90024 USA

[*]Institut für biologische Informationsverarbeitung
Kernforschungsanlage, Jülich
D-5170 Jülich, Federal Republic of Germany

INTRODUCTION

Yoshikami and Hagins (1971) first proposed that rod outer segments contained large amounts of Ca stored within the discs, and that this Ca was released by light and diffused to the plasma membrane to bind to and block the light-dependent conductance. This proposal, as well as the discovery of light-dependent G protein binding (Kühn, 1980), GTP turnover (Fung & Stryer, 1980), and phosphodiesterase activation (Wheeler & Bitensky, 1977) stimulated much research into the role of Ca and other second messengers in the production of the photoreceptor light response. As a result of this effort, it has now become clear that the central tenet of the "Ca hypothesis" of Yoshikami and Hagins is incorrect. The elegant experiments of Fesenko and his collaborators (Fesenko, Kolesnikov & Lyubarsky, 1985) and of Yau and his colleagues (Yau, Haynes & Nakatani, 1986) have shown beyond any reasonable doubt that the light-dependent conductance is controlled by changes in cyclic GMP within the rod. Although high concentrations of Ca at the cytoplasmic surface of the outer segment membrane do have some direct effects upon the light channels (Yau et al., 1986), these effects are small and possibly not of physiological significance (Matthews, Torre & Lamb, 1985)

Though Ca appears not to be directly involved in regulating the light-dependent conductance, there is considerable evidence that it nevertheless plays some important role in the physiology of photoreceptors. Rods have been shown to contain unusually large amounts of Ca concentrated within their outer segments. As Yoshikami and Hagins first postulated, the Ca is probably sequestered within the discs, is released in light, and is taken back up during dark adaptation. This "Ca cycle" is produced by a specific enzymatic machinery whose components are just now beginning to be identified. Furthermore, there is considerable experimental evidence that extrinsic changes in extracellular and cytosolic Ca concentration produce large effects on the rod dark current and membrane potential (see Fain & Lisman, 1981). Decreases in extracellular Ca also dramatically alter rod sensitivity (Bastian & Fain, 1982), and the introduction of Ca buffers into rods changes both the amplitude and kinetics of the light response

(Torre, Matthews & Lamb, 1986; Miller & Korenbrot, 1987a).

In order to elucidate the role of Ca in rod function, we have been attempting to understand the mechanisms responsible for Ca movements in rods, both in darkness and as the result of illumination (Schröder & Fain, 1984; Fain & Schröder, 1985a; Fain & Schröder, 1987). We have been using a new method of micro analysis called laser micromass analysis, or LAMMA. This method is more sensitive for Ca than the more commonly used energy-dispersive X-ray analysis (EDX) and has the additional advantage that it is possible to measure separately the concentrations of each of the isotopes. Cells normally containing almost exclusively ^{40}Ca (the most abundant naturally occurring isotope) can be incubated in solutions containing the stable isotopes ^{42}Ca or ^{44}Ca. After a given time, the amounts of each isotope can be measured, and in this way movements of Ca into or out of the cells can be determined.

THE CA ECONOMY OF RODS

As the result of experiments from many laboratories, including our own, it is possible to make some definitive statements about the Ca economy of rods.

1. Rods contain large amounts of total Ca mostly within their outer segments, but the free Ca concentration in the cytosol is quite low. Our own estimates of the total Ca content of rods are based on experiments like the one in Fig. 1 (Fain & Schröder, 1985a). In Fig. 1A we show a rod from which several LAMMA measurements were made using a LAMMA 500 (Leybold-Heraeus, Köln, FRG). A portion of an isolated retina from the toad Bufo marinus was fast frozen and the tissue was dehydrated either by freeze drying or by freeze substitution. A thin section from the retina was placed in a vacuum on the stage of a light microscope, and a portion of the outer segment was vaporized with a high-energy Nd:YAG laser. Ions formed during vaporization were accelerated and focused onto the detector of a mass spectrometer.

Sample spectra from such experiments are given in the rest of Fig. 1. Fig. 1B shows the ^{40}Ca peak for a rod from a dark-adapted retina frozen immediately after the retina was isolated from the animal. Using internal controls consisting of CaF_2 films vacuum-evaporated directly onto the tissue sections (Schröder & Fain, 1984), we have shown that this peak represents 4-5 mmol Ca per 1 tissue volume. This is equivalent to about 8 x 10^9 Ca atoms per rod outer segment or 1-2 Ca per rhodopsin molecule. Fig. 1C shows the spectrum of a rod from a retina which had been incubated for 30 min in 20 mM ^{40}Ca, Fig. 1D that of a rod incubated for 30 min in Ringer having a free Ca concentration of 10^{-9} M, and Fig. 1E the spectrum obtained when the LAMMA was aimed to one side of the rods into the embedding medium.

By aiming the laser of the LAMMA 500 up and down the photoreceptor (as in Fig. 1A), we were able to study the distribution of Ca within the rod (Schröder & Fain, 1984; Fain & Schröder, 1985a). We found that over 95% of the Ca was within the outer segment. Significant concentrations of Ca were found in the inner segment only within the mitochondrion-rich ellipsoid body. There was no significant difference in Ca content as a function of distance along the outer segment or (within the limit of our measurements) from one side of the outer segment to the other.

Our measurements were the first to be made from photoreceptors in intact retina. They were preceded by numerous attempts to measure the Ca content of isolated outer segments, the best of which were probably those of Schnetkamp, Kaupp, and co-workers (see Kaupp & Schnetkamp, 1982). Our measurements were followed by those of Somlyo and Walz (1985), who were unable to measure significant quantities of Ca in the frog outer segment using EDX, and by the recent study of Nicol, Kaupp, and Bownds (1987), who measured the Ca content of intact rods isolated in normal Ringer solution from frog retina and obtained an outer segment Ca content nearly identical to the one we obtained.

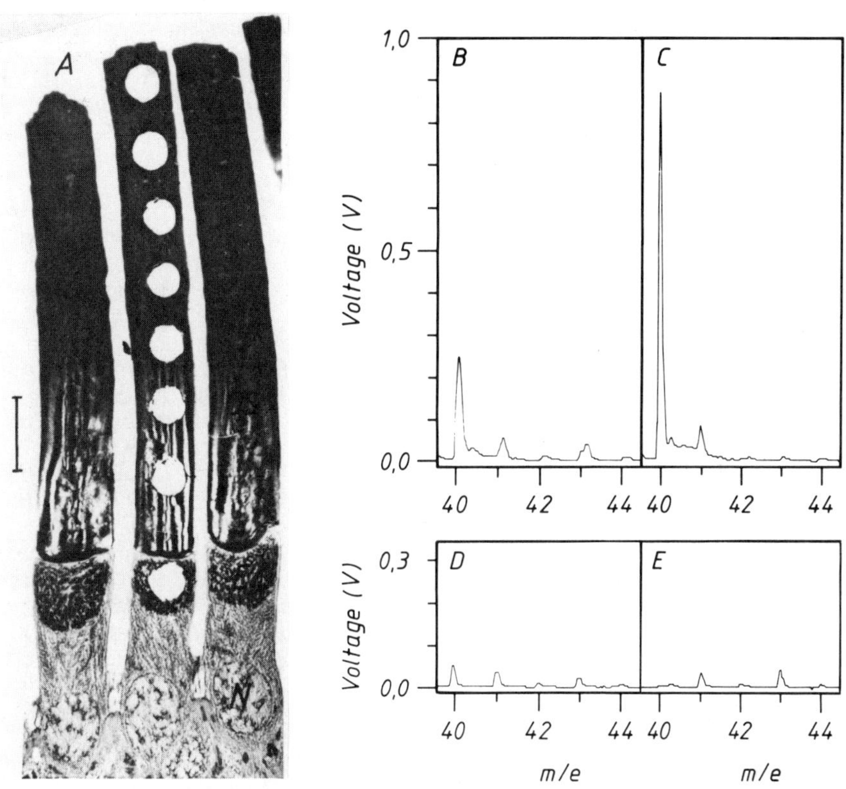

Fig. 1. LAMMA measurements from toad rods. A., Low-power electron-micrograph of rods showing the holes produced by the high-energy laser of the LAMMA apparatus. OS, outer segment; E, ellipsoid region containing mitochondria; N, nucleus. Scale bar is 10 um. B., control spectrum for m/e=40 to m/e=44 from dark-adapted retina frozen immediately after dissection. Ordinate gives the voltage signal from the secondary electron multiplier of the LAMMA after amplification. Abscissa gives mass in a.m.u. divided by charge, which is equivalent to mass in a.m.u. since the microplasma produced by the high-energy laser of the LAMMA consists almost exclusively of singly charged particles (see Fain & Schröder, 1985). C., spectrum from retina incubated in darkness for 30 min in Ringer containing 20 mM ^{40}Ca. Note that peak at m/e=40 is much larger than in B. D., spectrum from retina incubated in darkness for 30 min in Ringer for which the Ca was buffered with EGTA to produce a free Ca concentration of 10^{-9} M. Note reduction in peak at m/e=40. E., spectrum from same section as for B, but with laser aimed to one side of the photoreceptors into the embedding medium. All spectra are at same amplifier gain. Part A reprinted with permission from Schröder and Fain (1984), copyright Macmillan Journals, Ltd.

Although the total Ca content of rods seems therefore to be quite high, the free Ca concentration in the cytosol is low. The most reliable estimates of cytosolic free Ca concentration appear to be those of Lamb, Matthews and Torre (1986), of McNaughton, Cervetto and Nunn (1986), and of Miller and Korenbrot (1987a). Lamb et al. dialyzed solutions of the chelator BAPTA together with Ca at a fixed free concentration into rods using patch pipettes in the whole-cell recording mode. If the free Ca concentration in the dialysis solution was very low, they observed a pronounced increase in the outer segment Na conductance (probably produced indirectly by an increase in cyclic GMP). By varying the free Ca concentration until no change in Na conductance was observed, they estimated the normal free Ca concentration to be in the region of 1 uM. McNaughton et al. filled rods with the photoprotein aequorin and estimated the free Ca to be no greater than 0.6 uM. Miller and Korenbrot (1987a) measured free Ca by loading rods with the fluorescent Ca indicator Quin2 and estimated the free Ca concentration to be between 121 and 534 nM. These estimates are all within experimental error and show that in rods as in other neurons, the overwhelming majority of Ca is bound or sequestered. The free Ca content is only 0.01% of the total Ca and is below the limit of resolution of the LAMMA (about 50 uM per liter wet tissue volume).

2. **In darkness Ca moves rapidly into the rod through the light-dependent channels in the outer segment and through Ca channels in the inner segment, and it is rapidly expelled via Na/Ca exchange.** The permeability of the rod plasma membrane to Ca in darkness is quite high (Yau & Nakatani, 1984a; Hodgkin, McNaughton & Nunn, 1985). Yau and Nakatani (1985) have estimated that 10-15% of the current normally entering the outer segment in darkness is carried by Ca. This is equivalent to 10^7 Ca ions per rod per sec, or 15% of the total rod Ca content per min. Ca also enters voltage-dependent Ca channels in the inner segment (Fain, Gerschenfeld & Quandt, 1980). The value of this current at the rod resting membrane potential is difficult to estimate, since the amplitude of the Ca current is a steep function of potential in this region (Corey, Dubinsky & Schwartz, 1984). However, it is likely that the resting Ca current through the voltage-dependent channels in the inner segment is several pA in amplitude, which is of the same order as the Ca current through the light-dependent channels in the outer segment.

Though Ca enters the rod at a rapid rate in darkness, it is rapidly expelled primarily if not exclusively via Na/Ca exchange. Rapid Na/Ca exchange in rods was first demonstrated in bovine outer segments by Schnetkamp (1980, 1986), who measured the Ca efflux using ^{45}Ca or Arsenazo III and showed that Ca could be expelled at a rate of up to 5 x 10^6 Ca per outer segment per sec. In frog rods, which have a larger surface area, the maximal rate of Na/Ca exchange is larger (up to 6 x 10^7 Ca per outer segment per sec) and exceeds by several times the normal rate of Ca entry into the rod via the dark current and voltage-dependent Ca channels. Similarly high values for the maximal rate of Na/Ca exchange have been obtained by measuring the current produced by the exchanger (which is electrogenic) using electrophysiological methods (Yau & Nakatani, 1984b).

3. **Although Ca exchange across the plasma membrane is rapid, bulk exchange of Ca between the external medium and the outer segment is quite slow. Most of the Ca in the rods is not freely exchangeable, probably because it is sequestered within the discs.** It is possible to measure bulk exchange of Ca between the outer segment and the extracellular medium using the LAMMA method (Fain & Schröder, 1985a). An isolated retina is placed in Ringer for which all of the Ca (which is normally 97% ^{40}Ca) is

replaced with a stable isotope such as ^{42}Ca or ^{44}Ca. The retina is then incubated in this solution in darkness for a given period of time, fast-frozen, and analyzed. Using this method, we have shown (see Fig. 2) that the accumulation of bulk Ca from the extracellular medium into the rod outer segments occurs very slowly in intact photoreceptors at the normal (1.8 mM) Ca concentration (Fain & Schröder, 1985a). The rate of Ca exchange is of the order of 10^5 Ca per rod per sec, or about 10% of the total rod Ca per hour. Since, as we have said, the plasma membrane of the rod is quite permeable to Ca, the slow rate of exchange suggests that most of the Ca in the outer segment is either bound within the cytosol or sequestered. We have previously argued (Fain & Schröder, 1985a) that binding of Ca to proteins or lipids within the cytosol is an unlikely explanation for our results, since our data would then predict a mean life-time for the Ca-enzyme or Ca-lipid complex which is orders of magnitude greater than that for even the Ca-binding proteins calmodulin, parvalbumin, or troponin. It seems more likely that most of the Ca is sequestered.

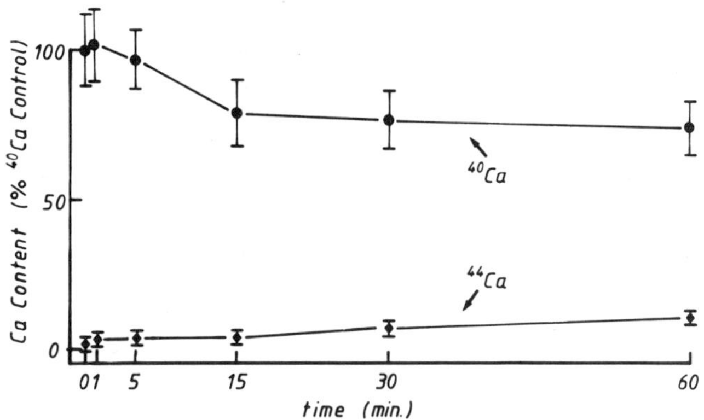

Fig. 2. Ca exchange in intact rod outer segments in darkness. Data points give the means and one standard deviation for ten measurements each from 4 to 6 retinas. Each retina was removed from a dark-adapted animal and cut into 6 pieces. Data points at time zero give the Ca content for control pieces, which for each retina were frozen directly after the dissection. The remaining pieces were incubated in darkness in Ringer containing the normal Ca concentration (1.8 mM) but with all of the ^{40}CaCl$_2$ substituted with ^{44}CaCl$_2$. Pieces were removed from this solution at 1, 5, 15, 30, and 60 min and immediately frozen. Data points give the ^{40}Ca (circles) and ^{44}Ca (diamonds) content, normalized to the mean ^{40}Ca content of control pieces. Reprinted with permission from Fain & Schröder (1985).

More recently we have examined the pooling of Ca within the outer segment by incubating the rods in solutions containing elevated concentrations of stable Ca isotopes. Fig. 3 shows the results of an experiment in which a retina was placed in 20 mM ^{44}CaCl$_2$ for 30 min in darkness. The spectrum to the left is from a control, showing the normal content of ^{40}Ca and ^{44}Ca in the outer segment. The spectrum to the right shows the result of the incubation in the high Ca solution. Note the large peak at m/e=44. The rate of ^{44}Ca accumulation in 20 mM CaCl$_2$ is about 10^7 Ca per rod outer segment per sec (Fain & Schröder, 1987), nearly 100 times larger than the rate of Ca accumulation at the normal Ca concentration (see Fig. 2).

Note however that, in spite of the large increase in the ^{44}Ca, the ^{40}Ca
peak in Fig. 3B is nearly as large as in the control spectrum in Fig. 3A.
Thus, even though large amounts of ^{44}Ca have accumulated within the cell,
this ^{44}Ca is not exchanging with the ^{40}Ca which was within the cell at the
beginning of the experiment.

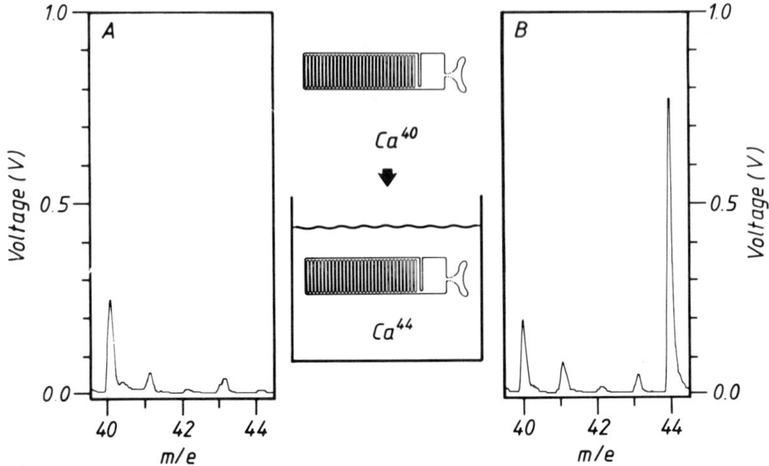

Fig. 3. Incubation of retina in darkness in high ^{44}Ca Ringer.
Representative spectra in same format as for Fig. 1 B.-E. (see
legend). A., spectrum from rod outer segment of control piece
of dark-adapted retina frozen immediately after the dissection. B., spectrum of rod outer segment from same retina as
for A but from another piece which had been incubated for 30
min in darkness in Ringer containing 20 mM ^{44}CaCl$_2$. Peak at
m/e=44 represents approximately 12–15 mmol ^{44}Ca/l tissue
volume which entered rods from external medium. Reprinted
with permission from Fain and Schröder (1987).

The results in Fig. 3 suggest that Ca in the rods can exist in two
pools. One pool, containing the great majority of Ca normally present
within the outer segment, exchanges slowly with extracellular Ca. This is
presumably the Ca within the discs. In addition, Ca can accumulate within
the rod in a second pool, so small under normal conditions that it is
below the limit of resolution of the LAMMA technique but much larger after
incubating the retina in high Ca. This pool presumably consists of Ca
free or loosely bound within the cytoplasm of the outer segment. In order
to see whether or not the Ca in this second pool is freely exchangeable
across the plasma membrane, we incubated the rods in darkness with two
different stable isotopes in succession, as outlined in Fig. 4A (Fain &
Schröder, 1987). The retinas were first placed for 15 min in 20 mM
^{42}CaCl$_2$ to load the rods (see Fig. 4C). The results of this initial
incubation were similar to those in Fig. 3B, except that the increase in
Ca content was registered predominantly at m/e=42. The small increases in
the peaks at m/e=40 and m/e=44 in Fig. 4C were the result of impurity in
the ^{42}Ca isotope. The retinas were then placed in 20 mM ^{44}CaCl$_2$. The
spectrum in Fig. 4D is taken from a rod fast-frozen only 1 min after the
change from ^{42}Ca to ^{44}Ca. The ^{42}Ca has nearly disappeared and is replaced

by a large peak at m/e=44. Notice that the peak at m/e=40 is nearly the
same as in the control spectrum (Fig. 4B).

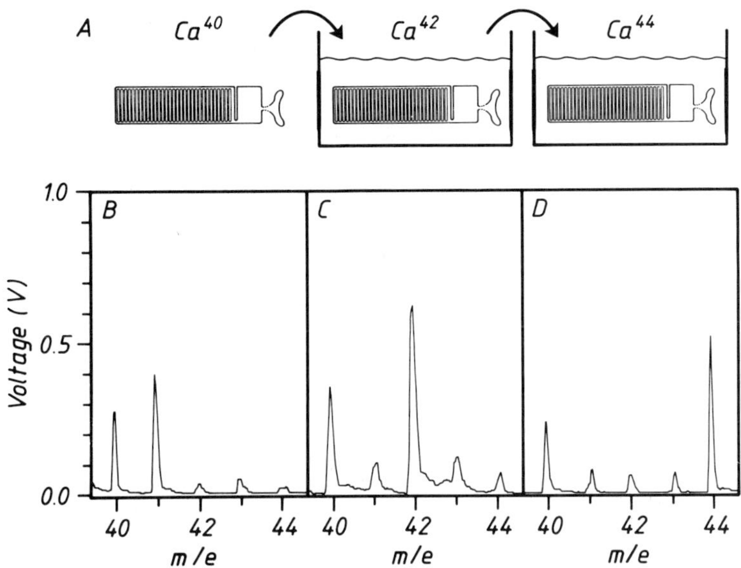

Fig. 4.   Double-isotope labelling experiments: protocol and
representative spectra.   A., flow chart for experiments.
Dark-adapted retinas with rod outer segments containing mostly
^{40}Ca were incubated in darkness first in 20 mM ^{42}CaCl$_2$ Ringer
to load rods with ^{42}Ca and then in 20 mM ^{44}CaCl$_2$ Ringer. B.-
D., sample spectra in same format as for Fig. 1 B.-E. (see
legend). B., spectrum from outer segment of control piece of
retina, frozen immediately after the dissection. C., spectrum
from outer segment of piece of retina incubated for 15 min in
20 mM ^{42}CaCl$_2$. Note large amplitude peak at m/e=42. Peaks at
m/e=40 and m/e=42 were larger than in B because of impurity of
the ^{42}Ca label. D., spectrum from outer segment of piece of
retina incubated first for 15 min in 20 mM ^{42}CaCl$_2$ and then
for 1 min in 20 mM ^{44}CaCl$_2$. Note that nearly all of the ^{42}Ca
left the rod apparently in exchange for ^{44}Ca. Reprinted with
permission from Fain and Schröder (1987).

Complete results of experiments of this kind are given in Fig. 5.
Here we show the means and standard deviations of the ^{40}Ca, ^{42}Ca, and ^{44}Ca
content, corrected for isotope impurity (Fain & Schröder, 1987) and
plotted as a percentage of the mean ^{40}Ca content of control outer segments. As in Fig. 4, the rods were first incubated for 15 min in 20 mM
^{42}CaCl$_2$ and then placed in 20 mM ^{44}CaCl$_2$.   During the incubation in
^{42}CaCl$_2$, the rods accumulated ^{42}Ca at a rate of about 2 x 10^7 Ca per rod
per sec. After the change to the 20 mM ^{44}CaCl$_2$ solution, there followed a
rapid exchange of ^{44}Ca for ^{42}Ca, whose rate can be estimated to be at
least 2-3 x 10^8 Ca per rod per sec. As soon as the ^{42}Ca pool was exhausted, the rate of ^{44}Ca accumulation decreased to about 1.5 x 10^7,
nearly the same as the rate of ^{42}Ca accumulation during the initial part
of the experiment. These results show directly that the Ca which enters
the outer segments in high Ca solutions remains accessible to exchange
with the extracellular solution. There is also some loss of ^{40}Ca from the
rods but at a much lower rate than the efflux of ^{42}Ca. Thus the ^{40}Ca

present in the rod under normal conditions clearly resides within a pool
different from the one occupied by the bulk of the Ca which enters the rod
from the extracellular space.

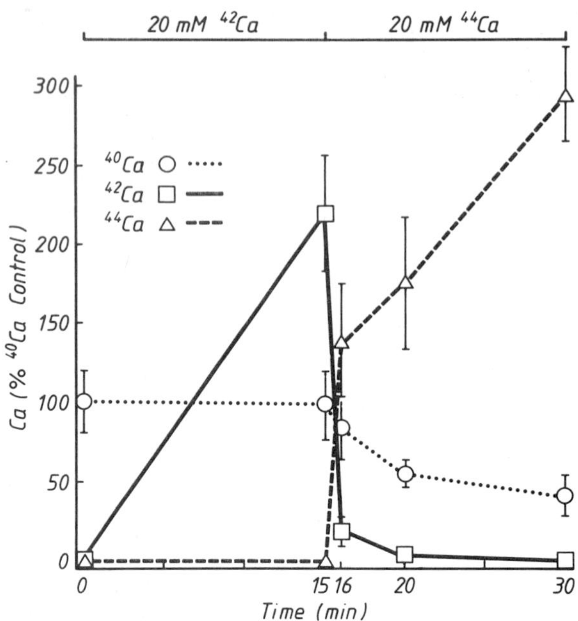

Fig. 5. Double-isotope labelling experiments: time course of
change in Ca content in rod outer segments from dark-adapted
retinas incubated in darkness first in 20 mM $^{42}CaCl_2$ and then
in 20 mM $^{44}CaCl_2$. Data points give ^{40}Ca (circles), ^{42}Ca
(squares), ^{44}Ca (triangles), corrected for isotope impurity.
Means with one standard deviation from 10 outer segments of
each of 4–5 retinas are given as percentage of the mean
corrected ^{40}Ca content of control pieces. Reprinted with
permission from Fain and Schröder (1987).

4. Like the plasma membrane, the discs contain a cGMP-dependent
conductance permeable to Ca^{2+} and perhaps also a Na/Ca exchanger. In
addition, discs contain an ATP-dependent Ca uptake mechanism, which may be
similar to the Ca ATPase in sarcoplasmic reticulum. The results from our
experiments suggest that nearly all of the Ca normally present in the rod
is sequestered within the discs. A similar conclusion has been reached by
Schnetkamp (1979), who measured the time course of release of Ca from
lysed outer segments. If the Ca within the discs were all free, it would
be present at a concentration of 30–100 mM. However, it seems likely that
most of the Ca is bound, either to lipids (McLaughlin & Brown, 1981) or to
rhodopsin (Tsui, Sundberg, & Hubbell 1985) and other proteins present in
the disc membrane.

There appear to be as many as three mechanisms for the transport of Ca
across the disc membrane. Cavaggioni and his collaborators first described a cyclic-GMP-dependent Ca permeability in discs (see Capovilla,
Caretta, Cavaggioni, Cervetto, & Sorbi, 1983), which has recently become
the subject of extensive investigation, in particular by Kaupp and his
collaborators (Koch & Kaupp, 1985; Cook, Zeilinger, Koch & Kaupp, 1986),
by Goldin and his co-workers (Puckett & Goldin, 1986; Pearce & Goldin,
1986), and by Schnetkamp (1986,1987). Cyclic-GMP-dependent Ca release

from bovine disks appears to consist of two components: a fast release with a high affinity for cyclic GMP, which can be selectively eliminated by pre-incubating the discs in high Na^+ (Schnetkamp, 1987); and a slow release with lower affinity, which is blocked by 1-cis-diltiazem. Both components are activated specifically by cyclic GMP (cyclic AMP is almost completely ineffective), and hydrolysis of the nucleotide is not necessary to produce Ca release. The two components of release are probably mediated by different proteins, and the one responsible for fast release has recently been solubilized and reconstituted into liposomes (Cook et al., 1986). The slower component accounts for the larger part of the Ca release from bovine disks and appears to be remarkably similar in its ion specificity, cyclic-GMP dependence, and pharmacology to the light-dependent conductance of the outer segment plasma membrane (Kaupp & Koch, 1986). Only this slower component has been observed in frog discs (Schnetkamp & Bownds, 1987).

It is possible to estimate the rate of Ca exchange across the discs in intact rods from our measurements of the rate of ^{44}Ca accumulation (see Fig. 2 and Fain & Schröder, 1985a, 1987). When this is compared to the rate of Ca movement across the plasma membrane, it is apparent that Ca

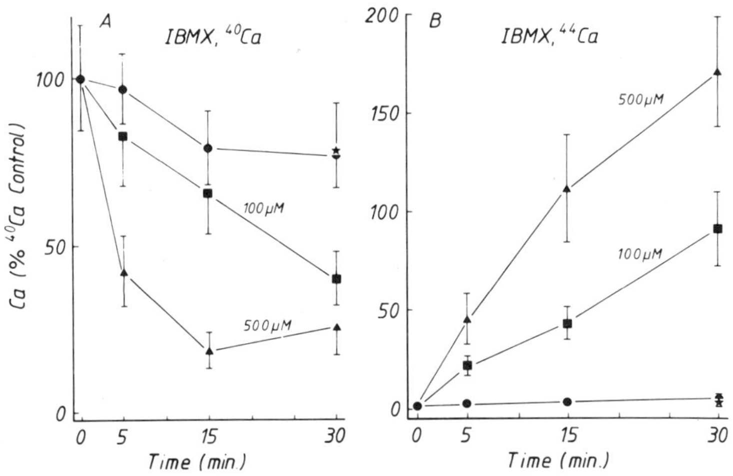

Fig. 6.   Effect of IBMX on Ca exchange in darkness. Retinas were incubated in Ringer containing 1.8 mM $^{44}CaCl_2$ and IBMX at a concentration of 100 uM (squares) or 500 uM (triangles). Data points give means with one standard deviation of 10 outer segments from 4-7 retinas for the ^{40}Ca content (A) and ^{44}Ca content (B) as a percentage of the mean ^{40}Ca content of control pieces. Circles are from Fig. 1 and show the decrease in ^{40}Ca (A) and increase in ^{44}Ca (B) in Ringer without IBMX. Stars give the ^{40}Ca and ^{44}Ca contents from rods in the present experiments (that is, for pieces from the same retinas as those treated with IBMX) after 30 min in Ringer containing only 1.8 mM $^{44}CaCl_2$. These are clearly within the range of the previous measurements. Reprinted with permission from Fain and Schröder (1987).

movement across discs is more than three orders of magnitude slower than across the plasmalemma (per cm^2 of membrane area). This is puzzling. If discs and plasma membrane both contain a similar cyclic-GMP-dependent permeability, and if the free cyclic GMP concentration within the outer segment is high enough in darkness to permit rapid movement of Ca across the plasma membrane, Ca should also move rapidly across discs. It is possible that the cyclic-GMP-dependent permeability is present at a much lower concentration in disc membranes than in plasma membranes. It is also possible that the permeability in discs is somehow inactivated in darkness (Puckett & Goldin, 1986).

If the permeability of the discs is experimentally augmented by increasing the cyclic GMP concentration within the photoreceptor, the rate of Ca exchange in the outer segment can be dramatically increased (Fain & Schröder, 1987). In Fig. 6, we show the results of experiments in which dark-adapted retinas were incubated in Ringer containing 100 or 500 uM of the phosphodiesterase inhibitor isobutylmethylxanthine (IBMX). IBMX has been shown to inhibit the light-activated phosphodiesterase of rod outer segments (Baehr, Devlin & Applebury, 1979) and to produce a large increase in the light-dependent permeability of the plasma membrane (Cervetto & McNaughton, 1986), presumably by increasing the free cyclic GMP concentration. The results in Fig. 6 show that IBMX also greatly increases the rate of bulk Ca exchange in the rods. We interpret these results as showing that rapid ^{40}Ca exit and ^{44}Ca entry can occur in outer segments if the disc membrane permeability is increased by increasing the cytosolic cyclic GMP concentration. They support our contention that most of the Ca in a rod is within the discs, and that the disc membrane is the major barrier to Ca exchange in rod outer segments in darkness.

In addition to the cyclic-GMP-dependent permeability, the membranes of disks may also contain a Na/Ca exchange transporter. Schnetkamp (1980, 1986) has shown that outer segments whose plasma membranes have been permeabilized can still take up and release Ca in a Na-dependent manner, though the maximum rate of Ca transport is smaller than in intact outer segments (Schnetkamp & Bownds, 1987). The significance of this observation is uncertain, since Na/Ca exchange has not yet been demonstrated in isolated discs. Furthermore, disc membranes appear not to contain a Na-K ATPase, and there may be no Na gradient across the discs to drive Ca entry or removal.

Finally, the recent experiments of Puckett, Aronson and Goldin (1985) have demonstrated an ATP-dependent high-affinity Ca uptake system in discs. Isolated, purified discs from bovine outer segments incubated in the presence of ATP take up Ca with an apparent K_m (for Ca) of 6 uM. Uptake is inhibited by vanadate but not by ruthenium red. The maximal rate of Ca uptake is only 0.013 Ca^{2+} per rhodopsin molecule per min, which for toad would be approximately 8×10^5 Ca per outer segment per sec. This is slow by comparison to the maximal rates of Ca movement across the plasma membrane through the light-dependent conductance or Na/Ca exchanger, but it may be fast enough to account for the normal rate of bulk Ca exchange observed in the outer segments in darkness (see Fig. 2). ATP-dependent Ca uptake is the only mechanism so far identified which seems capable of producing the sequestration of Ca into discs.

5. **Light produces a release of Ca from the sequestered pool and a decrease in the total Ca content of the outer segment. Ca is taken back up into the rods after the light is extinguished, during dark adaptation.** After the publication of the "Ca hypothesis" of Yoshikami and Hagins (1971), many attempts were made to measure light-induced Ca release from isolated discs or rod outer segments (summarized in Kaupp & Schnetkamp, 1982; Fain & Schröder, 1985b). These were largely unsuccessful, demon-

strating either a small release or none at all, probably because isolation of the outer segments has some deleterious effect either upon transduction or upon the mechanism responsible for Ca efflux. Attention then focused on intact photoreceptors in whole retina, and Gold and Korenbrot (1980) and Yoshikami, George, and Hagins (1980) nearly simultaneously discovered that light produces a transient increase in the Ca concentration in the extracellular space around the outer segments. Both of these groups interpreted this change as resulting from an increase in Ca efflux from the rods, due to light-dependent Ca release. However, it is now apparent that nearly all of the Ca increase which they measured was not due to an increase in Ca efflux but instead to a decrease in Ca influx, produced by the closing of the light-sensitive channels in the outer segment (Gold, 1986; Miller & Korenbrot, 1987b). As we have previously explained, Ca enters the rod at a rapid rate through the light-dependent channels in darkness and is rapidly expelled by Na/Ca exchange. When the channels are closed, Ca ceases to enter the rod but apparently continues to be expelled, and this produces the increase in extracellular Ca.

The continual entry and exit of Ca across the plasma membrane produces a large background Ca flux of the order of 10^7 Ca per rod per sec, which makes it difficult to detect Ca release from the discs in intact photoreceptors. Ca release has as yet only been detected with quantitative microanalysis (Schröder & Fain, 1984, 1987). Our LAMMA measurements of light-dependent Ca release are shown in Fig. 7. As in previous experiments, dark-adapted retinas were bathed in Ringer for which the $^{40}CaCl_2$ was replaced with $^{44}CaCl_2$. The retinas were then exposed to continuous

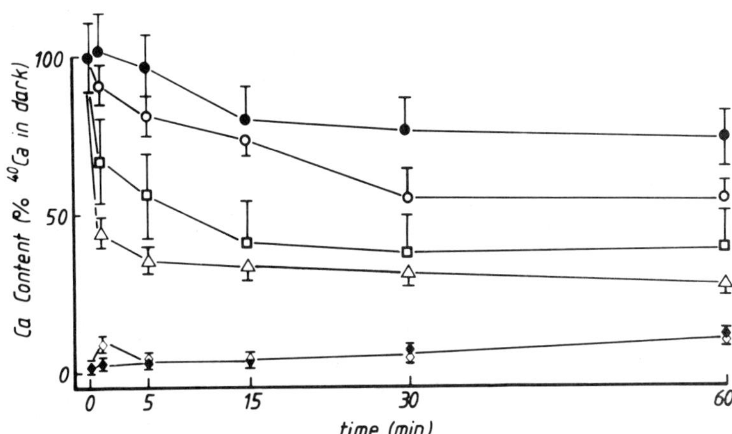

Fig. 7. Light-dependent Ca release. Retinas were cut into six pieces as for the experiment of Fig. 1 and placed in Ringer for which all of the $^{40}CaCl_2$ was replaced with $^{44}CaCl_2$. Closed circles and triangles are the ^{40}Ca and ^{44}Ca contents of retinas incubated in darkness and are taken from Fig. 1. Other symbols give mean Ca content in continuous illumination with one standard deviation, calculated from 10 outer segments each from 4 retinas. Intensities for ^{40}Ca were as follows: 68 Rh[*] per rod per sec (open circles), 8.1 x 10^3 Rh[*] per rod per sec (open squares), and 10^6 Rh[*] per rod per sec (open triangles). Open diamonds are ^{44}Ca content at 10^6 Rh[*] per rod per sec. Light stimulus was a continuous 503 nm field 50 mm in diameter and was uniform to within ±10%. Reprinted with permission from Schröder & Fain (1984), copyright Macmillan Journals, Ltd.

illumination at a given intensity and fast-frozen and analyzed. The closed circles and diamonds in Fig. 7 are the ^{40}Ca loss and ^{44}Ca accumulation in darkness, taken from Fig. 2 and reproduced here for ease of comparison. The three sets of open symbols in the upper part of Fig. 7 are the ^{40}Ca loss in light of three different intensities. The open circles are at the dimmest intensity of 68 rhodopsin molecules bleached (Rh*) per rod per sec, and after subtracting the ^{40}Ca loss in darkness, these data give a rate of light-induced ^{40}Ca decrease of about 10^6 Ca per rod per sec, or about 10^4 Ca per Rh*. At brighter intensities, the rate of ^{40}Ca release was greater, and at the highest intensity we used (10^6 Rh* per rod per sec, open triangles), the maximum rate of Ca release could be estimated to be of the order of 10^8 Ca per rod per sec. At this intensity, the release was transient and largely complete within 1 min.

Although continuous light produces a release of the ^{40}Ca, it has very little effect on the rate of ^{44}Ca accumulation. The open diamonds in Fig. 7 give the rate of ^{44}Ca uptake at the brightest illumination. There was a transient increase in ^{44}Ca content at 1 min which we have now seen in many experiments and believe to be real. At longer times, there was no significant difference in the rate of ^{44}Ca accumulation in continuous light and in darkness. Since the decrease in ^{40}Ca in continuous illumination is not compensated by a commensurate increase in ^{44}Ca, the total Ca concentration of the rod must have decreased. The data in Fig. 7 predict that the total Ca of outer segments from light-adapted animals should be about half that of dark-adapted animals, and this is in fact what we have found (Schröder & Fain, 1987).

Since a dark-adapted rod contains 4-5 mmol Ca per liter tissue volume and a light-adapted rod only 2-3 mmol, there must be some mechanism for replenishing the rods with Ca. We have recently discovered that, although ^{44}Ca uptake is not stimulated in continuous light, it is enhanced once the light is turned off, during dark adaptation (Schröder & Fain, 1987). The rate of ^{44}Ca accumulation in darkness after bright illumination is 2-3 times greater than in unstimulated rods. Even at this rate reuptake is rather slow, requiring 1-2 hours for fully recharging the outer segments with their full dark-adapted complement of Ca after the light-dependent pool has been exhausted in bright light. The rate of reuptake appears to be independent of the amount of rhodopsin bleached as long as the light-dependent pool has been fully depleted. It is as yet unclear whether the recovery of rod Ca has any relation to the recovery of sensitivity following bright light exposure.

At present, we have little information about the mechanism of Ca release or reuptake. Although, as we have said, the discs contain a cyclic-GMP-dependent Ca permeability, it is uncertain whether the gating of this permeability could occur in light, since light produces a decrease in cyclic GMP (Woodruff & Bownds, 1979). There is evidence suggesting a light-stimulated production of inositol triphosphate (IP$_3$) in rods (Ghalayini & Anderson, 1984; Brown, Blazynski & Cohen, 1987). Given the nearly ubiquitous role of IP$_3$ in producing release of Ca from internal stores, we are tempted to speculate that IP$_3$ may gate the opening either of the cyclic-GMP-dependent channels or of some other permeability or transport mechanism in the discs. The mechanism of Ca reuptake during dark adaptation is also unknown, but it seems possible that the Ca is pumped back into the discs by the ATP-dependent Ca pump (Puckett et al., 1985), whose activity may somehow be increased by prior illumination.

FUNCTIONAL SIGNIFICANCE OF Ca RELEASE

If Ca in the rods is sequestered within the discs, released in light, and taken back up after the light is turned off, then it is reasonable to suppose that Ca has some role in the physiology of the photoreceptor. The original "Ca hypothesis" of Yoshikami and Hagins (1971) seems clearly incorrect. Not only does cytosolic Ca seem to have little or no effect on the light-dependent conductance at physiological concentrations (see Yau et al., 1986), but the intracellular Ca concentration appears actually to decrease rather than increase over most of the working range of the rod (Yau & Nakatani, 1985; McNaughton, Cervetto & Nunn, 1986; Owen, Payne, Ratto & Tsien, 1987). The reason for this seems to be that the decrease in Ca influx into the rod caused by the closing of the light-dependent conductance is much larger at dim or moderate illuminations than the Ca released from the discs. Only in very bright light (for example 10^6 Rh[*] per rod per sec, see Fig. 7) might the efflux of Ca from the discs be expected temporarily to exceed the decrease in influx through the light dependent conductance. It remains to be shown whether release even in bright light leads to a functionally significant change in the free Ca concentration in the outer segment cytosol.

If Ca release from the discs is not responsible for the closing of the light-dependent conductance and if, within most of the working range of the photoreceptor, the intracellular free Ca concentration <u>decreases</u> in the light, one is left wondering what role Ca plays in transduction. It would be surprizing if rods contained specific mechanisms for Ca sequestration and light-dependent release, and these served no useful purpose. Of course, it is possible that Ca release has some role unrelated to signal transduction, such as triggering the shedding of the discs from the outer segments (Besharse, 1987). However, it seems also possible that the function of the release is not to change the cytosolic or extracellular Ca concentration but rather to change the Ca levels within the discs. Decreases in disc Ca may be responsible for regulating the sensitivity of the rod, for example during dark adaptation. At present, we know much more about the mechanisms for Ca uptake and release than about the function of Ca in photoreceptors. This is likely to change soon, as more information becomes available about the role of Ca in these cells.

ACKNOWLEDGEMENTS

We are grateful to A. Einerhand and J. Lauer for their excellent technical assistance. This research was supported in part by NIH grants EY 01844 and EY 00331 and N.S.F. grant INT 84-04028 to G.L.F. and by a S.F.B. 160 grant to W.H.S.

REFERENCES

Baehr, W., Devlin, M.J. and Applebury, M.L., 1979. Isolation and characterization of cGMP phosphodiesterase from bovine rod outer segments. J. Biol. Chem. 254:11669-11677.

Bastian, B.L. and Fain, G.L., 1982. The effects of low calcium and background light on the sensitivity of toad rods. J. Physiol. 330:307-329.

Besharse, J.C., 1987. Photosensitive membrane turnover: differentiated membrane domains and cell-cell interaction, in "The Retina. A Model for Cell Biology Studies, Part I," R. Adler and D. Farber, eds., Academic, New York, pp. 297-352.

Brown, J.E., Blazynski, C. and Cohen, A.I., 1987. Light induces a rapid and transient increase in inositol triphosphate in intact vertebrate rods. Invest. Ophthalmol. Vis. Sci., ARVO Abst., in press.

Capovilla, M., Caretta, A., Cavaggioni, A., Cervetto, L. and Sorbi, R.T., 1983. Metabolism and permeability in retinal rods. Prog. Retin. Res. 2:233-247.

Cervetto, L. and McNaughton, P.A., 1986. The effects of phosphodiesterase inhibitors and lanthanum ions on the light-sensitive current of toad retina rods. J. Physiol. 370:91-109.

Cook, N.J., Zeilinger, C., Koch, K.-W., and Kaupp, U.B., 1986. Solubilization and functional reconstitution of the cGMP-dependent cation channel from bovine rod outer segments. J. Biol. Chem. 261:17033-17039.

Corey, D.P., Dubinsky, J.M. and Schwartz, E.A., 1984. The calcium current in inner segments of rods from the salamander (Ambystoma tigrinum) retina. J. Physiol. 354:557-575.

Fain, G.L., Gerschenfeld, H.M. and Quandt, F.N., 1980. Calcium spikes in toad rods. J. Physiol. 303:495-513.

Fain, G.L. and Lisman, J.E., 1981. Membrane conductances of photoreceptors. Prog. Biophys. Mol. Biol. 37:91-147.

Fain, G.L. and Schröder, W.H., 1985a. Calcium content and calcium exchange in dark-adapted toad rods. J. Physiol. 368:641-665.

Fain, G.L. and Schröder, W.H., 1985b. Calcium content and light-induced release from photoreceptors: measurements with laser micro-mass analysis, in "Contemporary Sensory Neurobiology," M.J. Correia and A.A. Perachio eds., Alan R. Liss, New York, pp. 3-20.

Fain, G.L. and Schröder, W.H., 1987. Calcium in dark-adapted toad rods: evidence for pooling and cyclic-guanosine-3'-5'-monophosphate-dependent release. J. Physiol., in the press.

Fesenko, E.E., Kolesnikov, S.S. and Lyubarsky, A.L., 1985. Induction by cyclic GMP of cationic conductance in plasma membrane of retinal rod outer segment. Nature 313:310-313.

Fung, B.K.K. and Stryer, L., 1980. Photolyzed rhodopsin catalyzes the exchange of GTP in retinal rod outer segments. Proc. Natl. Acad. Sci. U.S.A. 77:2500-2504.

Ghalayini, A. and Anderson, R.E., 1984. Phosphatidylinositol 4,5-biphosphate: light-mediated breakdown in the vertebrate retina. Biochem. Biophys. Res. Comm. 124:503-506.

Gold, G.H., 1986. Plasma membrane calcium fluxes in intact rods are inconsistent with the "calcium hypothesis". Proc. Natl. Acad. Sci. U.S.A. 83:1150-1154.

Gold, G.H. and Korenbrot, J.I., 1980. Light-induced calcium release by intact retinal rods. Proc. Natl. Acad. Sci. U.S.A. 77:5557-5561.

Hodgkin, A.L., McNaughton, P.A., and Nunn, B.J., 1985. The ionic selectivity and calcium dependence of the light-sensitive pathway in toad rods. J. Physiol. 358:447-468.

Kaupp, U.B. and Schnetkamp, P.P.M., 1982. Calcium metabolism in vertebrate photoreceptors. Cell Calcium 3:83-112.

Koch, K.-W. and Kaupp, U.B., 1985. Cyclic GMP directly regulates a cation conductance in membranes of bovine rods by a cooperative mechanism. J. Biol. Chem. 260:6788-6800.

Kühn, H., 1980. Light- and GTP-regulated interaction of GTPase and other proteins with bovine photoreceptor membranes. Nature 283:587-589.

Lamb, T.D., Matthews, H.R. and Torre, V., 1986. Incorporation of calcium buffers into salamander retinal rods: a rejection of the calcium hypothesis of phototransduction. J. Physiol. 372:315-349.

Matthews, H.R., Torre, V. and Lamb, T.D., 1985. Effects on the photoresponse of calcium buffers and cyclic GMP incorporated into the cytoplasm of retinal rods. Nature 313:582-585.

McLaughlin, S. and Brown, J.E., 1981. Diffusion of calcium ions in retinal rods. A theoretical calculation. J. Gen. Physiol. 77:475-487.

McNaughton, P.A., Cervetto, L. and Nunn, B.J., 1986. Measurement of the intracellular free calcium concentration in salamander rods. Nature 322:261-263.

Miller, D.L. and Korenbrot, J.I., 1987a. Effects of a cytoplasmic Ca buffer on photoexcitation in retinal rods: a role for Ca in the control of gain and kinetics of phototransduction. J. Gen. Physiol., in press.

Miller, D.L. and Korenbrot, J.I., 1987b. Kinetics of light-dependent Ca fluxes across the plasma membrane of rod outer segments: a dynamic model of the regulation of cytoplasmic Ca concentration. J. Gen. Physiol., in press.

Nicol, G.D., Kaupp, U.B. and Bownds, M.D., 1987. Transduction persists in rod photoreceptors after depletion of intracellular calcium. J. Gen. Physiol. 89:297-319.

Owen, W.G., Payne, R., Ratto, G.M. and Tsien, R.Y., 1987. Use of fura2 to monitor changes in cytosolic free $[Ca^{2+}]$ of rods in the isolated retina of the bullfrog. J. Physiol. 383:41P.

Pearce, L.B. and Goldin, S.M., 1986. Rapid kinetic analysis of cyclic nucleotide-stimulated cation release from purified bovine rod outer segment discs. Soc. Neurosci. Abstr. 12, part 1:630.

Puckett, K.L., Aronson, E.T. and Goldin, S.M., 1985. ATP-dependent Ca uptake activity associated with a disc membrane fraction isolated from bovine retinal rod outer segments. Biochemistry 24:390-400.

Puckett, K.L. and Goldin, S.M., 1986. Guanosine 3',5'-cyclic monophosphate stimulates release of actively accumulated calcium in purified disks from rod outer segments of bovine retina. Biochemistry 25:1739-1746.

Schnetkamp, P.P.M., 1979. Calcium translocation and storage of isolated intact cattle rod outer segments in darkness. Biochem. Biophys. Acta 554:441-459.

Schnetkamp, P.P.M., 1980. Ion selectivity of the cation transport system of isolated intact cattle rod outer segments. Evidence for a direct communication between the rod plasma membrane and the rod disk membrane. Biochem. Biophys. Acta 598:66-90.

Schnetkamp, P.P.M., 1986. Sodium-calcium exchange in the outer segments of bovine rod photoreceptors. J. Physiol. 373:25-45.

Schnetkamp, P.P.M., 1987. Sodium ions selectively eliminate the fast component of cyclic GMP-induced Ca^{2+} release from bovine rod outer segment disks. Biochemistry, in press.

Schnetkamp, P.P.M. and Bownds, M.D., 1987. Na^{+}- and cGMP-induced Ca^{2+} fluxes in frog rod photoreceptors. J. Gen. Physiol., in press.

Schröder, W.H. and Fain, G.L., 1984. Light-dependent calcium release from photoreceptors measured by laser micro mass analysis. Nature 309:268-270.

Schröder, W.H. and Fain, G.L., 1987. Light-activated Ca release and re-uptake in rods. Invest. Ophthalmol. Vis. Sci., ARVO Abst., in press.

Somlyo, A.P. and Walz, B., 1985. Elemental distribution in Rana pipiens retinal rods: quantitative electron probe analysis. J. Physiol. 358:183-195.

Torre, V., Matthews, H.R. and Lamb, T.D., 1986. Role of calcium in regulating the cyclic GMP cascade of phototransduction in retinal rods. Proc. Natl. Acad. Sci. U.S.A. 83:7109-7113.

Tsui, F.C., Sundberg, S.A., and Hubbell, W.L., 1985. Analysis of surface electrical structure of photoreceptor membranes. Biophysical J. 47:106a.

Wheeler, G.L. and Bitensky, M.W., 1977. A light-activated GTPase in vertebrate photoreceptors: Regulation of light-activated cyclic GMP phosphodiesterase. Proc. Natl. Acad. Sci. U.S.A. 74:4238-4242.

Woodruff, M.L. and Bownds, M.D., 1979. Amplitude, kinetics, and reversibility of a light-induced decrease in guanosine 3',5'-cyclic monophosphate in frog photoreceptor membranes. J. Gen. Physiol. 73:629-653.

Yau, K.-W., Haynes, L.W. and Nakatani, K., 1986. Roles of calcium and cyclic GMP in visual transduction. Fortsch. der Zool. 33:343–366.

Yau, K.-W. and Nakatani, K., 1984a. Cation selectivity of light-sensitive conductance in retinal rods. Nature 309:352–354.

Yau, K.-W. and Nakatani, K., 1984b. Electrogenic Na-Ca exchange in retinal rod outer segment. Nature 311:661–663.

Yau, K.-W. and Nakatani, K., 1985. Light-induced reduction of cytoplasmic free calcium in retinal rod outer segment. Nature 313:579–582.

Yoshikami, S., George, J.S. and Hagins, W.A., 1980. Light-induced calcium fluxes from outer segment layer of vertebrate retinas. Nature 286:395–398.

Yoshikami, S. and Hagins, W.A., 1971. Light, calcium, and the photo-current of rods and cones. Amer. Biophys. Soc. Abs. 15:47a.

SMALL CONDUCTANCE Ca ACTIVATED K CHANNELS IN MOLLUSKS

A. Hermann[1], C. Erxleben[2] and D. Armstrong[3]

[1]University of Salzburg, Institute of Zoology
 Hellbrunnerstrasse 34, A-5020 Salzburg, Austria

[2]University of Konstanz, Faculty of Biology, D-7750
 Konstanz, FRG

[3]University of California, Department of Biology
 Los Angeles, CA 90024, USA

INTRODUCTION

In many cells intracellular Ca ions activate a conductance to K ions
in the plasma membrane. Among the cells where this property of Ca ions
was detected first were neurons from the marine mollusk, _Aplysia
californica_ (Meech, 1976). The biophysical, pharmacological and
physiological properties of that K conductance have now been studied in
detail in molluscan neurons. The K conductance is not only sensitive to
calcium ions but also intrinsically voltage-dependent. Probing the
internal Ca receptor site with different divalent cations has revealed
that ions such as cadmium or strontium also activate these K channels
effectively (Gorman and Hermann, 1979). Pharmacological studies further
showed that the Ca activated K current is highly sensitive to block by
external tetraethylammonium (TEA) and that other types of K currents in
these cells are not equally sensitive to TEA or 4-aminopyridine (4-AP)
(Hermann and Gorman, 1981a, b; Hermann and Hartung, 1986). The
physiological properties attributed to the Ca dependent conductance
include a contribution to the repolarization and after-hyperpolarization
of action potentials, post-tetanic hyperpolarization, frequency
adaptation, and the termination of bursting pacemaker potentials (Meech,
1978; Hermann and Hartung, 1983).

Further investigation of the Ca activated K conductance at the single
channel level in various cell types from vertebrates has revealed a
variety of Ca activated K channels with differences in their calcium- and
voltage-dependence and in their pharmacological properties (Marty, 1983;
Latorre and Miller, 1983; Levitan, this volume). A common feature of many
of these channels is their large unitary conductance, of the order of one
to three hundred picosiemens. Recently, charybdotoxin (ChTX) a, toxin
component isolated from the venom of the scorpion, _Leiurus quinquestriatus_,
has been shown to block the large conductance, Ca activated K channel
isolated from rat skeletal muscle and inserted into lipid bilayers (Miller
et al., 1985). Neurons from the terrestrial snail, _Helix pomatia_, express
another type of Ca activated K channel with a smaller unitary conductance

but similar pharmacological properties compared to the large conductance
channel (Lux et al., 1981; Hermann and Hartung, 1982; Ewald et al., 1985).

The experiments reported here were performed on nerve cells in the
abdominal ganglion of _Aplysia_ _californica_, using voltage-clamp and
patch-clamp techniques. Membrane current fluctuations and single channel
records were analyzed to investigate whether Ca activated K channels from
Aplysia neurons fall into the high or low conductance category. In
addition we studied the action of charybdotoxin on this channel. In
particular we were interested to determine the toxin's affinity,

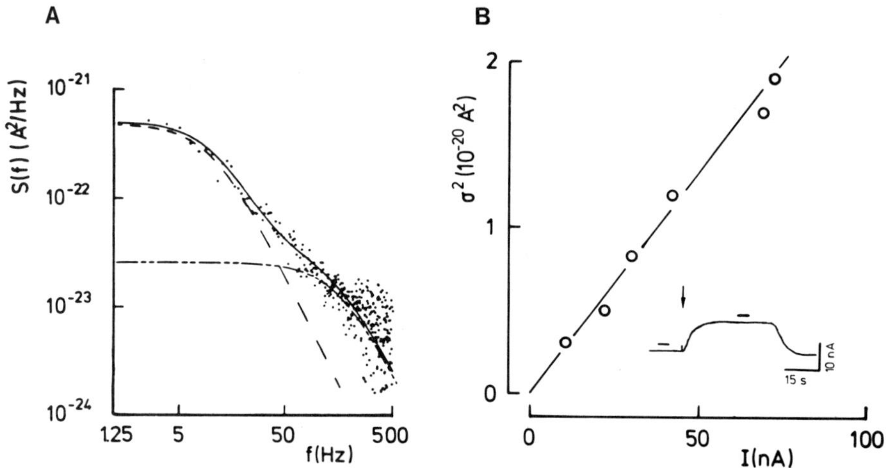

Fig. 1.   (A) Difference spectrum of Ca activated membrane current
          fluctuations. Spectral density, S(f), is plotted vs. the
          frequency, f, on double logarithmic scale. The difference
          spectrum is fitted by the sum of two Lorentzian components
          (continuous line) with corner frequencies of 10 and 150 Hz
          and zero frequency amplitudes of 4.5 x 10^{-22} and 2 x
          10^{-23} A^2/Hz (broken lines). Holding potential –40 mV.
          (B) Plot of the variance (σ^2) of Ca activated current
          fluctuations vs. the mean Ca activated K current (I). The
          insert shows the activation of the current by ionophoretic
          Ca injection (arrow). Power spectra were taken during the
          time indicated by bars (Hartung and Hermann, 1987).

specificity, and selectivity for different types of Ca activated K
channels, and its effect on action potentials and pacemaker activity in
molluscan neurons. The current through Ca activated K channels was
isolated by ion substitution and pharmacological procedures and was
activated in intact cells by ionophoretic injection of Ca ions. Details
of the experimental methods have been described elsewhere (Gorman and
Hermann, 1979; Hermann and Gorman, 1981a; Hermann and Hartung, 1982;
Hermann and Erxleben, 1987). For some experiments a cell line (GH_3)
derived from a rat pituitary tumor was used (Tashjian, 1979).

<u>Fluctuations of Ca activated K-current</u>

Fig. 1A shows the difference of power spectra of membrane current
fluctuations recorded before and during ionophoretic injection of Ca ions
(see insert, Fig. 1B). The difference spectrum can be attributed to the
fluctuation of Ca activated K channels between open and closed states and
is fitted by a minimum of two Lorentzian functions. Power spectra
obtained this way are very similar to those recorded previously in <u>Helix</u>
neurons (Hermann and Hartung, 1982). The single channel conductance,
estimated from the variance of the current fluctuations and the mean Ca
activated current (Fig. 1B), was about 11 pS at a holding potential of -30

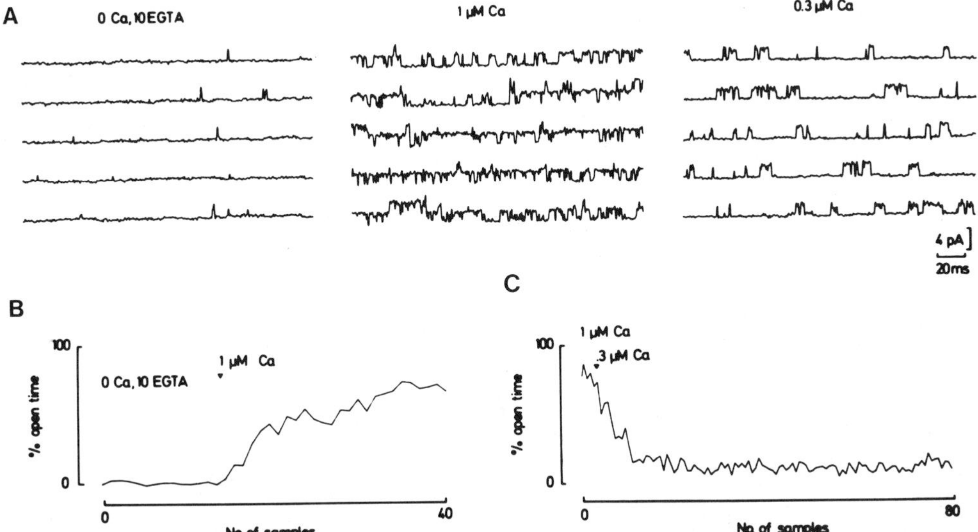

Fig. 2. (A) Single channel recordings from an inside-out patch in
solutions buffered with 10 mM EGTA to different
concentrations of free Ca ions. (B) and (C) Plots of
percent open time of channels vs. time, measured as the
number of 205 ms sample intervals. The concentrations of
free Ca ions was changed as indicated at the arrows.

mV. After correction for different internal K concentrations and the
non-linear current-voltage relationship, the single channel conductance
calculated from noise data is 12-14 pS (Hartung and Hermann, 1987). This
estimate of the single channel conductance from fluctuation analysis is
close to the value obtained from single channel recordings under similar
conditions (see Fig. 2A).

<u>Unitary currents through single Ca activated K channels</u>

Ca activated K channels were recorded in the cell-attached,
inside-out and outside-out configuration (cf. Hamill et al., 1981; Hermann
and Erxleben, 1987). Fig. 2A shows single channel recordings from an

inside-out patch over a range of internal Ca concentrations. In nominally
zero Ca solution containing 10 mM EGTA (free Ca concentration < 10^{-9} M)
the opening probability was greatly reduced but not eliminated (Fig. 2A
left panel). It appears that the channels can open at Ca ion
concentrations lower than those inside most cells. Under those
conditions, channel gating is characterized by openings of short duration,
reminiscent of the gating behavior of other Ca activated K channels in the
absence of Ca (Barrett et al., 1982; Pallotta, 1985). It is possible
therefore that in our preparation the brief openings observed in Ca free
solution reflect the voltage-dependent component of channel activation.
Increasing the internal Ca concentration to 0.3 µM and 1 µM increased
both the frequency of opening and the dwell time in the open state (Fig.
2B, C). The apparent probability of channel opening, uncorrected for the
number of active channels in the patch, increased from 1-2% in < 10^{-9} M
Ca to 11-19% in 3×10^{-7} M Ca and 79-93% in 10^{-6} M Ca solution.

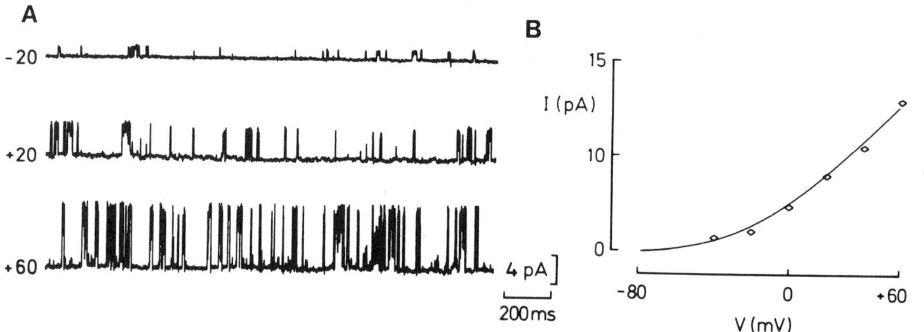

Fig. 3.   (A) Unitary Ca activated K currents in an inside-out patch
at membrane potentials from −20 mV to +60 mV. Bath Ca
concentration 0.3 µM. (B) Single channel current-voltage
relationship. The line drawn to the experimental points was
calculated using the Goldman-Hodgkin-Katz equation, with a
single channel permeability of 8×10^{-14} (cm^3s^{-1}), an
external K concentration of 10 mM and a K equilibrium
potential of −90 mV.

Fig. 3A shows single channel recordings at different membrane
potentials from −20 mV to +60 mV. The amplitude of the unitary current as
well as the probability of channel opening increased at more positive
membrane potentials. The current-voltage relationship from single channel
recordings (Fig. 3B) shows outward rectification as expected from the
asymmetric K ion distribution on either side of the patch. The line drawn
to the experimental points was calculated using the constant field
equation (Hodgkin and Katz, 1949) and follows the prediction for a K

selective ion channel. The reversal potential of single channel currents
was obtained by extrapolation of the current-voltage relationship to
negative potentials. It was found to be close to the theoretical value of
E_K, i.e. -90 mV (with K_i = 360 mM and K_O = 10 mM). The average
single channel conductance measured in asymmetrical K solution at zero
membrane potential was 35 ($\pm$5) pS (n=16). In contrast to the probability
of opening, the amplitude of single channel currents was not dependent on
the external (1-20 mM) or internal ($< 10^{-9} - 3 \times 10^{-6}$ M) Ca
concentration.

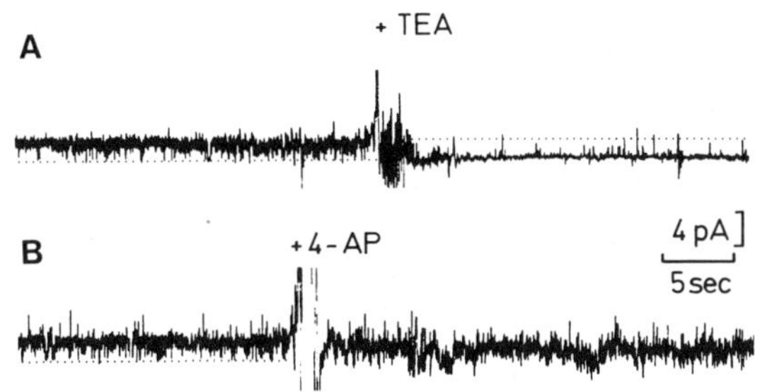

Fig. 4.   (A) Single channel recording from an outside-out patch
           before and after application of tetraethylammonium (TEA, 2
           mM) and (B) before and after application of 4-aminopyridine
           (4-AP, 10 mM). The current artifacts indicate the time at
           which the drug was added to the bath. In 4-AP containing
           solution the patch recording became more unstable. The
           patch electrode in (A) and (B) contained 1 µM Ca. Holding
           potential 0 mV.

     The dual dependence of gating on voltage and Ca, and a reversal
potential near E_K, distinguish this small conductance channel from most
other ion channels in molluscan neurons. The identification of these
small conductance channels was confirmed pharmacologically in outside-out
patches. The putative Ca activated K channels were identified by their Ca
and voltage-dependence and by their unitary conductance of approximately
32 pS (n=6). The open probability of these channels was reduced by
submillimolar concentrations of external tetraethylammonium (TEA) and was
completely blocked by 2-5 mM TEA (Fig. 4A). In contrast, 4-aminopyridine
(4-AP) did not block the channels at concentrations up to 10 mM (Fig.
4B). Thus, the pharmacological responses of these low conductance
channels are similar to those observed during macroscopic recordings of Ca
activated K current (cf. Hermann and Hartung, 1983, 1986).

<u>Charybdotoxin blocks the Ca activated K conductance</u>

The effect of the scorpion toxin, charybdotoxin (ChTX), on the
Ca activated K conductance was investigated with several different
experimental approaches: (1) by depolarizing voltage pulses under
voltage-clamp conditions (2) by intracellular injection of Ca ions (3) by
single channel recording and (4) by recording action potentials and
pacemaker activity under current-clamp conditions. Fig. 5 shows
current-voltage curves of isochronic outward currents measured at the end
of 150 ms pulses. Outward currents were recorded in a solution containing
10 mM 4-AP to suppress other voltage-activated K currents and to maximize
the visibility of the Ca activated K current component. The typically
N-shaped current-voltage relation of cell R-15 was suppressed in the
presence of ChTX in a dose dependent manner. The suppression was limited
to currents in the voltage range between +20 mV and +80 mV where the Ca
activated K current is predominant. The delayed, voltage-dependent K

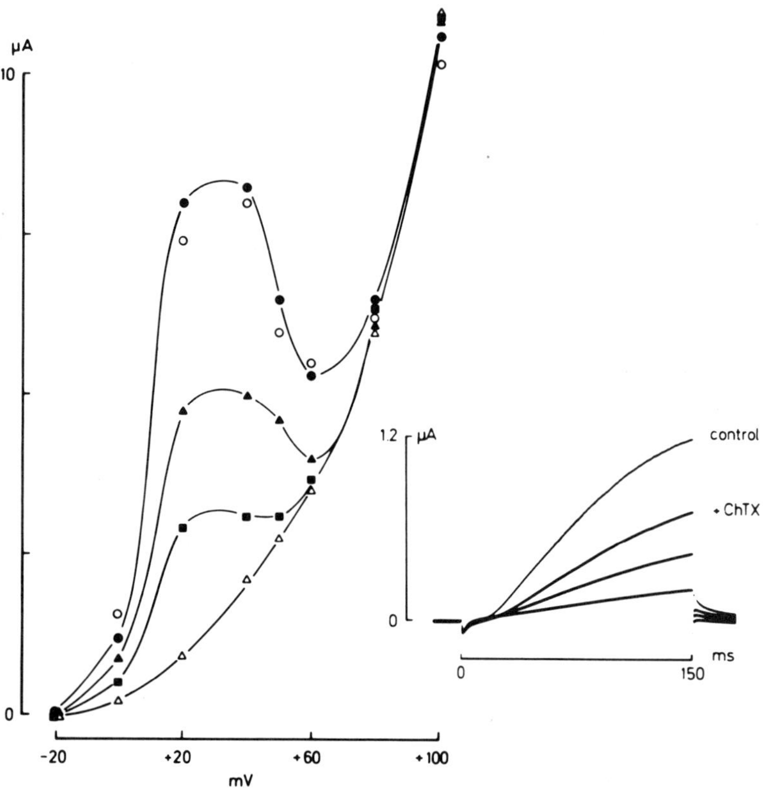

Fig. 5. Effect of charybdotoxin (ChTX) on membrane currents
         initiated by depolarizing voltage pulses. Current-voltage
         curve of isochronal K currents in response to voltage pulses
         lasting 150 ms. Currents were measured in artificial
         seawater solution containing 10 mM 4-aminopyridine, before
         (closed circles) and after application of ChTX, 100 nM
         (closed triangles), 200 nM (closed squares), 300 nM (closed
         triangles) and after washing out the toxin (open circles).
         The insert shows the outward current at 0 mV under control
         conditions (ASW solution containing 10 mM 4-AP) and the
         lower traces show the currents after application of 100, 200
         and 300 nM ChTX. Holding potential -40 mV; cell R-15.

current was not altered by the toxin at potentials more positive than +80
mV. Half—maximal inhibition of the Ca activated K current at Vm = +20 mV
(after correction for series resistance) was observed at 135–155 nM ChTX.
The effect of the toxin was reversible after washing with toxin—free
solution. The insert of Fig. 5 shows superimposed recordings of outward
currents before (control) and after addition of ChTX. Outward currents,
as well as tail—currents at the end of the voltage steps, were reduced in
a dose—dependent manner, but the time course of decay was not changed,
suggesting that the kinetics of gating had not been altered. The inward
current at the beginning of the voltage step was not altered significantly
by the toxin.

The effect of the toxin was also tested on the Ca activated K current
elicited by ionophoretically injecting Ca ions intracellularly (Gorman and
Hermann, 1979). The outward current, elicited at one minute intervals,
reached a peak shortly after the Ca injection was terminated and then
declined to the initial level of holding current. Fig. 6A shows
recordings of Ca activated K currents before and during application of
ChTX. The toxin reduced the peak amplitude of the currents but did not
alter the time course of their decay. The onset of inhibition by toxin
was studied in experiments where Ca ions were injected until the outward
current had reached a stationary level at which time ChTX was applied
(Fig. 6B). The apparent dissociation constant (KD) was obtained from
measurements of the inhibition produced by different toxin concentrations
at different membrane potentials. The KD for a 1:1 toxin—receptor complex
(indicated by the dose—response relationship) is about 30 nM at Vm = − 30

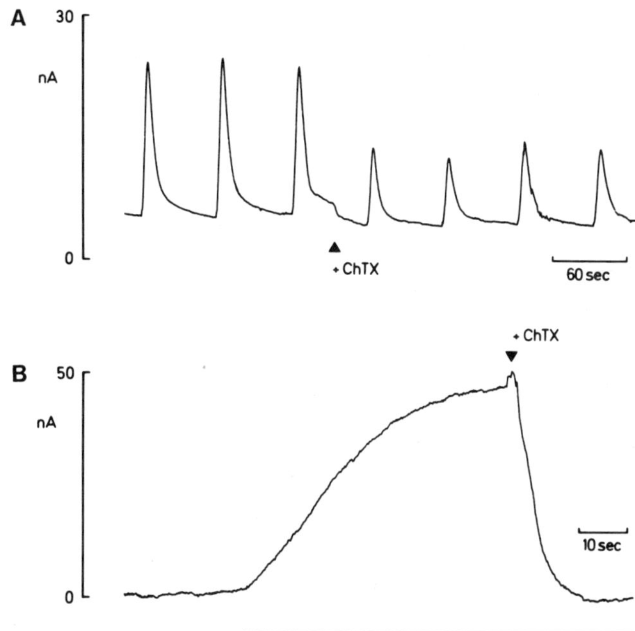

Fig. 6. (A) Effect of charybdotoxin (ChTX, 20 nM) on Ca activated K
         currents elicited by intracellular, ionophoretic injection
         of Ca ions under voltage—clamp conditions. Injection pulses
         were 400 nA for 2 sec, repeated every minute. (B) Outward
         current activated by a smaller Ca injection for the time
         indicated by the line beneath the current trace. After the
         current has reached a stationary level, ChTX (100 nM) was
         added to bath solution (arrowhead). Holding potential −30
         mV in (A) and (B). Cell R-15.

mV and 170 nM at Vm = + 20 mV. The weaker inhibition at more positive
membrane potentials suggests that the block may be voltage-dependent. The
voltage-dependence of inhibition was analyzed by estimating KD values at
different membrane potentials. The KD changed e-fold for a 63-69 mV
change in membrane potential implying that the toxin molecule senses a
fraction (about 35%) of the transmembrane electric field.

In addition to blocking peak outward currents, ChTX produced an
apparent inward current (see Fig. 6A), whose magnitude increased with the
toxin concentration. Its negative reversal potential, TEA sensitivity,
dependence on external Ca, and its insensitivity to 4-AP all suggest that
this current results from the blockade of a resting K conductance which is
activated by internal Ca ions. Such a contribution of a Ca activated K
conductance to the membrane's resting conductance has, in fact, been
indicated previously in _Aplysia_ neurons (Gorman and Hermann, 1982; Johnson
and Thompson, 1983). The specificity of the action of ChTX was tested on
various membrane currents under conditions which permitted optimal
isolation of the channel under investigation. The toxin had no effect on
Na or Ca currents or on other voltage-activated K currents, such as the
delayed rectifier current ($I_{K,V}$) or the transient outward current
($I_{K,A}$) (Hermann and Erxleben, 1987).

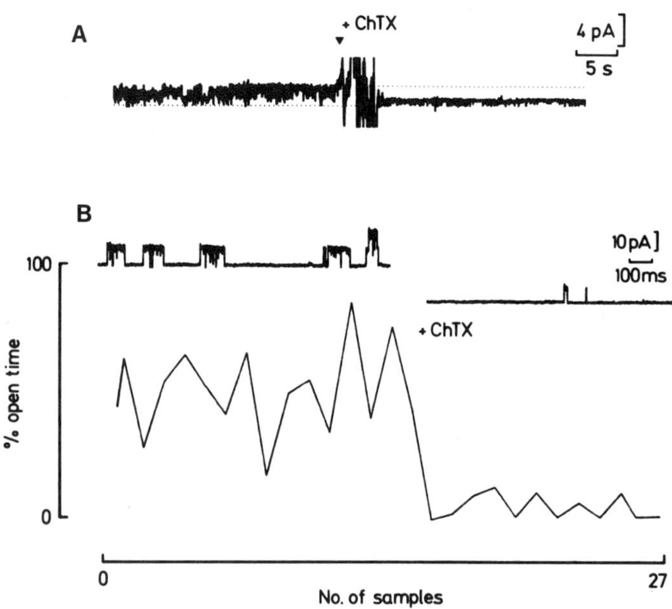

Fig. 7. (A) Single channel recording Ca activated K of channels in
an outside-out patch from an _Aplysia_ neuron, before and
after application of charybdotoxin (100 nM). The patch
electrode contained 10^{-6} M Ca. (B) Outside-out patch
recording of Ca activated K channels from a GH3 cell.
Percent of time in the open state plotted vs. time, measured
as the number of 205 ms sample intervals, before and after
application of ChTX (20 nM). The patch pipette contained 2
x 10^{-7} M Ca. The membrane potential in both (A) and (B)
was 0 mV.

Apamin, a component of bee venom, blocks the Ca activated K
conductance in a variety of preparations (cf. Cook and Haylett, 1985).
Our previous experiments have shown that apamin from various sources (a
gift from Dr. Habermann, University of Giessen and from Serva, Heidelberg,
FRG) has no significant effect on the Ca activated K current in <u>Helix</u> and
<u>Aplysia</u> neurons (Hermann and Hartung, 1983). Apamin obtained from a third
source (Sigma, St. Louis) had a minor blocking effect of about 10% at a
relatively high concentration of 2 x 10^{-5} M. Since the Sigma toxin is
only 50% pure, it is likely that some ingredient other than apamin caused
the effect.

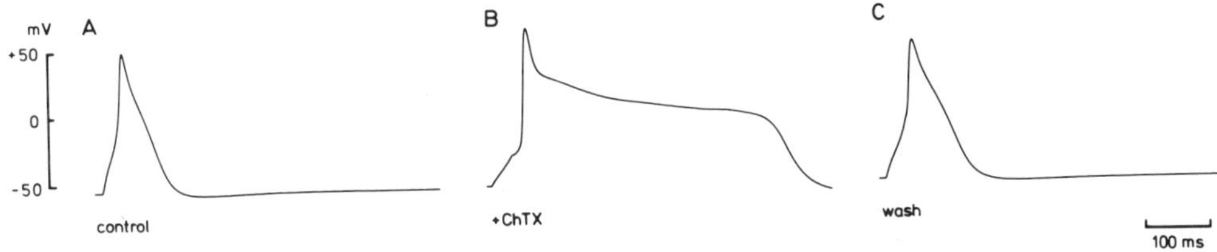

Fig. 8. (A) Effect of charybdotoxin (100 nM) on action potentials.
The artificial seawater contained 10 mM 4-aminopyridine.
Action potentials were elicited by brief (30 ms),
intracellular current injections.

<u>Toxin action on single Ca activated K channels</u>

In order to test whether ChTX can be used to discriminate between
different types of Ca activated K channels we have compared the effect of
the toxin on small conductance channels in <u>Aplysia</u> and large conductance
channels in a mammalian cell line. Fig. 7A shows an outside-out patch
recording of the small conductance Ca activated K channel from <u>Aplysia</u>
before and after ChTX (100 nM) was added to the bath solution. ChTX
completely blocked the Ca dependent channel but spared a Ca independent
channel of even smaller current amplitude. Fig. 7B shows a similar
recording of a large conductance K channel activated by Ca in GH3 cells, a
rat pituitary tumor cell line. After application of ChTX (20 nM) the open
probability of the channel was reduced almost to zero. It is obvious from
these experiments that ChTX blocks Ca activated K channels of both small
and large conductance. Like the ChTX sensitive channel in <u>Aplysia</u>, the

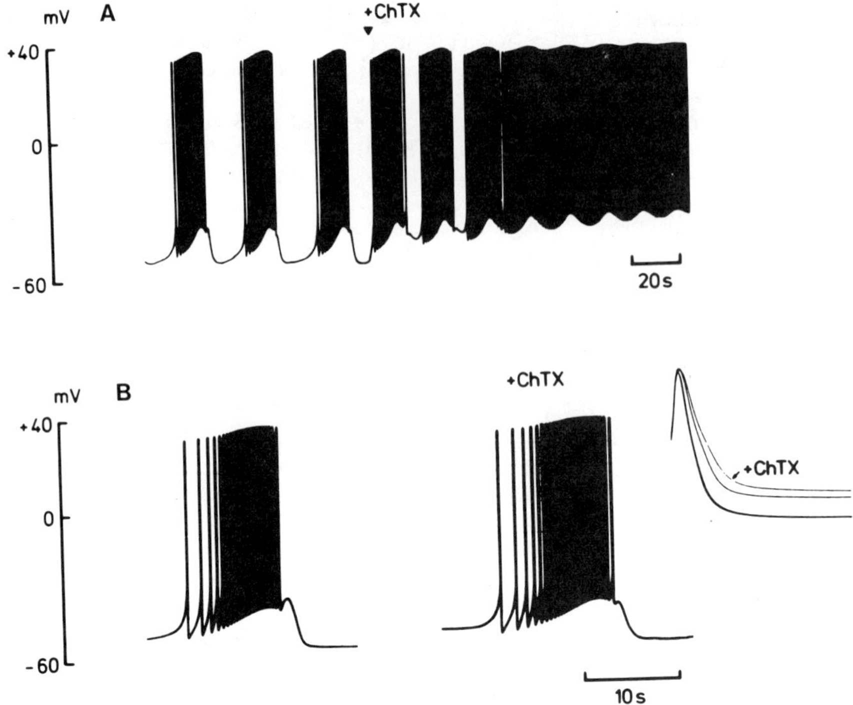

Fig. 9. Effect of ChTX (100 nM) on bursting pacemaker activity. (A)
Pacemaker activity recorded from cell R-15 before and after
application of ChTX (100 nM) (arrowhead). (B) Single bursts
displayed on an expanded time scale before and after toxin
application. To superimpose the first action potential of
each burst, the cell was hyperpolarized by current injection
in ChTX containing solution. The insert shows superimposed
action potentials before and during application of ChTX.
Cell R-15.

ChTX sensitive, Ca activated K channels in GH_3 cells were insensitive to
apamin (1 µM) or 4-AP (1 mM), but were blocked readily by TEA (1 mM)
(Hermann and Armstrong, unpublished results).

Toxin effects on action potentials and bursting pacemaker activity

In molluscan neurons ChTX prolonged the duration of action potentials
from spontaneously bursting cells but had no effect on action potentials
of beating cells. The effect of the toxin on the action potential was
observed most clearly in the presence of 4-AP (Fig. 8), which suppresses
other voltage-dependent K currents, but not the Ca activated K channels
(Hermann and Gorman, 1981a). The experiments indicate that under those
conditions a Ca activated K conductance contributes to the repolarization
of action potentials. In bursting pacemaker cells, ChTX caused a
depolarization, accompanied by an increase in the frequency of bursts,
which eventually led to tonic discharge of action potentials (Fig. 9A).
Repolarizing the cell by current injection reestablished bursting activity
but with some important changes in burst parameters: the number of action
potentials was increased by 20-25%; the final action potential of each
burst was prolonged; and the post-burst hyperpolarization was

diminished by 2-6 mV (Fig. 9B). It is concluded that ChTX blocks a Ca
activated K conductance that is present at negative membrane potentials
and therefore, contributes to the pace of action potential discharge. The
results further suggest that a Ca activated K outward current partially
contributes to the termination and after-hyperpolarization of bursts. The
finding that charybdotoxin alters but does not block bursting is in
accordance with both major hypotheses on the mechanism of pacemaker
activity, namely a Ca dependent inactivation of a persistent inward Ca
current (Adams and Levitan, 1985; Kramer and Zucker, 1985) which is not
affected by the toxin, and a Ca activated K current (cf. Gorman et al.,
1981, 1982; Smith and Thompson, 1987), which is blocked by the toxin.

In summary, the results from single channel recordings substantiate
the notion that in various types of cells at least two distinct classes of
Ca activated K channels exist: 1) channels with large unitary conductance
of 100-300 pS (big K or BK channels) found in vertebrate cells (Marty,
1983; Latorre and Miller, 1983) and 2) channels with small unitary
conductance of 30-60 pS (small K or SK channels). Although these channels
differ by almost an order of magnitude in their conductance, they appear
identical in their pharmacology. According to a hypothesis proposed by
Latorre and Miller (1983) these properties can be attributed to an
extracellular domain of the channel polypeptide which is quite similar in
both types of channels and which reacts with particular blocking
molecules. The difference in conductance may be explained by differences
in the selectivity filter or the length of the channel tunnel, with big
channels having a short tunnel and small channels having a longer tunnel.
High and low conductance Ca activated channels which are also sensitive to
charybdotoxin have been isolated recently from rat brain (see Levitan,
this volume). Thus, it appears that low conductance Ca activated K
channels are more common in cells from different preparations than
previously thought.

With the introduction of charybdotoxin a new and powerful
pharmacological tool is available for identifying Ca activated K
channels. In view of the high affinity of the toxin for the channel, it
may be feasible to use the toxin in a labelled form to localize Ca
activated K channels in cellular membranes and to estimate their density
and distribution. However, although ChTX blocks Ca activated K channels
at concentrations more than three orders of magnitude lower than TEA
(Hermann and Gorman, 1981b), it does not appear to discriminate between
different categories of Ca activated K channels since it antagonizes both
apamin-sensitive and apamin-insensitive channels (Abia et al., 1986).
Nevertheless, the effects of the toxin on cell excitability indicate that
a Ca activated K conductance contributes to the membrane resting
potential, the repolarization of action potentials, and the termination of
bursting pacemaker potentials.

ACKNOWLEDGEMENTS

Most of the experiments reported here were conducted at UCLA, in the
laboratory of the late Roger Eckert. A.H. was a Heisenberg Fellow on
leave of absence from the University of Konstanz, Faculty of Biology, FRG.
We thank C. Miller, Brandeis University, for his gift of charybdotoxin.

REFERENCES

Abia, A., Lobaton, C. D., Moreno, A., and Garcia-Sancho, J., 1986, Leiurus
     quinquestriatus venom inhibits different kinds of Ca2+-dependent
     K+channels. Biochem. et Biophys. Acta, 856:403.

Adams, W. B., and Levitan, I. B., 1985, Voltage and ion dependence of the
    slow currents which mediate bursting in _Aplysia_ neurone R15, J.
    Physiol., 360:69.
Barrett, J. N., Magleby, K. L., and Pallotta, B. S., 1982, Properties of
    single calcium-activated potassium channels in cultured rat muscle,
    J. Physiol., 331:211.
Colquoun, D., Neher, E., Reuter, H., and Stevens, C. F., 1981, Inward
    current channels activated by intracellular Ca in cultured cardiac
    cells, Nature, 294:752.
Cook, N. S., and Haylett, D. G., 1985, Effects of apamin, quinine and
    neuromuscular blockers on calcium-activated potassium channels in
    guinea-pig hepatocytes, J. Physiol., 358:373.
Ewald, D. A., Williams, A., and Levitan I. B., 1985, Modulation of single
    Ca2+-dependent K+-channel activity by protein phosphorylation,
    Nature, 315:503.
Gorman, A. L. F., and Hermann, A., 1979, Internal effects of divalent
    cations on potassium permeability in molluscan neurones, J. Physiol.,
    296:393.
Gorman, A. L. F., Hermann, A., and Thomas, M. V., 1981, Intracellular
    calcium and the control of neuronal pacemaker activity, Fed. Proc.,
    40:2233.
Gorman, A. L. F., and Hermann, A., 1982, Quantitative differences in the
    currents of bursting and beating molluscan pacemaker neurones, J.
    Physiol., 333:681.
Gorman, A. L. F., Hermann, A., and Thomas, M. V., 1982, The ionic
    requirements for membrane oscillations and their dependence upon the
    free intracellular calcium concentration in a molluscan pacemaker
    neurone, J. Physiol., 327:185.
Hamill, O. P., Marty, E., Neher, E., Sakmann, B., and Sigworth, F.
    J., 1981, Improved patch clamp techniques for high resolution current
    recording from cell and cell-free membrane patches, Pflueg. Arch.,
    391:85.
Hartung, K., and Hermann, A., 1987, Fluctuations of Ca2+-activated
    K+current in _Aplysia_ neurones, Biochim. et Biophys. Acta, 897:201.
Hermann, A., and Erxleben, C., 1987, Charybdotoxin selectively blocks Ca
    activated K channels in _Aplysia_ neurones, J. Gen. Physiol. (in press).
Hermann, A., and Hartung, K., 1982, Properties of a Ca2+ activated K+
    conductance in Helix neurones investigated by intracellular Ca2+
    ionophoresis, Pflueg. Arch., 393:248.
Hermann, A., and Hartung, K., 1983, Ca2+ activated K+ conductance in
    molluscan neurones, Cell Calcium, 4:387.
Hermann, A., and Hartung, K., 1986, Pharmacological aspects of Ca
    activated K conductance in molluscan neurones, in: "Calcium
    electrogenesis and neuronal functioning", U. Heinemann, M. Klee, E.
    Neher and W. Singer, ed., Exp. Brain Res. 14:124.
Hermann, A., and Gorman, A. L. F., 1981a, Effects of 4-aminopyridine on
    potassium currents in a molluscan neuron, J. Gen. Physiol., 78:63.
Hermann, A., and Gorman, A. L. F., 1981b, Effects of tetraethylammonium on
    potassium currents in a molluscan neuron, J. Gen. Physiol., 78:87.
Hodgkin, A. L., and Katz, B., 1949, The effects of sodium ions on the
    electrical activity of the giant axon of the squid, J. Physiol.,
    108:37.
Johnson, J. W., and Thompson, S. H., 1983, Calcium dependence of resting
    neuronal conductance, Soc. Neurosci., Abstr., 9:1187.
Kramer, R. H., and Zucker, R. S., 1985, Calcium-induced inactivation of
    calcium current causes the inter-burst hyperpolarization of _Aplysia_
    bursting neurones, J. Physiol., 362:131.
Latorre, R., and Miller, C., 1983, Conduction and selectivity in potassium
    channels, J. Memb. Biol., 71:11.
Lux, H. D., Neher, E., and Marty, A., 1981, Single channel activity
    associated with the calcium dependent outward current in Helix
    pomatia, Pflueg. Arch., 389:293.

Marty, A., 1983, Ca2+-dependent K+ channels with large unitary
    conductance, Trends in Neurosci., 6:262.
Meech, R. W., 1976, Intracellular calcium and the control of membrane
    permeability. In Calcium in Biological Systems. Symp. Soc. exp. Biol.
    no. 30. pp. 161-191. Cambridge University Press.
Meech, R. W., 1978, Calcium-dependent potassium activation in nervous
    tissues, Ann. Rev. Biophys. Bioeng., 8:1.
Miller, C., Moczydlowski, E., Latorre, R., and Phillips, M., 1985,
    Charybdotoxin, a protein inhibitor of single Ca2+-activated K+
    channels from mammalian skeletal muscle, Nature, 313:316.
Pallotta, B. S., 1985, N-bromoacetamide removes a calcium-dependent
    component of channel opening from calcium-activated potassium
    channels in rat skeletal muscle, J. Physiol., 86:601.
Siegelbaum, S. A., Camardo, J. S., and Kandel, E. R., 1982, Serotonin and
    cyclic AMP close single K+ channels in _Aplysia_ sensory neurones,
    Nature, 299:413.
Smith, S. J., and Thompson, S. H., 1987, Slow membrane currents in
    bursting pace-maker neurones of Tritonia. J. Physiol., 382:425.
    Korrektur am 26.3.87.
Tashjian, A.H., 1979, Clonal strains of hormone-producing pituitary
    cells. Methods Enzymol., 57:527-535.

Ca^{2+} DIFFUSION IN THE CYTOPLASM OF _APLYSIA_ NEURONS: ITS RELATIONSHIP TO LOCAL CONCENTRATION CHANGES

Douglas Tillotson and Enrico Nasi

Boston University School of Medicine
Department of Physiology
Boston, Mass.

The transient rise in intracellular free Ca^{2+}, which occurs in many nerve cells with excitation of the plasma membrane, has been shown to modulate the activity of a number of membrane conductance systems. Among these are the activation of a K$^+$ channel (Meech, 1978) and the inactivation of Ca^{2+} channels (Tillotson, 1979; Eckert and Tillotson, 1981). A precise understanding of any cellular mechanism mediated by Ca^{2+} requires a definition of the Ca^{2+} concentration in the vicinity of the relevant regulatory sites. In the case of plasma membrane conductances, the regulatory sites must be associated with the channels themselves; either directly or via local biochemical machinery. The time-varying Ca^{2+} concentration at any point in a nerve cell body is determined by a number of factors including: 1) the size of the Ca^{2+} load and the locus where it is applied 2) the cytoplasmic mechanisms responsible for regulating the free Ca concentration in the intracellular compartment 3) the rate of diffusion of Ca^{2+} in the cytoplasm. The validity of predictions of local Ca^{2+} changes associated with membrane excitation depends on having a reasonable estimate for each of these factors.

The intracellular Ca^{2+} transient produced by activation of Ca^{2+} channels at the plasma membrane is initially a steep concentration gradient that rapidly relaxes with time. A spatially and temporally resolved measurement of such a transient is not feasible with current methods of detection of intracellular Ca^{2+}. Considerations of speed alone preclude use of Ca^{2+} selective microelectrodes and radioisotope flux measurements. In addition, detection of such a steep gradient (spanning several orders of magnitude of concentration) requires a wide dynamic range. This eliminates the bioluminescent (Aequorin: Blinks et al. 1978) and currently available fluorescent indicators (Quin2, Fura2 and Indol: Grynkiewicz, Poenie, and Tsien, 1985). Only spectrophotometric measurement of Ca^{2+} using metallochromic indicators (eg. Arsenazo III and Antipyralazo III) offers sufficient speed and dynamic range to record the transient associated with membrane Ca^{2+} entry.

Unfortunately, the absorbance change signal recorded from a cell loaded with an optical indicator represents, in the best case, an integrated measure of the total free Ca^{2+} in the intracellular compartment. Such a signal provides no spatial information. One way to

obtain spatial information about Ca^{2+} using arsenazo III (AzIII)
involves restricting the region of the cell from which the optical
signal is recorded. A version of this method is shown in Figure 1. A
small collecting probe that records light absorbance from a restricted
region of the cell to the photodetector. This experimental arrangement
was used for measuring the diffusion of ions in the cell body of _Aplysia_
giant neruons. The selected nerve soma is held under voltage clamp and
is pressure injected with purified AzIII. The cell is then penetrated
with a multi-barrelled microelectrode containing $CaCl_2$ and $BaCl_2$
solutions (100mM) in separate barrels. This allowed ionophoretic

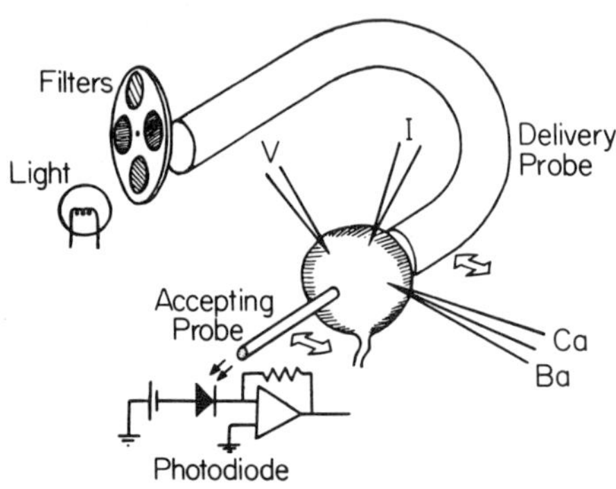

Figure 1. Schematic depiction of the experimental arrangement for
the _in vivo_ measurements. The cell soma is positioned between the
two fiber optics probes, which are mounted on a Huxley-type
micromanipulator, and can be displaced as a monolithic assembly in
any direction. Two standard microelectrodes are used to hold the
cell under voltage clamp (one of them serving also to
pressure-inject the metallochromic indicator Arsenazo III), and a
third, multibarrelled microelectrode is employed to apply minute
ionophoretic injections of either Ca^{2+} or Ba^{2+}. The probe assembly
is displaced while test injections are administered, until the
evoked AzIII signal reaches a maximum amplitude, presumably when
the tip of the injecting electrode is located near the center of
the measuring "beam". This point is taken as the 0-distance
reference point, and the probes are then moved in precise steps
away from it, repeating the injections and measuring the resulting
absorbance signal at each position. (From Nasi and Tillotson, 1985)

injections of each ion to be made at the same point within the nerve
cell. The optical microprobe assembly is positioned so that a small
test injection of Ca^{2+} produces a maximal absorbance change signal. This
position is considered to represent 0 distance between the source and
the optically sampled region. After a series of Ca^{2+} and Ba^{2+}
injections are presented and the resulting absorbance changes recorded,
the whole optical probe assembly is moved in measured steps, and the
test injections are repeated.

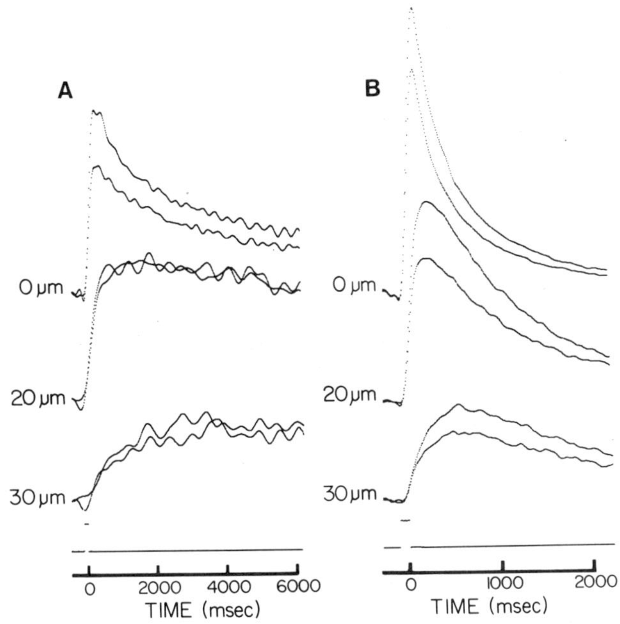

Figure 2. A. Differential absorbance signals (660-690 nm)
recorded in vivo at three distances from the ionophoretic
microelectrode tip (indicated near each pair of traces), in
response to injections of Ca^{2+} and Ba^{2+}. Each trace represents the
average of 16 records, and was low-pass filtered at a cutoff
frequency of 2 Hz with a digital algorithm with zero phase
characteristics in order to preclude signal distortion. The records
corresponding to 20 and 30 microns distance from the ionophoretic
electrode tip are scaled by a factor of 5 compared with the 0
distance record. The cell had been previously pressure-injected
with the dye to an estimated concentration of 1 mM.
B. Absorbance signals in response to Ca^{2+} and Ba^{2+} injections in
vitro. The procedure is essentially identical to the way the in
vivo measurements were performed, except that the probe assembly
was placed in a small volume bath (500 ul approx.) containing 0.2
mM AzIII, 100 mM KCl, 10 mM TRIS, pH =7.7. The cutoff frequency
for the low-pass digital filter was 6 Hz. The two lower pairs of
records are scaled by a factor of 2.5, with respect to the top
pair. (From Nasi and Tillotson, 1985)

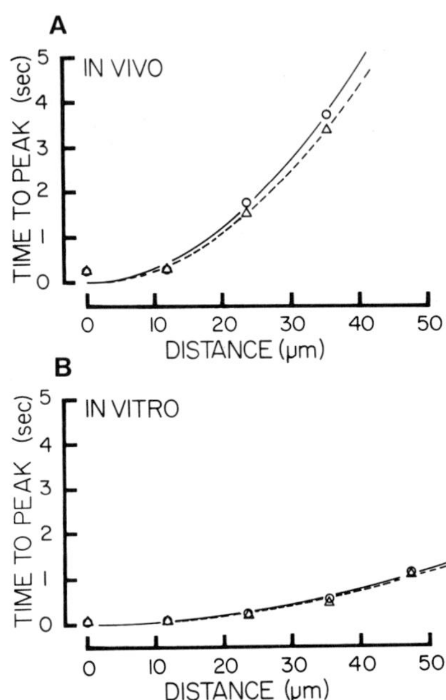

Figure 3. A. Time to peak of the Arsenazo III absorbance signal
    following a 100 ms, 50 nA injection of Ca (open circles) or Ba
    (open triangles) into a nerve soma. The two lines represent the
    best-fitting theoretical relations if time-to-peak as a function of
    distance from the source. The resulting value was 8.3×10^{-7}
    cm^2/sec for Ca (continuous line) and 9.2×10^{-7} cm^2/sec for Ba
    (dotted line).
    B. Time to peak for the AzIII signal _in vitro_. The procedure was
    similar to the _in vivo_ conditions, and yielded a value of the
    diffusion coefficient of 5.2×10^{-6} cm^2/sec for Ca (filled circles,
    continuous line) and 5.4×10^{-6} cm^2/sec for Ba (filled triangles,
    dotted line). (From Nasi and Tillotson, 1985)

The same apparatus was used to record absorbance changes by ion
injections _in vitro_. In these experiments both the probe assembly and
the injecting microelectrode were placed in a small volume of saline
solution containing AzIII. The injecting tip-to-probe separation
experiment described above could thus be repeated under conditions of
unimpeded diffusion. A comparison of the diffusion rate in the cell and
in free solution can then be used to estimate the impediments to
diffusion within the cytoplasm for different test ions that can be
detected with AzIII, with the aim of separating non-specific factors
(eg. viscosity) from factors such as binding to cytoplasmic sites.

Data from an _in vivo_ Ca^{2+}/Ba^{2+} comparison experiment are shown in
Figure 2A. Shown are absorbance change signals produced by a 100 nA,
100 msec duration Ca^{2+} and Ba^{2+} injections into a nerve cell body. Each
trace represents a digitally filtered, ensemble average of 16 raw
records. A computerized protocol determined the time at which the
maximum absorbance change occurred. It can be seen that at the distance
between the ion injecting tip and the measuring beam increased the
amplitude of the signals decreased and the time to peak of the signals
increased, as expected from diffusion theory (Crank, 1975). An _in vitro_
experiment using similar injections of Ca^{2+} and Ba^{2+} (50 nA for 100
msec) generated the averaged and filtered data shown in Figure 2B.

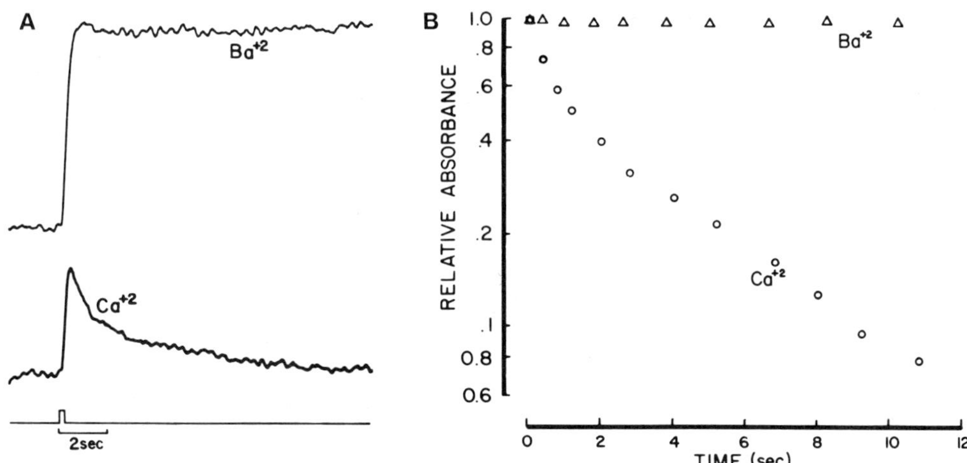

Figure 4. Comparison of AzIII signals associated with Ca^{2+} and
   Ba^{2+} entry through voltage-activated Ca^{2+} channels. (A) Dye
   absorbance records from an R15 cell with a 200msec voltage clamp
   pulse to +20mV in 10mM extracellular Ca^{2+} and Ba^{2+}. (B)
   Semilogarithmic plot of the falling phase of the normalized
   absorbance change records versus time. (Adapted from Fig. 2 of
   Tillotson and Gorman, 1983).

The peak of the absorbance signal measured in these experiments
occurs at the moment in time when the total free Ca^{2+} in the measuring
region reaches its maximum. Furthermore, the relationship between the
time of the peak concentration and the distance between the source and
the measuring beam is uniquely defined by the coefficient of diffusion.
The higher the diffusion rate, the shorter the time to peak at a given
distance. Figure 3A shows a plot of the time to peak of Ca^{2+}-AzIII and
Ba^{2+}-AzIII signals as a function of the optical probe-injecting tip
separation for _in vivo_ and _in vitro_ data. The smooth curves represent
the result of least-square fitting the theoretical relation between
distance from a point-source and the expected time-to-peak, in order to
determine the best estimate for the diffusion coefficient for each set

of data. The effective diffusion coefficients found in this experiment
were 8.3 x 10^{-7} cm^2/sec for Ca^{2+} and 9.2 x 10^{-7} for Ba^{2+} in the cell.
The _in vitro_ data is plotted in Figure 3B. The corresponding values
were 5.2 x 10^{-6} and 5.4 x 10^{-6} cm^2/sec, respectively. The measurements
in vitro were very consistent in different experiments, with a
variability of less than 5% (n=3), while a larger degree of variability
was observed in cell somas, covering a range of 7-12 x 10^{-7} cm^2/sec
(n=4).

  We and others have reported that Ba^{2+} is less well "buffered" than
Ca^{2+} (Connor et al., 1981; Tillotson and Gorman, 1983). This was based
in part on the observation that whole-cell Ba-AzIII signals decay much
more slowly than Ca-AzIII signals associated with ionophoretic
injections or with influx through the plasma membrane. This is
illustrated in Figure 4. If such difference in "buffering" reflected a
difference in reversible binding of Ca^{2+} and Ba^{2+} to intracellular
sites, we would predict that the reduction in mobility in the cell as
compared to a free solution should be more pronounced than for Ba^{2+}. In
addition, one would expect the actual size of the Ba-AzIII signal to be
many times greater than an equivalent Ca-AzIII signal. The similarity
between the effective diffusion coefficients found for these ions both
in vitro and _in vivo_ suggests that Ca^{2+} and Ba^{2+} are buffered (in the
sense of rapidly equilibrating, reversible binding) to the same extent.
Furthermore, equal injections of Ba^{2+} and Ca^{2+} produce AzIII signals
that are less than 50% different. in amplitude (Tillotson and Gorman,
1983). This further suggests that the difference in "buffering" noted
above is likely to be due to slower processes of uptake and/or
extrusion. Those processes would appear to be more strongly selective
for Ca^{2+} over Ba^{2+} than the reversible binding. This underscores the
importance of distinguishing between reversible binding and
uptake/extrusion mechanisms when considering neuronal regulation of
Ca^{2+}.

  At least two factors contribute to the reduction in the rate of
diffusion of Ca^{2+} in the nerve cell body compared to "free solution".
One is the physical impediment or tortuosity imposed by immobile
structures (ie. cytoskeleton, organelles) and the viscosity of the
cytoplasm. The second and perhaps more important factor is reversible
binding of Ca^{2+} to relatively immobile intracellular sites. Our method
for measuring Ca^{2+} diffusion gives no direct information about the
relative contribution of these two factors, although a comparison of the
diffusion rate _in vivo_ vs _in vitro_ for Ca and other ions that do not
bind to cytoplasmic constituents could clarify this point.

  If the assumption is made that the 6-10 fold reduction that has been
observed in cytoplasm can be attributed entirely to reversible binding
to a high capacity buffer system of immobile sites (Blaustein and
Hodgkin, 1969; Smith and Zucker, 1980), this would indicate that
cytoplasmic free calcium is 10-15% of the total intracellular calcium.
Such estimate appears to be substantially higher than the ratio of free
Ca to total Ca in squid axons (Keynes and Lewis, 1956; Di Polo et al.,
1976; Brinley et al., 1977). One factor which may contribute to this
discrepancy is that the assumption of the immobility of the binding
sites is likely to be an oversimplification since the mobility of Ca in
squid axoplasm as compared to free solution is reduced much more
dramatically than the self-diffusion coefficient (Hodgkin and Keynes,
1957).

  Both in the _in vivo_ and the _in vitro_ experiments, a finite time lag
before the occurrence of the peak is observed even at the "zero
distance" point. This appears to reflect momentary local saturation of
the AzIII in the vicinity of the injecting tip shortly after the pulse,

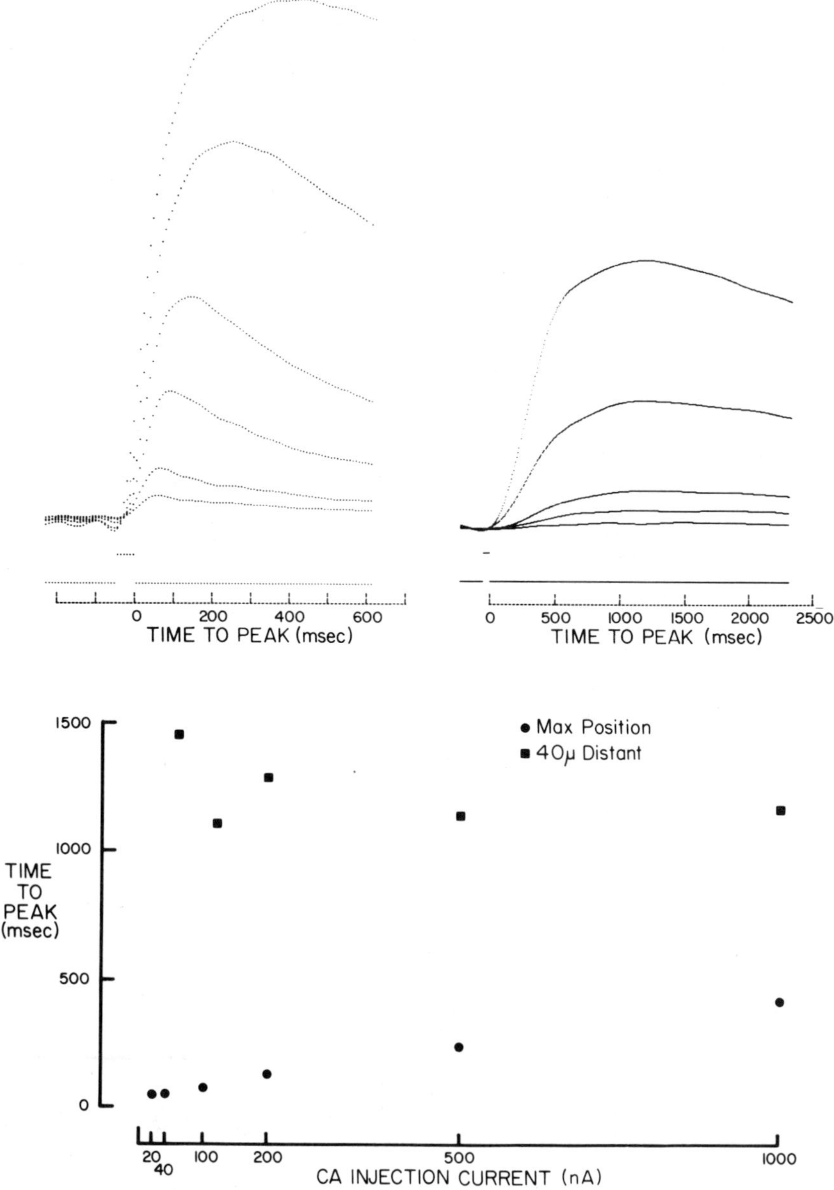

Figure 5. Effect of Ca^{2+} injection intensity on time-to-peak at two different optical microprobe - injection tip separations. Traces on left are absorbance change signals recorded at various injection intensities with the Ca^{2+} injector tip centered between the optical probes. The traces on the right were recorded at a optical probe -injection tip separation of 40 microns. The plot below is the time-to-peak vs. injection intensity for these data.

in such a way that the recorded absorbance signal continues to rise as the injected ions diffuse away from the saturated region. Such interpretation is consistent with the results shown in Figure 5 that at the zero-distance point the time to peak increases as injection intensity is increased. By contrast, measurements of the time-to-peak absorbance at a distance of 40 microns from the injection tip remained constant. This indicates that AzIII saturation does not distort the diffusion coefficients found.

Ca^{2+} indicator saturation used to determin local Ca^{2+} concentration:

It seems clear that shortly after an substantial Ca^{2+} load the concentration can rise locally to very high levels, saturating the Ca indicator in that region. This implies that some fraction of the Ca load initially is not detected. As time elapses Ca diffuses away from the region where the AzIII is saturated and therefore encounters free indicator molecules, forming additional complexes. Since the AzIII signal reflects the total quantity of the Ca-dye complex in the domain sampled by the spectrophotometer, this will result in a further increase of absorbance, in spite of the fact that no extra Ca has been introduced into the system. This notion is depicted graphically in Figure 6. The optical signal will reach its maximum value when diffusional redistribution will make the Ca concentration profile relax below saturating level, thereby allowing maximum Ca-dye interaction. At that moment the highest concentration (which according to diffusion-reaction theory is found at the locus of the source) will have dropped just below the range where saturation occurs. Such value can then be estimated empirically by referring to a suitable titration curve obtained in vitro. Such curves indicate that at AzIII concentrations normally used intracellularly, the level of saturating Ca^{2+} is in the hundreds of micromolar (Nasi and Tillotson, 1987).

It is a fact that if one carfully measures the absorbance signal of in Aplysia neuron filled with AzIII at high time resolution and a depolarizing voltage clamp step is administered, the trace is seen to rise <u>after</u> the offset of the clamp step. While the time lag to the peak of the signal with a clamp step is much shorter than that found with an ionophoretic injection, it was interesting to speculate that this delay could reflect local saturation of AzIII. If this were the case, it would be possible to determine the near-membrane Ca^{2+} level at one point in time (the time of the peak as discussed above).

The simplest way to test if AzIII saturation occurs following voltage clamp activation of Ca^{2+} channels is to vary the amount of Ca^{2+} which enters. The time of the peak AzIII signal would be predicted to increase with the Ca^{2+} load. The extent of calcium influx through membrane channels can be manipulated by varying the amplitude of a depolarizing step (Connor, 1979; Gorman and Thomas 1980). The amplitude of the AzIII signal provides adequate qualitative evidence that the manipulation is effective for varying the amount of Ca entering the cell.

The effect of varying the amplitude of a depolarizing voltage clamp step on the absorbance signal from a cell injected with AzIII is shown in Figure 7. Each trace represents the average of 16 raw records. An expanded view of the period of time near the end of the step reveals that the peak of the signal occurs many milliseconds <u>after</u> the termination of the step, and that such time lag becomes longer at voltages that induce a greater influx of Ca.

140

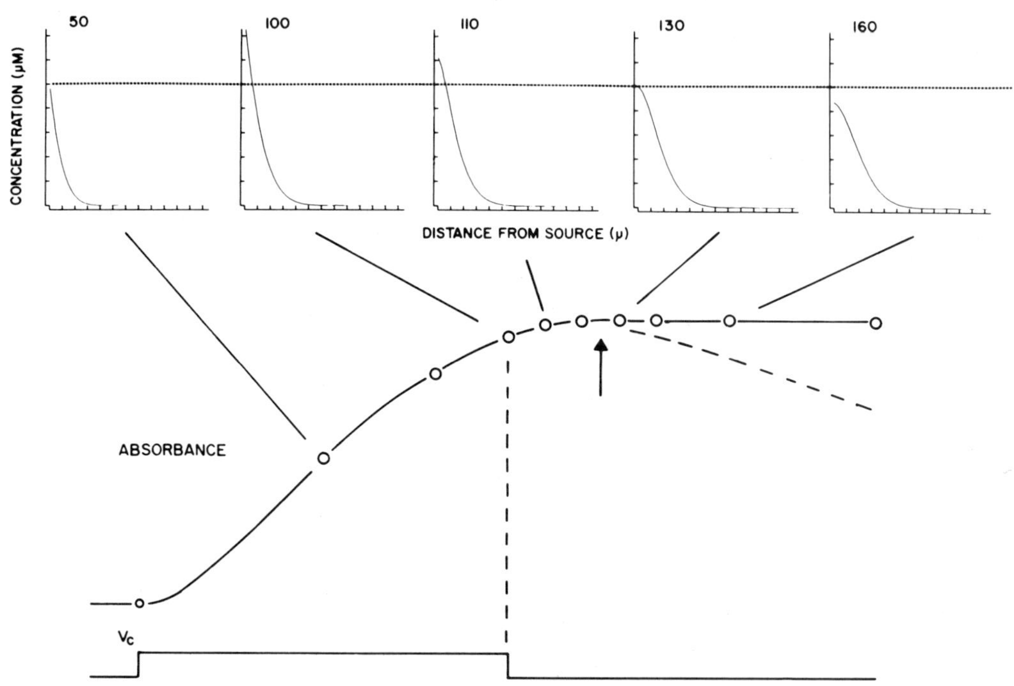

Figure 6. Simplified illustration of the notion that locally
saturating Ca concentration can result in an optical signal that
continues to rise beyond the time when Ca is introduced. An
idealized titration curve of the AzIII-Ca reaction is assumed,
changing abruptly from linear complexation to total saturation as a
function of $[Ca^{2+}]$. The dashed line represents the minimum
saturationg Ca concentration. A membrane Ca^{2+} load was simulated
(assuming a 100msec continuous entry at 7pA/$micron^2$, diffusion
coefficient = 8.3 x 10^{-7} cm^2/sec and "cytoplasmic" buffering
capacity of .85). Spatial concentration profiles are shown at the
top at the noted times with respect to the beginning of the Ca^{2+}
load (the concentration gradations are 50 micromolar, the distance
gradations are 2 microns). Note that at early times a substantial
fraction of the Ca load is not detected by the dye (the recorded
absorbance signal at any given time is proportional to the integral
of the spatial profile <u>below</u> the dashed line). As time elapses,
diffusional redistribution results in profiles that fall
progressively into the measurable range. After the first profile
that lies entirely below the saturating level, no further change
will occur in the optical signal (assuming for simplicity that
total Ca is conserved). The approximated AzIII signal is plotted
below

A possible confounding factor that could be responsible for the
delayed time-to-peak of the AzIII signal with a voltage clamp step is
the presence of Ca tail currents which could easily be masked by K tail
currents. To control for this possibility we used the voltage clamp to
"instantaneously" eliminate the driving force on Ca^{2+} rather than
relying on the deactivation kinetics. To accomplish this voltage clamp
steps were administered to different levels of depolarization, but then,
instead of repolarizing the membrane, the voltage was stepped to +160
mV, a potential presumed to be close to or more positive than the
equilibrium potential for Ca in these cells (Gorman & Thomas, 1980). In
these circumstances, even if the channels remain open, the influx of Ca

should stop abruptly (within the settling time of the clamp, i.e. in
the hundreds of microsecond range) as the driving force on Ca is
eliminated. Figure 8 shows that the delay in the time-to-peak was once
again observed, and so was its orderly relation to the magnitude of the
Ca load applied to the intracellular compartment.

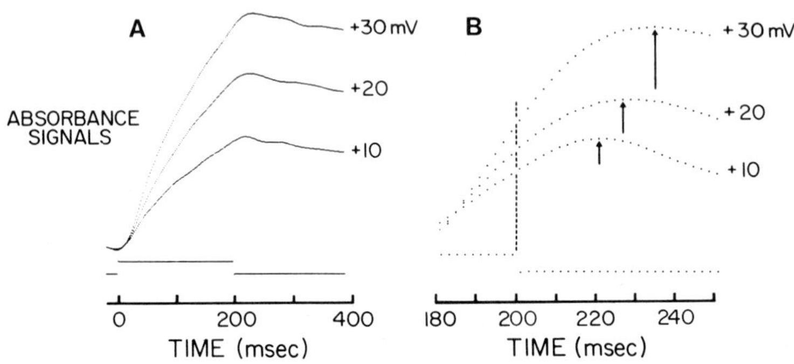

Figure 7. A. Effect of changing the Ca influx by varying the
      amplitude of a 200 msec voltage-clamp step to the potential
      indicated on the right hand side, on the AzIII absorbance signal.
      Each average record was digitally low-pass filtered at 20 Hz. B.
      Expanded view of the records shown in part A. The vertical dashed
      line marks the termination of the voltage-clamp step, while the
      arrows indicate the peak of the absorbance signal. It is evident
      that depolarization to more positive voltages causes a progressive
      displacement of the time at which the optical signal peaks. The
      cell was L-6. Holding potential was -50mV; the estimated AzIII
      concentration was 0.05 mM. (From Nasi and Tillotson, 1987)

      A second possible explanation for the time-to-peak delay is that,
although Ca influx from the extracellular compartment effectively stops
at the end of a depolarizing voltage clamp step, there could be internal
release from intracellular stores accounting for the further rise in the
absorbance signal. Evidence of Ca-induced Ca release has been presented
for cardiac muscle cells (eg. Fabiato & Fabiato, 1979), but has never
been demonstrated in nerve tissue. An indirect way to control for such
a potential problem was the replication of the saturation phenomenon
using Ba instead of Ca (unpublished observation). Ba ions permeate
through voltage-dependent Ca channels at least as readily as Ca ions do,
but have been shown ineffective in supporting a host of cellular
regulatory processes normally controlled by Ca, such as Ca-activated
potassium current (Gorman and Hermann, 1979), Ca-mediated Ca channel
inactivation (Tillotson, 1979), synaptic release (Augustine & Eckert,
1984), as well as some mechanisms of Ca-buffering (Tillotson & Gorman,
1983). It is very unlikely therefore that Ca^{2+} release could be used to
explain our results.

The time scale over which the time delays are seen approaches the speed of the Ca-AzIII reaction kinetics (published equilibration time constants range from 3 ms Scarpa et al., 1978; Palade and Vergara, 1981 to 10ms or greater Dorogi, 1984). Clearly, some portion of the delay in the time to peak is attributable to the Ca-AzIII kinetics. However, this should not vary with the size of the Ca^{2+} load. As a consequence, the significant difference in the time-to-peak absorbance seen at different membrane voltage levels is indicative of local AzIII saturation. However, the precise moment in time that this saturation is relieved at a given Ca load is uncertain.

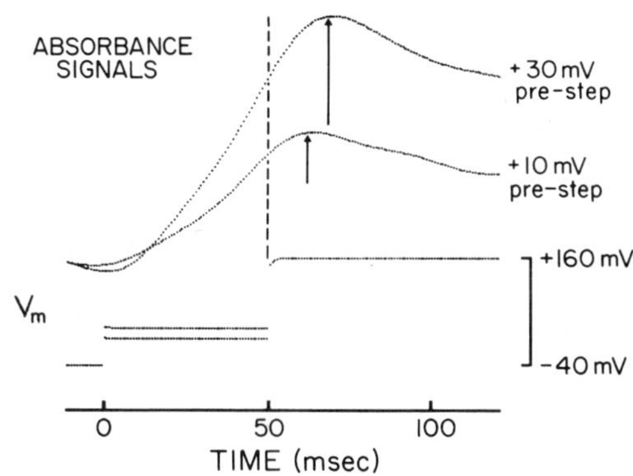

Figure 8. Control for potential contamination of the delayed peak absorbance change by calcium tail currents. The cell was held under voltage clamp at -40 mV, then a command step was administered to either +10 mV or +30 mV, following which the voltage was stepped to +160 mV. Since such potential is near the calcium equilibrium potential, Ca influx is abruptly stopped even though the channels may remain open. It can be seen that the peak of the AzIII signal still lags behind the termination of Ca entry, and that such delay is more pronounced for the step to +30 mV (top trace) which caused the greater Ca influx, as compared to the depolarization to +10 mV (second trace from the top). The voltages are shown in the bottom records. This R-15 cell was bathed in ASW containing 50mM TEA-Bromide substituted for NaCl to reduce the total outward current associated with the large depolarizations and improve the speed of voltage the transition. The intra-neuronal concentration of AzIII was estimated to be 0.22 mM. (From Nasi and Tillotson, 1987)

The idea that optical indicators may be susceptible to saturation by intracellular Ca during transients had been entertained as a potential nuisance for spectrophotometric measurements using the new generation of fluorescent compounds, that have a rather high affinity for Ca (see Thomas, 1982). However, it appeared to be an unlikely possibility in the case of Arsenazo III (Smith & Zucker, 1980), since it has an apparent dissociation constant in the micromolar range. Aside from

providing reasonable evidence for its existence under physiological
conditions, the novelty of the present report consists in exploiting the
occurrence of transient local saturation to derive information about the
extent of the accumulation of free Ca that can occur with membrane
excitation. By referring to the calibration curves obtained <u>in vitro</u>,
it can be concluded that the layer beneath the plasma membrane must
transiently experience free Ca concentrations in the 10^{-4} M range. While
such figure may appear shocking to some, it must be kept in mind that
previous estimates, suggesting a much more modest rise to 3-7 μM near
the membrane (Gorman & Thomas, 1980; Smith & Zucker, 1980) were
calculated on the basis of several assumptions concerning the behavior
of the Ca current, the Ca-AzIII interaction, and the buffering and
diffusion of Ca in the cytoplasm.

Our goal has been to empirically define the physical parameters
which determine the intracellular Ca^{2+} conentration following membrane
excitation. To this end the diffusion coefficient of Ca^{2+} in nerve
cytoplasm was measured. In addition, the near-membrane free calcium
level at one moment in time was estimated. It must be emphasized that
these inportant pieces of information are not sufficient for a complete
understanding of the spatio-temporal nature of transient calcium changes
in nerve cells following excitation. Since the diffusion equation is
linear, knowledge of one concentration value at a specified point in
space and time as well as of the time course of the Ca influx (on a
relative scale) and the effective diffusion coefficient makes it
possible to calculate the whole time-varying concentration profiles.
This simplified treatment holds true for brief times after the beginning
of stimulation, whereas at later times the slower uptake/extrusion
mechanisms play a significant role (Gorman, et al. 1984), and their
contribution has to be taken into account in the form of a
diffusion+irreversible reaction modelling scheme. When each relevant
parameter is fully characterized it should be feasible to develop the
desired comprehensive model of the fate of accumulated calcium.

REFERENCES

Augustine, G.J. & Eckert, R. 1984. Divalent cations differentially
support transmitter release at the squid giant synapse. J. Physiol.
(Lond.). 346: 257-271.

Blaustein, M.P., and Hodgkin, A.L. 1969. The effect of cyanide on the
efflux of calcium from squid axons. J. Physiol. (Lond.) 200:497-527.

Blinks, J.R., Mattingly, P.H., Jewell, B.R., van Leeuwen, M., Harrer,
G.C., and Allen, D.G. 1978. Practical aspects of the use of aequorin as
a calcium indicator: assay, preparation, microinjection, and
interpretation of signals. Meth. Enz, 58:292-328.

Brinley, F.J., Tiffert, T., & Mullins, L.J. 1977. Intracellular calcium
buffering capacity in isolated squid axon. J. Gen Physiol. 70:355-385.

Connor, J.A. 1979. Calcium current in molluscan neurones: measurement
under conditions which maximise its visibility. J. Physiol. (Lond.).
286: 41-60.

Connor, J.A., Ahmed, Z. & Ebert, G. 1981. Diffusion of Ca^{2+}, Ba^{2+}, H^+,
and Arsenazo III in neural cytoplasm. Soc. Neurosc. Abstr. 7: 15.

Crank, J. 1975. The Mathematics of Diffusion (2nd ed.). Oxford: Clarendon Press.

Di Polo, R., Requena, J., Brinley, F.J., Mullins, L.J., Scarpa, A., and Tiffert, T. 1976. Ionized calcium concentration in Squid Axons. J. Gen. Physiol. 67:433-467. Dorogi P.L. 1984. Kinetics and mechanism of Ca^{2+} binding to Arsenazo III and Antipyrylazo III. Biochem. Biophys. Acta. 799: 9-19.

Eckert, R. and Tillotson, D. 1981. Calcium-mediated inactivation of calcium conductance in caesium-loaded giant neurones of _Aplysia californica_. J. Physiol. (Lond.) 314:265-280.

Fabiato, A. & Fabiato, F. 1979. Use of chlortetracycline fluorescence to demonstrate Ca^{2+} - induced release of Ca^{2+} from the sarcoplasmic reticulum of skinned cardiac cells. Nature. 281: 146-148.

Gorman, A.L.F. & Hermann, 1979. Inaracellular effects of divalent cations on potassium permeability in molluscan neurones. J. Physiol. (Lond.) 296:393-410.

Gorman, A.L.F., Levy, S., Nasi, E. & Tillotson, D. 1984. Intracellular calcium measured with calcium-sensitive micro-electrodes and Arsenazo III in voltage-clamped Aplysia neurones. J. Physiol. (Lond.). 353: 127-142.

Gorman, A.L.F. & Thomas, M.V. 1978. Changes in the intracellular concentration of free calcium ions in a pace-maker neurone, measured with the metallochromic indicator dye arsenazo III. J. Physiol. (Lond.). 275: 357-376.

Gorman, A.L.F. & Thomas, M.V. 1980. Intracellular calcium accumulation during depolarization in a molluscan neurone. J. Physiol. (Lond.). 308: 258-285.

Grynkiewicz, G., Poenie, M., and Tsien, R.Y. (1985). A new generation of Ca^{2+} indicators with greatly improved fluorescence properties. J.Biol.Chem 260: 3440-3450.

Hodgkin, A.L., & Keynes, H.D. 1957. Movements of labelled calcium in squid giant axons. J Physiol. (Lond.) 138:253-281.

Keynes, H.D. & Lewis, P.R. 1956. The intracellular calcium content of some invertebrate nerves. J. Physiol. (Lond.) 134:399-407. Meech, R. W. 1978. Calcium-dependent potassium activation in nervous tissue. Rev. Biophys. Bioeng. 7:1-18.

Moczydlowski, E. & Latorre, R. 1983. Gating kinetics of Ca^{2+} -activated K^+ channels from rat muscle incorporated into planar lipid bilayers: evidence for two voltage-dependent Ca^{2+} binding reactions. J. Gen. Physiol. 82: 511-542.

Mullins, L.J. & Requena, J. 1979. Ca measurement in the periphery of an axon. J. Gen. Physiol. 74: 393-413.

Nachshen, D.A. 1984. Intracellular free Ca^{2+} activity in brain nerve terminals measured with Quin 2. Biophys. J. 45: 265a.

Nasi, E. & Tillotson, D. 1985. The rate of diffusion of Ca^{2+} amd Ba^{2+} in a nerve cell body. Biophys. J. 47:735-738.

Nasi, E. & Tillotson, D. 1987. Estimation of submembrane free calcium concentration following nerve stimulation. (submitted for publication),

Palade, P. & Vergara, J. 1981. Detection of Ca^{2+} with optical methods. In A.D. Grinnel & M. A.B. Brazier (Eds.) The regulation of muscle contraction: excitation-contraction coupling. New York: Academic Press.

Scarpa, A., Brinley, F.J., Jr & Dubyak, G. 1978. Antipyrylazo III, a "middle range" Ca^{2+} metallochromic indicator. Biochemistry. 17: 1378-1386.

Smith, S.J. & Zucker, R.S. 1980. Aequorin response facilitation and intracellular calcium accumulation in molluscan neurones. J. Physiol. (Lond.). 300: 167-196.

Thomas, M.V. 1982. Techniques in calcium research. London: Academic Press.

Tillotson, D. 1979. Inactivation of Ca conductance dependent on entry of Ca ions in molluscan neurons. Proc. Natl. Acad. Sci. USA. 76: 1497-1500.

Tillotson, D.L. & Gorman, A.L.F. 1983. Localization of neuronal Ca^{2+} buffering near plasma membrane studied with different divalent cations. Cell. Molec. Neurobiol. 3: 297-310.

FURA-2 IMAGING OF LOCALIZED CALCIUM ACCUMULATION WITHIN SQUID 'GIANT'

PRESYNAPTIC TERMINAL

Stephen J Smith, Luis R. Osses and George J. Augustine

Section of Molecular Neurobiology
Howard Hughes Medical Institute
Yale University Medical School
New Haven, CT 06510

Section of Neurobiology
Department of Biological Sciences
University of Southern California
Los Angeles, CA 90089-0371

INTRODUCTION

Calcium ions play a key role in synaptic transmission: a transmembrane flux of calcium, through voltage-gated calcium channels, is the
trigger that initiates secretion of neurotransmitters from presynaptic
terminals (Katz, 1969; Llinas, 1982; Augustine, et al., 1987a). The
central role of Ca channel-mediated Ca influx in transmission, coupled with
the well-documented compartmentalization of intracellular Ca signals (Rose
and Loewenstein, 1975; Gilkey, et al., 1978; Williams, et al., 1985; Tsien
and Poenie, 1986) makes it important to know where Ca channels are located
in the membrane of presynaptic terminals and to understand the propagation
of Ca signals within presynaptic cytoplasm.

Previous studies have established that Ca channels are not uniformly
distributed in the plasma membrane of neurons. Katz and Miledi (1969)
found that the amplitude and rate of rise of Ca-dependent action potentials
was smaller in the axonal region of squid neurons than in the terminal
region of these cells, which suggests that Ca channels are concentrated in
the synaptic regions of the plasma membrane. Studies with intracellular Ca
indicators revealed that depolarization-induced changes in intracellular Ca
concentration also are largest in synaptic regions, reinforcing the notion
that Ca channels are selectively localized in terminal regions of the
presynaptic membrane (Miledi and Parker, 1981; Stockbridge and Ross, 1984;
Llinas, 1985).

Thus far, there has been little definite information regarding
localization of Ca channels within presynaptic terminals. Theoretical
studies (Zucker and Stockbridge, 1983; Stockbridge and Moore, 1984; Simon
and Llinas, 1985; Zucker and Fogelson, 1986) have emphasized that the brief
delay between Ca entry and transmitter release (Llinas, et al., 1981;
Augustine, et al., this volume) and the finite (but unknown; see Augustine
et al., 1987a) rate of diffusion of Ca ions within presynaptic cytoplasm
require that sites of Ca entry be near sites of transmitter release.

Because exocytosis and transmitter release appearently occur only at restricted regions within presynaptic terminals (Heuser, et al., 1974, 1979; Torri-Tarelli, et al., 1986), it is possible that that Ca channels also may be restricted to relatively small areas within the terminal membrane. In fact, freeze fracture microscopy has identified populations of large intramembranous particles, restricted to small regions within presynaptic membranes, and it has been noted that they occur in approximately the correct number to be presynaptic Ca channels (Heuser, et al., 1974; Pumplin, et al., 1981).

This chapter describes our preliminary attempts to locate sites of Ca influx within presynaptic terminals. We have used the fluorescent Ca-indicator dye, fura-2 (Grynkiewicz, et al., 1985), and digital fluorescence microscopy (Williams, et al., 1985) to obtain 2-dimensional views of stimulus-induced changes in intracellular Ca in the unique 'giant' presynaptic terminals of squid. We present evidence that, immediately following electrical stimulation, these changes in intracellular Ca are preferentially localized to the region of the presynaptic terminal cytoplasm immediately adjacent to the synaptic cleft. At later times after stimulation the intracellular Ca increase is spatially dispersed, presumably due to rapid propagation of the Ca signal throughout presynaptic cytoplasm. The initial localization of intracellular Ca signals suggest that Ca channels indeed are concentrated in the region of the presynaptic membrane involved in transmitter release.

METHODS

Stellate ganglia were dissected and prepared for micropipette impalement by the procedure described by Augustine and Eckert (1984). A micropipette containing 1 mM fura-2 in aqueous solution was placed in the presynaptic fiber at a point approximately 0.5 mm proximal to the beginning of the presynaptic terminal, and 100 nA of current was passed for 0.5 hour to iontophoretically inject dye. Stimulating electrodes on the pallial nerve and a second micropipette in the postsynaptic fiber allowed verification of normal synaptic function during this procedure. The micropipettes were then withdrawn and the ganglion was removed from the injection chamber. This procedure resulted in dye loading that appeared uniform over the entire presynaptic terminal and a few mm of the presynaptic axon. Final dye concentrations were estimated to be near 100 micromolar by comparing terminals with fura-2 standards in a glass cannula. The preparation was maintained at $11^{\circ}C$ (+/- $3^{\circ}C$).

For fluorescence microscopy, the ganglion was placed dorsal surface down on the coverslip floor of a superfusion chamber designed to fit on the stage of an inverted compound microscope (Zeiss IM-35). This chamber was equipped with an array of platinum wire electrodes for grounding, bipolar stimulation of the pallial nerve and bipolar extracellular recording from both pallial and medial stellate nerves. This arrangement allowed stimulation of the presynaptic fiber and recording to confirm the occurence of the presynaptic action potential and synaptic transmission.

The optical arrangement for fluorescence consisted of a 75w Xenon arc lamp, a quartz collector lens, a changeable excitation filter, a dichroic mirror with a 420 nm excitation cutoff, a 63x 1.25 Neofluar objective and a barrier filter with a cutoff near 420 nm. The excitataion filters used were interference band-pass types with center wavelengths near 340 nm and 380 nm and bandwidths of approximately 12 nm.

Fluorescence images were collected using a SIT camera (Hamamatsu C-2400), processed using a real-time video processor (Imaging Technology,

Inc. series 151) and an IBM AT personal computer and stored on diskettes.
Information necessary for photometric calibration of the the fluorescence
microscope, camera and video digitizer was collected using a microscope
slide cuvette filled with a fura-2 solution and a 0.15 log unit neutral
density filter. All such calibration information was collected in the form
of video images, allowing for simple and routine correction for spatial
variations in fluorescence illumination and camera sensitivity over the
image field. A correction for tissue background fluorescence was made by
measuring fluorescence of a portion of the ganglion remote from the
injected presynaptic terminal.

Fluorescence recording runs consisted of three or four equally spaced
episodes of video frame summation in rapid succession. Each frame
summation episode involved summation of 8 or 16 video frames (1/30 sec
each) and two frame times for calculation and storage of images. The first
episode measured baseline fluorescence. The second episode began
simultaneously with a train of presynaptic stimulus pulses of approximately
the same duration as the summation episode (0.25 sec or 0.5 sec), and thus
measured fluorescence during the stimulus train. A third and sometimes a
fourth frame summation episode measured fluorescence after the stimulus
train. To collect spectral data, recording runs were repeated using
different excitation filters. A recovery period of at least 6 minutes was
always allowed between successive recording runs.

Processing of images to depict stimulus-induced changes in the
binding of Ca to fura-2 was begun by applying photometric and background
corrections to the images resulting from each summation episode. The
baseline fluorescence image for each run was then subtracted from the
images collected during and after stimulation, resulting in a raw
fluorescence change image. These images were normalized by dividing by the
baseline fluorescence image. An image normalized in this way should
express the chemical state of the indicator dye (i.e. the proportion
binding Ca ions) independent of variations in concentration of the dye or
optical path length. Such images should therefore convey accurate
information about localization of cytoplasmic Ca concentration increase,
the main goal for this study. These images do not directly represent Ca
concentration because they do not encompass the properties of the binding
reaction between Ca ions and fura-2 molecules. Schemes for calibration of
fura-2 signals for equilibrium conditions have been described (Grynkiewicz,
et al., 1985), but a calibration scheme adequate for the rapid kinetics and
large concentration gradients characteristic of the synaptic terminal
remains to be devised (see Neher, 1986).

The normalized fluorescence change images have been prepared for
presentation here by mapping to a color scale as indicated by the reference
scales included in the figures. The images were masked using the baseline
fluorescence to modulate brightness. This procedure was helpful in
visualizing the outlines of the injected presynaptic terminal and was
necessary to suppress distracting noise as the denominator for fluorescence
change normalization approaches zero.

RESULTS

Microinjection of fura-2 into presynaptic terminals, to an estimated
concentration of 100 uM, had little effect on synaptic transmission. This
contrasts with Arsenazo III, an earlier-generation Ca indicator dye, which
greatly attenuates synaptic transmission at dye concentrations required for
measurement of presynaptic Ca transients (Miledi and Parker, 1981;
Charlton, et al., 1982; Augustine, et al., 1985).

A light microscopic image of a living 'giant' synapse is shown in
Fig. 1A. The image represents a small portion of the left stellate
ganglion, viewed from the dorsal surface. The presynaptic fiber can been
seen to enter at the top right corner of the figure and then lie juxtaposed
to the right margin of the larger, curving postsynaptic fiber. The entire
region of contact between these two fibers (typically 0.5 to 1.0 mm in
length) is the site of synaptic transmission. The portion of the
presynaptic fiber leading up to the synaptic contact is designated as
axonal while along the region of synaptic contact the fiber is properly
designated a presynaptic terminal. The boundary between these two fibers
can be loosely designated the synaptic cleft, although microanatomy of the
synaptic contact is rather intricate (Young, 1973; Pumplin and Reese,
1978).

There is considerable individual variability in the relative
positions of the presynaptic and postsynaptic fibers within the stellate
ganglion (Martin and Miledi, 1986). When viewed from the dorsal surface of
the ganglion, the presynaptic element may appear above, below or beside the
postsynaptic element, and the two fibers often make a partial spiral as
they contact one another. The use of laterally-disposed presynaptic
terminals, such as the one shown in Figure 1A, was essential for
localization of sites of presynaptic Ca entry in our study. This
morphology allowed visualization of both longitudinal and lateral calcium
gradients, with the longitudinal axis corresponding to the transition from
axonal to presynaptic terminal regions and the lateral axis running from
the synaptic cleft to the oppposite, non-synaptic surface of the
presynaptic terminal (where transmitter release presumably does not occur).

An image of the same presynaptic terminal shown in Fig. 1A is shown
in Fig. 1B where digital processing has been used to depict changes in
fura-2 fluorescence related to intracellular Ca concentration changes.
This image represents the change in fluorescence during a train of 50
presynaptic action potentials fired at 100 Hz. As discussed below, the
coloration in images like Fig. 1B represents increases in cytoplasmic Ca
concentration. Such increases were detected only in response to electrical
stimulation and presumably reflect a response to opening of voltage-
dependent Ca channels in the presynaptic fiber membrane. Striking
variations in the colors representing magnitude of fluorescence change can
be observed along both longitudinal and lateral axes in Figure 1B.

Longitudinal gradients in stimulus-induced Ca accumulation

Stimulus-induced fluorescence changes are much larger in the terminal
as opposed to the axonal region of the presynaptic fiber. The peak
fluorescence change observed within the terminal is over ten times as large
as the change observed in the axon merely 250 micrometers from the
beginning of synaptic contact. This must correspond to at least a ten-fold
greater Ca concentration increase within corresponding terminal vs. axon
cytoplasmic volumes. Previous observations of enhanced Ca concentration
changes within terminal regions of presynaptic cells (Miledi and Parker,
1981; Stockbridge and Ross, 1984; Llinas, 1985) are extended here to a
finer level of spatial resolution. Images such as those in Figure 1
demonstrate that there can be a very distict boundary between presynaptic
axonal and terminal Ca regimes and that this boundary coincides closely
with the boundary of the synaptic contact between presynaptic and
postsynaptic fibers.

Lateral gradients in stimulus-induced Ca accumulation

The second important feature of the fluorescence changes shown in
Figure 1B is that these changes are not uniform laterally within the

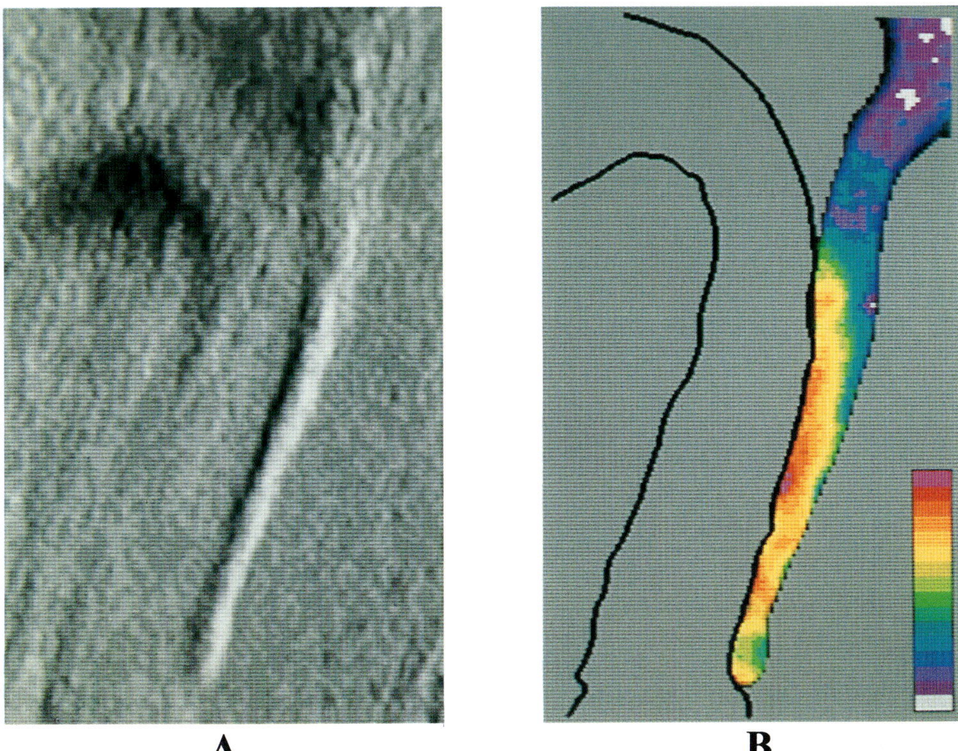

Fig. 1. Two images of the same small portion (approximately 0.44 mm wide by 0.72 mm high) of a squid stellate ganglion showing the 'giant' synapse. Presynaptic fiber enters from upper right. Larger, postsynaptic fiber enters at upper left and curves prior to the elogated region of synaptic contact where the two fibers run parallel. A. Single video image recorded using a combination of brightfield and fluorescence illumination. Presynaptic fiber is lightened due to fluorescence of fura-2 with which it has been injected. B. Result of digitally processing a series of video images to represent localization of stimulus-induced increases in presynaptic intracellular Ca concentration. Outlines of postsynaptic fiber, determined from image shown in A, are traced onto this image. The stimulus was a series of 50 presynaptic action potentials evoked by electrical stimulation at 100 Hz. Fluorescence of fura-2 in response to nominal 380 nm wavelength excitation was measured by integrating for 16/30 of a second immediately before and then during electrical stimulation. The color reference scale represents change in fluorescence emission divided by pre-stimulus baseline emission, from 0 (bottom, no change in Ca) to -0.18 (top, increase in Ca).

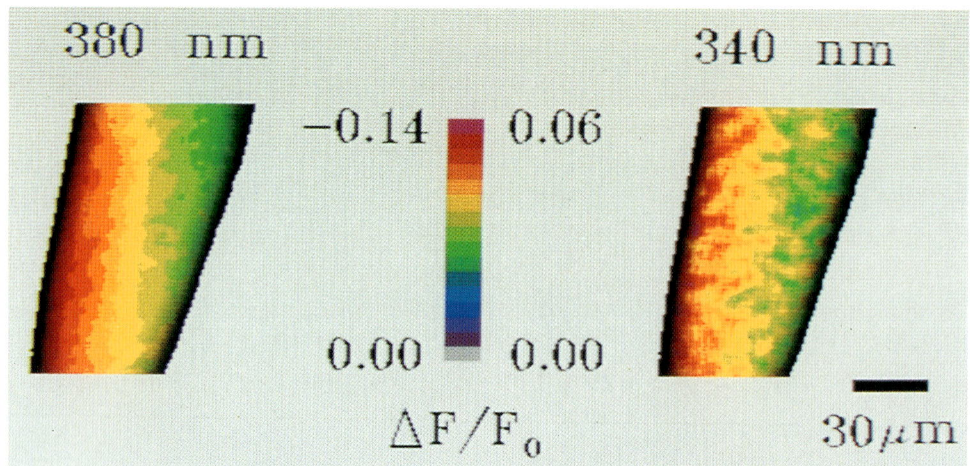

Fig. 2. Spectral properties of fura-2 fluorescence change induced by 25 presynaptic action potentials at 100 Hz are consistent with Ca increase preferentially localized to synaptic cleft surface (left side of each image). Images are from the same presynaptic terminal segment with differing nominal excitation wavelengths as indicated.

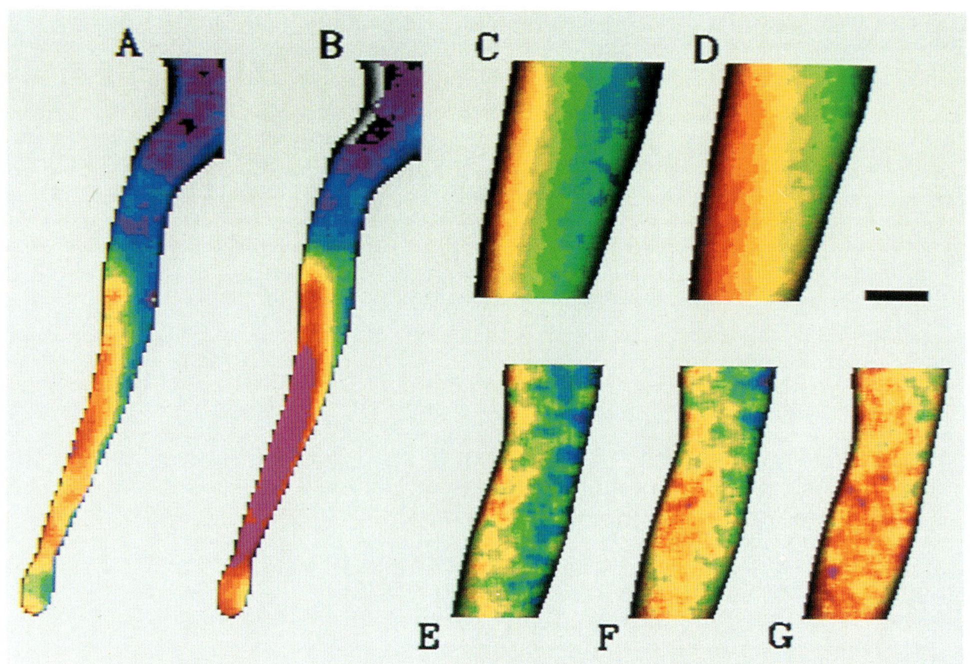

Fig. 3. Redistribution of fura-2 fluorescence changes following presynaptic action potentials. Three separate experiments. For one experiment (A,B), the stimulus was 50 spikes at 100 Hz and video frames were summed for 16/30 sec. In two others (C,D and E,F,G), the stimulus was 25 spikes at 100 Hz and frames were summed for 8/30 sec. Summation for images A, C and E began simultaneously with stimulation. Summation for B began 18/30 sec after stimulus onset, while for D and F it began 10/30 sec after stimulus onset. Summation for G began 20/30 sec after stimulus onset.

presynaptic terminal. It is apparent that the early changes in
fluorescence produced by stimulation are larger in the portion of the
presynaptic terminal closest to the postsynaptic cell. In Fig. 1B, this
can be see most clearly over the wider proximal segment of the presynaptic
terminal. This effect is somewhat blunted in the data of Fig. 1B, however,
due to the long duration of stimulation employed and because this
presynaptic terminal was not absolutely lateral to the postsynaptic fiber.

The lateral localization of the stimulus-induced fluorescence change
was much more obvious with shorter stimuli and in segments of presynaptic
terminals that lay precisely alongside the postsynaptic axon. Examples of
such data are shown in Figures 2, 3C-F. As in Figure 1, the postsynaptic
axon and synaptic cleft lie to the left of the presynaptic terminal. Only
a portion of the presynaptic terminal is shown in these figures because the
images were recorded at a higher magnification than in Figure 1. These
figures illustrate changes in fluorescence during trains of 25 presynaptic
action potentials fired at 100 Hz. In these figures it is clear that the
early fluorescence changes are larger by a factor of at least two or three
at the left edge of the presynaptic terminal, adjacent to the cleft, as
compared to the right edge, opposite from the cleft.

<u>Identification of fluorescence changes with Ca accumulation</u>

Figure 2 documents that fluorescence changes caused by electrical
stimulation have spectral characteristics consistent with increased binding
of Ca ions to fura-2. The two images were obtained from the same
presynaptic terminal by repeating the same stimulus with two different
excitation filters in the fluorescence optical pathway. With excitation at
380 nm (left), electrical stimulation induced a <u>decrease</u> in fluorescence
(note minus sign on color reference scale value). With 340 nm excitation,
stimulation induced an <u>increase</u> in fluorescence (positive reference scale
value). These are the directions of change in fluorescence expected from a
increase in the proportion of fura bound to Ca, and thus from an increase
in intracellular Ca concentration (Grynkiewicz, et al., 1985).

<u>Rapid redistribution of accumulated intracellular Ca</u>

The localized accumulation of Ca near presynaptic sites of
transmitter release is short-lived. This is evident from measurements of
fluorescence change at different times with respect to presynaptic
stimulation, as shown in Figure 3. The figure illustrates the time course
of fluorescence change in three different presynaptic terminals. Each of
the three series begins with an image from a measurement made coincident
with the 100 Hz train of presynaptic action potentials (Fig. 3A, 50 action
potentials, 3C and 3E, 25 action potentials). Second in each series is an
image from a measurement made beginning about 0.07 sec after the conclusion
of the stimulus train (Figs. 3B, 3D and 3F). In the third series, there is
a final image corresponding to a measurement made approximately 0.47 sec
after the stimulus train (Fig. 3G). In each case it is readily apparent
that the later fluorescence changes are less sharply localized to the
synaptic cleft region. Where the longest time intervals have elapsed
(Figs. 3B and 3G), fluorescence changes are nearly uniform across the width
of the terminal. This rapid redistribution of fluorescence changes
following stimulation is presumably due to diffusion of Ca ions away from
sites of entry and into the bulk of the presynaptic terminal.

It is immediately evident from each series in Fig. 3 that the later
images reflect fluorescence changes that are not only redistributed but are
also much larger than the initial changes measured coincident with
stimulation. In each case this seems primarily due to the fact that the
underlying spike-induced Ca transients persist for several seconds. Thus a

measurement made a fraction of a second after repetitive stimulation
integrates essentially undiminished contributions from all spikes within
the train, whereas the measurement made coincident with the train
integrates spikes responses later in the train for successively shortened
periods of time. For the simple case where fluorescence change is directly
proportional to Ca entry and the measurement episodes are negligibly short
in comparison to the duration of the spike-induced Ca transient, one would
expect exactly twice the fluorescence change from the measurement following
the train as compared to the measurement made coincident with the train.
This is approximately what was observed.

Some additional process must be operating, however, to explain the
difference between Figs. 3F and 3G. Here both images represent
fluorescence changes measured <u>after</u> <u>termination</u> of the stimulus, but the
total is again larger for the later time point. This phenomenon is not
readily explained in terms of current understanding of either presynaptic
Ca metabolism or fura-2 behavior, and will require further investigation.
It may reflect Ca-stimulated release of intracellularly stored Ca ions.

DISCUSSION

The use of fura-2 and digital fluorescence microscopy on laterally
disposed squid giant synapses has revealed both longitudinal and lateral
gradients in the intracellular Ca accumulation induced by brief electrical
stimuli. Longitudinally, observed Ca transients become much larger with
the transition from axon to presynaptic terminal. This observation follows
similar conclusions from earlier electrical and optical Ca measurement
studies (Katz and Miledi, 1969; Miledi and Parker, 1981; Stockbridge and
Ross, 1984; Llinas, 1985). The present study achieves a new degree of
spatial resolution, however, and demonstrates for the first time that the
axonal-terminal transition is extremely abrupt. Laterally, Ca accumulation
within the presynaptic terminal is initially larger near the synaptic cleft
and smaller at the opposite lateral margin. This gradient rapidly
dissipates, resulting in a more uniform elevation of Ca concentration. The
simplest way to explain these observations is to assume that Ca channels
are concentrated at the synaptic cleft, in the same presynaptic membrane
region responsible for transmitter release.

Alternatively, the observed spatial Ca gradients could result from
spatial heterogeneity of cytoplasmic Ca-handling properties, even though Ca
channels themselves might be uniformly distributed. For instance, there
might be weaker Ca buffering near release sites and stronger Ca buffering
distant from those sites. This might explain the early gradients, but
could not explain the redistribution to more uniform Ca elevation observed
with the passage of time (Fig. 3). Alternatively, it is possible that
intracellular Ca storage sites, triggered to release Ca by Ca influx
(Fabiato, 1985), are preferentially localized near sites of release. This
particular hypothesis merits further attention: it is difficult to reject
on the basis of our present data. Of course, it is possible that <u>both</u> Ca
channels and cytoplasmic Ca-handling properties are differentially
localized with respect to the synaptic cleft.

Indications that Ca channels may be highly localized at synaptic
contacts raise a question at the molecular level of how such localization
might be maintained. The possibilities include stable linkages between Ca
channel molecules and either extracellular or cytoskeletal molecules of
fixed localization. There is also the possibility of a more dynamic
localization maintained by patterns of membrane traffic. The latter
possibilty arises especially because synaptic release sites are
characterized by the very brisk membrane traffic related to the exocytotic

mechanism of transmitter release. Nonetheless, stable linkages to the Ca
channel surely constitute the more plausible basis for localization. A
linkage between the Ca channel (probably a glycoprotein) and specific
postsynaptic surface molecules would provide a particularly parsimonious
basis for the ultrastructural organization of the synapse. Such linkage
might occur by a direct affinity between presynaptic and postsynaptic
membrane glycoproteins or it might be mediated by a distinct lectin. With
progress in the molecular studies of the Ca channel now being carried out,
clues to understanding Ca channel localization may soon be forthcoming.

Stimulus-induced fluorescence changes redistribute with time. The
simplest interpretation of this observation would be in terms of passive
diffusion of Ca ions within presynaptic cytoplasm. Diffusion of complexes
formed between Ca and fura-2 or endogenous mobile Ca-binding molecules
might also contribute to redistribution of the accumulated Ca load. More
refined versions of measurements like those presented in Fig. 3 should
allow some testing of theories that have been put forth to account for
presynaptic Ca diffusion (e.g., Llinas, et al., 1981; Zucker and
Stockbridge, 1983; Stockbridge and Moore, 1984; Simon and Llinas, 1985;
Zucker and Fogelson, 1986). Comparison of measured and theoretically
predicted spatio-temporal Ca profiles may also provide a means of detecting
an "active" Ca propagation regime that would accompany a Ca-dependent Ca
release process. In fact, a preliminary mathematical model (SJ Smith,
unpublished) indicates that the Ca redistribution implicit in Fig. 3 is
substantially more rapid than expected from passive diffusion based on
parameters in the range of the published presynaptic Ca diffusion models
(see Augustine, et al., 1987a).

In summary, we have used fura-2 and digital fluorescence microscopy
to visualize Ca accumulation and diffusion within squid presynaptic
terminals. Our initial efforts appear to support the idea that Ca channels
are localized to presynaptic regions involved in transmitter release.
Further application of this approach promises to provide additional
insights into a number of questions related to presynaptic Ca action.

ACKNOWLEDGEMENTS

Supported by Howard Hughes Medical Institue, Whitaker Foundation and
NIH grant NS-16671 to SJS, Fogarty Foundation and Muscular Dystrophy
Association fellowships to LRO and NIH grant NS-21624 to GJA. We thank
Dr. Charles F. Stevens for comments on this manuscript.

REFERENCES

Augustine, G.J., Charlton, M.P., and Smith, S.J., 1985, Calcium entry into
     voltage-clamped presynaptic terminals of squid, J. Physiol., 367:143.

Augustine, G.J., Charlton, M.P., and Smith, S.J, 1987a, Calcium action in
     synaptic transmitter release, Ann. Rev. Neurosci., 10:633.

Augustine, G.J., Charlton, M.P., and Smith, S.J, 1987b, Toward a molecular
     understanding of synaptic transmitter release: Physiological clues
     from the squid giant synapse, in "Ion Channel Modulation", eds. D.
     Armstrong, A. Grinnell and M. Jackson, Plenum, New York.

Augustine, G.J. and Eckert, R., 1984, The divalent cations differentially
     support transmitter release at the squid giant synapse, J. Physiol.,
     346:257.

Charlton, M.P., Smith, S.J., and Zucker, R.S., 1982, Role of presynaptic calcium ions and channels in synaptic facilitation and depression at the squid giant synapse, J. Physiol., 323:173.

Fabiato, A., 1985, Time and calcium dependence of activation and inactivation of calcium-induced release of calcium from the sarcoplasmic reticulum of a canine cardiac Purkinje cell, J. Gen. Physiol., 85:247.

Gilkey, J.C., Jaffe, L.F., Ridgway, E.B., and Reynolds, G.T., 1978, A free calcium wave traverses the activating egg of the medaka, Oryzias latipes, J. Cell. Biol., 76:448.

Grynkiewicz, G., Poenie, M., and Tsien, R.Y., 1985, A new generation of Ca^{2+} indicators with greatly improved fluorescence properties, J. Biol. Chem., 260:3440.

Heuser, J.E., Reese, T.S., and Landis, D.M.D., 1974, Functional changes in frog neuromuscular junctions studied with freeze-fracture, J. Neurocytol., 3:109.

Heuser, J.E., Reese, T.S., Dennis, M.J., Jan, Y., Jan, L., and Evans, L., 1979, Synaptic vesicle exocytosis captured by quick freezing and correlated with quantal transmitter release, J. Cell Biol., 81:275.

Katz, B., 1969, "The Release of Neural Transmitter Substances," Liverpool University Press, Liverpool.

Katz, B., and Miledi, R., 1969, Tetrodotoxin-resistant electrical activity in presynaptic terminals, J. Physiol., 203:459.

Llinas, R.R., 1982, Calcium in synaptic transmission, Sci. Am., 247:56.

Llinas, R.R., 1985, The squid giant synapse, Curr. Topics Memb. Biol., 22:519.

Llinas, R., Steinberg, I.Z. and Walton, K., 1981, Presynaptic calcium currents in squid giant synapse, Biophys. J., 33:289.

Martin, R. and Miledi, R., 1986, The form and dimensions of the giant synapse of squids, Phil. Trans. Roy. Soc. Lond. B, 312:355.

Miledi, R., and Parker, I., 1981, Calcium transients recorded with arsenazo III in the presynaptic terminal of the squid giant synapse, Proc. Roy. Soc. Lond. B, 212:197.

Neher, E., 1986, Concentration profiles of intracellular calcium in the presence of a diffusible chelator, in: "Experimental Brain Research, Series 14", Springer-Verlag, Berlin.

Pumplin, D.W., and Reese, T.S., 1978, Membrane ultrastructure of the giant synapse of the squid Loligo pealei, Neurosci., 3:685.

Pumplin, D.W., Reese, T.S., and Llinas, R., 1981, Are the presynaptic membrane particles the calcium channels?, Proc. Natl. Acad. Sci. USA, 78:7210.

Rose, B., and Loewenstein, W.R., 1975, Calcium ion distribution visualized by aequorin: Diffusion in cytosol restricted by energized sequestering, Science, 190:1204.

Simon, S.M., and Llinas, R.R., 1985, Compartmentalization of the submembrane calcium activity during calcium influx and its significance in transmitter release, <u>Biophys. J.</u>, 48:485.

Stockbridge, N., and Moore, J.W., 1984, Dynamics of intracellular calcium and its possible relationship to phasic transmitter release and facilitation at the frog neuromuscular junction, <u>J. Neurosci.</u>, 4:803.

Stockbridge, N., and Ross, W.N., 1984, Localized Ca^{2+} and calcium-activated potassium conductances in terminals of a barnacle photoreceptor, <u>Nature</u> 309:266.

Torri-Tarelli, F., Grohovaz, F., Fesce, R., and Ceccarelli, B., 1985, Temporal coincidence between synaptic vesicle fusion and quantal secretion of acetylcholine, <u>J. Cell Biol.</u>, 101:1386.

Tsien, R.Y., and Poenie, M., 1986, Fluorescence ratio imaging: A new window into intracellular ionic signaling, <u>Trends Biochem. Sci.</u>, 11:450.

Williams, D.A., Fogarty, K.E., Tsien, R.Y., and Fay, F.S., 1985, Calcium gradients in single smooth muscle cells revealed by the digital imaging microscope using Fura-2, <u>Nature</u>, 318:558.

Young, J.Z., 1973, The giant fibre synapse of <u>Loligo</u>, <u>Brain Res.</u>, 57:457.

Zucker, R.S., and Fogelson, A.L., 1986, Relationship between transmitter release and presynaptic calcium influx when calcium enters through discrete channels, <u>Proc. Natl. Acad. Sci. USA</u>, 83:3032.

Zucker, R.S., and Stockbridge, N., 1983, Presynaptic calcium diffusion and the time courses of transmitter release and synaptic facilitation at the squid giant synapse, <u>J. Neurosci.</u>, 3:1263.

TOWARD A MOLECULAR UNDERSTANDING OF SYNAPTIC TRANSMITTER RELEASE:

PHYSIOLOGICAL CLUES FROM THE SQUID GIANT SYNAPSE

George J. Augustine, Milton P. Charlton and Stephen J. Smith

Section of Neurobiology
Department of Biological Sciences
University of Southern California
Los Angeles, CA   90089-0371

Department of Physiology
University of Toronto
Toronto, Ontario M5S 1A8 Canada

Section of Molecular Neurobiology and
Howard Hughes Medical Institute
Yale University Medical School
New Haven, CT   06510

INTRODUCTION

Although it has been known for more than 30 years that Ca ions
trigger the release of neurotransmitters at chemical synapses (Katz,
1969), the molecular basis for this important physiological event is
unclear. While biochemical and theoretical studies have proposed a
number of models to account for this presynaptic triggering action of Ca,
none has emerged as a consensus favorite (Augustine, Charlton and Smith,
1987). This problem arises, in part, because there have been few
attempts to test or evaluate these hypotheses.

We have been using the squid giant synapse to explore the mechanism
which permits Ca to activate release. The unique experimental advantage
of this system is that the presynaptic terminal of the synapse is so
large that it permits insertion of micropipettes directly into the
portion of the presynaptic neuron at which transmitter release takes
place. This permits application of two related strategies to address the
problem of how Ca triggers release. First, it is possible to **microinject**
inhibitors, activators, cofactors, etc. of molecules that have been
proposed to be involved in release. Second, physiological measurements
can be made and used to place **functional constraints** upon hypothetical
models of release. This paper will focus on the second approach,
summarizing several examples taken from our recent studies. The point
that we wish to make here is that while a definitive answer is not yet in
hand, the combined application of both of these approaches is likely to
be very valuable in determining how Ca triggers transmitter release.

HOW TO STUDY TRANSMITTER RELEASE AT THE SQUID SYNAPSE

One key experimental advantage of the squid giant synapse is that in this preparation it is possible to measure Ca currents in the presynaptic terminal (Llinas, Steinberg and Walton, 1981a; Charlton, Smith and Zucker, 1982; Augustine and Eckert, 1984; Augustine, Charlton and Smith, 1985a; Augustine and Charlton, 1986). This is optimally done with the 3-microelectrode voltage clamp technique developed by Adrian, Chandler and Hodgkin (1970), which permits measurement of currents from the properly-clamped portion of the long presynaptic terminal while not interfering with the labile molecular apparatus responsible for release. Ca currents in the presynaptic terminal can be pharmacologically separated from other presynaptic currents by using 1 uM tetrodotoxin to block currents flowing through Na channels, as well as internal and external tetraethylammonium ions and external 3,4-diaminopyridine to block K channel currents. Ca current measurements made under these conditions correspond closely with simultaneous measurements of intracellular Ca accumulation made with the Ca indicator dye Arsenazo III (Fig. 1), indicating the reliability of such Ca current measurements.

As is true for many other well-studied synapses, such as neuromuscular junctions (for example, Kriebel, this volume) it is possible to measure postsynaptic electrical events at the squid synapse and use these responses as a rapid, sensitive assay for transmitter release. In some of the experiments discussed here a single microelectrode has been used to measure transmitter-induced postsynaptic potential changes (p.s.p.s), but we typically use two microelectrodes to voltage clamp the postsynaptic cell and measure postsynaptic currents (p.s.c.s). We prefer to measure p.s.c.s because they are immune to certain complications, such as non-linear summation, activation of voltage-dependent conductances, and charging of membrane capacitance, that plague quantitative interpretation of experiments with p.s.p.s (Martin, 1955; Stevens, 1975).

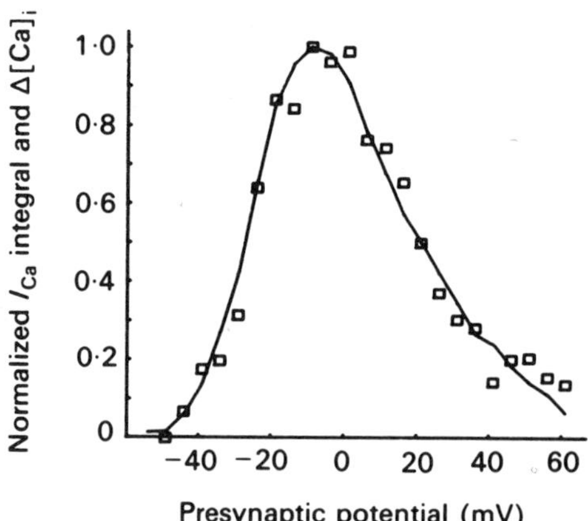

Fig. 1. Comparison between presynaptic Arsenazo III signals and Ca currents. The voltage dependence of Ca current integrals (continuous line) and peak Arsenazo III signal amplitudes (squares) elicited by 6-ms-long depolarizations closely correspond when scaled to the same peak value. From Augustine et al. (1985a).

   With these electrophysiological indicators, we can make detailed
measurements of both influx of Ca into the presynaptic terminal and its
relationship to transmitter release. The remainder of this paper will
use these measurements to consider (1) the **kinetics** of transmitter
release, (2) the **stoichiometry** of Ca action, and (3) the **selectivity** of
release for various divalent cations.

TRANSMITTER RELEASE HAS VERY RAPID KINETICS

   As first pointed out by Llinas, Steinberg and Walton (1981b), the
kinetics of transmitter release can be evaluated by examining the delay
between influx of Ca into the presynaptic terminal and the resultant
postsynaptic response. For depolarizations to negative potentials
there is a pronounced delay in the activation of Ca channels that
distorts the time course of transmission (Fig. 2A). This delay can be
circumvented by depolarizing to more positive potentials, such as +50 mV
(Fig. 2B). At such potentials Ca channels open, but no Ca enters the
presynaptic terminal because of the lack of electrochemical driving force
(Katz and Miledi, 1967). However, upon repolarization a large Ca influx
occurs, without delay, during the Ca 'tail' current. Under such
conditions (Fig. 2B) the delay between the onset of the tail current and
resultant p.s.c. is very brief, on the order of a few hundred usec
(Llinas et al., 1981b), although most of the postsynaptic response is
delayed by 1-2 msec (Augustine, Charlton and Smith, 1985b).

   In summary, this type of experiment reveals that the delay between
Ca influx and p.s.c., which includes not only the initiation of release,
but also diffusion of transmitter across the synaptic cleft and the
gating of postsynaptic, transmitter-activated channels, is short. This
means that the reactions that Ca activates to provoke release do not take
more than a fraction of a msec to occur. Because Ca has a finite rate of
diffusion, this also means that the site of Ca action must be very close
to the site of Ca entry.

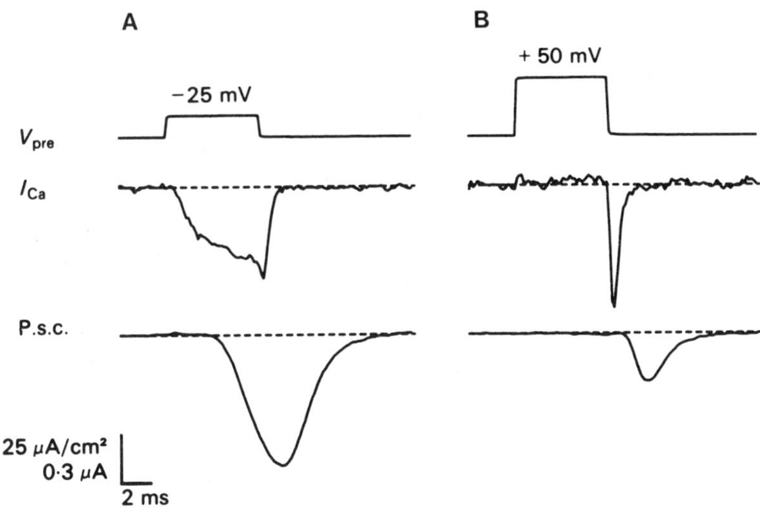

Fig. 2. Presynaptic Ca currents (I_{Ca}) and post-synaptic currents
        (p.s.c.) elicited by depolarizations of the presynaptic
        membrane (V_{pre}) to −25 mV (A) and +50 mV (B) from a holding
        potential of −70 mV. Currents were corrected for linear
        linear leakage and Ca-independent currents. Dotted lines
        indicate baselines measured prior to the voltage-clamp pulses.
        From Augustine et al. (1985b).

THERE IS A HIGHLY NON-LINEAR RELATIONSHIP BETWEEN Ca INFLUX AND RELEASE

A second important point that can be extracted from studies of
presynaptic Ca influx and postsynaptic response comes from consideration
of the stoichiometric relationship between the amount of Ca that enters a
presynaptic terminal and the quantity of transmitter release that
results. Modern analysis of this problem began with the work of Dodge
and Rahamimoff (1967). They found that transmitter release at the frog
neuromuscular junction is very sensitive to the external Ca
concentration, $[Ca]_o$. At very low $[Ca]_o$ they found that release
increased with the fourth power of $[Ca]_o$. This basic finding has since
been confirmed at many other types of synapses (Martin, 1977; Cohen and
Van der Kloot, 1985; Silinsky, 1985) and has been interpreted as
indicating that multiple Ca ions need to "cooperate" in order to trigger
release. In more contemporary terms, these results suggest that the
+receptor with which Ca must associate to trigger release may bind 4 or
more Ca ions to cause a single synaptic vesicle to fuse with the
presynaptic membrane.

Although this interpretation has proven popular, it suffers from the
limitation that many steps intervene between the change in $[Ca]_o$ and the
release of transmitter. Any (or all) of these steps could contribute to
the steep Ca dependence of release. In fact, the earliest attempts to
measure the relationship between presynaptic Ca current and postsynaptic
response at the squid synapse suggested a linear relationship between
these two parameters (Llinas et al., 1976, 1981b). If this were the
case, then the non-linear Ca dependence studied by Dodge and Rahamimoff
(1967) reflects a property of Ca entry into the presynaptic terminal,
rather than a property of the subsequent intracellular reactions that
lead to transmitter secretion. Alternatively, release could have a non-
linear dependence upon presynaptic Ca current, so that Ca entry would be
proportional to $[Ca]_o$.

We have performed two types of experiments to discriminate between
these two alternatives. First, we re-evaluated the relationship between
presynaptic Ca influx and transmitter release. We were particularly
concerned about voltage decrement within the long (ca. 1 mm) presynaptic
terminal of the squid synapse, since direct measurements indicated that
appreciable space clamp errors could arise during voltage clamp
experiments (Augustine et al., 1985a). This problem was avoided by using
a local Ca application method to eliminate Ca influx from all but the
well-clamped distal portion of the synapse (Fig. 3). This permitted
measurement of both presynaptic Ca currents and postsynaptic responses
from the portion of the synapse where space clamp problems should be
minimal.

Depolarization of presynaptic terminals under these conditions leads
to presynaptic Ca influx and resultant postsynaptic responses (Fig. 4).
The Ca currents elicited by the three smallest depolarizations shown in
this Figure increase by approximately equivalent increments, but the
p.s.c.s that they produce increase disproportionately. Thus there
appears to be a **non-linear** relationship between Ca influx and
postsynaptic response. To quantify this relationship, we attempted to
measure the presynaptic Ca currents and p.s.c.s in a number of ways
(Augustine, Charlton and Smith, 1985b). The basic conclusion of our
analysis is depicted in the synaptic transfer curve of Fig. 5. This
curve plots currents elicited with depolarizations ranging from −40 to
+50 mV, with currents resulting from depolarizations to −10 mV or less
indicated by open symbols and responses occurring at more positive
potentials indicated by closed symbols. Ca currents were integrated from
0.75 to 6 ms after the onset of the depolarization, while p.s.c.s were

integrated from 5 to 6 ms. On linear coordinates (Fig. 5, left), the
non-linear nature of the transfer curves is apparent. Plotting the same
data on logarithmic coordinates (Fig. 5, right) yields a straight line.
The solid line shown in the right graph is a power function with an
exponent of three. This function provides an adequate description of the
data points obtained from depolarizations both above and below -10 mV.

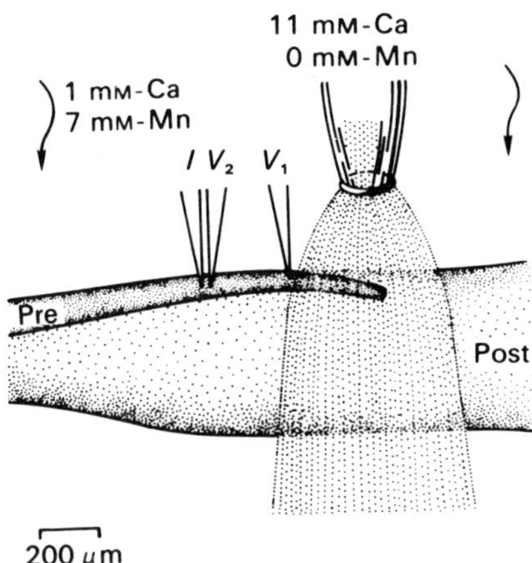

Fig. 3. Local application of Ca ions to voltage-clamped presynaptic
terminals. Presynaptic membrane potential was measured with
micro-electrodes V_1 and V_2, and current-passing electrode I
was connected to V_1 in voltage-clamp configuration to control
presynaptic membrane potential. The preparation was bathed in
a saline containing low (1mM) Ca and 7 mM-Mn. An extracellular
pipette was filled with saline containing 11 mM-Ca and no Mn
to restore Ca influx and transmitter release to the well-
clamped portion of the presynaptic terminal. From Augustine
et al. (1985a).

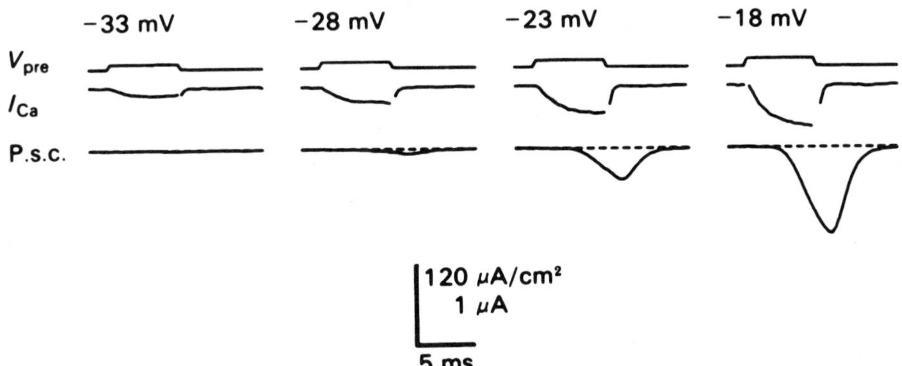

Fig. 4. Family of Ca currents and p.s.c.s elicited by presynaptic
depolarizations of 6 ms duration to the potentials indicated.
Roughly proportional increases in Ca currents (I_{Ca}) produced
by increasing depolarizations (V_{pre}) cause disproportionate
changes in p.s.c.s. From Augustine et al. (1985b).

The results of Fig. 5 allow us to make two conclusions regarding the relationship between presynaptic Ca influx and transmitter release. First, because the magnitude of postsynaptic responses obtained with similar calcium currents elicited by small and large depolarizations are comparable, transmitter release appears insensitive to presynaptic membrane potential (beyond effects of potential on Ca channel gating). This provides direct evidence against the proposal by Dudel, Parnas and Parnas (1983) that transmitter secretion has an intrinsic voltage dependence (see also Zucker and Lando, 1986). Second, the results demonstrate that there is a third-power relationship between presynaptic Ca influx and p.s.c. under these experimental conditions. This relationship may be as high as fourth-order under conditions where $[Ca]_o$ is lower (Augustine and Charlton, 1986).

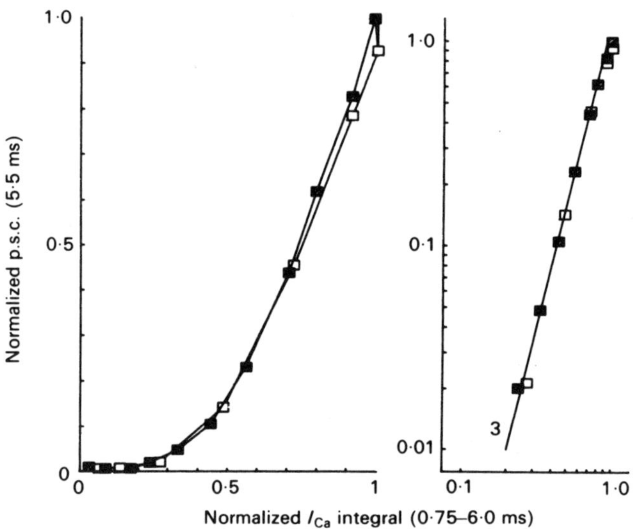

Fig. 5. Synaptic transfer curves relating Ca currents (I_{Ca}), integrated from 0.75–6 ms after initiating presynaptic depolarization, to p.s.c. integrated from 5–6 ms. Graphs are same data plotted on linear (left) and logarithmic (right) coordinates. From Augustine et al. (1985b).

Because of the fourth-order dependence of release upon $[Ca]_o$, the results of the experiments of Fig. 5 predict that presynaptic Ca currents should have a linear dependence upon $[Ca]_o$. We tested this prediction by varying $[Ca]_o$ at the squid synapse, while measuring both presynaptic Ca current and p.s.c. For these experiments we needed to know $[Ca]_o$ precisely, so the local Ca application method illustrated in Fig. 3 could not be applied. Instead, we used a local compression method (Charlton and Augustine, 1987) and/or careful positioning of presynaptic microelectrodes to minimize space clamp problems.

A different, but no less insidious, problem rears its ugly head when one attempts to change $[Ca]_o$ at the squid synapse. Because of extracellular diffusion barriers, changes in bulk Ca concentration may take hours to reach the synapse and affect transmission (Miledi and

Slater, 1966; Augustine and Eckert, 1984). We avoided this problem by
cannulating the circulatory system that provides the giant synapse with
blood and using this cannula to deliver Ca-containing saline to the
synapse (Fig. 6). This method, which is a variant of a method used by
Martin and Miledi (1975) to deliver fixatives to the synapse and Stanley
and Adelman (1984) to deliver a variety of physiological salines to the
synapse, improved the rate of delivery 10-100 fold, so that changes in
$[Ca]_o$ took minutes, rather than hours.

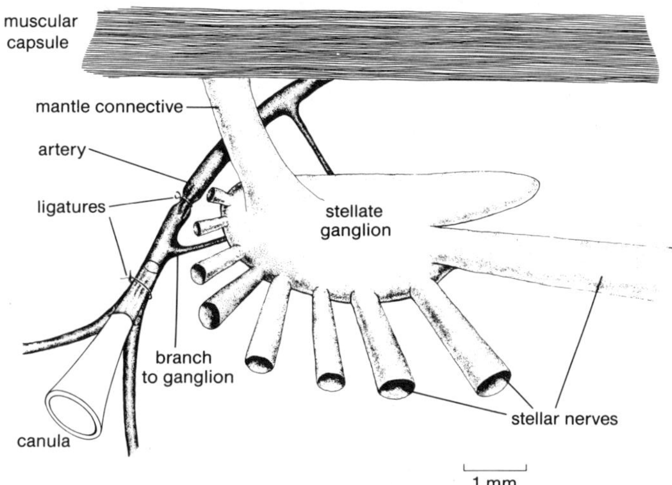

Fig. 6. Diagram of the cannulation method used to deliver solutions to
the giant synapse. A polyethylene cannula was inserted into
the artery, adjacent to the branch which provides the stellate
ganglion with blood. Solutions were gravity-fed into the
artery via the cannula. Ligatures were used to prevent the
cannula from slipping out of the artery and to prevent
shunting of solution flow. From Augustine and Charlton
(1986).

When we used these methods to change $[Ca]_o$ while measuring
presynaptic Ca current and p.s.c., we obtained the results shown in Fig.
7. In Fig. 7A both presynaptic Ca current and p.s.c. are plotted, on
linear coordinates, as a function of $[Ca]_o$. Presynaptic Ca current was
found to be roughly proportional to $[Ca]_o$, while p.s.c. varied in a
steep, non-linear way. On logarithmic coordinates (Fig. 7B), the data
could be described by the saturable function:

$$current \propto ([Ca]_o/(1+[Ca]_o/K_D))^n$$

where K_D is the apparent binding constant and n the order of the binding
reaction. The order, n, was 0.83 for presynaptic Ca current, while it
was 4 for p.s.c.s. Thus, the results of Figs. 5 and 7 confirm the notion
that there is a high-order relationship between presynaptic Ca influx and
release.

There are a number of ways in which these results may be
interpreted. For example,it is possible that the results are due to a
non-linear relationship between postsynaptic responses and rates of
transmitter release, but arguments presented in Augustine et al. (1985b)
and Augustine and Charlton (1986) suggest that this is not the case. We
think that the results indicate that transmitter release has a non-linear
dependence upon the intracellular Ca concentration and,

thus, that the reaction(s) responsible for triggering release require multiple Ca ions. Use of the Ca indicator dye, Fura-2, to make direct, high resolution measurements of presynpatic Ca transients during release might make it possible to determine whether or not this is true (Smith, Osses and Augustine, this volume).

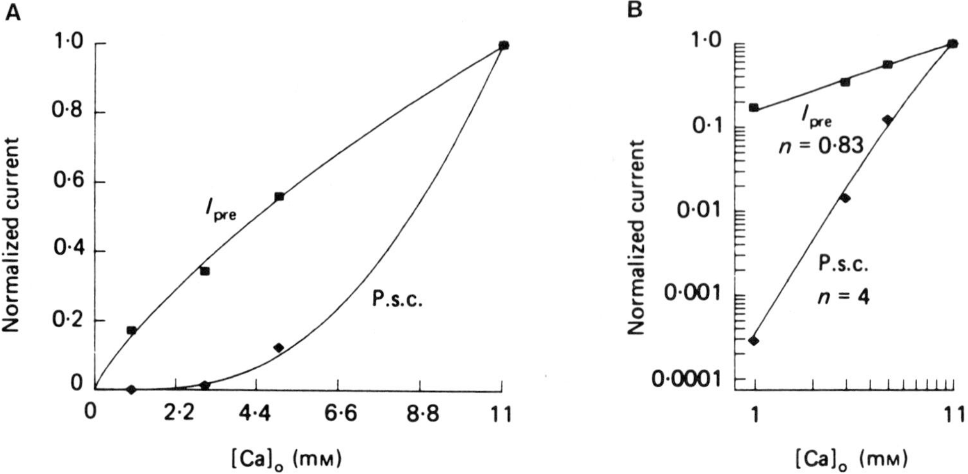

Fig. 7. Quantitative dependence of presynaptic Ca current (I_{pre}) and p.s.c. upon [Ca]$_o$. Responses were elicited by presynaptic depolarizations to -10 mV and were integrated from 2.5-3.5 ms after initiating presynaptic depolarization. A, plots on linear coordinates. B, plots on logarithmic coordinates. Lines connecting the data points are solutions of the equation listed in the text with n = 4 and K_D = 20 mM for p.s.c.s and n = 0.83 and K_D = 65 mM for I_{pre}.

TRANSMITTER RELEASE PREFERS Ca OVER OTHER DIVALENT CATIONS

The final set of physiological results which we will discuss addresses the question of the selectivity of the release process for divalent cations other than Ca. It has been known that divalent cations differ in their ability to support evoked transmitter release (e.g. Silinsky, 1985), but it was not clear whether this is due to differences in the ability of divalent cations to enter the presynaptic terminal or a difference in their efficacy in triggering exocytosis. We evaluated this question by using electrophysiological methods at the squid synapse to measure Ca channel currents and postsynaptic responses, while substituting Sr or Ba ions for Ca (Augustine and Eckert, 1984). We found that even though all three ions were rather similar in their ability to enter the presynaptic terminal, the postsynaptic responses which were elicited in solutions containing these three ions were quite different (Fig. 8). These differences in postsynaptic responses are not due to differences in sensitivity of the postsynaptic cell to transmitter, because miniature p.s.p.s were very similar in size in all three ions. Instead, the results indicate that the divalents differ in their efficacy in supporting transmitter release.

This difference in efficacy is seen more clearly in synaptic transfer curves measured in the presence of Ca, Sr and Ba (Fig. 9). These curves show that, for a given quantity of influx, postsynaptic responses are largest in Ca, smaller in Sr, and smallest in Ba. From this we conclude that upon entry into the presynaptic terminal these divalents differ in their ability to support transmitter release. This

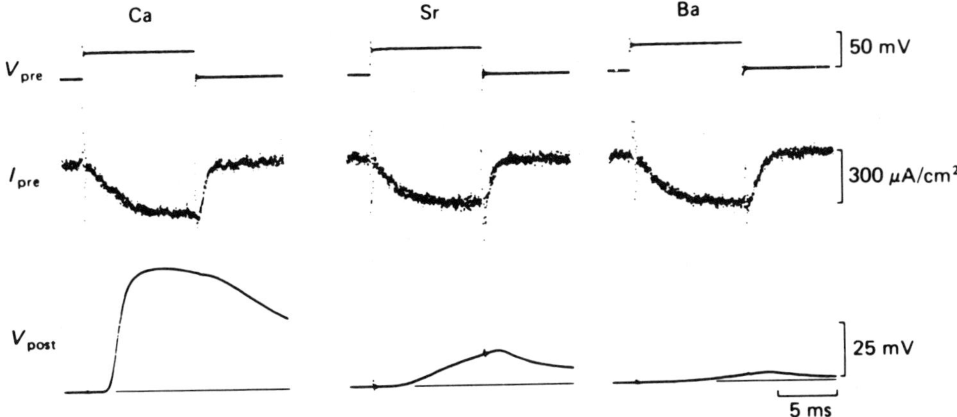

Fig. 8. Voltage-clamp measurements of presynaptic currents carried by
Ca, Sr, and Ba. Step changes in presynaptic potential (V_{pre})
elicit current transients (I_{pre}) which are similar in
amplitude in Ca, Sr, and Ba salines, although the postsynaptic potential (V_{post}) changes which they produce are
different. From Augustine and Eckert (1984).

could be due to differences in the ability of the presynaptic terminal to
remove these ions: for example, the terminal could remove Sr or Ba more
readily than Ca, so that the effective intracellular concentration of Sr
or Ba is lower than that of Ca. Studies on <u>Aplysia</u> neurons suggest just
the opposite, namely, that intracellular Sr or Ba transients are larger
than Ca transients (Tillotson and Nasi, this volume). Thus, we instead
propose that these divalents differ in their ability to initiate
transmitter release. This means that the reaction(s) responsible for
release are activated by divalents in the sequence Ca>Sr>Ba.

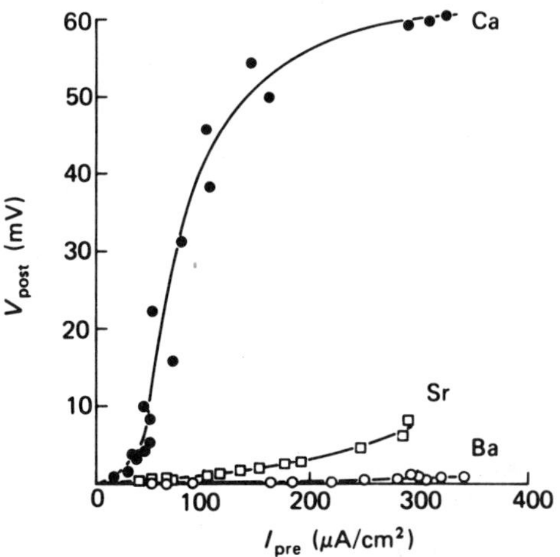

Fig. 9. Synaptic transfer curves in Ca, Sr and Ba. Curves are
sigmoidal, with postsynaptic responses (V_{post}) greatest in Ca
and smallest in Ba. From Augustine and Eckert (1984).

IMPLICATIONS FOR MOLECULAR MODELS OF TRANSMITTER RELEASE

The physiological results described above provide clues toward
understanding the molecular basis for transmitter secretion. The
kinetics of release indicate that the reactions underlying release are
very rapid, transpiring in as little as a few hundred usec. The
observation that the **stoichiometric** relationship between calcium influx
and postsynaptic response is a high-order power function suggests that
one or more Ca-dependent steps involved in release requires multiple Ca
ions. Finally, consideration of the **selectivity** of release for divalent
cations indicates that one or more steps in the process binds divalents
in the sequence Ca>Sr>Ba.

These physiological clues favor certain molecular models for
transmitter release. For example, calmodulin binds multiple Ca ions
(Huang, Chau, Chock, Wang and Sharma, 1981) in a relatively rapid fashion
(Klee, Crouch and Richman, 1980) and binds Ca with a higher affinity than
Sr or Ba (Teo and Wang, 1973). Thus it is possible that calmodulin, or
some other protein with similar "E-F hand" calcium-binding domains
(Kretsinger, 1980), is involved in the reactions which mediate release.
Further discussion of this hypothesis, which has not received universal
support, can be found in Augustine et al. (1987).

These clues also eliminate certain other molecular models of
release. The rapid kinetics of release excludes models requiring a large
number of intermediate reactions, and likely excludes single-step
reactions, such as protein phosphorylation, which have slow turnover
rates. The divalent selectivity of release excludes certain Ca-binding
proteins, such as synexin (Pollard, Creutz, Fowler, Scott and Pazoles,
1981), calelectrin (Walker, 1982), and probably other members of this
emerging family of Ca-binding proteins (Geisow, Fritsche, Hexham, Dash
and Johnson, 1986), because they do not exhibit the Ca>Sr>Ba selectivity
characteristic of release.

Hopefully, continued application of this approach will extract
additional physiological clues which may be used to eliminate other
classes of molecular hypotheses. Our eventual goal is to restrict the
list of hypotheses to the point where judicious application of the other
approach described at the beginning of this paper, namely microinjection
of inhibitors, substrates, cofactors, etc., can be used to home in on the
molecule that mediates Ca-dependent transmitter release.

ACKNOWLEDGEMENTS

We thank the organizers of this meeting for inviting us to participate,
C. Webb for reading a draft of this manuscript and L. Enriquez for typing
it. Supported by NIH grant NS-21624 to GJA, a MRC (Canada) grant to MPC,
and HHMI and NIH funds to SJS.

REFERENCES

Adrian, R.H., Chandler, W.K. and Hodgkin, A.L., 1970, Voltage clamp
     experiments in striated muscle fibres. J. Physiol. **208**:607.
Augustine, G.J. and Charlton, M.P., 1986, Calcium dependence of
     presynaptic calcium current and post-synaptic response at the squid
     giant synapse. J. Physiol. **391**:619.
Augustine, G.J., Charlton, M.P. and Smith, S.J., 1985a, Calcium entry
     into voltage-clamped presynaptic terminals of squid. J. Physiol.
     367:143.

Augustine, G.J., Charlton, M.P., and Smith, S.J., 1985b, Calcium entry and transmitter release at voltage-clamped nerve terminals of squid. J. Physiol. 367:163.

Augustine, G.J., Charlton, M.P. and Smith, S.J., 1987, Calcium action in synaptic transmitter release. Ann. Rev. Neurosci. 10:633.

Augustine. G.J. and Eckert, R., 1984, Divalent cations differentially support transmitter release at the squid giant synapse. J. Physiol. 346:257.

Charlton, M.P. and Augustine, G.J., 1987, Voltage clamping of geometrically complex cells: A method for improved voltage clamping of the squid giant synapse (submitted for publication).

Charlton, M.P., Smith, S.J. and Zucker R.S., 1982, Role of presynaptic calcium ions and channels in synaptic facilitation and depression at the squid giant synapse. J. of Physiol. 323:173.

Cohen, I. and van der Kloot, W., 1982, Calcium and transmitter release. Int. Rev. Neurobiol. 27:299.

Dodge, F.A. Jr., and Rahamimoff, R., 1967, Cooperative action of calcium ions in transmitter release at the neuromuscular junction. J. Physiol. 193:419.

Dudel, J., Parnas, I. and Parnas, H., 1983, Neurotransmitter release and its facilitation in crayfish muscle. VI. Release determined by both, intracellular calcium concentration and depolarization of the nerve terminal. Pflugers Arch. 399:1.

Geisow, M.J., Fritsche, U., Hexham, J.M., Dash, B. and Johnson, T., 1986, A consensus amino-acid sequence repeat in Torpedo and mammalian Ca^{2+}-dependent membrane-binding proteins. Nature 320:636.

Huang, C.Y., Chau, V., Chock, P.B., Wang, J.H. and Sharma, R.K., 1981, Mechanism of activation of cyclic nucleotide phosphodiesterase: Requirement of the binding of four Ca^{2+} to calmodulin for activation. Proc. Natl. Acad. Sci. USA 78:871.

Katz, B., 1969, "The Release of Neurotransmitter Substances", Liverpool University Press, Liverpool.

Katz, B. and Miledi, R., 1967, A study of synaptic transmission in the absence of nerve impulses. J. Physiol. 192:407.

Klee, C.B., Crouch, T.H. and Richman, P.G., 1980, Calmodulin. Ann. Rev. Biochem. 49:489.

Kretsinger, R.H., 1980, Structure and evolution of calcium-modulated proteins. CRC Crit. Rev. Biochem. 8:119.

Kriebel, M., 1987, Quantal classes and subunits of quanta in the neuromuscular junction, in "Ion Channel Modulation", eds. D. Armstrong, A. Grinnell and M. Jackson, Plenum, New York.

Llinas, R., Steinberg, I.Z. and Walton, K., 1976, Presynaptic calcium currents and their relation to synaptic transmission voltage clamp study in squid giant synapse, and theoretical model for the calcium gate. Proc. Natl. Acad. Sci. USA 73:2918.

Llinas, R., Steinberg, I.Z. and Walton, K., 1981a, Presynaptic calcium currents in squid giant synapse. Biophys. J. 33:289.

Llinas, R. Steinberg, I.Z.and Walton, K., 1981b, Relationship between presynaptic calcium current and post-synaptic potential in squid giant synapse. Biophys. J. 33:323.

Martin, A.R., 1955, A further study of the statistical composition of the end-plate potential. J. Physiol. 130:114.

Martin, A.R., 1977, Junctional transmission. II. Presynaptic mechanisms. in "Handbook of Physiology", section 1, The Nervous System, ed. Kandel, E.R., American Physiological Society, Bethesda, MD.

Martin, A.R. and Miledi, R., 1975, A presynaptic complex in the giant synapse of the squid. J. Neurocytol. 4:121.

Miledi, R. and Slater, C.R., 1966, The action of calcium on neuronal synapses in the squid. J. Physiol. 184:473.

Pollard, H.B., Creutz, C.E., Fowler, V., Scott, J. Pazoles, C.J., 1981, Calcium-dependent regulation of chromaffin granule movement, membrane contact, and fusion during exocytosis. <u>Cold Spring Harbor Symp. Quant. Biol.</u> **46**:819.

Silinsky, E.M., 1985, The biophysical pharmacology of calcium-dependent acetylcholine secretion. <u>Pharmacol. Rev.</u> **37**:81.

Smith, S.J., Osses, L. and Augustine, G.J., 1987, Fura-2 imaging of localized calcium accumulation within squid ´giant´ presynaptic terminals, in "Ion Channel Modulation", eds. D. Armstrong, A. Grinnell and M. Jackson, Plenum, New York.

Stanley, E.F. and Adelman, W.J. Jr., 1984, Direct access of ions to the squid stellate ganglion giant synapse by aortic perfusion: effects of calcium-free medium, lanthanum, and cadmium. <u>Biol. Bull.</u> **167**:467.

Stevens, C.F., 1976, A comment on Martin´s relation. <u>Biophys. J.</u> **16**:891.

Teo, T.S., Wang, J.H., 1973, Mechanism of activation if a cyclic adenosine 3´:5´-monphosphate phosphodiesterase from bovine heart by calcium ions. <u>J. Biol. Chem.</u> **248**:5950.

Tillotson, D. and Nasi, E., 1987, Ca^{2+} diffusion in the cytoplasm of <u>Aplysia</u> neurons: its relationship to local concentration changes. in "Ion Channel Modulation", eds. D. Armstrong, A. Grinnell and M. Jackson, Plenum, New York.

Walker, J.H., 1982, Isolation from cholinergic synapses of a protein that binds to membranes in a calcium-dependent manner. <u>J. Neurochem.</u> **39**:815.

Zucker, R.S. and Lando, L., 1986, Mechanism of transmitter release: Voltage hypothesis and calcium hypothesis. <u>Science</u> **231**:574.

QUANTAL CLASSES AND SUBUNITS OF QUANTA IN THE NEUROMUSCULAR JUNCTION

Mahlon E. Kriebel

State University of New York Health Science Center
Department of Physiology
Syracuse, New York 13210 USA

<u>Classes</u> <u>of</u> <u>quanta</u>:

There are three classes of spontaneous miniature endplate potentials (MEPPs) in the normal vertebrate preparation and the ratios vary
during synaptogenesis and with various treatments. In the unstressed
adult preparation (frogs and mammals) most MEPPs (96%) contribute to a
bell-shaped distribution (bell-MEPPs) which has a coefficient of
variance of about 30% (Fatt and Katz, 1952). Del Castillo and Katz
(1954) showed that the unitary evoked potentials are the same size as
MEPPs and they proposed that the larger endplate potentials are composed
of quanta the size of MEPPs. Liley (1956) and Boyd and Martin (1956)
published amplitude histograms of small endplate potentials that were
reduced with a low calcium-high magnesium saline that have distinct
peaks that are integral multiples of the mean MEPP. These results show
that both spontaneous and evoked release are from the same pool of
quanta. With conditions of low quantal content or with relatively low
rates of spontaneous release the size of a quantum remains very constant
(Katz, 1978). However, with high rates of evoked release (Kriebel,
1978) or induced spontaneous release, the average MEPP is greatly
reduced (Kriebel and Gross, 1974; Kriebel and Stolper, 1975; Kriebel, et
al., 1976; Kriebel and Florey, 1983; Kriebel and Pappas, 1987). These
changes reflect differences in the ratio of two classes of MEPPs that
are found in the normal preparation; and, do not usually reflect a
change in the percentage of giant MEPPs as found with chronic botulinum
toxin poisoning (see Thesleff and Molgó, 1983, for a review). Giant
MEPPs are calcium insensitive and Liley (1957) found that giants were
two or three times larger than the mean MEPP so he concluded that they
represented a class of multiquantal events. Giant MEPPs are almost
temperature independent, not influenced by electrotonic depolarization
and unresponsive to elevated potassium levels (Liley, 1957). Their
frequencies are potentiated with 4-aminoquinoline (Molgó and Thesleff,
1982) and vinblastine (Pecot-Dechavassine, 1976). They may have rapid
rising phases, have notches or breaks on their rising phase or show a
slow rising phase (Kriebel and Stolper, 1975; Thesleff and Molgó, 1983).
Giants are normally found in both amphibians and mammals but are not
evoked with nerve stimulation.

The occurrence of a small class of MEPPs which skew into the noise
(<u>skew-MEPPs</u>) was first reported by Dennis and Miledi (1971,1974) during

reinnervation and by Harris and Miledi (1971) in the botulinum toxin
treated preparation. Cooke and Quastel (1973) observed small MEPPs in
the normal mammal preparation and they found that the frequency of small
MEPPs was increased with focal depolarization which demonstrates that
the small MEPPs are generated at the same postsynaptic membrane as the
larger MEPPs. Gross and Kriebel (1973) reported small MEPPs in the
normal unstressed frog preparation that form a distinct mode which is 7
times smaller than the larger MEPPs (Fig. 1).

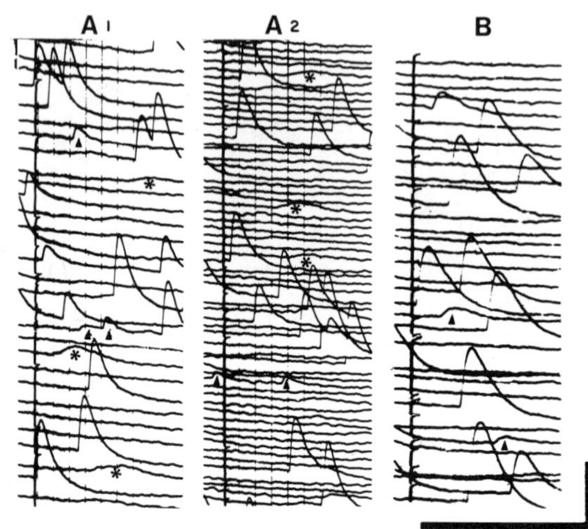

Fig. 1. Sample records of MEPPs: (A) Doubly innervated
          muscle cell. Camera film speed is slower in A_2
          than in A_1. MEPPs generated by a remote junction
          have longer time characteristics and are indicated
          with an asterisk. Sub-MEPPs are indicated with
          arrowheads. 4 x calcium saline. Calibration: 200
          ms and 4 mV; -92 mV resting potential. 12 μm cell.
          (B) Sub-MEPPs indicated with arrowheads. Mean of
          small mode MEPPs is 1/7 that of the major mode
          (3.8 mV). Calibration of high gain AC coupled
          trace is 100 ms and 4 mV. Resting potential is
          -90 mV. 10 μm cell. From Kriebel and Gross, 1974.

These were termed sub-MEPPs but since we have found them to belong to
the same class first described by Dennis and Miledi (1971), they will be
referred to as skew-MEPPs. During morphogenesis the ratio of skew- to
bell-MEPPs changes from mainly skew-MEPPs in the neonate to mainly MEPPs
that form a bell distribution in the adult (Muniak, et al., 1982). In
between stages show equal numbers of skew- and bell-MEPPs. Since the
distribution of the larger MEPP class is usually normal as first
described by Fatt and Katz (1952) these are referred to as bell-MEPPs in
order to distinguish them from skew-MEPPs. In the unstressed adult
preparation, the MEPP amplitude distribution shows a discontinuity
between the skew-MEPPs and the bell-MEPPs so there are two normal
classes of normally occurring MEPPs based on amplitude (Kriebel and
Gross, 1974; Bevan, 1976; Kriebel, et al., 1976; Carlson and Kriebel,
1985). Both classes have the same time characteristics indicating that
both are generated by the same postsynaptic sites and released from the
same presynaptic sites (Kriebel, et al., 1976; Erxleben and Kriebel,
1987).

After denervation of the amphibian preparation, the Schwann cell
forms an extensive Schwann cell-muscle fiber junction and spontaneous
miniature potentials reappear after an initial quiescent period (Birks,

et al., 1960; Bevan, et al., 1976; Kriebel, et al., 1980). These
<u>Schwann cell MEPPs</u> have about the same time characteristics as those
from nerve and occur at a surprisingly high frequency when studied in
small muscle fibers (10µ dia) with low noise electrodes (Kriebel, et
al., 1980). They form a skew distribution with a mode about one-third
that of the skew-MEPPs recorded from similarly sized innervated muscle
fibers. The Schwann cell MEPP frequency is up to 100/min in a small
muscle fiber that would generate only 10 nerve MEPPs/min (Kriebel, et
al., 1980). Identified Schwann cell muscle fiber junctions with high
MEPP frequencies contain few vesicles (Fig. 2).

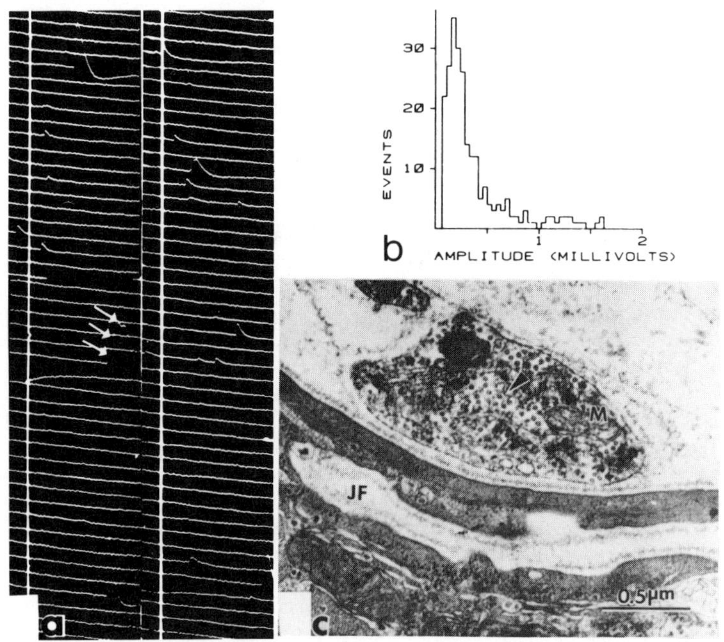

Fig. 2. Schwann cell miniature endplate potentials from a
10µ dia. muscle cell. Six days postoperative
preparation. (a) Schwann cell MEPPs (resting
potential, -83 mV). Note three breaks on the
rising phase of a large MEPP (three arrows)
indicating 'coupled' MEPPs. Calibration: each
trace 500 ms long, 0.25 mV between traces. (b)
Schwann cell MEPP amplitude histogram showing a
peak. This histogram indicates that few MEPPs
were lost to the noise. Frequency, 1/s. (c) The
Schwann cell-muscle junction from which (a) was
recorded. The Schwann cell contains numerous
microtubules (arrowhead) and a mitochondrion (M).
JF, junctional fold. From Kriebel, et al., 1980.

The Schwann cell MEPPs have entirely different release characteristics
than nerve MEPPs and do not appear to be released in the normally
innervated neuromuscular junction even though the Schwann cell has
processes between the muscle fiber and the nerve terminal. Schwann cell
MEPPs are little influenced by changes in extracellular calcium
(Kriebel, et al., 1980; cf. Bevan, et al., 1976). Ito and Miledi (1977)
observed that a calcium ionophore decreased the Schwann cell MEPP
frequency and since Kriebel, et al. (1982) noted that skew-MEPP
frequencies were greatly increased with a calcium ionophore we can
conclude that the skew-MEPPs in the normally innervated preparation are

not the product of Schwann cell release. Bevan, et al. (1976) noted
that actinomicin D prevented the appearance of Schwann cell MEPPs which
demonstrates that following nerve degeneration the Schwann cell is
induced to synthesize and/or release packets of acetylcholine. With
long recording periods we found that the Schwann cell MEPP frequencies
waxed and waned which probably explains some of the discrepancies
reported by Bevan, et al., (1973,1976) and Kriebel, et al. (1980) in
regards to lanthanum ions, calcium ions, temperature and tonicity. The
amplitude of Schwann cell MEPPs shows a slight skew to the larger ones
and many have breaks or notches on their rising phase indicating that
they are composed of smaller units (Fig. 2).

<u>Sub-unit hypothesis of the quantum of transmitter release</u>:

     Fatt and Katz (1952) constructed an amplitude histogram of 800
MEPPs which has a coefficient of variation of 30% and shows no secondary
shoulders or peaks so they concluded there is normally one class of
spontaneous MEPPs. With the improved electronics available, Kriebel and
Gross (1974) found a small class of MEPPs (skew-MEPPs) that formed a
distinct peak from the noise and which forms an amplitude distribution
separate from the bell-MEPPs. With signal to noise ratios of 50 to 1,
Kriebel and Gross (1974) also found that amplitude distributions have
multiple peaks that are multiples of the smaller class. Subsequent
studies demonstrated that amplitude histograms of both unitary evoked
potentials and MEPPs had the same multiple peaks (Fig. 3); and, when
MEPP and unitary evoked potential amplitude histograms were not similar
they still showed the same peak intervals (Kriebel, et al., 1982).
Usually, quanta the size of skew-MEPPs are not evoked (Bevan, 1976;
Kriebel, et al., 1982) but after high rates of release the skew class
may be evoked as unitary evoked potentials (Kriebel, 1978). The two
classes are also very different pharmacologically and have different
temperature coefficients so that the "control" ratio of skew- to bell-
MEPPs depends on the experimental temperature (Fig. 4; Kriebel, et al.,
1976; Carlson, et al., 1982). Moreover, ratios of skew- to bell-MEPPs
are difficult to quantitate because they can change spontaneously (Fig.
5). The effect of the ionophore X-537A is of interest because it
usually changes the MEPP amplitude distribution such that most MEPPs are
of the skew class (Fig. 6; Kriebel, et al., 1980).

     We have found that changes in the average MEPP amplitude are
usually ascribed to a change in the ratio of the skew- to bell-MEPP and
initially do not reflect a change in bell-MEPP size (Kriebel and Gross,
1974; Kriebel, et al., 1976; Kriebel and Florey, 1983). It is necessary
to record from the same muscle fiber to make these observations and to
maintain the signal-to-noise ratio such that the mode of the skew-MEPPs
is not lost into the background noise level. With these experimental
criteria, we have found that botulinum toxin treatment of the <u>in vitro</u>
preparation does not gradually change the bell-MEPP amplitude but first
blocks the bell-MEPP leaving the skew class of MEPPs. This change
occurs within a few hours and after the release of as few as 200 MEPPs
(Kriebel, et al., 1976). At a time when the bell-MEPPs are blocked, the
unitary evoked potentials are of the bell-MEPP size. β-bungarotoxin has
a similar effect but there is first a few minutes of high frequency release, then the skew class remains but at an elevated frequency (Fig. 7;
Llados, et al., 1980). After intensive nerve stimulation, (Kriebel,
1978), black widow spider venom treatment (Kriebel and Stolper, 1975),
intensive rates of release with lanthanum ions (Kriebel and Florey,
1983)or hypertonic saline (Kriebel and Pappas, 1987) mainly skew-MEPPs
remain and there is little, if any, change in the modal value of
skew-MEPPs.

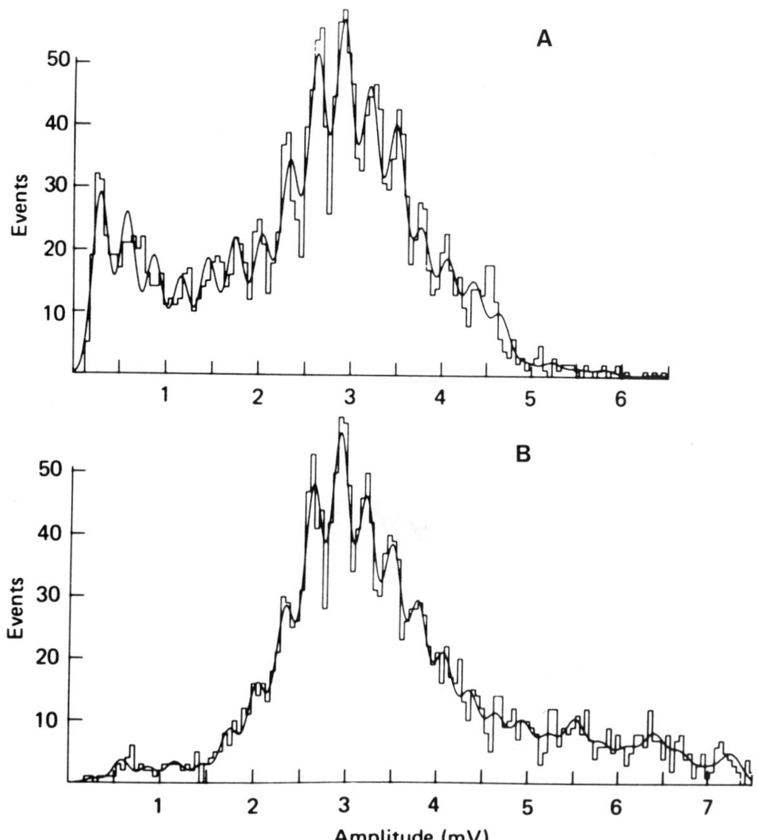

Fig. 3. MEPP and unitary EPP amplitude distributions from a 14 day old mouse diaphragm junction at room temperature. 25 C. No neostigmine. The phrenic nerve was stimulated at 2 Hz and all MEPPs and EPPs were measured during the 70 min of continuous recording. The continuous lines show the predicted distributions based on the subunit hypothesis. Membrane potential was −64 mV. Co^{++} was ca. 4 mM. A, MEPPs. Note that there are two general classes. The smaller ones form a skew distribution and, in this case, show two distinct peaks. 2.4 x 10 MEPPs (34/min) compose this histogram of which 20% are skew and 5% are sub-MEPPs. The estimated value of subunit size is 0.29 mV and of subunit standard deviation is 0.016 mV. B, EPPs. Since most stimuli resulted in failures (75%), most EPPs correspond to the classical unitary evoked potential. The EPP histogram shows multiple peaks which are in register with those in the MEPP amplitude distribution. 2.5 x 10 EPPs. From Matteson, et al., 1982.

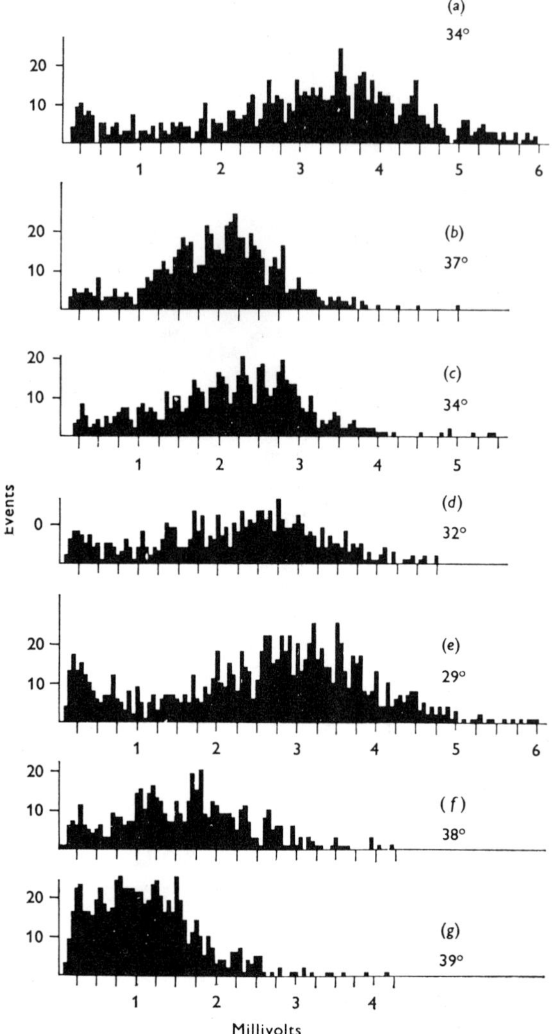

Fig. 4. Successive amplitude histograms showing effect of
small changes in temperature challenges on the mean
MEPP amplitude and the distribution of amplitudes.
These plots show all recorded MEPPs. Temperatures
were changed in about 1 min, which was relatively
rapid compared to the duration of each change.
Preparation (10-day-old mouse) was equilibrated for
1 hr at 34 C and frequency and mean MEPP amplitude
were stationary. Muscle cell resting potential was
−72 mV at the start and −70 mV at the end of 268 min
of continuous recording. Neostigmine, 10^{-7} g/ml.;
normal Ca saline. (a) 34 C, 824 MEPPs/60 min, 14
MEPPs/min; (b) 37 C, 659 MEPPs/16 min, 41 MEPPs/min;
(c) 34 C, 591 MEPPs/40 min, 15 MEPPs/min; (d) 32 C,
558 MEPPs/56 min, 10 MEPPs/min; (e) 29 C, 1005
MEPPs/80 min, 12 MEPPs/min; (f) 38 C, 500 MEPPs/12
min, 41 MEPPs/min; (g) 39 C, 686 MEPPs/4 min, 71
MEPPs/min. From Kriebel, et al., 1976.

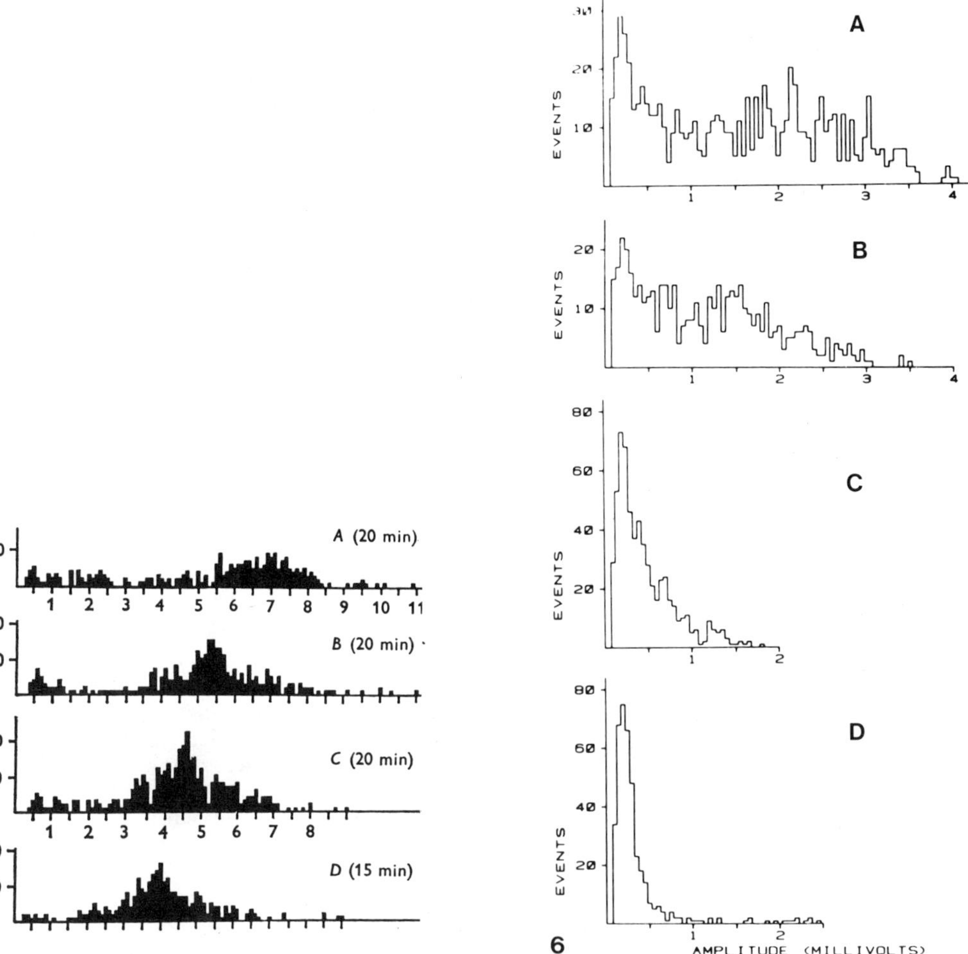

Fig. 5. Successive amplitude histograms showing a
spontaneous decrease in MEPP amplitude. These
histograms show all recorded MEPPs during
continuous monitoring. The muscle cell resting
potential was -65 mV at the start and -64 mV at
the end. Neostigmine, 10^{-6} g/ml; Nine-day-old
mouse. In D, the mean MEPP amplitude was
stationary. A, 269 MEPPs/20 min, 13 MEPPs/min; 20
min elapsed between the start of dissection and
the start of recording;.B, 329 MEPPs/20 min, 16
MEPPs/min; C, 398 MEPPs/20 min, 20 MEPPs/min; D,
293 MEPPs/15 min, 20 MEPPs/min. From Kriebel, et
al., 1976.

Fig. 6. Effect of 10 μM ionophore on MEPP distribution.
Twenty day old mouse. Temperature 28 C. Resting
potential -65 mV throughout the experiment. A,
Control, 20 min, frequency 0.6/s . B-D,
Contiguous 10 min recording sections after the
addition of the ionophore. (All MEPP amplitudes
during this 30 min challenge are shown).
Respective MEPP frequencies were 0.8, 1.0 and
0.9/s. From Kriebel, et al., 1980.

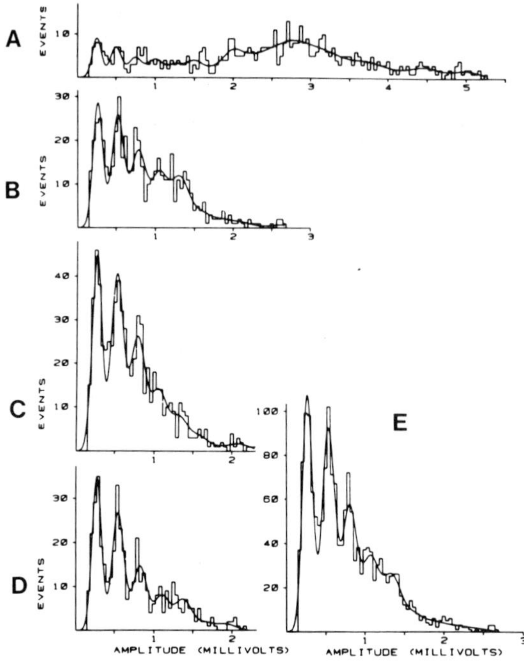

Fig. 7. Effect of β-BuTX on MEPP amplitude distribution.
A: control histogram, 20 min of recording. After
the addition of 20 μg/ml β-BuTX there was a period
of depression lasting approximately 5 min followed
by a 4-fold increase in frequency that reached its
peak 40 min after the drug was added. B,C and D:
successive 12 min periods 1 h after the addition
of β-BuTX. E: B, C and D combined. Temperature
was kept at 26 C during the experiment. Resting
potential 70 mV at the start and 66 mV at the end.
21-day-old mouse. The curves are predicted from
the subunit modal. From Llados, et al., 1980.

The skew class has the same time characteristics as the bell class
in regards to voltage (Kriebel, et al., 1976) or current (Erxleben and
Kriebel, 1987a). Skew-MEPPs are normal released quanta and must be
considered in attempts to correlate morphological findings to the
physiological state and history of a terminal (Rose, et al., 1978).
During synaptogenesis, in the developing junction as well as in the
reinnervating junction, most MEPPs are of the skew class (Dennis and
Miledi, 1971,1974; Muniak, et al, 1982). We have studied identified
junctions that generate mainly skew-MEPPs in the tadpole leg muscle
(Hanna, unpub.) and identified regions in the mouse diaphragm and found
neither differences in the synaptic vesicle mean diameter nor variance
(Kriebel, et al., 1986).

The ultrastructural relationship of skew-MEPPs to bell-MEPPs is not
clear. The bell-MEPP distribution and mean is changed with various
treatments but the mode of the skew-MEPP remains remarkably constant
(within experimental error of 25%). Even though the bell-MEPP mean can
be changed by a factor of up to 7 the mode of the skew remains within
experimental boundaries and the discontinuity between the skew-MEPP
class and the bell-MEPP class is maintained with certain treatments
until the bell-MEPPs disappear (La ions, BTX, hypertonic treatment).
There is little experimental evidence that the skew class represents
partially formed quanta that are prematurely released before reaching

bell-MEPP size. This possibility does not explain the discontinuity in amplitude histograms between the two classes, the constant mode size of the skew-MEPP nor the observation that quanta of the skew-class are usually not evoked (Bevan, 1976; Kriebel, 1978; Kriebel, et al., 1982).

The skew-MEPP class and bell-MEPP class of quanta have one property in common in that the amplitude histograms may show the same equally spaced peaks on amplitude distributions (Fig. 8; Llados, et al., 1980).

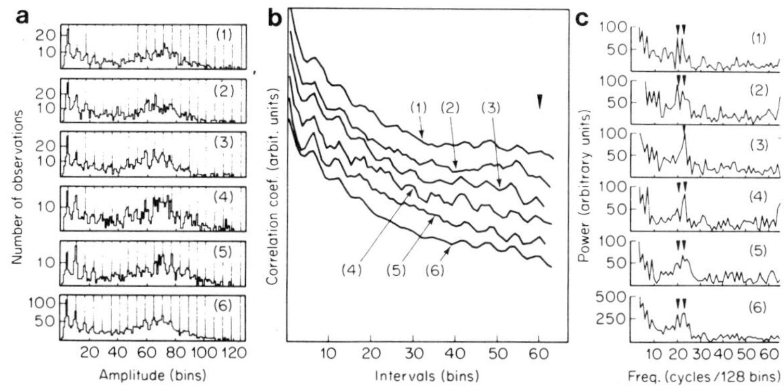

Fig. 8. Frequency analysis showing stationarity of peak intervals throughout sequential plots of MEPP amplitude distribution, according to Matteson, et al. (1981). (a) Sequential plots of successive series of 612 (1), 734 (2), 725 (3), 634 (4) and 591 (5) MEPP amplitudes. The total of 3297 MEPPs is plotted in (6). (b) Respective autocorrelograms of the histograms in (a). The oscillation period is constant throughout all the histograms of this experiment. As the tenth wave of the autocorrelograms corresponds to an interval of 59.5 bins (arrowhead), the calculated interpeak is 5.95 bins (c) Respective PDS of the histograms in (a). An increase in power in all the PDS between 20 and 23 cycles for 128 bins (arrowheads) indicates a constant frequency in the sequential plots (a). The interpeak observed was 5.56-6.40 bins or 0.28-0.32 mV since bin size was 0.05 mV. The size of the subunit noted by Matteson, et al. was 0.30 mV. The vertical lines appearing over the histograms in (a) represent 21.5 cycles/128 bins, as calculated, and correspond to the peaks in all the sequential plots. From Vautrin, 1985.

Thus, we have proposed that both the skew-MEPPs and the bell-MEPPs are composed of the same subunits (Matteson, et al., 1981). We have shown that the multiple peaked amplitude histograms of both the MEPPs and the unitary evoked potentials can be fit with distributions based on a subunit model with three assumptions:

1) The subunit amplitude is normally distributed with a mean and variance (10-16%). The measured variance of sub-MEPPs (first peak on skew-distribution) is composed of the actual variance plus the noise and measurement errors.

2) Larger amplitude MEPPs result from the synchronous release of two or more subunits. Subunits sum linearly and independently; therefore, the release of j subunits would produce a normally

distributed population of MEPP amplitudes with a mean of j times mean
subunit size (μ) and variance of j times the actually subunit variance
plus the noise-measurement variance.

3) The overall amplitude distribution is, therefore, composed of a
mixture of subpopulations. The numbers in each subpopulation are found
with a weighting factor which represents the probability that a MEPP
belongs to a given population. For any observed histogram estimates of
the subunit mean and variance and number of MEPPs in each subpopulation
were obtained from maximum likelihood estimates. The estimates were
independent for each histogram and a probability density function was
used to produce fits to the observed histograms.

The model was tested statistically on amplitude histograms obtained
under a variety of experimental conditions and the accuracy of the model
was tested by a chi-square goodness of fit test (Matteson, et al.,
1981). The generalized likelihood ratio test was used to test the
subunit model against a reduced bimodal mode (Fig. 9).

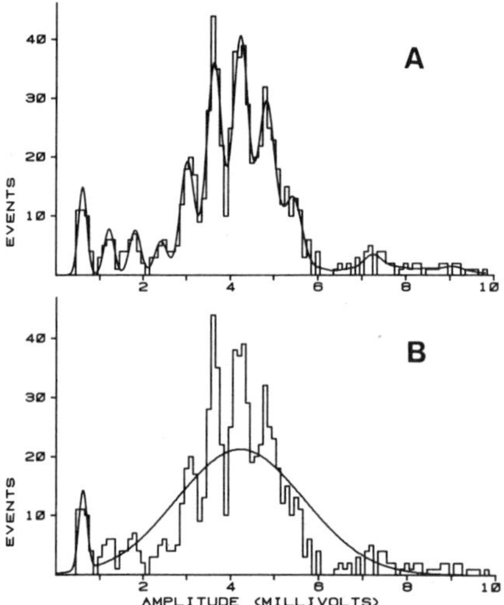

Fig. 9. (a) The fit produced by the multimodal model. The
number of subpopulations in this case was 16. For
each histogram this number was simply determined
by the amplitude range of the histogram, under the
assumption that the weighting factors of any
larger amplitude subpopulations are zero. (b) The
fit produced by the bimodal model. See Matteson,
et al., 1981 for details of maximum likelihood
estimates.

The fit produced by the subunit model was accurate for 90% of the 68
histograms tested and was significantly better than the fit produced by
the bimodal model for 93% of the 68 histograms. The model fit the peaks
of the unitary evoked potential amplitude histograms in those cases in
which the overall profiles of the unitary evoked histograms were similar
to the bell-MEPPs as well as those cases in which there was a great
discrepancy between the MEPP and the unitary evoked potential
distributions (Kriebel, et al., 1982).

The variance of the subunit used in the model was determined from
the noise-measurement error which was experimentally measured and from
the variance of all the peaks in the amplitude histograms (Matteson, et
al., 1981; Kriebel, et al., 1982). With large sample sizes we also know

that the variance of the sub-MEPP (first peak on skew-MEPP distribution), which was calculated by subtracting the measured noise-measurement errors from the sub-MEPP distribution, is the same value which was calculated for the subunit of the peaks in the amplitude histograms (10 to 16%) (Wernig and Motelica-Heino, 1978; Carlson and Kriebel, 1985; Kriebel and Motelica-Heino, 1987). The measured variance of the sub-MEPPs in amplitude histograms is usually 30 to 60% depending on the noise level. Thus, if this value is used for the subunit variance in the model no more than 5 or 6 peaks could be found as predicted by Katz (1978) (Fig. 10).

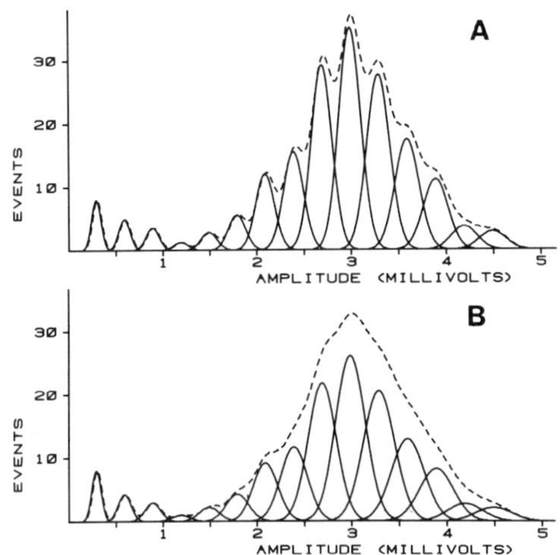

Fig. 10. (a) These curves represent theoretical MEPP amplitude distributions based on the multimodal model. The ordinate is scaled in number of events to be compatible with the scale used for amplitude histograms in the following figures. The values used to plot these curves were computed by multiplying the appropriate probability density function by the total number of observed events. Each solid curve represents a subpopulation of MEPP amplitudes resulting from the synchronous release of a constant number of subunits. The dashed curve is a mixture of all subpopulations; i.e., the overall theoretical MEPP amplitude distribution predicted by the multimodal model. Mean and standard deviation (S.D.) values used were: $\mu=0.3$ mV; $\sigma s=\sigma m=0.035$ mV. (b) These curves were also generated from the multimodal model using the same total s-MEPP variance as in (a). However, measurement and noise errors were assumed to be zero. Mean and S.D. values used were $\mu=0.3$ mV; $\sigma s=0.049$ mV. From Matteson, et al., 1981.

The small variance of the subunit places constraints on the molecular mechanism or the morphological container for the subunit.

We have demonstrated that the peaks in amplitude histograms remain in the same position with successive data samples and that the peaks become more prominent as the sample size increases (Figs. 7,8 and 11; Kriebel, et al., 1976,1982; Carlson and Kriebel, 1985). Moreover, Wernig and Motelica-Heino (1978) and Carlson and Kriebel (1985) report that the number of peaks remains constant but the peak interval is increased with an anticholinesterase agent (Fig. 11).

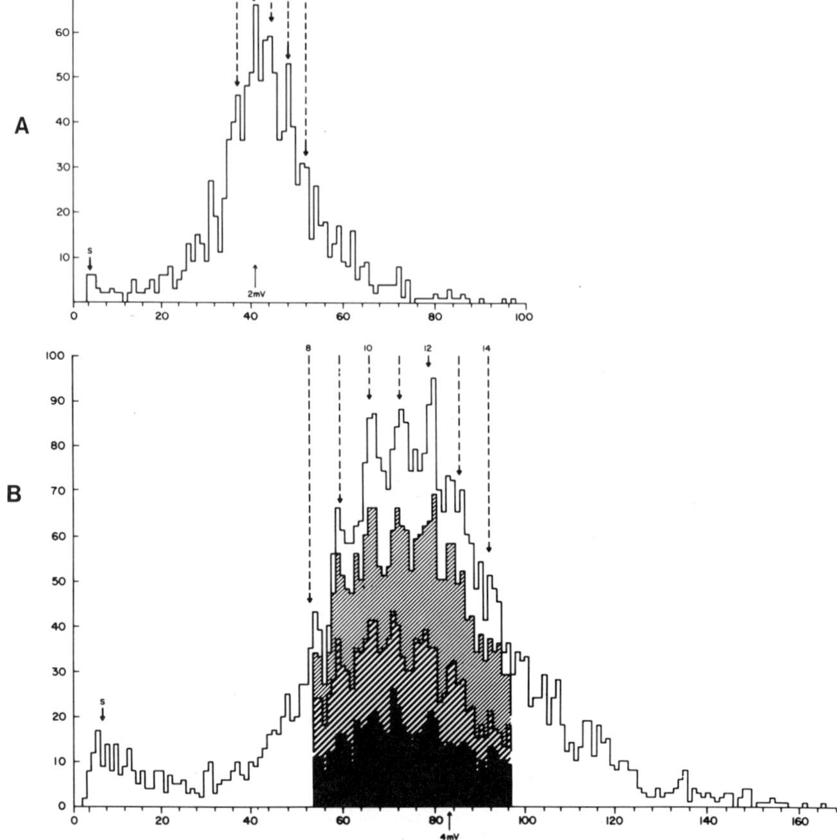

Fig. 11. A, control MEPP amplitude histogram. 1.2×10^3
MEPPs; 4.6/s. 36 C, ca. 22-day-old mouse. Peak
intervals are at 3.7 bins. The arrow labelled (s)
indicates the subunit size equal to the peak
intervals. The first blank bins represent noise.
Bell MEPP is 2.06 mV. B, MEPP amplitude histogram
after neostigmine. 3.7×10^3 MEPPs. Note that
central peaks are noticeable with the first 950
MEPPs (filled histogram) and that the peaks become
more pronounced with larger sample sizes. The
average bell MEPP (3.55 mV) increased by a factor
of 1.70 times the control value, and the peak
interval increased by a factor of 1.73 times the
control value. From Carlson and Kriebel, 1985.

Thus, the peaks do not result from spurious groupings of events by
chance which can result from small sample sizes (see Magleby and Miller,
1981). Moreover, Vautrin (1986) and Csicsaky, et al. (1985) found peaks
with the autocorrelation test using the histograms of Kriebel and
co-workers (Figs. 8 and 12). Csicsaky, et al. (1985) report that in 9
out of 11 of their experiments they found substantial evidence for the
existence of preferred amplitudes. Tremblay, et al. (1983, 1985) also
reported evidence for subunits. Vautrin and Mambrini (1981) found not
only integral amplitudes of the unitary evoked responses but also
unitary evoked potential classes based on latency. Erxleben and Kriebel
(1987) report that the number of peaks in MEP current amplitude
histograms remained constant with treatments that altered the

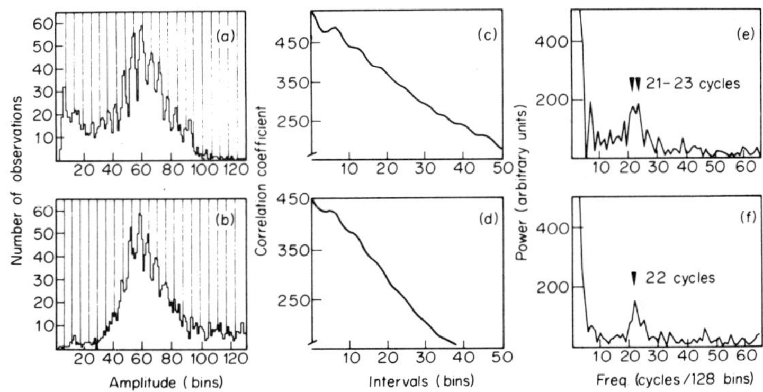

Fig. 12. Frequency analysis showing constant peak intervals throughout the amplitude distributions of miniature endplate potentials (a) and unitary evoked quantal releases (b) (same as Fig. 3; from Kriebel, et al., 1982). (c) and (d) autocorrelograms of histograms (a) and (b) respectively. Successive increases in the correlation coefficient indicate periodic oscillations in the number of events in the successive bins of the histograms. The periods indicated in both distributions correspond to 5-6 bins. (e) and (f) power density spectra (PDS) of histograms (a) and (b) respectively. The increase in power from 21 to 23 cycles for 128 bins in (e), and at 22 cycles for 128 bins in (f) indicate the same frequency in the histograms (a) and (b), which correspond to an interpeak of 5.81 bins, or 0.29 mV. The vertical lines in (a) and (b) over the histograms represent 22 cycles/128 bins, as calculated, and correspond to the peaks in both distributions. From Vautrin, 1986.

postsynaptic membrane characteristics and that the spacing of the peak intervals changed as predicted. For example, increasing the bath temperature, increasing the holding potential (or both), or adding an anticholinesterase agent increased the peak interval (subunit size). The MEPC amplitudes were determined with a computer program from magnetic tape and the peaks were independent of bin size and were clarified with a curve smoothing routine.

There is another, independent line of evidence for the subunit nature of MEPPs. Many bell- and skew-MEPPs have notches on their rising phases and these appear too often to be accounted for by chance coincidences of two smaller MEPPs occurring at nearly the same time. Short periods of elevated temperature, excess calcium ions, lanthanum ions, hypertonic treatment and calcium ionophores greatly increase the number of breaks on MEPPs (Kriebel and Stolper, 1975; Kriebel and Gross, 1974; Kriebel, et al., 1976; Kriebel and Florey, 1983; Kriebel and Pappas, 1987). Most breaks appear as a foot on the MEPP and many of these are the size of the sub-MEPPs. Amplitude histograms of the breaks do not reflect the overall MEPP amplitude distribution of all MEPPs with smooth rising phases but are of the sub-MEPP size which also demonstrates that MEPPs with breaks are not chance coincidences of two events. After high rates of release resulting from elevated temperature, lanthanum ions, black widow spider venom and hypertonic saline treatments, mainly sub-MEPPs remain and these are clustered (Kriebel and Stolper, 1975; Kriebel and Florey, 1983).

In conclusion, electrophysiological studies demonstrate that there
are normally two classes of transmitter quanta and that both classes are
composed of subunits. The bell-MEPP is composed of 9 to 11 subunits and
the average number is readily changed with high rates of release. The
same subunits compose MEPPs and the unitary evoked potential. Atypical
and slow bell- and skew-MEPPs probably represent asynchrony of the
subunits. Supported by NSF 19694.

References

Bevan, S., 1976, Sub-miniature end-plate potentials at untreated frog
    neuromuscular junctions, J. Physiol., 258:145-155.
Bevan, S., Grampp, W., and Miledi, R., 1973, Further observations on
    Schwann cell min.e.p.p.s, J. Physiol., 232:88-89P.
Bevan, S., Grampp, W., and Miledi, R., 1976, Properties of spontaneous
    potentials at denervated motor end-plates of the frog: Proc. Royal
    Soc. London B, 194:195-210.
Birks, R., Katz, B., and Miledi, R., 1960, Physiological and structural
    changes at the amphibian myoneural junction, in the course of nerve
    degeneration, J. Physiol., 150:145-168.
Carlson, C. G., and Kriebel, M. E., 1985, Neostigmine increases the size
    of subunits composing the quantum of transmitter release at the
    mouse neuromuscular junction, J. Physiol. 367:489-502.
Cooke, J. D., and Quastel, D. M. J., 1973, Transmitter release by
    mammalian motor nerve terminals in response to focal polarization,
    J. Physiol. 228:377-405.
Csicsaky, M., Papadopoulos, R., and Wiegand, H., 1985, Detection of
    sub-miniature endplate potentials by harmonic analysis, J.
    Neuroscience Methods, 15:113-129.
del Castillo, J., and Katz, B., 1954, Quantal components of the
    end-plate potential, J. Physiol., 124:560-573.
Dennis, M. J., and Miledi, R., 1971, Lack of correspondence between the
    amplitudes of spontaneous potentials and unit potentials evoked by
    nerve impulses at regenerating neuromuscular junctions, Nature New
    Biology, 232:126-128.
Dennis, M. J., and Miledi, R., 1974, Characteristics of transmitter
    release at regenerating frog neuromuscular junctions, J. Physiol.,
    239:571-594.
Erxleben, C., and Kriebel, M. E., (in press), a. Characteristics of
    spontaneous miniature and subminiature end-plate currents at the
    neonate and adult mouse neuromuscular junction, J. Physiol.
Erxleben, C., and Kriebel, M. E., (in press), b. Subunit composition of
    the spontaneous miniature end-plate currents at the mouse
    neuromuscular junction, J. Physiol.
Fatt, P., and Katz, B., 1952, Spontaneous subthreshold activity at motor
    nerve endings, J. Physiol., 117:109-128.
Gross, C. E., and Kriebel, M. E., 1973, Multimodal distribution of MEPP
    amplitudes: the changing distribution with denervation, nerve
    stimulation and high frequencies of spontaneous release, J. Gen.
    Physiol., 62:658-659a.
Harris, A. J., and Miledi, R., 1971, The effect of type D botulinum
    toxin on frog neuromuscular junctions, J. Physiol., 217:497-515.
Ito, Y., and Miledi, R., 1977, The effect of calcium-ionophores on
    acetylcholine release from Schwann cells, Proc. Royal Soc. London
    B, 196:51-58.
Katz, B., 1978, The release of the neuromuscular transmitter and the
    present state of the hypothesis, in: "Studies in Neurophysiology,"
    R. Porter, ed., Cambridge University Press. pp. 1-21.
Kriebel, M. E., 1978, Small mode miniature endplate potentials are
    increased and evoked in fatigued preparations and in high Mg^{++}
    saline, Brain Res., 148:381-388.

Kriebel, M. E., and Florey, E., 1983, Effect of lanthanum ions on the amplitude distributions of miniature endplate potentials and on synaptic vesicles in frog neuromuscular junctions, Neuroscience, 9:535-547.

Kriebel, M. E., and Gross, C. E., 1974, Multimodal distribution of frog miniature endplate potentials in adult, denervated, and tadpole leg muscle, J. Gen. Physiol., 64:85-103.

Kriebel, M. E., Hanna, R., and Muniak, C., 1986, Synaptic vesicle diameters and synaptic cleft widths at the mouse diaphragm in neonates and adults, Devel. Brain Research, 27:19-29.

Kriebel, M. E., Hanna, R. B., and Pappas, G. D., 1980, Spontaneous potentials and fine structure of identified frog denervated neuromuscular junctions, Neuroscience, 5:97-108.

Kriebel, M. E., and Pappas, G. D., 1987, Effect of hypertonic saline on quantal size and synaptic vesicles in identified neuromuscular junction of the frog, Neuroscience, in press.

Kriebel, M. E., Llados, F., and Carlson, C. G., 1980, Effect of the Ca^{2+} ionophore X-537A and a heat challenge on the distribution of mouse MEPP amplitude histograms, J. de Physiol. Paris, 76:435-441.

Kriebel, M. E., Llados, F., and Matteson, D. R., 1976, Spontaneous subminiature end-plate potentials in mouse diaphragm muscle: evidence for synchronous release, J. Physiol., 262:553-581.

Kriebel, M. E., Llados, F., and Matteson, D. R., 1982, Histograms of the unitary evoked potential of the mouse diaphragm show multiple peaks, J. Physiol., 322:211-222.

Kriebel, M. E., and Motelica-Heino, I., (in press), Description of the sub-MEPP distribution, determination of subunit size and number of subunits in the adult frog neuromuscular bell-MEPP, Neuroscience, in press.

Kriebel, M. E., and Stolper, D. R., 1975, Non-Poisson distribution in time of small- and large-mode miniature end-plate potentials, Amer. J. Physiol., 229:1321-1329.

Llados, F., Kriebel, M. E., and Matteson, D. R., 1980, β-bungarotoxin preferentially blocks one class of miniature endplate potentials, Brain Research, 192:598-602.

Liley, A. W., 1956, The quantal components of the mammalian end-plate potential, J. Physiol., 133:571-587.

Liley, A. W., 1957, Spontaneous release of transmitter substance in multiquantal units, J. Physiol., 136:595-605.

Magleby, K. L., and Miller, D. C., 1981, Is the quantum of transmitter release composed of subunits? A critical analysis in the mouse and frog, J. Physiol., London, 311:267-287.

Matteson, D. R., Kriebel, M. E., and Llados, F., 1981, A statistical model indicates that miniature end-plate potentials and unitary evoked end-plate potentials are composed of subunits, J. Theor. Biol., 90:337-363.

Molgó, J., and Thesleff, S., 1982, 4-aminoquinoline-induced 'giant' miniature endplate potentials at mammalian neuromuscular junctions, Proc. Royal Soc. London B, 214:229-247.

Muniak, C. G., Kribel, M. E., and Carlson, C. G., 1982, Changes in mepp and epp amplitude distributions in the mouse diaphragm during synapse formation and degeneration, Devel. Brain Res., 5:123-138.

Pécot-Dechavassine, M., 1976, Action of vinblastine on the spontaneous release of acetylcholine at the frog neuromuscular junction, J. Physiol. London, 261:31-48.

Rose, S. J., Pappas, G. D., and Kriebel, M. E., 1978, The fine structure of identified frog neuromuscular junctions in relation to synaptic activity, Brain Res., 144:213-239.

Thesleff, S., and Molgó, J., 1983, A new type of transmitter release at the neuromuscular junction, Neuroscience, 9:1-8.

Tremblay, J. P., Laurie, R. E., and Colonnier, M., 1983, Is the MEPP due
     to the release of one vesicle or to the simultaneous release of
     several vesicles at one active zone?, Brain Res. Rev., 6:299-314.
Tremblay, J. P., Robitaille, R., and Grenon, G., 1985, Miniature
     endplate potential amplitudes corrected for spatial decay are not
     normally distributed, Brain Res., 328:170-175.
Vautrin, J., 1986, Subunits in quantal transmission at the mouse
     neuromuscular junction: tests of peak intervals in amplitude
     distributions, J. Theor. Biol., 120:363-370.
Vautrin, J., and Mambrini, J., 1981, Caractéristiques du potentiel
     unitaire de plaque motrice de la grenouille, J. Physiol. Paris,
     77:999-1010.
Wernig, A., and Motelica-Heino, I., 1978, On the presynaptic nature of
     the quantal subunit, Neuroscience Lett., 8:231-234.

SECTION 3

ION CHANNEL MODULATION BY NEUROTRANSMITTERS AND SECOND MESSENGERS

CYTOPLASMIC MODULATION OF ION CHANNEL

FUNCTIONING IN THE NEURONAL MEMBRANE

Platon G. Kostyuk

Bogomoletz Institute of Physiology
Ukrainian Academy of Sciences
Kiev, USSR

INTRODUCTION

According to a general scheme that has emerged from extensive studies
carried out during recent decades, membrane ion channels and cell
metabolism represent operationally independent systems involved in the
maintenance of cellular reactivity. Metabolism creates the necessary
prerequisites for the generation of ionic currents forming transmembrane
electrochemical gradients by active ion transport. Ion channels use these
gradients but not metabolism as a source of energy for producing
transmembrane electric currents. This scheme seemed to be universal for a
long time; however now more and more experimental data have become
available indicating that in certain cases the function of ion channels
may be directly influenced by intracellular metabolic processes. The
present paper presents a survey of such data obtained in our laboratory
and discusses their significance for the understanding of cell functioning.

ELECTRICALLY-OPERATED CALCIUM CHANNELS

The first suggestion that the function of electrically-operated
membrane channels might be under direct metabolic control was formulated
on the basis of experiments carried out on cardiac muscle fibers. It is
known that calcium inward currents are potentiated under the action of
catecholamines; in parallel, an increase in $3',5'$-cAMP content takes place
in the fibers. Direct introduction of cAMP into the fiber produces
similar calcium current potentiation. Using modern ideas about cyclic
nucleotide activated phosphorylation of cellular proteins, these data have
been interpreted as an indication of the role of phosphorylation of
calcium channel proteins in the maintenance of channel activity; they seem
to have an additional gating mechanism operated directly by cellular
metabolic processes (Reuter, 1974, 1983). Quite recently it has been
directly shown that calcium channels isolated from heart and striated
muscle fibers and incorporated into a phospholipid bilayer can be
phosphorylated via a cAMP-dependent mechanism, this being followed by
modulation of their transition from a closed to an open state (Flokerzi et
al., 1986; Hosey et al., 1986).

Calcium channels in the somatic membrane of nerve cells possess a
property which has served us as a convenient basis for a study of the

metabolic dependence of their function. In the course of intracellular
perfusion by saline solution, the transmembrane calcium currents rapidly
decline in amplitude – within several minutes in large snail neurons and
within several dozens of seconds in smaller rat neurons (Fedulova et al.,
1981; Doroshenko et al., 1982; Byerly and Hagiwara, 1982). The decline
also continues despite measures taken to prevent an increase in the
intracellular calcium concentration; the speed of decline slows down
considerably at lower temperatures, as illustrated in Fig. 1. This
implies that the normal functioning of calcium channels requires some
cytoplasmic factor(s) washed out or destroyed during cell perfusion by
saline solution. In their absence the channels pass into an inactive
("sleeping") state.

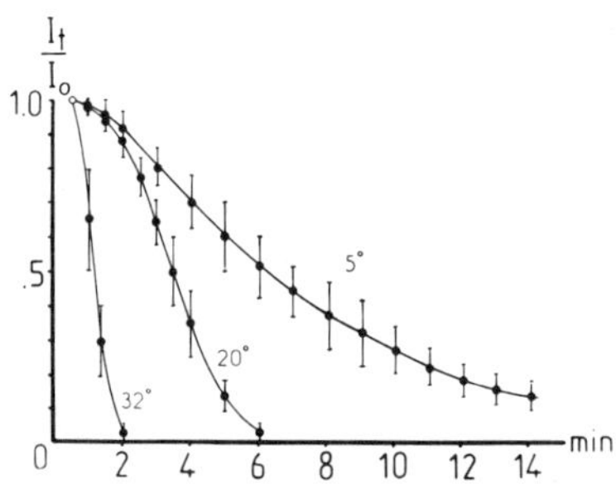

Fig. 1.   Changes in calcium membrane conductance of perfused rat
sensory neuron depending on duration of perfusion at
different temperatures (Doroshenko et al., 1982).

Introduction of cAMP together with ATP and Mg^{2+} into the perfusing
solution not only prevented the progressive decline of calcium conductance
but sometimes restored it to the initial level. Separate introduction of
either ingredient also exerted some stabilizing effect. In perfused snail
neurons the maximum effect was observed at about 10^{-4}M cAMP (see Fig.
2); similar concentration dependences have been obtained for restoring
calcium conductance in perfused rat sensory neurons (Fedulova et al.,
1981). Intracellular introduction of cGMP had no detectable effect on the
calcium conductance decline. Addition of fluoride to the perfusate in
quantities capable of activating membrane adenylate cyclase (several mM)
(together with Mg and ATP) also restored calcium conductance; this means
that the activity of membrane-bound enzymes is preserved in the perfused
neuron. In contrast, the activation of cell phosphodiesterase (resulting
in accelerated reduction of intracellular cAMP levels), as expected,
speeded up the decline, while inhibition of the activity of this enzyme by

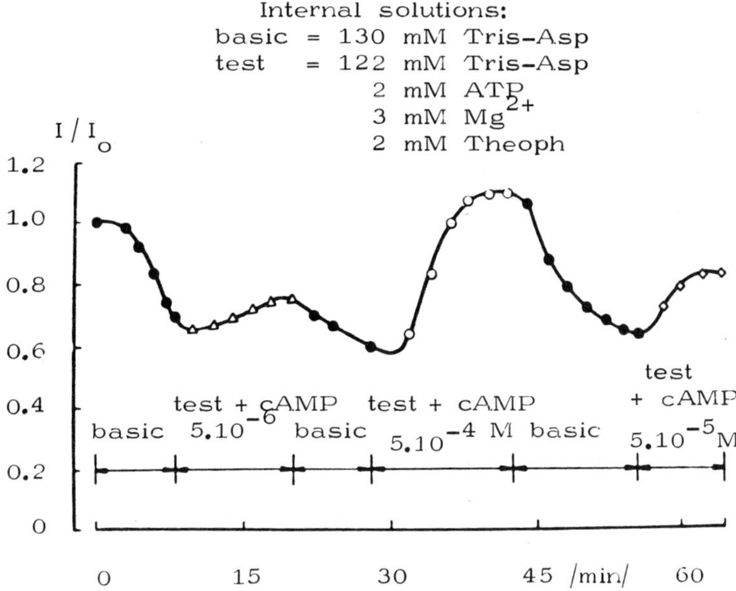

Fig. 2   Restoration of calcium membrane conductance of perfused
snail neuron upon intracellular injections of cAMP at
different concentrations, together with ATP and Mg ions
(Doroshanko et al., 1982).

theophylline (2 mM) or calmodulin blockers (triphtazine and R24571)
reduced the rate of calcium conductance decline (Doroshenko and Martynyuk,
1984). Finally, intracellular introduction of the catalytic subunit of
the cAMP-dependent protein kinase (obtained from rabbit myocardium)
together with ATP also stopped the decline of calcium conductance and
slowly restored it to the initial level (Doroshenko et al., 1984).

The described experiments confirm the idea that electrically-operated
calcium channels in the neuronal membrane, like similar channels in
cardiac fibers, depend in some way on the phosphotransferase activity of
cAMP-dependent protein kinase. The data from our laboratory have been
recently verified in several other laboratories on different types of
neurons. Most extensively, this problem has been investigated by R.
Eckert and his colleagues who not only confirmed the necessity of
cAMP-dependent phosphorylation for supporting the function of calcium
channels, but also demonstrated that the opposite process –
dephosphorylation by Ca-dependent phosphatase – speeds up the "washout" of
calcium currents and their inactivation in the course of sustained
membrane depolarization (Chad and Eckert, 1986). The "washout" can also
be partly due to activation of proteolysis, as it can be slowed down by
leupeptin – an inhibitor of Ca-dependent protease (Chad and Eckert,
1985). The stabilizing action of phosphorylation processes on calcium
channels has also been shown on GH$_3$ cells (Armstrong and Eckert, 1985),
in excitable cells from the intermediate zone of the hypophysis (Cota,
1986) and others. However, in some cases such stabilization could be

achieved by introduction of only ATP without cAMP (Byerly and Yazejian,
1986); this was considered as evidence against the leading role of
phosphorylation in the maintenance of the activity of calcium channels.
But keeping in mind the complexity of cell interior, one might assume that
under some perfusion conditions no effective control over the region of
protein kinase and phosphatase location can be achieved (see also Levitan,
1985).

In recent experiments the present scheme has been supplemented by
several details. The most important of these concerns the separation of
calcium channels into two subgroups - low-threshold and high-threshold
ones (Carbone and Lux, 1984; Fedulova et al., 1985; Matteson and
Armstrong, 1986 and others), which differ in their potential-dependence,
activation and deactivation kinetics and, especially, in the mechanism of
inactivation. While inactivation of high-threshold calcium channels
depends on the quantity of calcium ions carried into the cell by the
inward current (Kostyuk and Krishtal, 1977; Eckert and Tillotson, 1981),
the low-threshold channels are inactivated in a "classical" potential-
dependent way. The described metabolic control is manifested only in
high-threshold channels; the functioning of low-threshold channels did not
change much during intracellular perfusion (Fedulova et al., 1985).
Similar observations were also made on GH_3 cells (Armstrong and
Matteson, 1985).

Peculiar results were obtained upon investigation of high-threshold
calcium currents in the same cells (rat sensory neurons) at different
stages of their ontogenetic development. In cells from neonatal animals
the "washout" of calcium currents could always be prevented effectively by
the intracellular introduction of cAMP, but in cells from adult rats the
effectiveness of such introduction was greatly diminished (Veselovsky et
al., 1986). Also, it is not clear why cAMP injections into nonperfused
cells via an intracellular microelectrode depress calcium inward currents
(Kononenko et al., 1983); probably, in this case uncontrolled release of
calcium from intracellular stores takes place.

A comparison of the properties of calcium channels in the neuronal
and myocyte membranes shows an important difference in their metabolic
regulation. In cardiomyocytes the activation of adenylate cyclase induces
a considerable increase in calcium conductance compared with resting
level, and this effect is used for the mediation of the stimulating action
of catecholamines. In the neuronal membrane such an increase cannot be
detected. On the contrary, norepinephrine depresses calcium conductance
in some neurons via a mechanism not related to second messengers (Forscher
and Oxford, 1985). However, recently it has been detected in our
laboratory in intact snail neurons that specific activation of adenylate
cyclase by cholera toxin induces a considerable prolongation of the
calcium action potential followed by depression at high toxin
concentrations (Nalivaiko, 1986). An example of this effect is presented
in Fig. 3.

CHEMICALLY-OPERATED ION CHANNELS

It was found in 1975 that intracellular injection of cAMP into snail
neurons produces membrane depolarization associated with an increase in
ionic conductance (Liberman et al., 1975). Further studies of this effect
in our laboratory showed that membrane depolarization is produced by the
development of an inward current (Kononenko, 1980). Addition of
theophylline (a phosphodiesterase blocker) potentiated the effect, while
imidazole (a phosphodiesterase activator) inhibited it. The action of
tolbutamide (a protein kinase inhibitor) indicates that the cAMP effect is

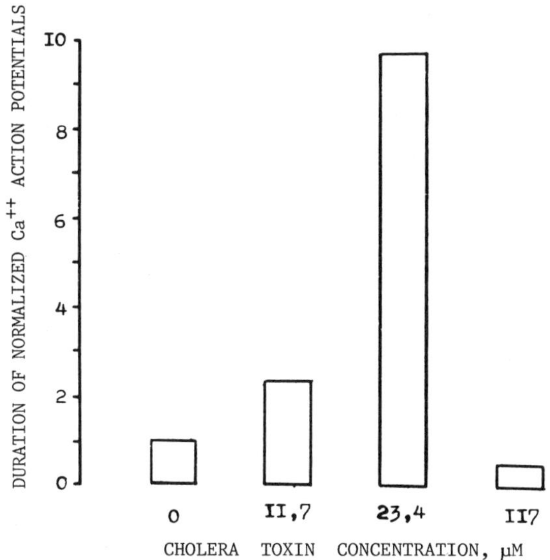

Fig. 3.   Changes in plateau duration of calcium action potential in
nonperfused snail neuron under the action of various
concentrations of cholera toxin.

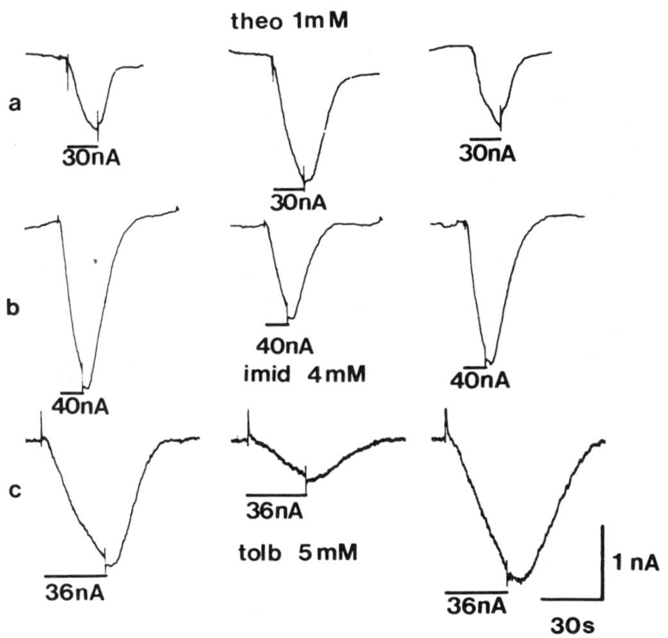

Fig. 4.   Changes in cAMP-induced currents of snail neuron under the
action of inhibitors of cAMP metabolism: theophylline (a),
imidazole (b) and tolbutamide (c)   (Kononenko et al., 1983)

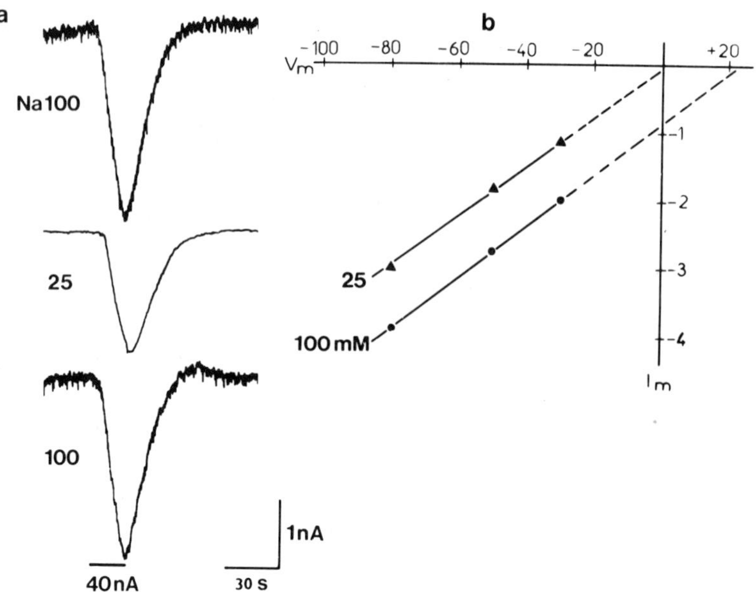

Fig. 5.   cAMP-induced inward currents in snail neuron. Changes in
the current upon varying Na concentration in the medium are
shown on the left. Right part shows current-voltage
characteristics for these currents (Kononenko et al., 1983)

mediated through the activation of cAMP-dependent protein kinase (Fig. 4):
at 5 mM concentration it induced a 4-fold reduction of the rate of rise
and decline of the cAMP-induced current and a decrease of its amplitude
(Kononenko et al., 1983, 1986).

As can be seen from Fig. 5, the described inward current is not
carried via electrically-operated ion channels; thus, the neuronal
membrane possesses ion channels which can be opened by a chemical
influence (phosphorylation) from inside the cell. The selectivity
properties of such channels are close to those of chemically-operated
channels activated by excitatory transmitters from outside the cell.
Obviously, the described effect of injected cAMP represents the final step
in a complex cellular response initiated by primary activation of the
secondary messenger system, especially, synthesis of cAMP. In the same
neurons generation of similar inward current could be produced by external
application of serotonin; this current could be affected in a similar way
by the described activators or inhibitors of the cyclic nucleotide
metabolism (Shcherbatko, 1985). It is quite possible that the action of
serotonin (and also possibly of some other neurotransmitters and
neuromodulators) on these neurons is mediated via such metabolically
operated ion channels.

CONCLUSIONS

Thus, cellular enzymatic systems cannot be regarded only as a source

of energy for ionic pumps creating the corresponding electrochemical
gradients. In many cases the function of ion channels is itself under
direct metabolic control via phosphorylation of their protein components.
In one case phosphorylation appears to play the role of "gating"
mechanism, _per se_, rendering the channel from a closed state into an open
one; in another case it only modulates the gating of the channels. Apart
from ion channels investigated in our laboratory, other studies have
revealed more metabolically dependent channels, among them those of a
calcium-dependent potassium current (dePeyer et al., 1982) and of a
special potassium current depressed by serotonin (Klein et al., 1982).

It should be stressed that the presence of direct metabolic control
is most characteristic for those ion channels whose activation or current
transfer is determined by calcium ions (calcium and calcium-dependent
potassium channels). The leading role of calcium in triggering or
regulation of practically all intracellular processes - starting from the
synthesis of substances necessary for cellular activity, their
intracellular transport and incorporation into the corresponding cellular
structures and ending with their release from the cell is well known.
Therefore, the injection of a certain amount of these ions during every
excitatory cycle should be considered in the first place as a factor
coupling membrane activity to intracellular processes. Accordingly, to
ensure continuous normal functioning of the excitable cell, just such an
injection must be, in turn, under recurrent control from the intracellular
metabolic systems.

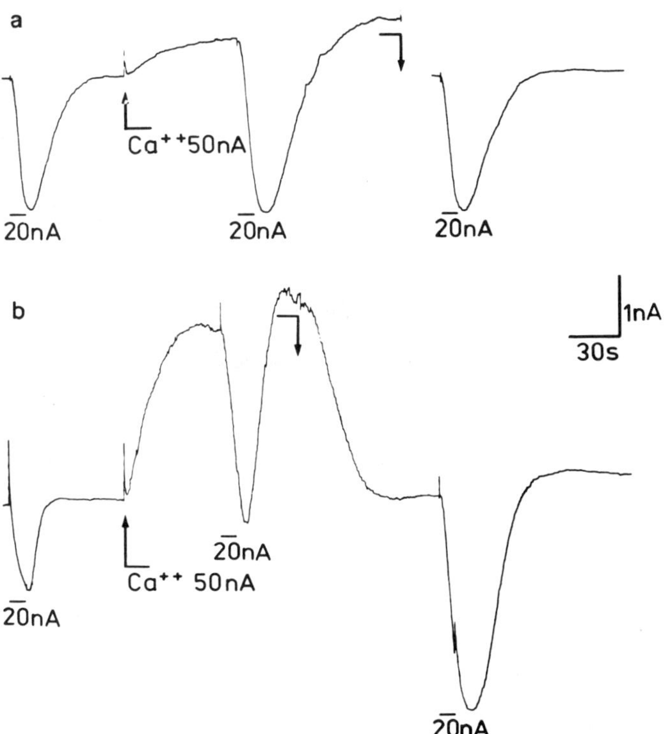

Fig. 6. Potentiation of cAMP-induced current in snail neuron upon
increasing the intracellular Ca concentration. Arrows show
the onset and the end of Ca injection by 50 nA iontophoretic
current (Kononenko et al., 1986)

The presence in the nerve cell (as in other excitable cells) of a
mechanism mediating external influences via a system of intracellular
metabolic messengers which, in turn, alter the function of membrane ion
channels creates vast possibilities for continuous regulation of the
effectiveness of these influences, depending on current activity of the
cell. For instance, we have found that the amplitude and duration of the
inward current induced by intracellular cAMP are considerably increased by
preceding elevation of the intracellular calcium concentration. Most
important is the fact that such potentiation of the cAMP-induced current,
as can be seen in Fig. 6, is retained for a long time after the return of
intracellular calcium to normal levels (Kononenko et al., 1986). A
biochemical basis for this effect could be the presence in the
corresponding channels of two allosterically connected phosphorylation
centers: one cAMP-dependent and one Ca-calmodulin-dependent. From the
functional viewpoint it may represent the possibility of an intracellular
link between two sets of ion channels – the electrically-operated and
chemically operated ones. As a result, the generation of spike discharge
in a cell would proceed to long-lasting changes in the effectiveness of
chemical (transmitter or hormonal) influences upon it, i.e. to some kind
of memory about its previous activity.

REFERENCES

Armstrong, D., Eckert, R., 1985, Phosphorylating agents prevent washout of
    unitary calcium currents in excised membrane patches, J. Gen.
    Physiol., 86:25a.
Byerly, L., and Yazejian, B., 1986, Intracellular factors for the
    maintenance of calcium currents in perfused neurones from the snail,
    Lymnaea stagnalis, J. Physiol. (London), 370:631.
Carbone, E., and Lux, H. D., 1984, A low voltage-activated, fully
    inactivating Ca channel in vertebrate sensory neurones, Nature,
    310:501.
Chad, J.E. and Eckert, R., 1986, An enzymatic mechanism for calcium
    current inactivation in dialyzed Helix neurones. J. Physiol.
    378:31–51.
Chad, J. E., and Eckert, R., 1985, Ca current inactivation is slowed in
    dialysed snail neurons by the substitution of ATP-γ-S for internal
    ATP, J. Gen. Physiol., 86:27a.
Cota, G., 1986, Calcium channel currents in pars intermedia cells of the
    rat pituitary gland. Kinetic properties and washout during
    intracellular dialysis, J. Gen. Physiol., 88:83.
Doroshenko, P. A., Kostyuk, P. G., and Martynyuk, A. E., 1982,
    Intracellular metabolism of adenosine 3',5'-cyclic monophosphate and
    calcium inward current in perfused neurones of Helix pomatia,
    Neuroscience, 7:2125–2134.
Doroshenko, P. A., Kostyuk, P. G., and Martynyuk, A. E., Kursky, M.D., and
    Vorobetz, Z.D., 1984, Intracellular protein kinase and calcium inward
    currents in perfused neurones of the snail Helix pomatia,
    Neuroscience, 11:263–267.
Doroshenko, P. A. and Martynyuk, A. E., 1984, Effect of calmodulin
    blockers on inhibition of potential-dependent calcium conductance by
    intracellular calcium ions in nerve cell, Dokl. AN SSSR (Moscow),
    274: 471 (in Russian).
Eckert, R., and Tillotson, D., 1981, Calcium-mediated inactivation of the
    calcium conductance in caesium-loaded giant neurones of Aplysia
    californica, J. Physiol. (London), 314:265.
Fedulova, S. A., Kostyuk, P. G., and Veselovsky, N. S., 1981, Calcium
    channels in the somatic membrane of the rat dorsal root ganglion
    neurons, effect of cAMP, Brain Res., 214:210.

Fedulova, S. A., Kostyuk, P. G., and Veselovsky, N. S., 1985, Two types of
    calcium channels in the somatic membrane of newborn rat dorsal root
    ganglion neurones, J. Physiol. (London), 359:431.
Flokerzi, V., Oeken, H. J., Hofmann, F., Pelzer, D., Cavalie, A., and
    Trautwein, W., 1986, Purified dihydrophyridine-binding site from
    skeletal muscle t-tubules in a functional calcium channel, Nature,
    323:66.
Forscher, P., and Oxford, G. S., 1985, Modulation of calcium channels by
    norepinephrine in internally dialysed avian sensory neurons, J. Gen.
    Physiol., 85:743.
Hosey, M. M., Borsotto, M., and Lazdunski, M., 1986, Phosphorylation and
    dephosphorylation of dihydropyridine-sensitive voltage-dependent Ca
    channel in skeletal muscle membranes by cAMP- and Ca-dependent
    processes, Proc. Natl. Acad. Sci. USA, 83:3733.
Kalman, D., and Eckert, R., 1985, Injection of the catalytic and
    regulatory subunits of protein kinase into Aplysia neurons alters
    calcium current inactivation, J. Gen. Physiol., 86:26a.
Klein, M., Camardo, J., and Kandel, E. R., 1982, Serotonin modulates a
    specific potassium current in the sensory neurons that show
    presynaptic facilitation in Aplysia, Proc. Natl. Acad. Sci. USA,
    79:5713.
Kononenko, N. I., 1980, Ionic mechanisms of the transmembrane current
    evoked by injection of cyclic AMP into identified Helix pomatia
    neurons, Neurophysiology (Kiev), 12:526 (in Russian).
Kononenko, N. I., Kostyuk, P. G., and Shcherbatko, A. D., 1983, The effect
    of intracellular cAMP injections on stationary membrane conductance
    and voltage- and time-dependent ionic currents in identified snail
    neurons, Brain Res., 268:321.
Kononenko, N. I., Kostyuk, P. G., and Shcherbatko, A. D., 1986, Properties
    of cAMP-induced transmembrane current in mollusc neurons, Brain Res.,
    376:239.
Kostyuk, P. G., and Krishtal, O. A., 1977, Effects of calcium and
    calcium-chelating agents on the inward and outward currents in the
    membrane of mollusc neurones, J. Physiol. (London), 270:569.
Levitan, I. B., 1985, Phosphorylation of ion channels, J. Membrane Biol.,
    87:177.
Liberman, E. A., Minina, S. V., and Golubtsov, K., 1975, Study of the
    metabolic synapse. I. Effect of intracellular microinjection of
    3',5'-AMP, Biofizika (Moscow), 20:451 (in Russian).
Matteson, D. R., and Armstrong, C. M., 1986, Properties of two types of
    calcium channels in clonal pituitary cells, J. Gen. Physiol., 87:161.
Peyer, J. E. de, Cachelin, A. B., Levitan, I. B., and Reuter, H., 1982,
    Ca-activated K conductance in internally perfused snail neurons is
    enhanced by protein phosphorylation, Proc. Natl. Acad. Sci. USA,
    79:4207.
Reuter, H., 1974, Localization of beta adrenergic receptors and effects of
    noradrenaline and cyclic nucleotides on action potentials, ionic
    currents and tension in mammalian cardiac muscle, J. Physiol.
    (London), 242:429.
Reuter, H., 1983, Calcium channel modulation by neurotransmitters,enzymes
    and drugs, Nature, 301:569.
Shcherbatko, A. D., 1985, On the possible participation of the cyclase
    system in indirect electrical reactions of the neurone induced by
    serotonin, Dokl. Akad. Nauk SSSR (Moscow), 281:1014 (in Russian).
Veselovsky, N. S., Fedulova, S. A., and Kostyuk, P. G., 1986, Changes in
    ionic mechanisms of electrical excitability of the somatic membrane
    of rat's dorsal root ganglion neurons during ontogenesis.
    Relationship between calcium channel functioning and intracellular
    metabolism, Neurophysiology (Kiev), 18:827.

CONTROL OF THE GENERATION AND REMOVAL OF CALCIUM-MEDIATED INACTIVATION

OF THE CALCIUM CURRENT IN <u>HELIX</u> <u>ASPERSA</u> NEURONS

John Chad

Department of Neurophysiology
University of Southampton, SO9 3TU, U.K.

INTRODUCTION

The level of intracellular calcium is the hub of many control
processes, providing a transduction of electrical activity into
biochemical control. Depolarizing activity in excitable cells activates
voltage-dependent calcium currents (Hagiwara and Byerly, 1981; Tsien 1983:
McCleskey et al., 1986), which can be regenerative, resulting in a rapid
increase in the activity of intracellular calcium ions in the vicinity of
the channel. The elevation in intracellular calcium is closely controlled
by buffering processes and subsequent extrusion of the calcium (Baker,
1976, 1987) maintaining a low resting level of calcium; uncontrolled and
generalized increases in calcium being toxic (Hajos et al., 1986). Under
physiological conditions spatially and temporally restricted elevations in
the level of free calcium (Chad and Eckert, 1984; Simon and Llinas, 1985)
are responsible for control of ion channels and the transduction of
electrical activity into responses such as exocytosis and contraction.
The increased level of calcium modulates calcium-dependent enzymes, such
as those linked to calmodulin (Klee et al., 1980; Stoclet et al., 1986),
to alter their activity and hence generate a diversity of responses. The
intracellular calcium ions can act as second messengers to control and
modulate ionic channels, producing activation (Meech 1978; Colquhoun et
al., 1981; Yellen, 1982) or inactivation (Chad and Eckert, 1984), and play
a pivotal role in rhythmic activity (Hermann et al., this volume). It is
the calcium-dependent enzyme within the cell which gives the divalent
specificity of action while the channel conductance properties give the
specificity of the ionic current measured. In this chapter we will
discuss one such process which can be studied in detail in isolated,
dialyzed <u>Helix aspersa</u> neurons.

The particular response we have investigated is the calcium-mediated
inactivation of the calcium current in <u>Helix aspersa</u> neurons. In
molluscan neurons the inactivation depends specifically upon the nature of
the entering divalent ion (Tillotson, 1979; Eckert and Tillotson, 1981;
Plant and Standen, 1981; Plant et al., 1983; Eckert and Ewald, 1983 a,b).
Any intervention that reduces the intracellular accumulation of calcium
ions causes a reduction in the inactivation. Thus, substitution of barium
ions for calcium ions in the bathing solution decreases inactivation of
the current despite an overall increase in the current size. Similarly
increasing the calcium buffering capacity of the cell with EGTA reduces

the inactivation. Calcium-dependent inactivation of calcium currents can
also be distinguished from purely voltage-dependent inactivation by
analysis of current-voltage relations. The calcium current response to a
voltage step initially increases as the depolarization increases, the
current amplitude reaches a maximum and further increases in
depolarization decrease the calcium current amplitude as the voltage
approaches calcium's reversal potential and the driving force is reduced.
The inactivation of the calcium current also follows this bell shape,
rising with depolarization to a maximum, coincident with the peak current,
and decreasing as the peak current decreases at very depolarized
potentials. The pattern of the inactivation mirrors the current
amplitudes rather than increasing in direct proportion to the increased
depolarization as would be expected for voltage-dependent inactivation
(Eckert and Ewald, 1983a). Calcium currents have been categorized on the
basis of experimental evidence as having either calcium-dependent or
voltage-dependent inactivation (reviewed by Eckert and Chad, 1984);
however these are not necessarily mutually exclusive properties.
Calcium-dependent inactivation was originally analysed in _Paramecium_ by
Brehm and Eckert (1978) and has since been described in a wide range of
cell types, including cardiac muscle (Mentrard et al., 1984; Lee et al.,
1985; Nilius and Roder, 1985).

The possible role of phosphorylation in modulation of calcium channel
activity was suggested by experiments on cardiac muscle cells (Tsien,
Giles and Greengard, 1972) in which it was observed that cyclic AMP was
the second messenger involved in producing beta-adrenergic stimulation.
The elevation in cyclic AMP activates a cyclic AMP-dependent protein
kinase, leading to increased peak calcium currents and a slowing of
inactivation. This appears to be produced by increased probability of
channel opening in response to depolarization (Osterrieder et al., 1982;
Reuter et al., 1983; Bean et al., 1984; Kameyama et al., 1985). Our
understanding of the processes underlying modulation of calcium currents
has been advanced by experiments involving intracellular dialysis of
single neurons, which allows the isolation of calcium currents free from
contamination by potassium, sodium, chloride or proton currents, and
permits the introduction into the cell of exogenous enzymes (Lee et al.,
1978; Kostyuk et al., 1977). A full account of the methods we have
employed has been published (Chad and Eckert, 1986).

An important but unhelpful initial observation was that the isolated
calcium current rapidly decreased in size when the cell was dialyzed with
a simple ionic solution. This loss of peak current has been termed
washout. Washout occurs whenever the cytoplasmic face of the membrane is
exposed to a simple ionic solution, whether by whole cell dialysis
(Hagiwara and Nakajima, 1966; Byerly and Hagiwara, 1982; Doroshenko et
al., 1982) or after isolation of an excised patch of cell membrane
(Fenwick et al., 1982; Cavalie et al., 1983; Nilius et al., 1985). The
loss of activity is complex in that it can be slowed or arrested but not
reversed by interventions which prevent elevation of calcium ions at the
intracellular face of the membrane. Thus, dialysis of whole cells with
calcium buffering solutions slows washout (Byerly and Hagiwara, 1982;
Fenwick et al., 1982). A further reduction in washout can be achieved by
the use of substances which promote cyclic AMP-dependent phosphorylation
(Doroshenko et al., 1982, 1984; Forscher and Oxford, 1985; Byerly and
Yazejian, 1986; Chad and Eckert, 1986).

Our interest in the activity and control of calcium-dependent
processes precluded the use of intracellular calcium buffers as they would
interfere with the very processes we wished to examine. Therefore, our
initial problem was to produce conditions in which a stable calcium
current could be maintained without supplementation of calcium buffering,

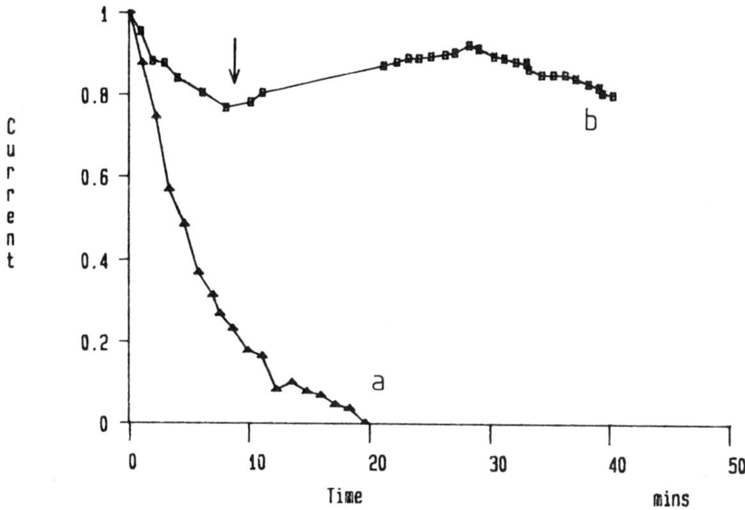

Figure 1.   Dependence of Ca current on kinase activity. Dialysis of
<u>Helix</u> neurons leads to washout of the Ca current.
Normalized peak current amplitudes of Ca currents (leak
corrected) are plotted against time during dialysis for
two different cells. The holding potential was -40 mV and
the extracellular Ca concentration was 50 mM in each
case. (a) Dialysis of a cell with a Cs-aspartate solution
that contained TEA (10 mM), HEPES (100 mM, pH 7.8), ATP-Mg
(7 mM) and leupeptin (0.1 mM) led to a loss of Ca
current. (b) Dialysis of another cell with the same
solution produced an initial loss of current, but the
addition of CS (20 µg/ml) of the cAMP-dependent protein
kinase at the arrow produced recovery and maintenance of
the Ca current.

and it is this difference in our experimental protocol which has led to
the dissection of the processes underlying calcium current washout.

PROTEIN KINASE ACTIVITY PREVENTS WASHOUT

Calcium currents were recorded from voltage-clamped dialyzed <u>Helix</u>
neurons. In the absence of contaminating currents the leak-corrected
inward current response to depolarization accurately reveals the
macroscopic kinetics of calcium channel activation and inactivation. In
Figure 1 the peak calcium current, recorded in response to a depolarizing
voltage clamp step from a holding potential of -40 mV to a potential of
+10 mV, is normalized to the current recorded at time zero, and plotted
against time after the establishment of perfusion. In these experiments
calcium was used as the charge carrier (extracellular solution, 50 mM Ca)
and no intracellular calcium buffers were included. When a cell was
perfused with a solution containing cesium aspartate, ATP-Mg and HEPES,
the peak calcium current rapidly decreased (Fig 1a). After washout there
was no evidence of any remaining asymmetrical, contaminating currents
which might have concealed the true kinetics of the current recorded under

these conditions. We have observed that the inclusion of either EGTA or
the catalytic subunit of cyclic AMP dependent protein kinase will slow but
not substantially reverse the washout. However, it was found that the
washout of calcium current could be prevented or reversed by the inclusion
of the catalytic subunit of cAMP dependent protein kinase and ATP Mg in
the dialysate. Complete reversals were only observed when the dialysate
contained 0.1 mM leupeptin, an inhibitor of calcium-activated
intracellular proteases (Aoyagi et al., 1969; Murachi, 1983), throughout
the dialysis. The progress of the peak calcium current through time is
shown in Figure 1b. Initially there is washout of the calcium current,
leupeptin alone having little effect. When catalytic subunit of cyclic
AMP dependent protein kinase (20 µg/ml) was added to the dialysate
(arrow), washout was reversed and the current increased to a steady
value. Thus it appears that the calcium current in these cells can be
maintained by protein kinase activity in the absence of intracellular
calcium buffers, provided that the calcium-dependent protease activity is
inhibited. We have identified two components in washout: channels can be
inactivated in a calcium-dependent manner by loss of kinase activity, and
channels can be degraded by proteolytic activity leading to irreversible
channel loss. The degradation of dephosphorylated channels appears more
rapid than that of phosphorylated channels as currents can be maintained
by protein kinase in the absence of leupeptin but can not be fully
restored. The action of protein kinase required the presence of a
hydrolyzable form of ATP and was not exhibited by heat-inactivated
catalytic subunit (Chad and Eckert, 1986), implying that a phosphorylation
of the channel or a closely associated site was required for activity.

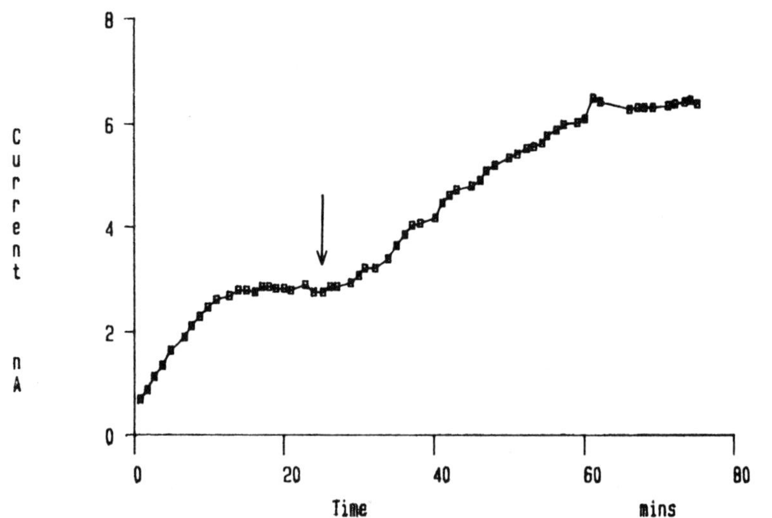

Figure 2.   Dependence of calcium current on intracellular calcium
             activity. Dialysis of Helix neurons with Cs-aspartate,
             TEA, HEPES and ATP-Mg solution buffered to 1 µM Ca with
             EGTA/Ca blocked the Ca current. The presence of leupeptin
             (0.1 mM) prevented irreversible washout. Peak current
             amplitudes for Ca current are plotted against time after
             changing dialysate from a Ca buffer set to 1 µM to Ca
             buffer of 0.1 µM. The decreased intracellular Ca level
             leads to a progressive increase in the Ca current,
             presumably due to endogenous kinase activity. The
             addition of Ca/CaM kinase (500 units/ml) at the arrow
             produced a further increase in the Ca current.

We would predict that in the presence of leupeptin to block irreversible loss of channels we should be able to control the level of calcium channel activity by alterations of the level of intracellular calcium concentration. Figure 2 shows a plot of the peak current against time in an experiment where the dialysis was begun with a solution buffered to 1 μM Ca using Ca/EGTA. Initially there was no net inward calcium current. The dialysate was then changed to one buffered to 0.1 μM Ca. As the calcium level within the cell decreased there was a steady increase (wash-in) of the calcium current which reached a plateau. The dialysate was then supplemented with Ca-calmodulin kinase (500 units/ml at arrow) and a further increase in calcium current was observed which could be maintained by catalytic subunit. Thus it appears that when proteolysis is prevented by leupeptin the calcium current can be completely inactivated, in a reversible manner, by raised intracellular calcium. Restoration can be produced by a reduction in the intracellular calcium level, presumably by a decrease in inactivation and hence the increased net effect of endogenous kinase activity. Exogenous Ca-calmodulin kinase can affect calcium channel activity and this raises the possibility that a calcium-dependent increase in channel activity could occur _in_ _vivo_ under appropriate conditions.

The discovery of multiple calcium channel types in mammalian cells (Nowycky et al., 1984, 1985; Carbone and Lux, 1984; Tsien, 1986) has raised the question of whether we are dealing with more than one type of calcium current. These channel types have been distinguished at a macroscopic level by a number of criteria, including voltage-dependent inactivation and divalent ion selectivity. We have routinely used a holding potential of -40 mV to inactivate the potassium A current (Connor and Stevens,1971), but changing the holding potential did not change either the amplitude or the characteristics of inactivation of the calcium currents we recorded (Chad and Eckert, 1986). Single channel studies on mammalian cells have shown that the class of channels with the largest conductance in barium require metabolic maintenance and rapidly become inactive when cell-attached patches are excised. In contrast the other class of channels with a smaller conductance, which are largely inactivated at -40 mV, remain active in cell-free patches (Armstrong and Eckert,1987; Nilius et al., 1985). The experiments reported here are concerned with a current which is not inactivated at these relatively depolarized holding potentials, has a greater conductance to barium than calcium and can be fully washed out by intracellular dialysis. Thus, this current appears from the macroscopic data to be attributable to the activity of a single dominant type of calcium channel.

DIVALENT SPECIFICITY OF INACTIVATION

A purely voltage-dependent mechanism of inactivation would be unaffected by the nature of the ion carrying the current. A current-dependent mechanism of inactivation, however, would be expected to depend on the species of divalent involved. Recordings of inward current were obtained in equimolar divalent solutions (50 mM) for calcium, strontium and barium. The currents recorded in response to steps to +10 mV from a holding potential of -40 mV are shown in Figure 3A, and have been scaled by the factor indicated to match the peak currents for the three divalents. The rate of inactivation of the inward current response is much slower in Ba solutions than in calcium, even though the barium current is much larger. This precludes the possibility that intracellular divalent accumulation is having its effect directly by reducing the driving force, for if this were the case, the larger current would be predicted to inactivate more rapidly. The strontium currents are of intermediate magnitude and inactivation. For each divalent species

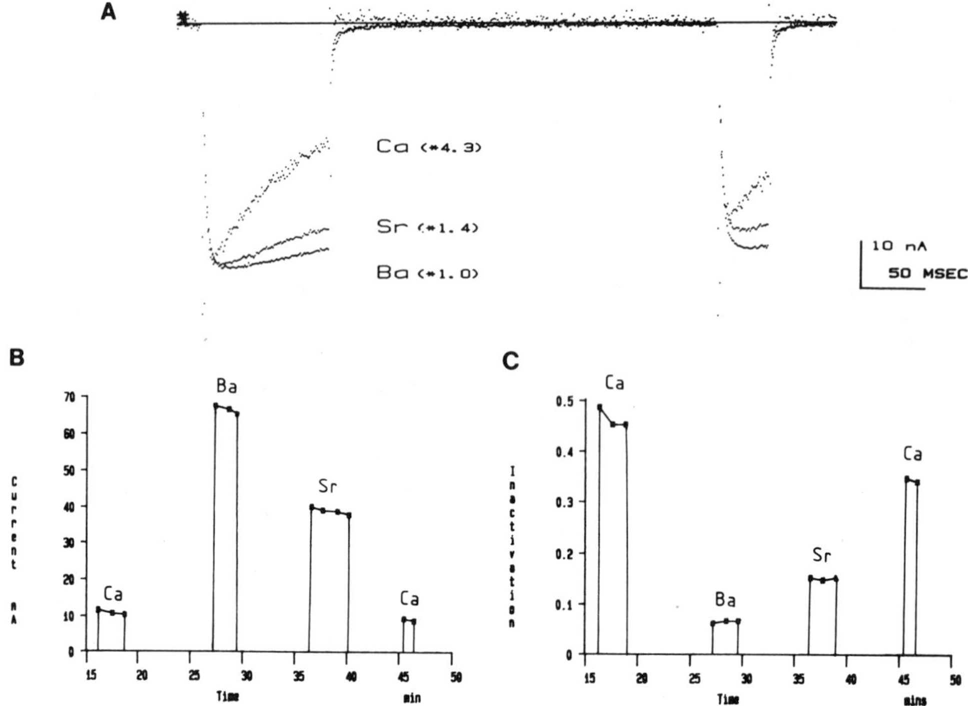

Figure 3. Divalent dependence of inactivation. A) Currents recorded from dialyzed <u>Helix</u> neurones during double pulses to +10 mV in 50 mM Ca, Ba or Sr. The first pulse peaks have been normalized to that of Ba using the multiplicative factor shown. The inactivation is most pronounced for Ca and least for Ba. B) The peak inward currents are plotted for the different divalents which were superfused sequentially over the cell. The order of current magnitude is Ba>Sr>Ca. C) The inactivation during the first pulse is plotted for the three divalents tested. The order of inactivation is Ca>Sr>Ba. The current-dependent inactivation depends upon the species of divalent rather than the absolute magnitude of the current.

carrying current, the inactivation increases during the response in proportion to the integral of the current (Chad et al., 1983,1984; Standen and Stanfield, 1982). However, there is a strong dependence on the nature of the divalent ion. The sequence for generation of inactivation is Ca>Sr>Ba, which matches the selectivity of calmodulin (Cheung, 1984). In these experiments measurements were made sequentially on the same cell and show that the order of peak current amplitudes (Fig. 3B) is the reverse of the order of inactivation measured at the end of the pulse (Fig. 3C). This suggests that the mechanism of inactivation is a separate and distinct process from that which confers selectivity on ion permeation through the channel (Hess et al., 1986; Tsien et al., 1987).

A further difference between the divalents is that the I/V curves are moved to more negative potentials by the substitution of barium for calcium. Thus, the effective depolarization is greater for barium than calcium currents in these experiments. However the inactivation is less in barium, again suggesting that inactivation is dependent on the divalent current rather than membrane potential.

202

CALCINEURIN ENHANCES Ca-DEPENDENT INACTIVATION

The apparent loss of voltage-dependent calcium channel activity
during washout, which can be restored by phosphorylation, appears to be
similar to the loss or inactivation of current during a single response.
Both washout and inactivation have a calcium-dependence, and both can be
slowed by increased kinase activity (Chad and Eckert, 1986). One can
surmise that there would be a system to remove the phosphorylation
produced by kinase activity and that this would require a phosphatase (cf.
Reuter, 1983). Thus, a plausible mechanism for the development of
calcium-dependent inactivation would be the calcium-dependent increase in
activity of a phosphatase. If this were the case the properties of the
phosphatase would confer the divalent specificity upon the system.

Phosphatase IIb, calcineurin, is a calcium-dependent phosphatase
originally isolated from bovine brain, which under physiological
conditions is closely associated with calmodulin. Calcium can directly
stimulate phosphatase activity at micromolar concentrations via binding
sites within calcineurin itself, and also indirectly via calmodulin (Klee
et al., 1979; Stewart et al., 1982). Thus, calcineurin fulfills the
postulated requirements for the linkage between calcium entry and the
inactivation. The ability of exogenous calcineurin to enhance
calcium-dependent inactivation was tested by its inclusion in the
dialysate of isolated <u>Helix</u> neurons. The basic dialysate included CsCl,
Hepes, leupeptin, ATP-Mg and catalytic subunit to stabilize the calcium
current, but no calcium buffers that would interfere with the elevation of
intracellular calcium ion concentration (Chad and Eckert, 1985b, 1986).

The effect of adding calcineurin (40 µg protein/ml) and calmodulin
(10 µM) during the course of dialysis with a fixed concentration of
catalytic subunit and stable calcium current is shown in Figure 4A. The
stimulus protocol consisted of paired steps to +10 mV from a holding
potential of −40mV. The control currents (i) show the typical activation
to a peak followed by an inactivation during the prolonged first pulse.
The second pulse elicits a calcium current of similar shape but with a
decreased peak amplitude due to residual inactivation from the first
pulse. Measurements of the response to the second pulse at different
interpulse intervals were used to monitor the recovery from inactivation.
After dialysis with calcineurin/calmodulin there was little change in the
initial current peak (ii) but an increased rate of inactivation during the
pulse. This can be seen in more detail in currents recorded on a faster
time scale (B). In this experiment pulses with a single stimulus at
varying intervals were repeated to monitor the changes in the calcium
current through time. The control trace (i) was obtained when the calcium
current was stabilized as before. Addition of calcineurin to the
dialysate led to a progressive increase in the inactivation (ii) and some
reduction in peak current. Inactivation occurring during the rising phase
of the calcium current will tend to decrease the peak current and also
decrease the time to peak (Chad et al., 1984). After some time the
dialysate was replaced with a simple ionic solution and the current was
allowed to washout (iii). There was little net current remaining after
washout, indicating the absence of any washout-resistant channel types.
The absence of net outward current after washout implies that
Ca-insensitive contaminating currents do not interfere with the
measurements of inactivation

The rate of recovery from inactivation, as indicated by the peak
current response to the second of the pair of command steps, seems to be
little affected. It is noteworthy that the addition of increased calcium
buffering capacity normally gives rise to a decreased rate of
calcium-dependent inactivation (Eckert and Tillotson, 1981; Chad et al.,

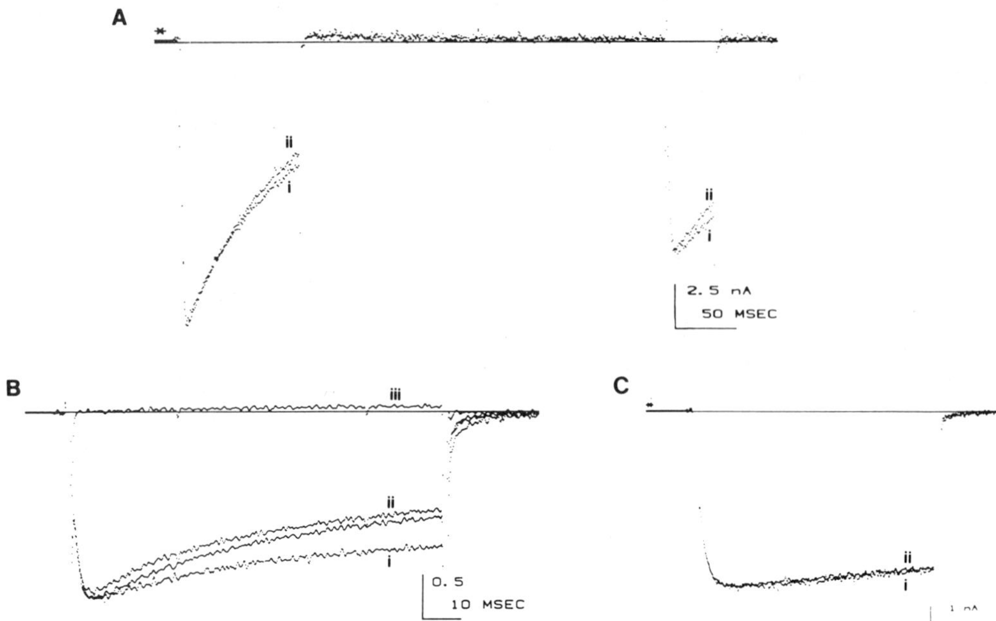

Figure 4. Calcineurin enhances calcium-dependent inactivation. A) The calcium currents (leak-corrected) recorded from a dialyzed cell during a double-pulse experiment with steps to +10 mV. The cell was dialyzed with Cs-aspartate, TEA, HEPES, ATP-Mg, leupeptin and CS to stabilize the calcium current. i. control currents show inactivation and recovery from inactivation during the interpulse interval. ii. After the addition to the dialysate of calcineurin/calmodulin (40 μg/ml, 10 μM) both pulses produced currents of similar peak size but with accelerated inactivation. Calcineurin promotes inactivation but has little effect on the rate of recovery from inactivation. B) Dialysis as before of another cell. i. Control current showing inactivation. ii. Addition of calcineurin/calmodulin had little effect on peak current but caused a progressive increase in inactivation. iii. Further dialysis with a solution lacking in kinase led to a complete washout of calcium current and revealed only an insignificant possible contamination or leak correction error. C) Dialysis of a neuron as before but with the solution supplemented with EGTA (1mM) gives a current with greatly reduced inactivation. i. Control current. ii. The addition of calcineurin/calmodulin produces no change in the current recorded in the presence of intracellular EGTA.

1984); however, addition of the extra calcium buffering of calcineurin/calmodulin leads to an enhanced inactivation. Thus the activity of calcineurin appears to be distinct from its simple role as a calcium buffer. The properties of the inactivation produced by calcineurin appear to be the same as the endogenous inactivation, dialysis of the cell with EGTA reduces the endogenous inactivation and also prevents any change on the introduction of calcineurin (Fig.4C). Similarly replacement of external calcium with barium blocks the inactivation produced by calcineurin (Chad and Eckert, 1986). This form

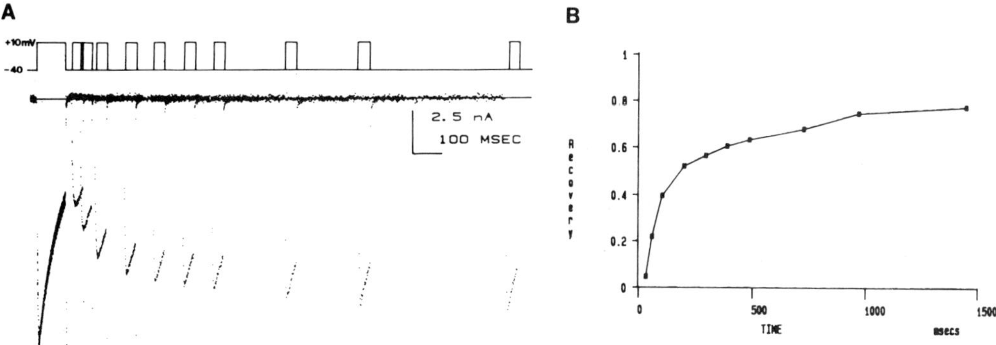

Figure 5. Recovery from inactivation. Experiments on Helix neurones
         dialyzed as before and stepped to +10 mV in a double pulse
         protocol. A) Ca currents recorded to double pulses with
         varying interpulse intervals are shown. The protocol is
         illustrated in the inset. The responses to the first
         pulse overlay each other showing full recovery between
         trials, the amplitude of the second, test, pulse is
         reduced by inactivation remaining from the first pulse.
         B) The recovery from inactivation is quantified as the
         ratio (peak of the test pulse – end maximally inactivated
         current of the first pulse) / (peak of first pulse – end
         of first pulse). This ratio is plotted against interpulse
         interval to show the recovery from inactivation, which is
         initially rapid but then slows and will finally be unity
         at 1 minute (time between trials).

of inactivation can also be modulated by altering the activities of
endogenous enzymes in intact neurons of <u>Aplysia</u> (Chad et al., 1987).

RECOVERY FROM INACTIVATION

     The hypothesis that calcium-dependent inactivation results from
dephosphorylation of a site on the calcium channel, or closely associated
with it, leads to the conclusion that recovery from inactivation requires
rephosphorylation of this site. We have used twin-pulse experimental
protocols on <u>Helix</u> neurons to investigate the properties of the recovery
from inactivation. The cells were voltage-clamped at –40 mV and were then
stepped to +10 mV for a 100 ms prepulse, which produces extensive
inactivation. After various periods of time at the holding potential, the
cell was stepped back to +10 mV in a test pulse to measure the extent of
recovery from calcium-mediated inactivation. This was repeated at minute
intervals for a number of interpulse intervals, and the resultant clamp
currents are shown in Figure 5A. The currents during the prepulse are
superimposed as the time between runs is adequate for complete recovery.
The test pulses show a progressive recovery from inactivation which can be
plotted as the ratio of the increase in test-pulse current peak above that
at the end of the pre-pulse, to the total inactivation during the
pre-pulse (Fig 5B). Thus, recovery is complete when this recovery ratio
is unity. This ratio does not measure the exact ratio of inactivated

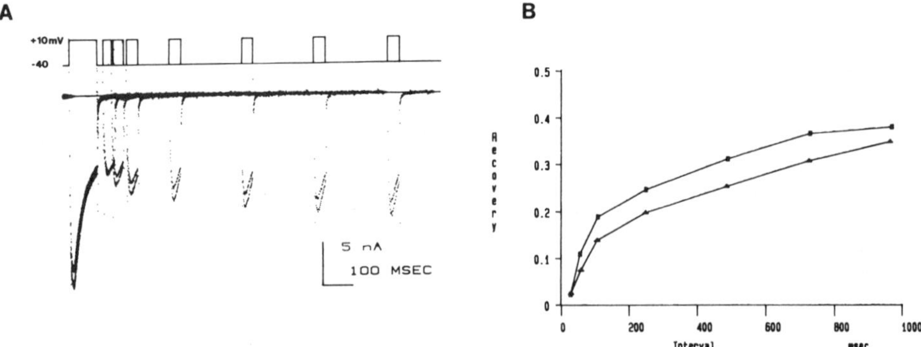

Figure 6.   Effect of increased kinase activity on recovery from
inactivation. A) Ca currents recorded during double pulse
experiments as before but for two different concentrations
of CS of cAMP-dependent protein kinase. Dialysis with
80 µg/ml CS gave a larger peak current in the first
pulse and in the second pulse at all interpulse
intervals. Dialysis with 8 µg/ml CS gave a smaller peak
current in each case. There was a full recovery to the
initial current size between trials for both
concentrations of CS. B) Recovery from inactivation
(calculated as before) was plotted against interpulse
interval for 8 µg/ml CS (triangles) and for 80 µg/ml
CS (squares).

channels at the various times as some inactivation occurs during the
activation phase of the recorded currents (Chad et al., 1984), which
distorts the measurements.

In order to see whether the kinetics of recovery were dominated by
the phosphorylating activity of the protein kinase or by the decrease in
dephosphorylating activity caused by the decrease in intracellular calcium
ion activity, we repeated these experiments on cells dialyzed with two
different concentrations of the catalytic subunit of cAMP-dependent
protein kinase, with fixed concentrations of ATP-Mg (7 mM) and
calcineurin/calmodulin (24 µg/ml). The cell was held at a potential of
-40 mV and calcium currents were elicited by paired steps to +10 mV. The
first pulse had a duration of 100 ms which allowed inactivation to
develop; the subsequent test pulse was briefer and measured the degree of
recovery of the peak current. These pairs of test pulses were repeated at
60 s intervals, which permitted complete recovery of the first peak
current. The alteration of the interpulse interval allowed us to follow
the time course of recovery from the inactivation produced by the first
pulse. The superposition of the resulting currents is shown in Figure
6A. A ten fold increase in the concentration of catalytic subunit (8 µg
protein/ml and 80 µg protein/ml) produced a slight increase in the peak
current and more inactivation during the first pulse. The pattern of
recovery observed from the envelopes of test pulses was very similar in
both cases. The recovery from inactivation was calculated and has been
plotted against interpulse interval in Figure 6B. This value will be
unity at 60 s in both cases because of the full recovery between tests.

The calculated recovery is not a direct measure of rephosphorylation as
the peak currents used in the calculation are not an accurate measure of
all the calcium channels available to be phosphorylated. Increased kinase
levels produce some increased recovery at all interpulse intervals but
little change in the rate of recovery from inactivation, despite a
ten-fold difference in the concentration of kinase. Reductions in the
level of catalytic subunit below the lower value lead to progressive loss
of the calcium current when tested at minute intervals and so were not
used in these experiments. Previous experiments have shown that the
addition of intracellular EGTA not only decreases the extent of
inactivation, but also speeds the recovery from the remaining
inactivation. Thus it appears that it is the decreasing dephosphorylating
activity produced by the reduction in intracellular calcium ion activity

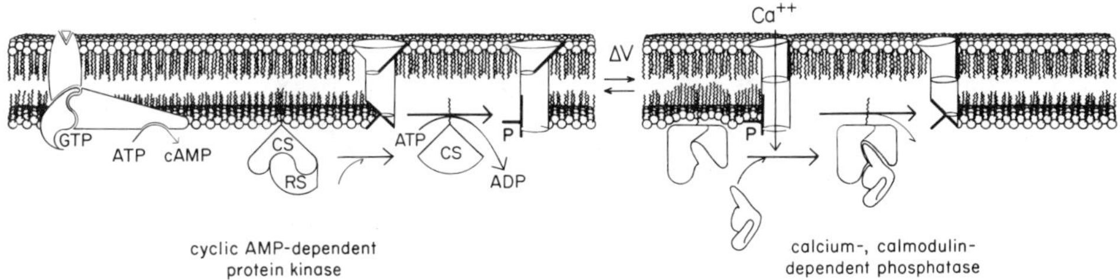

Figure 7. Depiction of the hypothesized enzymatic mechanism of
           Ca-dependent inactivation. The cAMP-dependent protein
           kinase is shown as a membrane bound complex. In the
           presence of cAMP the soluble catalytic subunit catalyses
           the phosphorylation of the calcium channel, removing
           inactivation and allowing the channel to be activated by
           depolarization. Depolarization opens the channel
           permitting entry of calcium ions which activate the
           membrane bound calcineurin/calmodulin complex to catalyse
           the dephosphorylation and hence inactivation of the
           calcium channel. The calmodulin is thought to be
           permanently complexed with the calcineurin under
           physiological conditions.

which dominates recovery from inactivation, rather than the level of
kinase activity, although the latter obviously has some effect. Kinase is
required to maintain channel activity in the long term, and also slows the
rate of inactivation during a single depolarizing pulse (Chad and Eckert,
1986). Similar experiments have also been conducted on intact _Aplysia_
neurons (Chad et al., 1987), with essentially similar results.

DISCUSSION

     The data we have presented is consistent with the overall scheme for
channel maintenance and modulation outlined in Figure 7. The predominant
assumptions of this scheme are that for the calcium channel to be

activated by depolarization it must be in the phosphorylated state,
dephosphorylation rendering it inactivated, i.e. greatly reducing the
probability of the channel opening in response to depolarization. Single
channel studies in GH3 cells also show that kinase activity can maintain
calcium channel activity in isolated membrane patches (Armstrong and
Eckert, 1987), increasing the likelihood that the phosphorylation site is
on the channel itself or on a closely associated protein. The proportion
of the channels which are in the inactivated state will be determined by
the balance between the activities of the phosphatases and kinases
present. It appears that in intact _Aplysia_ and _Helix_ neurons the resting
level of kinase activity predominates, maintaining most of the calcium
channels in the phosphorylated and hence activatable state. However, when
the channels are opened, the influx of calcium ions leads to activation of
a calcium-dependent phosphatase such as calcineurin and hence to an
increased rate of phosphatase activity. The shift in the balance towards
dephosphorylation increases the proportion of dephosphorylated,
inactivated channels producing the calcium-dependent inactivation
observed. The plausibility of this scheme is enhanced by the
immunocytochemical localization of calcineurin in GH3 cells and _Aplysia_
neurons (Farber et al., 1985; Saitoh and Schwartz, 1983), and also
identification of soluble cAMP-dependent protein kinases in _Aplysia_ and
Helix neurons (Bandle and Levitan, 1977; Novak-Hofer et al., 1985). This
system also appears to function in the intact _Aplysia_ neuron, and in the
GH3 cell (Kalman et al., 1987; Armstrong and Eckert, 1987; Chad et al.,
1987).

Advances in the study of the nature of the molecular identity of the
calcium channel have revealed interesting parallels between the enzymatic
maintenance and modulation of calcium channel activity and the properties
of the putative channel. The isolation of a dihydropyridine-binding
protein from skeletal muscle has revealed a multisubunit complex, one
subunit of which is very sensitive to proteolysis (Curtis and Caterall,
1984). One or more of these subunits can be phosphorylated by the
catalytic subunit of cAMP dependent protein kinase and also by
Ca-calmodulin kinase (Curtis and Caterall, 1985; Hosey et al., 1986). The
phosphorylated protein can also be dephosphorylated in a calcium-dependent
manner by calcineurin (Hosey et al., 1986). All these properties are
reflected in the dependence of calcium channel activity observed in our
experiments: sensitivity to proteolysis, dependence on phosphorylation by
a kinase, and the inactivation in a calcium-dependent manner by a
phosphatase.

This scheme also accounts for the washout of calcium current observed
upon dialysis and its prevention by the cAMP-dependent protein kinase.
The initial phase of washout can be slowed by supplementation with ATP-Mg
but not prevented. Exogenous catalytic subunit is necessary for long term
maintenance of the current when no calcium buffers are added to the
dialysate and calcium is used as the charge carrier. When dialysis is
established, the more rapidly diffusing cellular constituents are
removed. Initially the monovalent ions exchange, Cs replacing K and
blocking outward currents. As dialysis continues larger molecular weight
constituents such as ATP are lost, followed by the soluble catalytic
subunit of cAMP-dependent protein kinase. Kinase activity decreases and
elevated levels of intracellular calcium can activate a phosphatase to
give unopposed dephosphorylation and hence inactivation of the calcium
channels. There also seems to be a calcium-dependent process causing
irreversible loss of calcium channels which can be blocked by leupeptin.
A family of calcium-dependent proteases, the calpains, have been described
which are blocked by leupeptin and also EGTA (Murachi et al., 1983) and
which are normally controlled by an endogenous, soluble protein,
calpastatin (Aoyagi et al., 1983; Sasaki et al., 1984). This endogenous

regulator may be removed by dialysis permitting calcium to activate
calpain. We have observed that the irreversible loss of calcium current
appears to be slowed by kinase activity (Chad and Eckert, 1986), and it
has been shown that phosphorylation can reduce the susceptibility of
proteins to degradation by proteolytic enzymes (Holzer and Heinrich,
1980). Thus dialysis of neurons with EGTA containing solutions or using
Ba as a charge carrier may serve to prevent activation of both the
calcium-dependent phosphatase and protease systems and hence account for
the ability of ATP-Mg to substantially reduce washout (Byerly and
Yazejian, 1986). The degradation of calcium channels by a
calcium-dependent process could subserve a role in the control of the
lifetime of the channel itself, and hence in channel number and turnover.

The dependence of the calcium channel upon phosphorylation can
explain the divalent specificity of inactivation (Ca>Sr>Ba) if it is
assumed that the divalents have different abilities to stimulate the
phosphatase, whereas the channel conductance properties will determine the
divalent specificities of the current (Ba>Sr>Ca). It is interesting to
note that it has been recently reported that in the leech retzius cell the
order for inactivation of the current is the same as in _Helix_; however the
order for the divalent specificity of the current is different (Sr>Ba>Ca;
Bookman and Liu, 1987), providing further evidence for the separation of
the properties of permeation from those of inactivation. The data we have
presented cannot exclude the involvement of a cascade of phosphatase
activities in which calcineurin may dephosphorylate and hence activate
phosphatase 1 (Cohen, 1982; Hemmings et al., 1984) which could then
dephosphorylate the calcium channel or a closely associated site.
Phosphatase 1 has indeed been shown to interfere with the beta-adrenergic
stimulation of the calcium current in cardiac cells (Kameyama et al.,
1986); however, calcineurin can directly dephosphorylate the putative
calcium channel from skeletal muscle (Hosey et al., 1986). Thus the
scheme we have presented appears to be the simplest explanation for the
available data.

The fact that the calcium-mediated inactivation itself does not
appear to be removed, or washed out, by dialysis is of great interest,
particularly in the light of evidence that calcineurin and the catalytic
subunit of cAMP-dependent protein kinase can both be membrane-bound, often
in close association with one another (Aitken et al., 1982; Hathaway et
al., 1981). These properties may be of fundamental importance in that
they will be involved in the spatial control of calcium-dependent
processes. The location of elevations in calcium activity with respect to
the calcium sensors and their substrates will govern the responses
observed to calcium influx (Chad and Eckert, 1984; Haiech and Dmaille,
1981).

In summary it appears that the phenomenon of calcium-mediated
inactivation of calcium current is part of a channel regulation process
involving phosphorylation/dephosphorylation mechanisms. The
phosphorylation can be produced by the activity of the cAMP-dependent
kinase to phosphorylate the channel and hence allow it to be activated by
depolarization. Elevations of intracellular calcium activate a
phosphatase to dephosphorylate and hence inactivate the calcium channel
producing an autoregulation of calcium influx.

ACKNOWLEDGEMENTS

The original direction and inspiration for this work came from Roger
Eckert and was based on the pioneering studies conducted in his laboratory
by Paul Brehm and Doug Tillotson. I am very grateful for Roger's

friendship and guidance, and I would like to thank his family for their
continued affection.

I also thank D. Armstrong, G. Augustine, D. Kalman and H. Wheal for
their helpful comments and suggestions, and I am very grateful for the
gifts of calmodulin and calcineurin from C. Klee, catalytic subunit from
S. Halegoua, A. Nairn and I. Levitan, and calcium/calmodulin kinase from
T. Bartfai. This work was supported by a Javits Neuroscience Investigator
award to Roger Eckert from NINCDS, USPHS NS 08364. Attendance at this
Symposium was made possible by the generous provision of travel funds from
the Guarantors of Brain, and the Wellcome Trust.

REFERENCES

Aitken, A., Cohen, P., Santikarn, S., Williams, D.H., Calder, A.G., Smith,
    A. and Klee, C.B., 1982, Identification of the NH2-terminal blocking
    group of calcineurin B as myristic acid. FEBS Letters 150:314-318.
Aoyagi, T, Miyatu, S., Nanbo, M., Koyima, F., Matsuyaki, M., Isizuka, M.
    Takeuchi, T. and Umezawa, H., 1969, Biological activities of
    leupeptins. Journal of Antibiotics 22:558-561.
Armstrong, D. and Eckert, R., 1987, Voltage-activated calcium channels
    that must be phosphorylated to respond to membrane depolarization.
    Proceedings of the National Academy of Sciences of the USA. (in press)
Baker, P., 1976, The regulation of intracellular calcium. In: "Calcium in
    Biological Systems", Cambridge University Press, Cambridge.
Baker, P., 1987, In: "Calcium and the Cell", CIBA Symposium series.
Bandle, E. F. and Levitan, I. B., 1977, Cyclic AMP-stimulated
    phosphorylation of a high molecular weight endogenous protein
    substrate in sub-cellular fractions of molluscan nervous systems.
    Brain Research 125:325-331.
Bean, P. B., Nowycky, M. C. and Tsien, R. W., 1984, Beta-adrenergic
    modulation of calcium channels in frog ventricular heart cells.
    Nature 307:371-375.
Bookman, R. J. and Liu, Y., 1987, Calcium channel currents in cultured
    retzius cells of the leech. Biophysical Journal 51:226a.
Brehm, P. and Eckert, R. O., 1978, Calcium entry leads to inactivation of
    the calcium current in Paramecium. Science N.Y. 202:1203-1206.
Brezina, V., Eckert, R. and Erxleben, C., 1987, Suppression of calcium
    current by an endogenous neuropeptide in neurones of Aplysia
    Californica. Journal of Physiology. (in press)
Byerly, L. and Hagiwara, S., 1982, Calcium currents in internally perfused
    nerve cell bodies of Lymnaea stagnalis. Journal of Physiology
    322:503-528.
Byerly, L. and Yazejian, B., 1986, Intracellular factors for the
    maintenance of calcium currents in perfused neurons of the snail
    Lymnaea stagnalis. Journal of Physiology 370:631-650.
Carbone, E. and Lux, H. D., 1984, A low voltage-activated, calcium
    conductance in embryonic chick sensory neurons. Biophysical Journal
    46:413-418.
Cavalie, A., Ochi, R., Pelzer, D. and Trautwein, W., 1983, Elementary
    currents through calcium channels in guinea-pig myocytes. Pflugers
    Archiv 398:284-297.
Chad, J. E. and Eckert, R. O., 1984, Calcium "domains" associated with
    individual channels may account for anomalous voltage relations of
    Ca-dependent responses. Biophysical Journal 45:993-999.
Chad, J. and Eckert, R., 1985, Calcineurin, a calcium-dependent
    phosphatase, enhances Ca-mediated inactivation of Ca current in
    perfused snail neurons. Biophysical Journal 47:266a.
Chad, J. E. and Eckert, R., 1986, An enzymatic mechanism for calcium
    current inactivation in dialyzed Helix neurons. Journal of
    Physiology 378:31-51.

Chad, J.,Eckert, R. and Ewald, D., 1983, Kinetics of calcium current inactivation simulated with an heuristic model. Biophysical Journal 41:61a.

Chad, J. E., Eckert, R. and Ewald, D., 1984, Kinetics of Ca-dependent inactivation of calcium current in neurones of Aplysia californica. Journal of Physiology 347:279-300.

Chad, J., Kalman, D. and Armstrong, D., 1987, The role of cyclic AMP-dependent phosphorylation in the maintenance and modulation of voltage-activated calcium channels. Journal of General Physiology. (in press)

Cheung, W. Y., 1984, Calmodulin: its potential role in cell proliferation and heavy metal toxicity. Federation proceedings 43:2995-2999.

Cohen. P., 1982, The role of protein phosphorylation in neural and hormonal control of cellular activity. Nature 296:613-620.

Colquhoun, D., Neher, E. Reuter, H. and Stevens C. F., 1981, Inward current channels activated by intracellular Ca in cultured cardiac cells. Nature 294:752-757.

Connor, J. and Stevens, C., 1971, Voltage clamp studies of a transient outward membrane current in gastropod neural somata. Journal of Physiology 213:21-30.

Curtis, B. M. and Caterall, W. A., 1984, Purification of the calcium antagonist receptor of the voltage-sensitive calcium channel from skeletal transverse tubules. Biochemistry 23:2113-2118.

Curtis, B. M. and Caterall, W. A., 1985, Phosphorylation of the calcium antagonist receptor of the voltage sensitive calcium channel by cAMP-dependent protein kinase. Proceedings of the National Academy of Sciences of the USA. 82:2528-2532.

Doroshenko, P. A., Kostyuk, P. G., Martynyuk, A. I., 1982, Intracellular metabolism of adenosine 3'-5'-cyclic monophosphate and calcium inward current in perfused neurones of Helix pomatia. Neuroscience 7:2125-2134.

Doroshenko, P. A., Kostyuk, P. G., Martynyuk, A. I., Kursky, M. D. and Vorobetz, Z. D., 1984, Intracellular protein kinase and calcium inward currents on perfused neurones of the snail Helix pomatia. Neuroscience 11:263-267.

Eckert, R. and Chad, J. E., 1984, Inactivation of calcium currents. Progress in Biophysics and Molecular Biology 44:215-267.

Eckert, R. and Ewald, D., 1983a, Calcium tail currents in voltage-clamped intact nerve cell bodies of Aplysia californica. Journal of Physiology 345:533-548.

Eckert, R. and Ewald, D., 1983b, Inactivation of calcium conductance characterised by tail current measurements in neurones of Aplysia californica. Journal of Physiology 345:549-565.

Eckert, R. and Tillotson, D., 1981, Calcium-mediated inactivation of the calcium conductance in caesium-loaded giant neurones of Aplysia californica. Journal of Physiology 314:265-280

Farber, L., Ianetta, F., Kirby, T. and Wolff, D. J., 1985, Calmodulin dependent phosphatase of PC-12, C-6-Glioma and GH3 pituitary adenoma cell lines. Neuroscience Abstracts 11:855

Fenwick, E. M., Marty, A. and Neher, E., 1982, Sodium and calcium channels in bovine chromaffin cells. Journal of Physiology 331:599-635.

Forscher, P. and Oxford, G. S., 1985, Modulation of calcium channels by norepinephrine in internally dialyzed avian sensory neurons. Journal of General Physiology 85:743-763.

Hagiwara, S. and Byerly, L., 1982, Calcium channel. Annual Review of Neuroscience 4:69-125.

Hagiwara, S. and Nakajima, S., 1966, Effects of the intracellular Ca ion concentration upon the excitability of the muscle fiber membrane of a barnacle. Journal of General Physiology 49:807-818.

Haiech, J. and Dmaille, J. G., 1981, Supramolecular organisation of
     regulatory proteins into calcisomes: a model of the concerted
     regulation by calcium ions and cyclic adenosine 3'-5'-monphosphate in
     eukaryotic cells. In: "Metabolic interconversion of Enzymes", Proc.
     in Life Sciences:303-313 Springer Verlag, Berlin.
Hajos, F. Garthwaite, G. and Garthwaite, J., 1986, Reversible and
     irreversible neuronal damage caused by excitatory amino acid
     analogues in rat cerebellar slices. Neuroscience 18:417-436.
Hathaway, D. R., Adelstein, R. S. and Klee, C. B., 1981, Interaction of
     calmodulin with myosin light chain kinase and cyclic AMP-dependent
     protein kinase in bovine brain. Journal of Biological Chemistry
     230:8183-8189.
Hemmings, H. C., Greengard, P., Lim Tung, H. Y. and Cohen, P., 1984,
     DARPP-32, a dopamine-regulated neuronal phosphoprotein, is a potent
     inhibitor of protein phosphatase-1. Nature 310:503-505.
Hess, P., Lansman, J. B. and Tsien, R. W., 1986, Calcium channel
     selectivity for divalent and monovalent cations. Voltage- and
     concentration-dependence of single channel current in guinea pig
     ventricular heart cells. Journal of General Physiology 88:293-319.
Holzer, H. and Heinrich, P. C., 1980, Control of proteolysis. Annual
     review of Biochemistry 49:63-91.
Hosey, M. M., Borsetto, M. and Lazdunski, M., 1986, Phosphorylation and
     dephosphorylation of dihydropyridine sensitive voltage-dependent
     calcium channel in skeletal muscle membranes by cAMP- and
     Ca-dependent processes. Proceedings of the National Academy of
     Sciences of the USA. 83:3733-3737.
Kalman, D., Erxleben, C. and Armstrong, D., 1987, Inactivation of the
     dihydropyridine-sensitive calcium current in GH3 cells is a
     calcium-dependent process. Biophysical Journal 51:432a.
Kameyama, M., Hofmann, F. and Trautwein, W., 1985, On the mechanism of
     beta-adrenergic regulation of the Ca channel in the guinea pig
     heart. Pflugers Archiv 405:285-293.
Kameyama, M., Hescheler, J., Mieskes, G. and Trautwein, W., 1986, The
     protein-specific phosphatase 1 antagonizes the beta-adrenergic
     increase in the cardiac Ca current. Pflugers Archiv 407:461-463.
Klee, C. B., Crouch, T. H. and Krinks, M. H., 1979, Calcineurin: A calcium
     and calmodulin-binding protein of the nervous system. Proceedings of
     the National Academy of Sciences of the USA. 76:6270-6273.
Klee, C. B., Crouch, T. H. and Richman, P. G., 1980, Calmodulin. Annual
     Review of Biochemistry 49:489-515.
Kostyuk, P. G. and Krishtal, O. A., 1977, Separation of sodium and calcium
     currents in the somatic membrane of mollusc neurons. With an
     Appendix by Shakhovalov,Yu A. Journal of Physiology 270:545-568.
Lee, K. S., Akaike, N. and Brown, A. M., 1978, Properties of internally
     perfused, voltage-clamped, isolated nerve cell bodies. Journal of
     General Physiology 71:489-507.
Lee, K. S., Marban, E. and Tsien, R. W., 1985, Inactivation of calcium
     channels in mammalian heart cells: joint dependence on membrane
     potential and intracellular calcium. Journal of Physiology
     364:395-411.
McCleskey, E. W., Fox, A. P., Feldman, D. and Tsien, R. W., 1986,
     Different types of calcium channels. Journal of Experimental
     Biology 124:177-190.
Meech, R. W., 1978, Calcium dependent-potassium activation in nervous
     tissues. Annual Review of Biophysics and Bioengineering 7:1-18.
Mentrard, D., Vassort, G. and Fischmeister, R., 1984, Calcium-mediated
     inactivation of the calcium conductance in caesium-loaded frog heart
     cells. Journal of General Physiology 83:105-131.
Murachi, T., 1983, Intracellular Ca protease and its inhibitor protein:
     Calpain and calpastatin. In: "Calcium and Cell function", vol. IV.
     ed. Cheung,W.Y. pp. 377-410, Academic Press.

Nilius, B., Hess, P., Lansman, J. B. and Tsien, R. W., 1985, A novel type
    of cardiac calcium channel in ventricular cells. Nature 316:443-446.
Nilius, B. and Roder, A., 1985, Direct evidence of Ca-sensitive
    inactivation of slow inward channels in frog atrial myocardium.
    Biomedical and Biochemical Acta 44:1151-1161.
Novak-Hofer, I., Lemof, S., Villemain, M. and Levitan, I. B., 1985,
    Calcium and Cyclic nucleotide-dependent protein kinases and their
    substrates in the Aplysia nervous system. Journal of Neuroscience
    5:151-159.
Nowycky, M. C., Fox, A. P. and Tsien, R. W., 1984, Two components of
    calcium channel current in chick dorsal root ganglion cells.
    Biophysical Journal 45:36a.
Nowycky, M. C., Fox, A. P. and Tsien, R. W., 1985, Three types of neuronal
    calcium channel with different calcium agonist sensitivity. Nature
    316:440-443.
Osterrieder, W., Brum, G. Hescheler, J., Trautwein, W., Flockerzi, V. and
    Hofman, R., 1982, Injection of subunits of cyclic AMP-dependent
    protein kinase into cardiac myocytes modulates calcium current.
    Nature 298:576-578.
Plant, T. D. and Standen, N. B., 1981, Calcium current inactivation in
    identified neurones of _Helix aspersa_. Journal of Physiology
    321:273-285.
Plant, T. D., Standen, N. B. and Ward, T. A., 1983, The effects of
    injection of calcium ions and calcium chelators on calcium channel
    inactivation in Helix neurones. Journal of Physiology 334:189-212.
Reuter, H., 1983, Calcium channel modulation by neurotransmitters, enzymes
    and drugs. Nature 301:569-574.
Saitoh, T. and Schwartz, J. H., 1983, Serotonin alters the subcellular
    distribution of a calcium/calmodulin-binding protein in neurons of
    Aplysia. Proceedings of the National Academy of Sciences of the
    USA. 80:6708-6712.
Sasaki, T., Kikuchi, T., Yomoto, N., Yoshimura, N. and Murachi, T., 1984,
    Comparative specificity and kinetic studies on porcine calpain I and
    calpain II with naturally occurring peptides and synthetic
    fluorogenic substrates. Journal of Biological Chemistry
    259:12489-12494.
Simon, S. M. and Llinas, R. R., 1985, Compartmentalization of the
    submembrane calcium activity during calcium flux and its significance
    in transmitter release. Biophysical Journal 48:485-498.
Standen, N. B. and Stanfield, P. R., 1982, A binding-site model for
    calcium channel inactivation that depends on calcium entry.
    Proceedings of the Royal Society of London. B 217:101-110.
Stewart, A. A., Ingbretsen, T. S., Manalan, A., Klee, C. B. and Cohen, P.,
    1982, Discovery of a Ca- and calmodulin-dependent protein
    phosphatase: probable identity with calcineurin, CaM-BP 80). FEBS
    Letters 137:80-84.
Stoclet, J. C., Gerard, D., Kilhoffer, M-C., Lugnier, C. Miller, R. and
    Schaeffer, P., 1986, Calmodulin and its role in intracellular calcium
    regulation. Progress in Neurobiology. (in press)
Tillotson, D., 1979, Inactivation of Ca conductance dependent on entry of
    Ca ions in molluscan neurons. Proceedings of the National Academy of
    Sciences of the USA. 77:1497-1500.
Tsien, R. W., Hess, P., McCleskey, E. W. and Rosenberg, R. L., 1987,
    Calcium channels: Mechanisms of selectivity, permeation and block.
    Annual Review of Biophysics and Biophysical Chemistry. (in press)
Tsien, R. W., 1983, Calcium channels in excitable cell membranes. Annual
    Review of Physiology 45:341-358.
Tsien, R. W., Giles, W. and Greengard, P., 1972, Cyclic AMP mediates the
    effects of adrenaline on cardiac purkinje fibres. Nature New
    Biology 240:181-183.
Yellen, G., 1982, Single calcium-activated non-selective cation channels
    in neuroblastoma. Nature 296:357-359.

THE ROLE OF PROTEIN PHOSPHORYLATION IN THE RESPONSE OF DIHYDROPYRIDINE-
SENSITIVE CALCIUM CHANNELS TO MEMBRANE DEPOLARIZATION IN MAMMALIAN
PITUITARY TUMOR CELLS

By David Armstrong* and Daniel Kalman

Department of Biology, UCLA, Los Angeles, California, and
*Laboratory of Cellular and Molecular Pharmacology,
 National Institute of Environmental Health Sciences,
 Research Triangle Park, North Carolina

INTRODUCTION

The phosphorylation and dephosphorylation of serine and threonine
alters the activity of a wide variety of proteins that contribute to the
characteristic structure and function of cells in the brain (Nestler &
Greengard, 1986). In fact, since electric fields act only on charged
molecules, it may seem obvious in retrospect that adding or removing a
densely charged phosphate group should alter the response of ion channels
to changes in the voltage across the membrane. In that regard it is
interesting to note that, despite the striking structural homology in the
putative membrane spanning portions of voltage-activated sodium and
calcium channel proteins (Tanabe et al., 1987), only sodium channels
continue to respond to depolarization when the cell's cytoplasm is
replaced with artificial saline solutions lacking the ingredients to
support protein phosphorylation (Baker et al., 1962; Hagiwara & Nakajima,
1966). The recent advances in electrophysiological techniques which allow
one to observe the current passing through individual ion channels in
cell-free patches of native membrane (Hamill et al., 1981) have made it
possible to investigate such differences directly by studying the
phosphorylation-dependent alterations in ion channel activity produced by
purified kinases and phosphatases and inhibitors of their native
counterparts. We have used this approach to study the role of protein
phosphorylation in the regulation of a prominent class of voltage-
activated calcium channels in cells derived from a rat pituitary tumor
(GH_3; Tashjian, 1979).

RESULTS AND DISCUSSION

Two kinds of voltage-activated calcium channels in GH_3 cells

The two classes of unitary inward barium currents observed during depolarizations of cell-free patches from GH$_3$ cells (Armstrong & Eckert, 1985; 1987) are illustrated in Figure 1A. One class (traces at lower left and right) has a smaller conductance (~10 pS) and a lower threshold of activation near -40 mV. The other class (traces at upper left) has a larger conductance (~25 pS), a higher threshold of activation near -20 mV, and shorter mean open times. Both classes are blocked by 2 mM cobalt and have extrapolated reversal potentials more positive than +40 mV. No other voltage-activated channels are observed in cell-free patches from GH$_3$ cells with cesium chloride solutions on the cytoplasmic side of the patch and tetrodotoxin (TTX,2 μM) and tetraethylammonium ions (TEA, 20 mM) on the extracellular side.

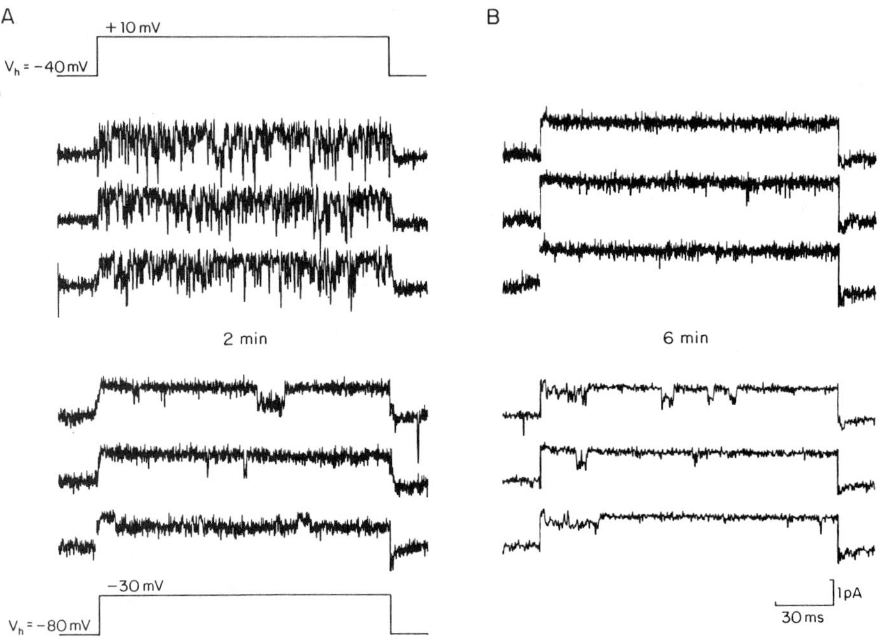

Figure 1. Unitary barium currents through voltage-activated calcium channels in an outside-out patch from a GH$_3$ cell exposed to 90 mM Ba, 20 mM TEA and 2 μM TTX on the extracellular side of the membrane and a CsCl solution in the pipette, buffered with HEPES (pH 7.2) and EGTA (pCa 8). A. Current records taken during the indicated voltage steps 2 minutes after excising the patch; B. the response to identical voltage steps 4 minutes later still.

Six minutes after excising the outside-out patch in Figure 1 into a standard physiological saline solution, the larger conductance channel stopped responding to depolarization altogether (Fig. 1B, traces at upper right). In contrast, the smaller conductance channel continued to respond to depolarization as long as the gigaohm seal remained intact (traces at lower right). Thus, unlike the larger conductance channels, they do not appear to require exogenous metabolic maintenance to remain active. In the absence of the larger conductance openings at more positive potentials, it can also be seen that the smaller conductance channels undergo nearly complete steady-state inactivation at the holding potential of -40 mV (Fig. 1B). Although the larger conductance channels do not inactivate in barium, they do require some cytoplasmic factor for activity. Furthermore, the larger conductance channels, but not the

216

smaller conductance ones, are modulated by dihydropyridines (Armstrong &
Eckert, 1987). Many of these differences in the properties of the two
kinds of calcium channels are summarized in Table 1.

TABLE 1.

Two Types of Calcium Channels in GH3 cells

L, large & long lasting	nomenclature	T, tiny & transient
$\sim$23 pS	conductance in 90mM Ba^{2+}	$\sim$10pS
high: -20mV	threshold of activation	low: -40mV
fast (FD) $\sim$1ms	rate of deactivation open channel lifetime	slow (SD) 2-5 ms
calcium-dependent	inactivation	voltage-dependent
washes out	activity in cell-free patches	persists
yes	dihydropyridine-sensitive	no

Two classes of voltage-activated calcium channels with very similar
properties have been described in a wide variety of cells (Carbone & Lux,
1984; Deitmer, 1984; Armstrong & Matteson, 1985; Bean, 1985; Fedulova et
al., 1985; Nilius et al., 1985; Nowycky et al., 1985). A third class has
been described in excitable cells derived from the neural crest (Nowycky
et al., 1985; Fox et al., this volume); however, we have found no evidence
of such channels in GH$_3$ cells. Given the postulated role of such
channels in neurotransmitter release (Fox et al.; Umbach & Gundersen;
Ewald et al., this volume), it is important to note that GH$_3$ cells
synthesize and secrete prolactin in a manner that depends largely on
calcium influx across the plasma membrane through dihydropyridine-
sensitive channels (Enyeart et al., 1985; Luini et al., 1985).

This paper is concerned primarily with these dihydropyridine-
sensitive calcium channels that require some form of exogenous metabolic
maintenance to remain active in cell-free patches or dialyzed cells. They
have been given various nicknames by other investigators, including "Type
II," "L" for their large conductance and lack of inactivation in barium,
"HVA" for their high threshold of voltage activation, and "FD" because
their brief openings produce fast deactivating inward tail currents, and
they are also almost certainly the same class of channels that are
modulated by cyclic AMP-dependent phosphorylation in cardiac muscle cells
(Reuter, 1983; Tsien et al., 1986). In addition to the loss of activity,
or "wash-out," in minimal saline solutions, these channels can be
distinguished from other voltage-activated calcium channels in GH$_3$ cells
by their lack of inactivation in barium and by their sensitivity to
modulation by dihydropyridines (Cohen & McCarthy, 1987; Kalman et al.,
1987).

Figure 2A illustrates these two properties of the pharmacologically
isolated calcium current before it washed out in a GH$_3$ cell dialyzed
internally through a patch pipette in the whole-cell configuration. The
inactivation of the inward current during a sustained depolarization is
eliminated by substituting barium for calcium as the charge carrier in the
external solution. Thus, inactivation is judged to be a calcium-dependent
process. Furthermore, the calcium current recorded during voltage steps
from -40 mV was completely blocked by 1 uM nimodipine, a dihydropyridine
antagonist (Cohen & McCarthy, 1987). We chose -40 mV as the holding
potential because the dihydropyridine-insensitive calcium channels are
largely inactivated at that potential (cf. Fig. 1). Therefore,
properties like inactivation of these dihydropyridine-sensitive calcium
channels can be studied in isolation.

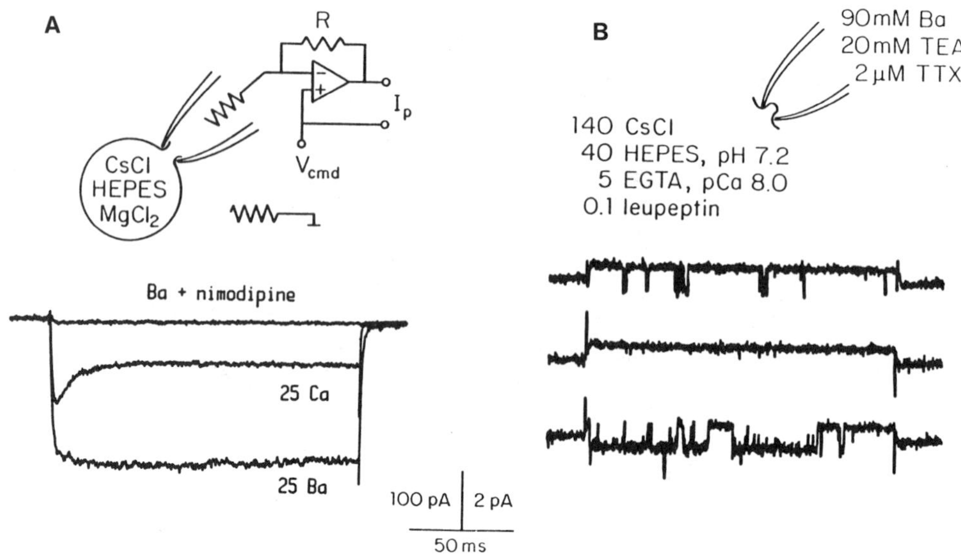

Figure 2. Two views of dihydropyridine-sensitive calcium channel gating in response to voltage steps from −40 to 0 mV. <u>A.</u> Macroscopic currents from dialyzed cells, voltage-clamped with patch pipettes in the whole-cell configuration. The inward current in 25 mM Ba does not inactivate and is completely blocked by 1 µM nimodipine. <u>B.</u> Representative traces of unitary barium currents in a cell-attached patch illustrating three distinct modes of gating.

Individual dihydropyridine-sensitive calcium channels respond to depolarization in one of three characteristic gating patterns or "modes" (Hess et al., 1984; Nowycky et al., 1985) that are illustrated in Figure 2B with records from a cell-attached patch on a GH$_3$ cell. They respond most frequently with clusters of very brief openings, but often they do not respond at all, as if they were inactivated when the membrane was depolarized. Much more rarely the channel responds with openings of very long duration. Transitions between these different modes occur every few seconds. Consequently, Tsien and his collaborators have suggested that each mode of gating reflects a different functional conformation of the channel and that dihydropyridines modulate these channels by stabilizing one or the other of the less frequent conformations, antagonists stabilizing the inactive state and agonists, like BAY K 8644 (Schramm et al., 1983), stabilizing the state with longer mean open times. We have tentatively identified the change in calcium channel structure that underlies the transitions between the silent and normal modes of activity.

<u>Phosphorylation-dependence of dihydropyridine-sensitive channel activity</u>

A number of observations (Armstrong, 1986; Armstrong & Eckert, 1985; 1987; Armstrong et al., 1987; Chad et al., 1987; Kalman et al., 1987a,b) have led us to conclude that wash-out, calcium-dependent inactivation and dihydropyridine sensitivity all depend on the same structural feature of this predominant class of voltage-activated calcium channels: when the channel or a closely associated regulatory site in the membrane is not phosphorylated by the cyclic AMP dependent protein kinase, the probability

218

of the channel being activated by physiological depolarization in its
native membrane is so close to zero that it cannot be detected.
Therefore, we refer to these dihydropyridine-sensitive channels as being
phosphorylation-dependent in the same spirit that acetylcholine-activated
channels at the vertebrate neuromuscular junction are called
acetylcholine-receptor channels even though they have been demonstrated to
open very infrequently in the absence of acetylcholine (Jackson, 1984).

The rapid cessation of activity by the large conductance channel in
cell-free patches exposed to simple saline solutions is a widespread
phenomenon (Fenwick et al., 1982; Cavalie et al., 1983; Carbone & Lux,
1984; Nilius et al., 1985; Armstrong & Eckert, 1987). This loss of
activity can be slowed by buffering the calcium ion concentration below
10^{-8} M with EGTA (Byerly and Hagiwara, 1982; Fenwick et al., 1982) or by
cooling the preparation significantly (Carbone & Lux, 1984), but not
prevented (Byerly & Yazejian, 1986; Chad & Eckert, 1986; Armstrong &
Eckert, 1987). However, when ATP-Mg and the catalytic subunit of the
cyclic AMP-dependent protein kinase are included in the solution bathing
the cytoplasmic side of the membrane, activity persists in cell-free
patches for as long as the gigaohm seal remains intact (Armstrong &
Eckert, 1987). In the absence of ATP, the purified kinase has no effect,
so it is reasonable to conclude that channel activity depends on
phosphorylation. Because activity can be restored by introducing ATP and
the kinase after several minutes of inactivity in inside-out patches that
have been continuously perfused (Armstrong & Eckert, 1987), the
phosphorylated site that regulates the response of these calcium channels
to depolarization must be closely associated with the molecules forming
the channel in the membrane. That conclusion is supported by the
demonstration that cyclic AMP-dependent protein phosphorylation
dramatically increases the activity of calcium channels purified from
mammalian skeletal muscle and reconstituted in lipid bilayers (Flockerzi
et al., 1986) or those expressed in _Xenopus_ oocytes after injections of
mammalian cardiac muscle messenger RNA (Dascal et al., 1986).

Figure 3 illustrates an important control experiment that we have
carried out to ensure that the lack of activity observed in Figure 1B

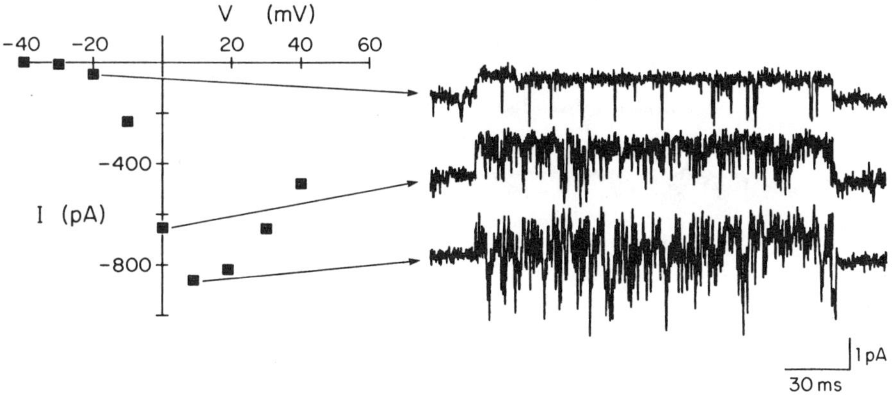

Figure 3. Current-voltage relation of macroscopic inward current in
whole cell exposed to 90 mM Ba, 20 mM TEA and 2 μM TTX. Subsequently,
the patch pipette was withdrawn from the cell to form an outside-out patch
in the same solution. On the right are displayed representative records
of unitary barium current activity during depolarizations from −40 mV to
the voltage indicated.

accurately reflects the probability of channel opening. On the left we
have plotted the current-voltage relation of a GH$_3$ cell voltage-clamped
in the whole-cell configuration at a holding potential of -40 mV. The
peak current elicited by depolarization in 90 mM Ba occurs at
approximately +10 mV. After obtaining the data for the I-V curve, the
patch pipette was withdrawn from the cell to form an outside-out patch.
Representative responses of the individual channels in that patch during
identical depolarizations are shown on the right. Thus, the channel is
activated maximally by depolarizations to +10 mV, so the absence of
activity at that voltage after wash-out truly reflects a dramatic
reduction in the probability of opening.

Figure 4 illustrates the effect of ATP alone on calcium channel
activity in cell-free patches in the absence of exogenous kinase. These
representative traces were obtained at 0.15 Hz during successive
depolarizing steps from -40 to +10 mV, 15 minutes after forming an
outside-out patch. In the experiment shown on the left (Fig. 4A), the
channel was still remarkably active although periods of activity were
often interrupted by several seconds of inactivity. In contrast, such
activity never persists in GH$_3$ cells for longer than 15 minutes at 20° C
in the absence of ATP (Armstrong & Eckert, 1987). The effect of ATP was
not mimicked by a nonhydrolyzable analogue of ATP, AMP-PCP (Armstrong,
1986). The typical result of including the Walsh inhibitor protein of the
cyclic AMP dependent kinase with ATP is illustrated in the experiment
shown on the right (Fig. 4B). Although a calcium channel did open
infrequently, proving that there was a functional channel remaining in the
patch, the probability of opening was more than two orders of magnitude

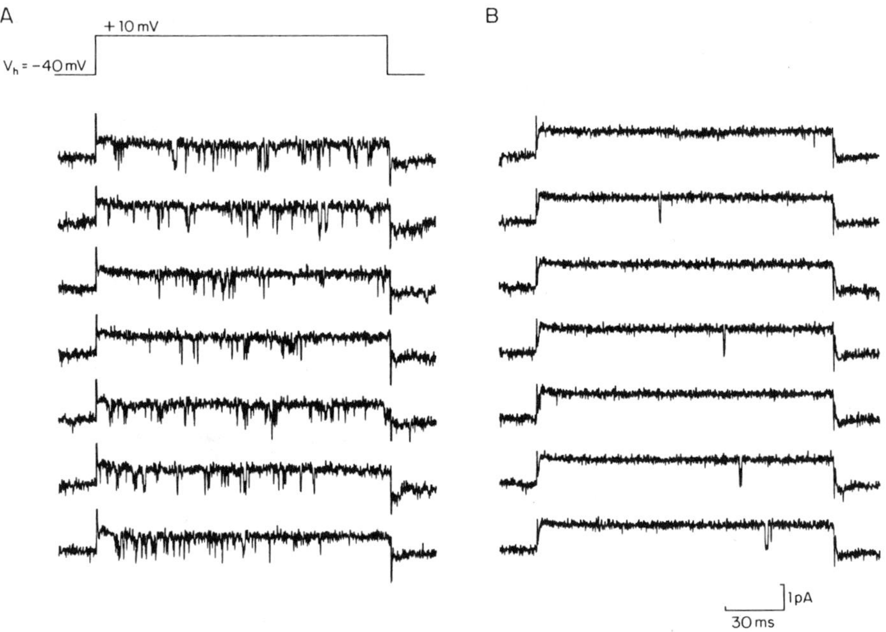

Figure 4. Calcium channel activity during depolarizations from -40 to
0 mV 15 minutes after forming an outside-out patch. Consecutive traces
taken at 6 sec intervals. A. In addition to the minimal CsCl solution
buffered with HEPES and EGTA, the pipette solution contained 2 mM ATP-Mg.
B. In a separate experiment the pipette solution used in A was
supplemented with approximately 0.2 µg/ml of the specific catalytic
subunit inhibitor protein, originally purified from skeletal muscle by
Walsh et al., 1971).

lower in the presence of the kinase inhibitor. This result strongly
suggests that an endogenous cyclic AMP dependent kinase is often bound to
the plasma membrane in GH$_3$ cells in sufficient proximity to the channel
to regulate its activity on a physiological time scale (Armstrong, 1986).

Calcium-dependent inactivation of phosphorylation-dependent calcium channels

Unlike voltage-activated sodium channels which exhibit voltage-
dependent inactivation, the predominant calcium channels in a wide variety
of cells throughout the animal kingdom inactivate in a calcium-dependent
manner (Eckert & Chad, 1984). Different mechanisms of inactivation are
appropriate for the different roles of those two voltage-activated
channels in nerve cell physiology. Sodium channels function to depolarize
the membrane by rapidly injecting charge into the cell, but because the
internal concentration of sodium is high, the sodium current does not
alter it. In contrast, the internal concentration of calcium is
submicromolar, and a variety of cellular processes are triggered by higher
concentrations. The calcium current through dihydropyridine-sensitive
calcium channels in GH$_3$ cells can produce a marked increase in free
calcium inside the cell (Schlegel et al., 1987), so it is not surprising
that mechanisms have evolved to inhibit calcium influx when the
intracellular concentration of calcium rises. Calcium-dependent
inactivation of calcium channels provides an important mechanism for such
feedback inhibition of calcium influx.

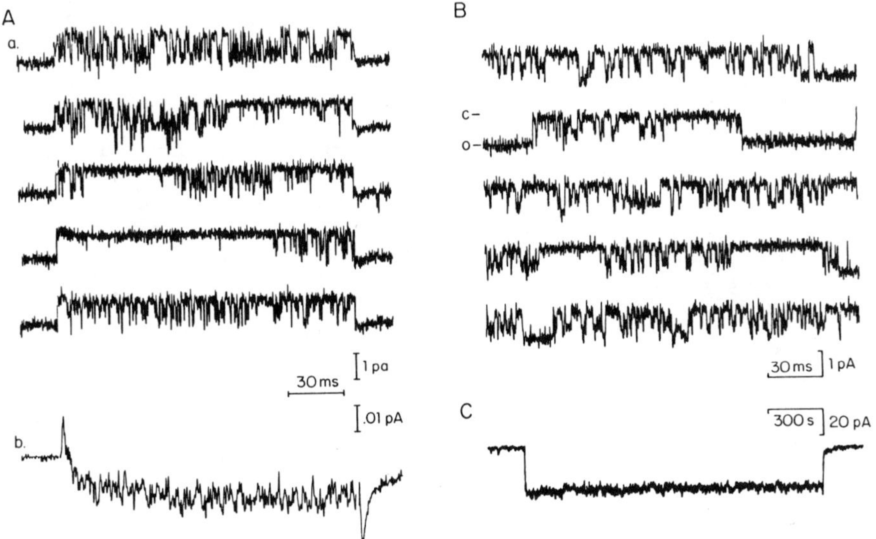

Figure 5. Barium currents through calcium channnels maintained by
phosphorylation do not inactivate. A. a, selected records of unitary
barium currents during voltage steps from -40 to 0 mV; b, ensemble average
of more than 100 records taken at 6 sec intervals from the same
outside-out patch in which 2 mM ATP-Mg and approximately 1.0 µg/ml of
the purified catalytic subunit of the cyclic AMP-dependent protein kinase
were added to the pipette solution bathing the cytoplasmic side of the
membrane. B. Consecutive records from another outside-out patch with ATP
and kinase present, in which the membrane was clamped to 0 mV and activity
was sampled at 3 sec intervals. Although the channel continued to
alternate between open (o-) and closed (c-) states, it did not
inactivate. C. Macroscopic inward current recorded from a GH$_3$ cell
dialyzed with CsCl and bathed in 25 mM Ba, 20 mM TEA and 2 µM TTX. Note
the different time scale; nevertheless, no inactivation is evident.

Historically, two mechanisms have been described for calcium channel inactivation: a voltage-dependent mechanism and a calcium-dependent one (Eckert & Chad, 1984). Eckert and his colleagues devised three criteria for distinguishing calcium-dependent mechanisms from voltage-dependent ones: substituting barium for calcium as the charge carrier in the external solution, depolarizing sufficiently close to the calcium equilibrium potential to reduce calcium influx, and reducing the rise in intracellular calcium with exogenous buffers. Each of these experimental manipulations reduced or eliminated the inactivation of the predominant calcium current in <u>Paramecia</u> and in molluscan neurons (Brehm, Eckert & Tillotson, 1980; Eckert & Tillotson, 1981).

When GH_3 cells are voltage-clamped in the whole-cell configuration without adding exogenous calcium buffers to the solution dialyzing the inside of the cell and held at -40 mV to minimize contamination by dihydropyridine-insensitive calcium channels, the inactivation of the dihydropyridine-sensitive, phosphorylation-dependent calcium current in GH_3 cells is strictly calcium-dependent by the criteria outlined above (Kalman et al., 1987a). When barium is substituted for calcium as the charge carrier (cf. Fig. 2A), we find no evidence for any additional voltage-dependent component of inactivation. That conclusion is underscored in Figure 5 which illustrates unitary barium currents through a dihydropyridine-sensitive calcium channel maintained by exogenous kinase and ATP in a cell-free patch from a GH_3 cell. Figure 5A contains five current traces taken during depolarizations from -40 to 0 mV and an ensemble average (Fig. 5Ab) of all the records from this patch, which illustrate that the probability of channel opening with barium as the charge carrier does not diminish during the depolarization. Figures 5B & 5C make the same point on a much longer time scale. Figure 5B shows five successive traces taken at 3 second intervals while the membrane was held continuously at 0 mV. Although the channel regularly enters periods of prolonged inactivity on this time scale, activity always resumes in an equally robust manner despite holding the membrane continuously at 0 mV (cf. Cavalie et al., 1986). Macroscopic barium currents on that time scale (Fig. 5C) confirm that conclusion, and exhibit no evidence for voltage-dependent inactivation of the phosphorylation-dependent calcium channels in GH_3 cells.

In view of the many parallels between the calcium-dependent loss of activity associated with both inactivation and wash-out and the ability to reverse wash-out by cyclic AMP-dependent phosphorylation, Eckert and Chad (1984) proposed that inactivation of the phosphorylation-dependent channels results when an endogenous calcium-dependent phosphatase is stimulated to dephosphorylate the channel by the increase in intracellular calcium associated with channel activation. In experiments on dialyzed molluscan neurons (Chad & Eckert, 1986; Chad, this volume), they have reconstituted calcium-dependent inactivation with calcineurin, a calcium- and calmodulin- dependent phosphatase (IIB) purified from mammalian brain (Klee et al., 1979; Stewart et al., 1982). Phosphatase I has been reported to reduce peak calcium currents induced by β-adrenergic stimulation in cardiac myocytes (Kameyama et al., 1986a), but it is unlikely that phosphatase I plays a role in calcium-dependent inactivation <u>in</u> <u>vivo</u> because its activity is neither calcium-dependent nor membrane-bound.

GH_3 cells also contain calcineurin (Wolff et al., 1987) and its basal, calcium-independent activity may be responsible for the loss of activity we observe in cell-free patches in the absence of ATP, even though we buffer the calcium ion concentration below 10 nM on the cytoplasmic side of the patch (Armstrong & Eckert, 1987). The dephosphorylation of one channel every few minutes in a cell-free patch by a calcium-dependent phosphatase in the absence of calcium is not inconsistent with the dephosphorylation of hundreds of channels every millisecond when macroscopic calcium currents raise intracellular calcium

levels in the absence of exogenous buffers (cf. Fig. 2B). Figure 6
demonstrates a clear connection between cyclic AMP-dependent
phosphorylation and calcium-dependent inactivation of the
dihydropyridine-sensitive calcium channels in GH$_3$ cells: inactivation
was largely eliminated when cyclic AMP production was stimulated maximally
by forskolin or vasoactive intestinal peptide (Kalman et al., 1987b).

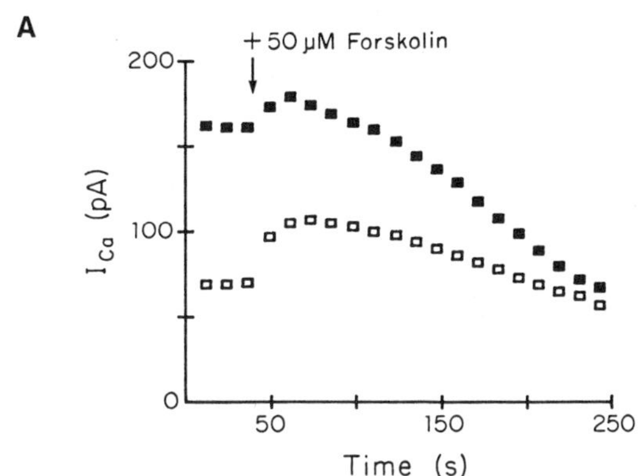

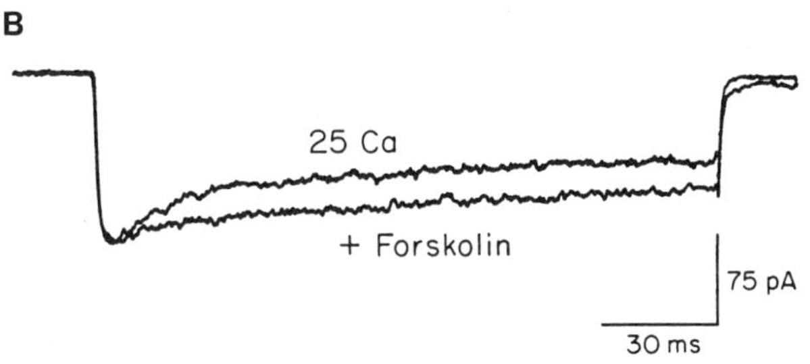

Figure 6. Stimulating cyclic AMP production with forskolin (Seamon
et al., 1981) slows calcium dependent inactivation. <u>A.</u> Peak (solid
squares) and steady-state (open squares) current recorded during voltage
steps from -40 to 0 mV from a dialyzed GH$_3$ cell versus time. Forskolin
(50 µM) was added at the arrow. Peak current continues to decline in
the absence of exogenous calcium buffers in the solution dialyzing the
cell interior. <u>B.</u> Representative records of currents with equal amplitude
peaks before and after adding forskolin.

Thus, the acceleration of both wash-out and inactivation of the
dihydropyridine-sensitive channels by calcium, and the reversal of both
processes by cyclic AMP-dependent phosphorylation, may reflect a single
structural alteration of the calcium channel protein. It is possible that
addition of a phosphate group to a single serine or threonine in the
cytoplasmic domain of the channel protein (Tanaba et al., 1987) shifts the
voltage-activation curve into the physiological voltage range. In this
view of the dihydropyridine-sensitive calcium channels, the dependence of
activity on phosphorylation is absolute, but multiple phosphorylation sites
may provide an explanation for additional effects of cyclic AMP-dependent
phosphorylation on calcium channel gating (Reuter et al., 1986; Tsien et al.,
1986; Kameyama et al., 1986b). However, unless the activity of the
dihydropyridine-sensitive calcium channels can be reconstituted in cell-
free patches of native membrane under conditions that preclude
phosphorylation of the channel, more complicated hypotheses will have to
await the construction of mutant proteins lacking various putative sites
of phosphorylation.

Acknowledgements. This work was supported by a Public Health Service
Grant (NS 8364) to the late Roger Eckert, the Los Angeles affiliate of the
American Heart Association and the Molecular Biology Training Grant (GM
07185) at UCLA. We are also grateful for the encouragement and support of
Drs. Paul O'Lague, Dennis O'Connor, Alan Grinnell and Francisco Bezanilla
at UCLA after Roger's death, and the generous gifts of pure reagents from
A. Nairn, K. Diltz and A. Scriabine.

REFERENCES

Armstrong, D., 1986, An endogenous cyclic AMP-dependent protein kinase
    modulates the activity of voltage-dependent calcium channels, Journal
    of General Physiology, 88:11a.

Armstrong, D. and Eckert, R., 1985, Phosphorylating agents prevent
    wash-out of unitary calcium currents in excised membrane patches.
    Journal of General Physiology, 86:25a-26a.

Armstrong, D. and Eckert, R., 1987, Voltage activated calcium channels
    that must be phosphorylated to respond to membrane depolarization.
    Proceedings of the National Academy of Sciences, 84:2818-2522.

Armstrong, D., Erxleben, C. and Kalman, D., 1987, Calcium channels
    modulated by BAY K 8644 appear less susceptible to dephosphorylation,
    Biophysical Journal, 51:233a.

Armstrong, C. M., and Matteson, D. R., 1985, Two distinct populations of
    calcium channels in a clonal line of pituitary cell. Science,
    227:65-67.

Baker, P. F., Hodgkin, A. L., and Shaw, T.I., 1962, Replacement of the
    axoplasm of giant nerve fibres with artificial solutions, Journal of
    Physiology, 164:330-337.

Bean, P. B., Nowycky, M. C., and Tsien, R. W., 1984, Beta-adrenergic
    modulation of calcium channels in frog ventricular heart cells,
    Nature, 307:371-375.

Bean, P. B., 1985, Two kinds of calcium channels in canine atrial cells,
    Journal of General Physiology, 86:1-30.

Brehm, P., Eckert, R., and Tillotson, D., 1980, Calcium-mediated
    inactivation of calcium current in Paramecium, Journal of Physiology,
    306:31-51.

Byerly, L., and Hagiwara, S., 1982, Calcium currents in internally
    perfused nerve cell bodies of Lymnaea stagnalis, Journal of
    Physiology, 322:503-528.

Byerly, L., and Yazejian, B., 1986, Intracellular factors for the
    maintenance of calcium currents in perfused neurons from the snail,
    Lymnaea stagnalis, Journal of Physiology, 370:631-650.

Carbone, E., and Lux, H. D., 1984, A low voltage-activated fully
    inactivating Ca channel in vertebrate sensory neurons, Nature,
    310:501-502.

Cavalie, A., Ochi, R., Pelzer, D., and Trautwein, W., 1983, Elementary
    currents through Ca^{2+} channels in guinea pig myocytes, Pflugers
    Archiv, European Journal of Physiology, 398:284-297.

Cavalie, A., Pelzer, D., and Trautwein, W., 1986, Fast and slow gating
behaviour of single calcium channels in cardiac cells. Pflügers
Archiv, European Journal of Physiology, 406:241-258.

Chad, J. E., and Eckert, R., 1986, An enzymatic mechanism for calcium
current inactivation in dialyzed Helix neurons, Journal of
Physiology, 378:31-51.

Chad, J. E., Kalman, D., and Armstrong, D., 1987, The role of cyclic AMP
dependent phosphorylation in the maintenance and modulation of
voltage activated calcium channels, in: "Cell Calcium and the
Control of Membrane Transport," D.C. Eaton and L.J. Mandel, editors.
Society of General Physiologists Series, Vol. 42, The Rockefeller
University Press, New York.

Cohen, C. J. and McCarthy, R. T., 1987, Nimodipine block of calcium
channels in rat anterior pituitary cells, Journal of Physiology,
387:195-225.

Deitmer, J., 1984, Evidence for two voltage-dependent calcium currents in
the membrane of the ciliate Stylonychia mytilus, Journal of
Physiology, 355:137-159.

Deitmer, J., 1986, Voltage-dependence of two inward currents carried by
calcium and barium in the ciliate Stylonychia mytilus, Journal of
Physiology, 380:551-574.

Doroshenko, P. A., Kostyuk, P. G., and Martynyuk, A. I., 1982,
Intracellular metabolism of adenosine 3'-5'-cyclic monophosphate and
calcium inward current in perfused neurones of Helix pomatia,
Neuroscience, 7:2125-2134.

Eckert, R., and Chad, J. E., 1984, Inactivation of calcium channels,
Progress In Biophysics and Molecular Biology, 44:215-267.

Eckert, R., and Tillotson, D., 1981, Calcium-mediated inactivation of the
calcium conductance in caesium-loaded giant neurones of Aplysia
californica, Journal of Physiology, 314:265-280.

Enyeart, J. J., Aizawa, T., and Hinkle, P. M., 1985, Dihydropyridine
Ca^{2+} antagonists: potent inhibitors of secretion from normal and
transformed pituitary cells, American Journal of Physiology,
248:C510-C519.

Farber, L. H., Wilson, F. J., and Wolff, D. J., 1987, Calmodulin-dependent
phosphatases of PC12, GH_3, C_6 cells: physical, kinetic and
immunochemical properties, Journal of Neurochemistry, 49:404-414.

Fedulova, S. A., Kostyuk, P. G., and Veselovsky, N. S., 1985, Two types of
calcium channels in the somatic membrane of new-born rat dorsal root
ganglion neurones, Journal of Physiology, 359:431-446.

Fenwick, E. M., Marty, A., and Neher, E., 1982, Sodium and calcium
channels in bovine chromaffin cells, Journal of Physiology,
331:599-635.

Hamill, O. P., Marty, A., Neher, E., Sakmann, B., and Sigworth, F. J.,
1981, Improved patch clamp techniques for high-resolution current
recording from cells and cell-free membrane patches, Pflugers Archiv.
European Journal of Physiology. 398:284-297.

Hess, P., Lansman, J. B., and Tsien, R. W., 1984, Different modes of Ca channel gating favored by dihydropyridine Ca agonists and antagonists, Nature, 311:538-544.

Jackson, M. B., 1984, Spontaneous openings of the acetylcholine receptor channel. Proc. Natl. Acad. Sci. USA, 81:3901-3904.

Kalman, D., Erlexben, C., and Armstrong, D., 1987a, Inactivation of the dihydropyridine-sensitive calcium current in GH_3 cells is a calcium-dependent process, Biophysical Journal, 51:432a.

Kalman, D., O'Lague, P. H., and Armstrong, D., 1987b, Increasing the intracellular concentration of cyclic AMP reduces Ca-dependent inactivation of Ca channels, Neuroscience Abstracts, (in press).

Kameyama, M., Hescheler, J., Mieskes, G., and Trautwein, W., 1986, The protein-specific phosphatase 1 antagonizes the β-adrenergic increase of the cardiac Ca current, Pflügers Archiv, European Journal of Physiology, 407:461-463.

Kameyama, M., Hescheler, J., Hofmann, F., and Trautwein, W., 1986, Modulation of Ca current during the phosphorylation cycle in the guinea pig heart, Pflügers Archiv, European Journal of Physiology, 407:123-128.

Kostyuk, P. G., 1984, Metabolic control of ionic channels in the neuronal membrane, Neuroscience, 13:983-989.

Klee, C. B., Crouch, T. H., and Krinks, M. H., 1979, Calcineurin: a calcium- and calmodulin-binding protein of the nervous system, Proceedings of the National Academy of Sciences, 76:6270-6273.

Luini, A., Lewis, D., Guild, S., Corda, D., and Axelrod, J., 1985, Hormone secretagogues increase cytosolic calcium by increasing cAMP in corticotropin-secreting cells, Proc. Natl. Acad. Sci. USA, 82:8034-8038.

Matteson, D. R., and Armstrong, C. M., 1986, Properties of two types of calcium channels in clonal pituitary cells, Journal of General Physiology, 87:161-182.

Nestler, E. J., and Greengard, P., 1984, "Protein Phosphorylation In The Nervous System," Wiley-Interscience, A Neurosciences Institute Publication, New York.

Nilius, B., Hess, P., Lansman, J. B., and Tsien, R. W., 1985, A novel type of cardiac calcium channel in ventricular cells, Nature, 316:443-446.

Nowycky, M. C., Fox, A. P., and Tsien, R. W., 1985, Three types of neuronal calcium channel with different calcium agonist sensitivity, Nature, 316:440-443.

Schlegel, W., Winiger, B. P., Mollard, P., Vacher, P., Wuarin, F., Zahnd, G. R., Wollheim, C. B., and Dufy, B., 1987, Oscillations of cytosolic Ca^{2+} in pituitary cells due to action potentials, Nature, 329:719-721.

Seamon, K. B., Padgett, W., and Daly, J. W., 1981, Forskolin: a unique diterpene activator of adenylate cyclase in membranes and in intact cells, Proceedings of the National Academy of Sciences, 78:3363-3367.

Stewart, A. A., Ingbretsen, T. S., Manalan, A., Klee, C. B., and Cohen, P., 1982, Discovery of a Ca- and calmodulin-dependent protein phosphatase: probable identity with calcineurin(CaM-BP$_{80}$), _FEBS Letters_, 137:80-84.

Tanabe, T., Takeshima, H., Mikami, A., Flockerzi, V., Takahashi, H., Kangawa, K., Kojima, M., Matsuo, H., Hirose, T., and Numa, S., 1987, Primary structure of the receptor for calcium channel blockers from skeletal muscle, _Nature_, 328:313-318.

Tashjian, A. H., 1979, Clonal strains of hormone-producing cells, _Methods in Enzymology_, 58:527-535.

Tsien, R. W., Bean, B. P., Hess, P., Lansman, J.B., Nilius, B., and Nowycky, M. C., 1986, Mechanisms of calcium channel modulation by β-adrenergic agents and dihydropyridine calcium agonists, _Journal of Molecular and Cellular Cardiology_, 18:691-710.

Walsh, D. A., Ashby, C. D., Gonzalez, C., Calkins, D., Fisher, E., and Krebs, E. G., 1971, Purification and characterization of a protein kinase inhibitor of adenosine 3',5' monophosphate-dependent protein kinases, _Journal of Biological Chemistry_, 246: 1977-1985.

MODULATION OF THE POTASSIUM CONDUCTANCE IN THE SQUID GIANT AXON

Eduardo Perozo*+, Christina K. Webb* and Francisco Bezanilla*

*Department of Physiology
Ahmanson Laboratory of Neurobiology and
Jerry Lewis Neuromuscular Research Center
University of California
Los Angeles, California 90024

+Instituto Venezolano de Investigaciones
Cientificas, Apartado 21827
Caracas 1020A, Venezuela

Marine Biological Laboratory
Woods Hole, Massachusetts 02543

INTRODUCTION

The potassium conductance of the squid giant axon is considered the
classical example of the delayed rectifier-type of voltage dependent
potassium permeability (Hodgkin and Huxley, 1952). In this preparation a
large number of experiments have been done on K ionic currents using
intact axons and under conditions of internal perfusion to study the
selectivity and voltage dependence of the conductance. It has also been
possible to record single channel events (Conti and Neher, 1980; Llano and
Bezanilla, 1985; Llano, Webb and Bezanilla, 1987) and gating currents
related to the movement of the charge responsible for the opening and
closing of the K conductance (White and Bezanilla, 1985). The squid giant
axon is then an ideal preparation for a detailed analysis of K channel
gating because all types of electrophysiological recordings can be made in
the same preparation, an important prerequisite for formulating a complete
model of channel gating. We review here evidence that the K conductance
in the squid axon is modulated by ATP-dependent phosphorylation. The
changes induced by ATP in the macroscopic currents appear to be the result
of a combined effect on more than one type of K channel. We also provide
evidence for a Ca-activated component of the K conductance.

INTERNAL ATP MODULATES K CONDUCTANCE

When a squid axon is dialyzed with a solution containing no ATP its
potassium current decreases to values as low as 50% of that observed
before dialysis is initiated (Bezanilla, Caputo, Dipolo and Rojas, 1986;
Perozo, Dipolo, Caputo, Rojas and Bezanilla, 1986). A typical family of
potassium currents recorded after dialyzing the axon with an ATP-free

solution is shown in Figure 1A. The dialysis and voltage clamp of the axon
was performed using techniques described previously (Brinley and Mullins,
1967; DiPolo, Bezanilla, Caputo and Rojas, 1985). The dialysis capillary
had a molecular weight cut-off of 9000 daltons. The internal solution
usually contained 20 mM phosphate. In this experiment the membrane
potential was held at -60 mV and pulses were given to more depolarized
potentials. Several features of the K currents were sensitive to internal
ATP. In the absence of ATP, the currents turned on more rapidly than in
the presence of ATP. At test potentials more positive than -40mV, the
currents recorded in the absence of ATP were smaller; at potentials more
negative than this value, the currents were actually larger in the absence
of ATP. When the axon was then dialyzed with a solution containing 2 mM
ATP (Fig 1B), the currents recovered most of their original amplitude and
kinetics recorded before initiating the dialysis.

Using perfused squid axons, ATP was found to change the K currents in
a manner similar to that seen in the dialyzed axon (Bezanilla et al., 1986;
Webb and Bezanilla, 1986). The axons were perfused and voltage clamped
using the modified Tasaki canulation technique (see Bezanilla, Vergara and
Taylor, 1982). In most of the experiments the internal perfusate contained
fluoride to increase the membrane resistance. The addition of 2 mM ATP to
the internal solution stimulated current amplitude and changed the kinetics
of both turn-on and turn-off; the turn-on kinetics were slower and the
turn-off kinetics were faster (Bezanilla et al, 1986; Webb and Bezanilla,
1986). It is interesting to note that the ATP effect in the perfused axon
does not reverse when ATP is washed out. This presumably is because
perfusing the axon removes soluble molecules which are normally responsible
for reversing the effect of ATP.

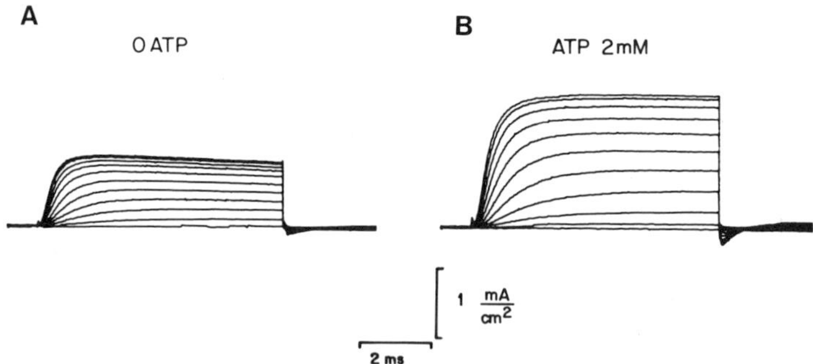

Fig. 1. Family of potassium currents recorded from a dialyzed axon.
A, after dialyzing with an ATP-free solution. B, In the
presence of ATP. The holding potential was -60 mV and test
pulse duration was 8 msec. The internal solution contained
310 mM K+, 3 mM Mg2+, 400 mM aspartate, 20 mM phosphate, 1
mM EGTA. The external solution was artificial sea water and
contained 300 nM TTX and 1 mM NaCN. Leakage and capacitive
currents have been subtracted.

ATP EFFECT IS MEDIATED BY A PHOSPHORYLATING STEP

The effects of ATP on K currents in both dialyzed and perfused axons could be due to the phosphorylation of a channel-related protein (Levitan, 1985) or simply could be a direct influence of ATP on K channel gating (Kakei, Noma and Shibasaki, 1985; Findlay, Dunne and Peterson, 1985). Several lines of evidence indicate that the effect is the result of phosphorylation. The effect of ATP requires magnesium and is not mimicked by AMP, ADP or cAMP (Bezanilla et al., 1986; Perozo, Bezanilla, Caputo and Dipolo, 1987). The ATP concentration which produces a half-maximal effect is 10 uM, close to the Km found for many other phosphorylation reactions (Hidaka, Inagaki, Kawamoto and Sasaki, 1984). Many other nucleotides were tested and none produced the same effect as ATP. These included hydrolysis-resistant analogs of ATP such as AMP-PCP and AMP-PNP (Perozo et al., 1986, 1987).

A further line of evidence that ATP is acting via a phosphorylation reaction is seen in perfused axons. When the internal perfusate contains an alkaline phosphatase the effects of ATP are partially reversible (Webb and Bezanilla, 1986). Furthermore, while the thio derivative ATP-gamma-S is effective in changing the conductance in the dialyzed axon, its effect is not reversible (Perozo et al., 1987). This result, which contrasts with the reversible action of ATP in dialyzed axons, presumably is due to the ATP-gamma-S inserting a thio-phosphate group which is resistant to hydrolysis by endogenous phosphatases (Sherry, Goreka, Kosoy, Debrouska and Hartshore, 1978). When axons are dialyzed with fluoride, a potent phosphatase inhibitor, the effects of ATP become irreversible (Perozo et al., 1987).

The differences in the reversibility of the ATP effect in the perfused and dialyzed axons support the notion that the ATP effect is due to phosphorylation. Unlike the dialyzed axon, the perfused axon is free of most soluble constituents because these are washed out with the axoplasm at the onset of perfusion. Consequently, soluble phosphatases are expected to be eliminated. In addition, since perfusion experiments often are done in the presence of fluoride, any remaining phosphatases are likely to be inhibited. This could explain the irreversible nature of the ATP effect in the perfused axon. The presence of endogenous phosphatases, then, would explain the reversibility of the ATP effect in the dialyzed axon.

Although the addition of a kinase to the internal perfusate is not required to see the ATP effect in perfused axons, the presence of the catalytic subunit of cAMP-dependent protein kinase enhances the ATP effect (Bezanilla et al., 1986; Webb and Bezanilla, 1986). This implies that the kinase responsible for the ATP effect on the K conductance is bound to, or very closely associated with, the membrane. At present we do not know whether the kinase normally involved in this phosphorylation is a cAMP-dependent kinase or one of the numerous other protein kinases found in nervous tissue (Nestler and Greengard, 1985).

THE EFFECT OF ATP ON K CONDUCTANCE IS VOLTAGE DEPENDENT

In the presence of ATP, the kinetics of K current turn-on are slower while the currents at turn-off are faster (Bezanilla et al., 1986; Webb and Bezanilla, 1986). Figure 2a illustrates the voltage dependence of activation time constants. In the presence of ATP these time constants are shifted by about 15 mV towards more positive potentials. In addition, there is a pronounced lag in the turn-on of the current in the presence of ATP (Fig. 2b). These kinetic changes suggest that ATP alters the gating

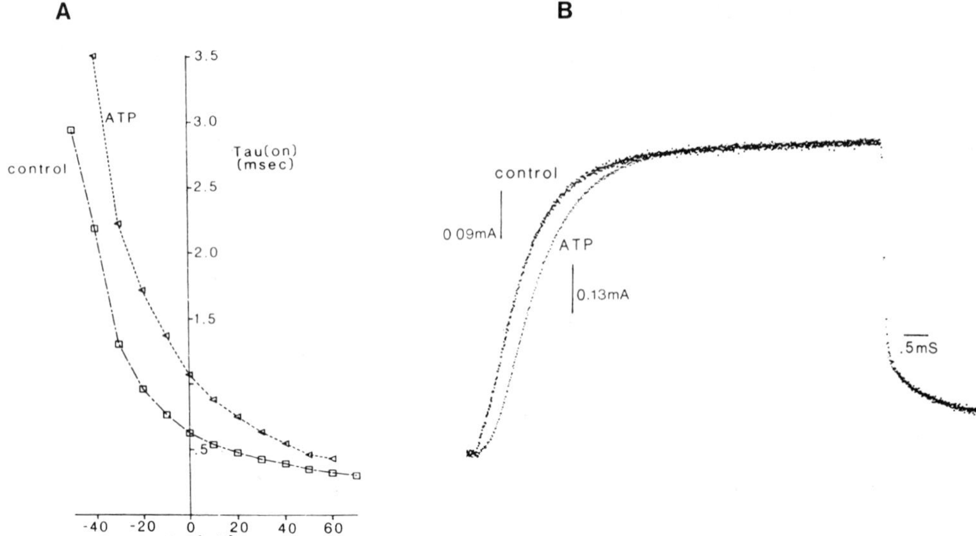

Fig. 2.   Effects of ATP-dependent phosphorylation on the kinetics of
K current activation.   A, Activation time constants plotted
against test pulse potential.   The holding potential was -60
mV.   The internal solution contained 200 mM K, 160 mM
glutamate, 20 mM phosphate, 2 mM Mg-ATP; pH 7.4.   B, Current
responses for a test pulse to 0 mV.   The ATP record has been
scaled down to superimpose with the control record.   The
holding potential was -50 mV.   The internal solution
contained 200 mM K, 160 mM glutamate, 40 mM fluoride, 2 mM
Mg-ATP, 75 nM catalytic subunit of cAMP-dependent protein
kinase (Sigma); pH 7.4.   Leakage and capacitive currents
have been subtracted.

machinery of the K conductance.   ATP might therefore be expected to affect
the kinetics and steady state properties of the charge movement related to
the K conductance (White and Bezanilla, 1985).

In order to detect K gating currents, axons must be internally
perfused with a solution containing no potassium ions.   We normally used
cesium as a K substitute.   We also used cesium in the external solution to
avoid the irreversible loss of K conductance caused by the lack of
permeant ions (Almers and Armstrong, 1980).

Gating currents recorded under these conditions were affected by
internal ATP (Webb and Bezanilla, 1987).   The kinetics of charge movement
were slower at the turn-on and faster at the turn-off of the currents.   In
addition, the voltage dependence of charge movement was shifted to more
depolarized potentials.   The total charge displaced at high
depolarizations, however, was not significantly altered by ATP.   These
results are consistent with the changes observed in the ionic currents.   A
quantitative comparison between ionic and gating currents is not yet
possible because the ionic conditions used to measure these two types of
currents are not the same.

A detailed analysis of ionic currents suggests that ATP also alters K channel inactivation. Most striking is the influence of holding potential (Vh) on ATP effects. When Vh was −60 mV the conductance potentiation was as high as 2.5 times the control value seen in the absence of ATP. However when Vh was −70 mV or more negative the current amplitude potentiation in the presence of ATP was usually quite small or sometimes not present at all. An investigation of the effect of holding potential on the amplitude of the K currents in the presence and absence of ATP yielded the results shown in Figure 3. In this figure, the ATP−dependent potentiation of the K current is plotted as a function of the holding potential. Potentiation is calculated as the ratio of currents recorded in the presence of ATP to those recorded in the absence of ATP. At potentials more negative than −70 mV there was little change in the

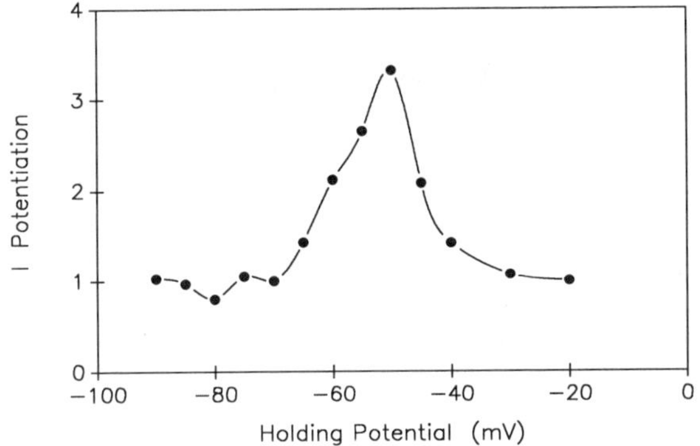

Fig. 3.   ATP−dependent potentiation of the K current at various
          holding potentials. A test pulse to 0 mV was given at each
          holding potential. Each experimental point corresponds to
          the ratio of current recorded in the presence of ATP to that
          recorded in the absence of ATP at a given holding potential.
          The curve drawn was fit by eye.

current amplitude in the presence of ATP. Between −60 mV and −45 mV the current amplitude increased remarkably in the presence of ATP. At potentials more positive than −40 mV, there is again little change in current amplitude with ATP. Direct measurements of the steady state slow inactivation of K currents indicates that ATP produces a positive shift as large as 15 mV (Perozo et al, 1986, 1987). Thus, when Vh is −60 mV and ATP is not present, about one half of the channels are inactivated; upon addition of ATP, inactivation at this holding potential is virtually removed and the same depolarizing pulse can elicit a larger K current. However, when Vh is −70 mV, most of the channels, in the absence of ATP, are not inactivated; a depolarizing pulse can then elicit a K current nearly as large as that observed in the presence of ATP. At holding

potentials more positive than about −40 mV, the steady state inactivation
in the presence of ATP is similar to that in the absence of ATP and,
again, the currents recorded before and after addition of ATP are nearly
equal.

THE EFFECT OF ATP ON K CONDUCTANCE IS MORE THAN A SIMPLE VOLTAGE SHIFT

Thus, ATP appears to be shifting the voltage dependence of both
activation and inactivation parameters along the voltage axis. The
direction of the voltage shift is consistent with the addition of a
negative charge near the voltage sensing domains of the channel proteins,
as expected from the phosphate group introduced by the phosphorylation
(Perozo and Bezanilla, 1987b). This explanation makes several testable
predictions. A simple shift implies that a current trace recorded at V1
mV in the absence of ATP should superimpose with a current trace recorded
at V1 + Vs mV in the presence of ATP, where Vs is the value of the shift
along the voltage axis. Attempts to superimpose traces with different
degrees of voltage shift were unsuccessful. Invariably, the trace recorded
in the presence of ATP had a more pronounced lag in the turn−on of the
current. This implies that a simple voltage shift is not adequate to
explain the effect of ATP on the K conductance.

Another test of the idea that the phosphorylation is adding a
negative charge near the voltage sensor and thus introducing a voltage
shift is to try to produce such an effect on the gating charge by
modifying the divalent cation concentration or by changing the internal
pH. When the internal pH is increased to about 9 the K conductance is
potentiated and the activation kinetics are slower. Conversely, the
conductance is drastically reduced when the pH is decreased to about 6
(Perozo and Bezanilla, 1987b). These results are consistent with the idea
that ATP is changing the potential sensed by the gating machinery and lend
some support to the hypothesis that ATP adds a negative charge near the
voltage sensor.

On the other hand, elevating internal free calcium in dialyzed axons
produces effects on the K conductance directly opposite to those seen when
the pH is decreased. At a calcium concentration of 40 uM and in the
absence of ATP, the current amplitude is increased and the turn−on
kinetics are slower (Perozo and Bezanilla, 1987a, 1987b). The calcium
concentration which produces a half−maximal effect is 1 uM, a
concentration which is in the physiological range. These effects seen
with calcium are strikingly similar to the effects seen with ATP.
Considering that a calcium ion has the opposite charge of a phosphate
group, it seems unlikely that the effects of both ATP and calcium are the
result of a change in the electrostatic charge.

THE EFFECTS OF INTERNAL CALCIUM ON THE K CONDUCTANCE

The changes seen in the K currents following the addition of calcium
to the internal solution could be explained by either an effect of calcium
on the delayed rectifier or the activation of a Ca−activated K channel.
Our results suggest that calcium has both effects.

In dialyzed axons we compared K currents, under conditions where the
internal calcium concentration was less than 1 nM, with currents recorded
in the presence of high calcium concentrations (up to 50 uM). When the
axon was held at −60 mV and pulses given to potentials more positive than
−30 mV, the current amplitude was increased up to 2−fold in the presence
of calcium. Much of this potentiation was blocked by the external
application of 100 nM charybdotoxin (CTX) (Perozo and Bezanilla, 1987a,

1987b). We found, in addition, that 50 mM tetraethylammonium (TEA)
applied externally also eliminates much of the potentiation produced by
internal calcium. External application of both CTX and TEA have been
shown to be specific blockers of a certain class of calcium-activated K
channels (Miller, Moczydlowski, Latorre and Phillips, 1985,; Latorre,
1986). Thus, we take these results as evidence that part of the K
conductance in the squid axon is a calcium-activated K channel.

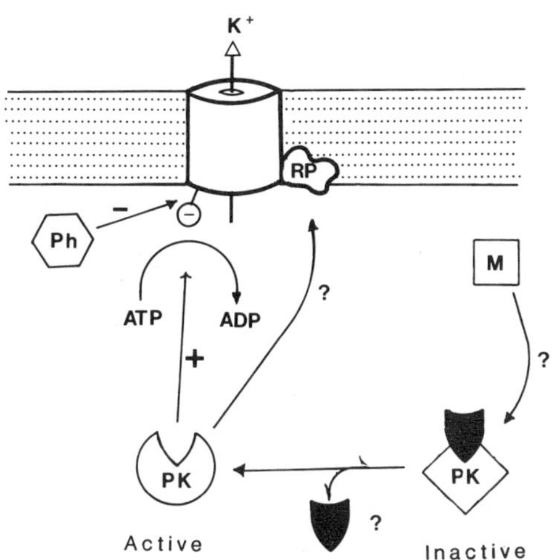

Fig. 4.   Schematic drawing of potassium channel regulation by ATP-
          dependent phosphorylation. The channel shown depicts any
          one of the three or more axonal potassium channels. The
          negative charge on the channel represents the phosphate
          group produced when ATP is split. PK, one of many possible
          protein kinases. The kinase is shown in both an active and
          inactive form; the regulatory subunit, as drawn, is only
          associated with the inactive form. This would be the case
          if the cAMP-dependent kinase was involved in K channel
          modulation. Other kinases would have different forms of
          regulation. RP, channel-associated regulatory protein; this
          could be the site of phosphorylation which results in
          changes in K channel gating. Ph, protein phosphatase,
          presumably soluble. M, the second-messenger system which
          normally activates the protein kinase.

     The effect of internal calcium is not exclusive in stimulating the
Ca-activated conductance. When internal Ca is increased and the
Ca-activated K current is blocked by CTX, the voltage dependence of the
conductance is shifted to more positive potentials and, depending on the
holding potential, the current amplitude is increased. A study of the
steady state K inactivation shows that calcium is shifting the
inactivation curve to more positive potentials (Perozo and Bezanilla,
1987b). Again, this effect is contrary to that expected from the addition
of a surface charge. Internal calcium appears to be affecting the delayed
rectifier in a manner which is remarkably similar to the effects of ATP.

MORE THAN ONE TYPE OF K CHANNEL

The above results suggest that the K channel population in the squid
axon is not homogeneous. Single channel events recorded with a patch
pipette applied to the internal surface of cut-open axons support this
interpretation (Llano and Bezanilla, 1985; Llano et al., 1987). These
recordings revealed a population of channels with a conductance of about 20
pS; this was identified as the delayed rectifier. Another channel,
observed less frequently, had a conductance of about 40 pS. Recent
experiments on the cut-open axon have revealed yet another channel with a
conductance of about 60 pS (Webb and Bezanilla, unpublished). Thus squid
axons contain a multitude of channels with different conductance properties.
It is not yet clear which of these are modulated by ATP or calcium.

DISCUSSION

In summary, regulation of K channels in the squid axon can be pictured
as in Fig. 4 where the participation of several components leads to the
phosphorylation of the channel itself or a channel-related regulatory
protein (RP). The phosphorylation is carried out by the catalytic region
of a protein kinase (PK). This protein kinase requires ATP and magnesium
and is highly selective for ATP over other nucleotides. At this time, we
do not know the nature of this protein kinase or the second-messenger
system which triggers the kinase. Also illustrated is a protein
phosphatase (Ph) which acts to down regulate the process. It is clear from
the evidence presented that a dephosphorylation process is occurring in the
axon (Webb and Bezanilla, 1986; Perozo et al, 1986, 1987); however the
nature of this phosphatase is not known. The channel depicted is a generic
one in that it could be any one of the three or more channels present in
the axon.

It is possible to speculate about the components of the K conductance
in the squid axon and their modulation by both calcium and ATP. The
experimental results demonstrate that calcium and ATP have similar effects
on the conductance. Since a fraction of the conductance appears to be a
calcium activated K channel, it is tempting to suggest that ATP is
increasing the sensitivity of the calcium binding site. Alteration of the
calcium sensitivity of calcium-dependent K channels by phosphorylation has
been observed in channels incorporated into bilayers (Ewald, Williams and
Levitan, 1985). In addition, both ATP and calcium affect the delayed
rectifier. These changes are at the level of the gating sensor thus
affecting both activation and inactivation in a voltage dependent fashion.
The effect of ATP seems to be to shift the voltage dependence of the gating
parameters to more depolarized potentials, as shown by the ionic and gating
current results. Likewise, calcium introduces a shift in the voltage
dependence of ionic current activation and inactivation. At present it is
not clear how both calcium and ATP-dependent phosphorylation produce the
similar effects. One possible suggestion is that phosphorylation changes
the gating structure of the channel and calcium either modulates the same
structure or the effectiveness of the phosphorylation. In addition, the
phosphorylation could perhaps be mediated by a calcium-dependent protein
kinase. Elevation of calcium would then increase the activity of the
protein kinase thus making it more readily available for phosphorylation.

REFERENCES

Almers, W., and Armstrong, C. M., 1980, Survival of K+ permeability and
    gating currents in squid axons perfused with K+-free media, J. Gen.
    Physiol., 75:61.

Bezanilla, F., Caputo, C., Dipolo, R., and Rojas, H., 1986, Potassium conductance of squid giant axon is modulated by ATP, Proc. Nat. Acad. Sci. U.S.A., 83(8):2743.

Bezanilla, F., Vergara, J., and Taylor, R. E., 1982, Voltage clamping of excitable membranes, in: "Methods of Experimental Physics: Biophysics," vol. 20, G. Ehrenstein and H. Lecar, eds., Academic Press Inc., London.

Brinley, F. J., and Mullins, L.M., 1967, Sodium extrusion by internally dialyzed squid axons, J. Gen. Physiol., 50:2303.

Conti, F., and Neher, E., 1980, Single channel recordings of K+ currents in squid axons, Nature (Lond), 285:140.

Dipolo, R., Bezanilla, F., Caputo, C., and Rojas, H., 1985, Voltage dependence of the Na/Ca exchange in voltage-clamped, dialyzed squid axons, J. Gen. Physiol., 86:457.

Ewald, D. A., Williams, A., and Levitan, I. B., 1985, Modulation of single Ca2+-dependent K channel activity by protein phosphorylation, Nature, 315:503.

Findlay, I., Dunne, M. J., and Petersen, O. H., 1985, ATP-sensitive inward rectifier and voltage and Calcium-activated K channels in cultured pancreatic islet cells, J. Memb. Biol., 88:165.

Hidaka, H., Inagaki, M., Kawamoto, S., and Sasaki, Y., 1984, Isoquinolinesulfonamides, novel and potent inhibitors of cyclic nucleotide dependent protein kinase and protein kinase C, Biochemistry, 23:5036.

Hodgkin, A. L., and Huxley, A.F., 1952, A quantitative description of membrane current and its application to conduction and excitation in nerve, J. Physiol. (Lond), 117:500.

Kakei, M., Noma, A., and Shibasaki, T., 1985, Properties of adenosine-triphosphate-regulated potassium channels in guinea-pig ventricular cells, J. Physiol., 363:441.

Latorre, R., 1986, The large Ca-activated potassium channel, in: "Ion channel reconstitution," C. Miller, ed., Plenum, New York.

Llano, I., and Bezanilla, F., 1985, Two types of potassium channels in the cut-open squid giant axon, Biophys. J., 47:221a.

Llano, I., Webb, C. K., and Bezanilla, F., 1987, Two types of unitary potassium channel currents in the squid giant axon, In preparation.

Miller, C., Moczydlowski, E., Latorre, R., and Phillips, M., 1985, Charybdotoxin, a protein inhibitor of single Ca2+-activated K+ channels from mammalian skeletal muscle, Nature, 313:316.

Perozo, E., and Bezanilla, F., 1987a, Intracellular calcium modulates the potassium conductance in dialyzed squid axons, Biophys. J., 51:547a.

Perozo, E., and Bezanilla, F., 1987b, The modulation of K channels in dialyzed squid axons: Effect of intracellular calcium, In preparation.

Perozo, E., Bezanilla, F., Caputo, C., and Dipolo, R., 1987, The modulation of K channels in dialyzed squid axons: ATP mediated phosphorylation, In preparation.

Perozo, E., Dipolo, R., Caputo, C., Rojas, H., and Bezanilla, F., 1986, ATP modification of K currents in dialyzed squid axons, Biophys. J., 49: 215a.

Sherry, S., Goreka, A., Kosoy, M., Debrouska, R., and Hartshore, D., 1978, Roles of calcium and phosphorylation in the regulation of the activity of Gizzard myosin, Biochemistry 17:4411.

Webb, C. K., and Bezanilla, F., 1986, K currents in perfused axons are modified by ATP, Biophys. J., 49:215a.

Webb, C. K., and Bezanilla, F., 1987, Potassium gating currents in the perfused axon are modified by ATP, Biophys. J., 51:547a.

White, M. M., and Bezanilla, F., 1985, Activation of squid axon K channels: Ionic and gating current studies, J. Gen. Physiol., 85:539.

PHYSIOLOGICAL INTERACTION BETWEEN CALCIUM AND CYCLIC AMP

IN AN APLYSIA BURSTING PACEMAKER NEURON

Richard H. Kramer, Edwin S. Levitan[1], and Irwin B. Levitan

Graduate Department of Biochemistry, Brandeis University
Waltham, MA 02254

INTRODUCTION

Cells respond to a variety of extracellular stimuli by utilizing
intracellular messenger systems. In recent years several intracellular
messengers have been identified as mediators of changes in the electrical
activity of excitable cells. The best understood messengers in nerve cells
are Ca^{2+} and cyclic AMP. An increase of free Ca^{2+} is a common by-product of
action potentials, while a variety of neurotransmitters and hormones
increase internal cyclic AMP. The study of what these messengers do, and
their mechanisms of action, has extended our understanding of neuronal
function. Now cellular neurophysiology is confronted with a new
challenge: to understand how intracellular messenger systems interact within
a neuron. Interactions between intracellular messenger systems may have a
special importance in neurons, allowing for the biochemical integration of
intrinsic electrical activity such as action potentials with synaptic and
hormonal influences. It is likely that studying interactions between
intracellular messenger systems will elucidate basic mechanisms involved in
generating long-term changes in neuronal function.

ACTIONS OF CALCIUM AND CYCLIC AMP IN APLYSIA NEURONS

The study of the intracellular messenger roles of Ca^{2+} and cyclic AMP
has been especially fruitful in identified neurons from Aplysia. Ca^{2+} and
cyclic AMP have been most thoroughly studied and perhaps best understood in
the endogenously bursting pacemaker cell R15 (for a review see Adams and
Benson, 1985). Changes in the concentration of Ca^{2+} due to bursts of action
potentials (Gorman and Thomas, 1978) and changes in the concentration of
cyclic AMP due to application of serotonin (5-HT; Levitan and Drummond,
1980) and egg-laying hormone (ELH; Levitan et al., 1987), have been detected
in cell R15. Hence R15 is an ideal cell in which to study interactions
between spike activity and neurotransmitter effects mediated by interactions
between the respective intracellular messenger systems.

----

[1]Present address: MRC Molecular Neurobiology Unit, University of
 Cambridge Medical School, Cambridge, England

Physiological effects of both Ca^{2+} and cyclic AMP are well known in R15.
Table 1 shows the diverse assortment of ionic currents that are regulated by
internal Ca^{2+}, cyclic AMP, or both substances in this neuron. Each
intracellular messenger has divergent actions on several ionic currents.
Ca^{2+} regulates at least three ionic currents (I_{Ca}, $I_{cat(Ca)}$, and $I_{K(Ca)}$)
whose participation in generating bursting pacemaker activity has been
studied in great detail (Adams, 1985; Adams and Levitan, 1985; Kramer and
Zucker, 1985a,b). Intracellular Ca^{2+} also regulates an inwardly rectifying
K^+ current, I_R, by a process that operates over a time course much longer
than the period of the burst cycle (see below). In addition, there is
evidence that an increase in intracellular Ca^{2+} also leads to the activation
of an inwardly rectifying Cl^- current, I_{Cl} (Chesnoy-Marchais, 1983;
R. Kramer, unpublished observations).

The effect of intracellular Ca^{2+} on the voltage-gated Ca^{2+} current,
I_{Ca}, is of particular importance for the physiology of _Aplysia_ bursting
neurons, such as R15. The pioneering work of Eckert and Tillotson (1981)
established that internal Ca^{2+} causes the inactivation of the I_{Ca} in neuron
R15. Molluscan bursting neurons are somewhat unusual in that I_{Ca} begins to
activate well below the threshold for action potentials (e.g. at membrane
potentials more positive than -60 mV). Recent experiments suggest that the
voltage-dependent activation, and the Ca^{2+}-dependent inactivation of the
subthreshold portion of I_{Ca} are the key factors that are responsible for the
oscillation of the membrane potential during bursting (Adams and Levitan,
1985; Kramer and Zucker, 1985b).

Cyclic AMP also regulates multiple ionic conductances in R15 (Lotshaw
et al., 1986). The best characterized of these is the inward or "anomalous"
rectifying K^+ current, I_R. Because of the inward rectification of I_R, the
current appears to exhibit the unusual property of activating upon
hyperpolarization of the membrane potential instead of upon depolarization.
Convincing evidence has been obtained over the years demonstrating that the
activation of I_R by 5-HT is mediated by cyclic AMP. The effect of 5-HT on
I_R is accentuated by applying phosphodiesterase inhibitors, and mimicked by
applying membrane permeable cyclic AMP analogs or activators of adenylate
cyclase, such as forskolin (Drummond et al., 1980). In addition, injection
of the free catalytic subunit of cyclic AMP-dependent protein kinase leads
to the activation of I_R (I. Levitan and colleagues, unpublished results),
and injection of a protein inhibitor of the kinase blocks the action of 5-HT
on I_R (Adams and Levitan, 1982). More recently, 5-HT has been shown to
affect two other ionic currents in R15, also via cyclic AMP (Lotshaw et al.,
1986; Levitan and Levitan, 1987). 5-HT, as well as other agents which
increase the level of cyclic AMP, causes an increase in the voltage-gated
Ca^{2+} current, I_{Ca}, and causes a decrease in an inwardly rectifying Cl^-
current, I_{Cl}.

Are the divergent effects of cyclic AMP physiological?

The myriad of Ca^{2+} and cyclic AMP effects on ionic currents in R15 is
suprising and somewhat bewildering. How do we know that the multiple
effects of these intracellular messengers are physiological? In the case of
Ca^{2+}, there is a natural event that causes the influx of Ca^{2+}, and leads to
modulation of each of the ionic currents described above; that is the
endogenous bursting activity for which the cell is held in high esteem. In
the case of the cyclic AMP-dependent actions of 5-HT, the physiological
relevence of modulation of multiple ionic currents has been less clear.
Unfortunately, there are no known serotonergic neurons that are presynaptic
to R15 (although there are serotonergic neurons in _Aplysia_), and possible
circulating 5-HT in the hemolymph has not been measured. Therefore we do
not know for sure that 5-HT is a neurotransmitter that affects R15 _in situ_.
However, we have found that a second neuromodulator with known actions on

240

Table 1. Ionic currents regulated by Ca^{2+} and/or cyclic AMP in R15

Name	Type	Ca^{2+} Effect	Cyclic AMP Effect
$I_{K(Ca)}$	Calcium-dependent K^+ current	Activation[1,2]	Activation[3]
$I_{cat(Ca)}$	Non-selective cation current	Activation[4]	?
I_{Ca}	Voltage-gated Ca^{2+} current	Inactivation[5-7]	Activation[8]
I_R	Inward rectifier K^+ current	Inactivation[9]	Activation[10]
I_{Cl}	Inward rectifier Cl^- current	Activation(?)[11]	Inactivation[8]

References: [1]Meech (1972), [2]Gorman and Hermann (1979), [3]Ewald and Eckert (1983), [4]Kramer and Zucker (1985a), [5]Eckert and Tillotson (1981), [6]Adams and Levitan, 1985, [7]Kramer and Zucker (1985b), [8]Lotshaw et al. (1986), [9]Kramer and Levitan (1987), [10]Benson and Levitan (1983), [11]Chesnoy-Marchais (1983)

R15 _in situ_ affects the same ionic currents as does 5-HT, and also operates via cyclic AMP. This second neuromodulator is the _Aplysia_ neuropeptide egg-laying hormone (ELH).

ELH causes burst augmentation in R15 via a cyclic-AMP mediated increase in two ionic currents

ELH is released along with other peptides during repetitive firing or "afterdischarge" of the bag cells neurons of the _Aplysia_ abdominal ganglion (Kupfermann and Kandel, 1970). Secreted ELH acts both on reproductive tissues and on the nervous system to induce egg-laying and associated behaviors (for a review see Mayeri and Rothman, 1985). Both afterdischarge of the bag cell neurons _in situ_ (Mayeri et al., 1979), and extracellular application of ELH (Mayeri et al., 1985) result in an augmentation of bursting activity in neuron R15. The burst augmentation is characterized by an increase in the frequency and number of spikes during the depolarizing phase of bursting, and an increase in the amplitude of the interburst hyperpolarization (Fig. 1A). The burst augmentation lasts for many minutes after a brief application of ELH. The persistence of burst augmentation in the absence of the neurotransmitter suggests the continued action of an intracellular messenger.

We have found that both local application of synthetic ELH (Fig. 2A) and bag cell afterdischarge (Fig. 2B) affect two distinct regions of the steady-state current-voltage (I-V) curve of R15. Both treatments cause an increase in the inward current at membrane potentials more hyperpolarized than -75 mV, which is near the K^+ equilibrium potential of neuron R15. This voltage range (-75 mV and below) corresponds to that at which I_R becomes fully activated; hence the enhanced inward current below -75 mV could be due to an increase in I_R. Both afterdischarge and ELH also cause an increase in

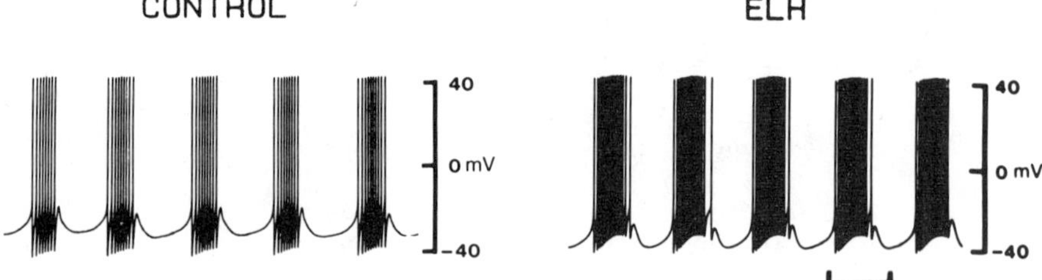

Fig. 1.   Effect of ELH on bursting pacemaker activity in neuron R15 from the
          abdominal ganglion of <u>Aplysia</u>. A micropipette situated 400 μm from
          the soma of R15 was used to locally "puff" 1 μl of 40 μM ELH at the
          cell. Note the augmentation of both the burst and the interburst
          hyperpolarization. The trace marked "ELH" was recorded 5 minutes
          after ELH application.

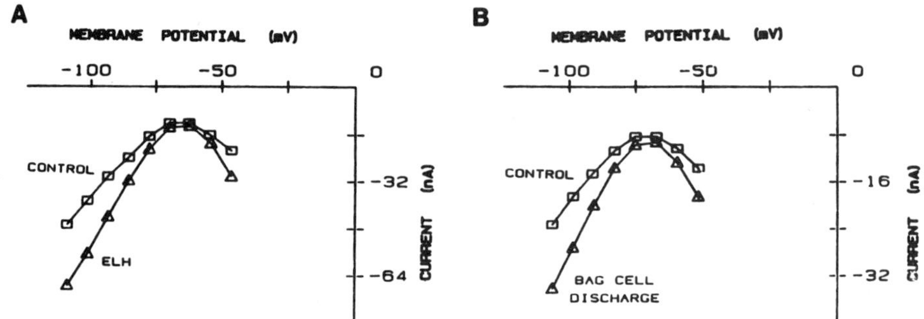

Fig. 2.   Effect of ELH and bag cell afterdischarge on the I-V curve of
          R15. A. I-V curves from an R15 before (squares) and 16 min after
          (triangles) local ELH application to R15. Steady-state currents
          were evoked with 1 sec voltage clamp pulses from -80 mV. B. I-V
          curves from another R15 before (squares) and 8 min after
          (triangles) bag cell afterdischarge. Steady state currents were
          evoked with 0.5 sec pulses from -60 mV. Bag cell afterdischarge
          was elicited by electrically stimulating the left pleuroabdominal
          connective.

a second inward current at membrane potentials more depolarized than about
-60 mV. The subthreshold portion of I_{Ca} begins to activate at these
depolarized voltages; hence the enhanced inward current above -60 mV could
result from an increase in I_{Ca}.

     Several types of evidence support the hypothesis that the effect of ELH
at potentials more negative than -75 mV is due to an increase in I_R. First,
ELH fails to modulate the membrane current below -75 mV if substances that
specifically block I_R (i.e 1mM Ba^{2+} or 5 mM Rb^+) are added to the bathing
medium. Secondly, the ELH-induced conductance at hyperpolarized potentials
increases as the external K^+ concentration is increased, and decreases as
the external K^+ concentration is decreased, as expected for I_R. Finally,
the effect of ELH below -75 mV is identical to the effect of 5-HT, and is
occluded by application of a satuarating concentration of 5-HT (i.e. 100
μM). This implies that ELH and 5-HT act on the same ionic conductance, and

since previous work has established that 5-HT enhances I_R (Benson and Levitan, 1983; Gunning, 1987), then ELH must enhance I_R as well.

Evidence also supports the hypothesis that ELH enhances I_{Ca}. The effect of ELH on membrane current at potentials more depolarized than -60 mV is blocked by substances that block Ca^{2+} currents, such as Co^{2+} and Mn^{2+}. ELH enhances the Ca^{2+}-dependent inactivation of I_{Ca} initiated by depolarizing voltage-clamp pulses, suggesting that ELH causes an enhanced influx of Ca^{2+}. The ELH-induced enhancement of inward current at depolarized membrane potentials is unaffected by addition of K^+ current blockers, such as tetraethylammonium or 4-aminopyridine, suggesting that ELH does not affect a K^+ current at depolarized voltages. Finally, the ELH affect on the inward current above -60 mV is also mimicked and occluded by application of a saturating concentration of 5-HT (100 μM), and there is previous evidence that 5-HT enhances I_{Ca} (Lotshaw et al., 1986; Levitan and Levitan, 1987).

Since ELH, like 5-HT, appears to enhance both I_R and I_{Ca}, it seems likely that ELH also operates via cyclic AMP as an intracellular messenger. We have measured the content of cyclic AMP in single R15 somata, and have found that ELH exposure results in an increase in cyclic AMP of 72% over untreated cells. In addition, the effects of ELH on I_R and I_{Ca} are accentuated by a phosphodiesterase inhibitor, as expected for effects that are mediated by cyclic AMP. Hence the evidence leads us to conclude that ELH affects multiple ionic currents (I_R and I_{Ca}) via cyclic AMP. We propose that the enhancement of both I_R and the subthreshold I_{Ca} by ELH causes an augmentation of the interburst hyperpolarization and the depolarizing phase of bursting, respectively, thereby augmenting bursting pacemaker activity.

PHYSIOLOGICAL INTERACTION BETWEEN CALCIUM AND CYCLIC AMP IN NEURON R15

Since so much is known about the physiological effects of Ca^{2+} and cyclic AMP in R15, and since we can induce physiological changes in the intracellular levels of both of these intracellular messengers, we are presented with the unique opportunity to examine physiological interactions between the two messenger systems. We have focussed on one particular phenomenon that seems to result from an interaction between Ca^{2+} and cyclic AMP in R15: the inactivation of I_R, which appears to be due to intracellular Ca^{2+} interfering with the cyclic AMP-dependent activation of the current (Kramer and Levitan, 1987; Kramer et al., 1987).

Calcium-dependent inactivation of I_R in neuron R15

We have discovered a novel action of intracellular Ca^{2+} on an ionic current in R15. Ca^{2+} influx, caused either by bursts of action potentials or by voltage-clamp depolarizations, results in a profound, long-lasting inactivation of I_R (Kramer and Levitan, 1987). Figure 3A shows the effect of bursting activity on membrane current elicited by hyperpolarizing voltage pulses from -90 to -130 mV. Most of the membrane conductance at these hyperpolarized membrane potentials is due to I_R. The cell was released momentarily from voltage clamp and allowed to fire a spontaneous burst of spikes; then the cell was clamped once again and the hyperpolarizing pulses were resumed. The burst is followed by a decrease in the amplitude of the inward currents induced by the hyperpolarizing voltage pulses; hence, there is a decrease in membrane conductance.

The kinetics of the conductance decrease induced by bursts has several interesting properties (Fig 3B). The conductance decrease is long-lasting, requiring more than 10 min to recover after a single burst of spikes. The conductance decrease exhibits a delay, such that the maximum effect of the burst does not occur until 60-90 sec after the end of the burst. Following

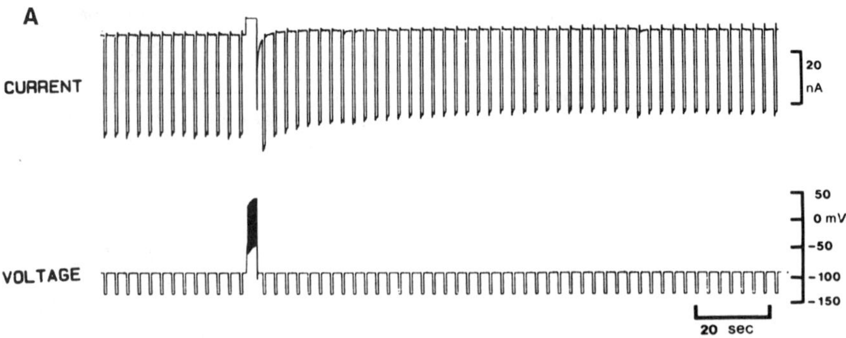

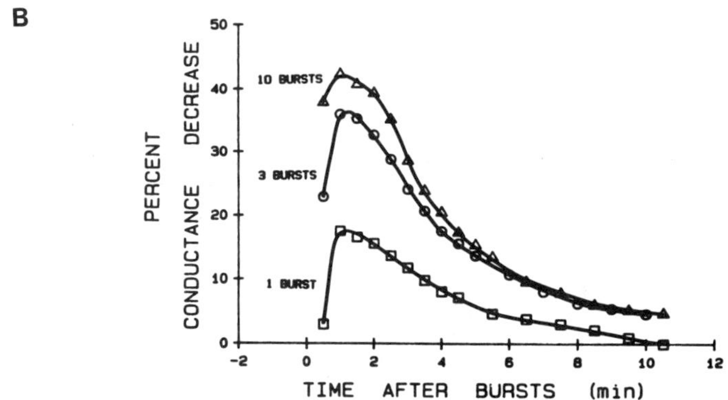

Fig. 3. Effect of bursting activity on membrane conductance of R15. **A**. A
single burst of action potentials (12 spikes) reduces the membrane
conductance of a voltage-clamped R15 cell. Currents were elicited
by repeated 400 msec hyperpolarizing voltage pulses from -90 to
-130 mV. The burst occurred spontaneously when the voltage-clamp
was momentarily turned off. **B**. Time course of the conductance
decrease following bursts of spikes. The percent conductance
decrease is the difference between the conductance at a given time
after the bursts and the conductance before the bursts, divided by
the conductance before the bursts. The steady-state membrane
conductance was measured by applying 400 msec voltage pulses from
-75 to -115 mV in a voltage-clamped R15 neuron.

multiple bursts of spikes, the decrease in conductance is larger and
longer-lasting, suggesting that the factor responsible for reducing the
membrane conductance accumulates over several bursts. In addition, the
delay of the maximal effect of the burst seems to diminish as additional
bursts are added, suggesting that the process that causes the conductance
decrease approaches equilibrium after repeated bursts of action potentials.

In order to examine the ionic basis of the conductance decrease, the
I-V curve of R15 was measured before and after a "burst" of voltage-clamp
depolarizations was applied (Fig. 4A). The change in the I-V curve induced
by the burst indicates that the decrease in membrane conductance is due to
the inactivation of I_R. The conductance decrease occurs at hyperpolarized
membrane potentials below about -75 mV. I_R exhibits steep activation with

hyperpolarization below this membrane potential. The conductance decrease is blocked if I_R is blocked by addition of 1 mM Ba^{2+} (Fig 4B) or 5 mM Rb^+ (not shown) to the saline. In addition, the conductance decrease shifts to more hyperpolarized membrane potentials as external K^+ is reduced, in accord with the expected shift of I_R. Hence the burst results in an inactivation of I_R.

The inactivation of I_R is dependent upon the influx and intracellular accumulation of Ca^{2+}. The inactivation of I_R fails to occur if Ca^{2+} is removed from the bathing medium (Fig. 4C), or if EGTA is iontophoretically injected into R15 (Fig. 4D). In addition, bursts of depolarizing voltage clamp pulses that approach the Ca^{2+} equilibrium potential (e.g. pulses to +100 mV) fail to cause inactivation of I_R. Hence the inactivation of I_R requires the influx and action of intracellular Ca^{2+}.

How does Ca^{2+} cause the inactivation of I_R?

There are a number of possible mechanisms whereby intracellular Ca^{2+} could lead to the inactivation of I_R. One possibility is that Ca^{2+} acts directly to allosterically regulate I_R channels, analogous to the direct activation of $I_{K(Ca)}$ channels by Ca^{2+} (Moczydlowski and Latorre, 1983). Alternately, Ca^{2+} may act indirectly to influence the activity of I_R channels. For example, Ca^{2+} could regulate the phosphorylation state of either the channel or an associated protein that controls its activity. It is also possible that Ca^{2+} inactivates I_R by opposing some step in the cyclic AMP cascade, which activates I_R.

There is circumstantial evidence that Ca^{2+} does not directly interact with and cause the inactivation of I_R channels (Kramer et al., 1987). If the Ca^{2+}-dependent inactivation of I_R were a direct result of Ca^{2+} interaction with the channels, then as the magnitude of I_R is increased, the Ca^{2+}-dependent inactivation of I_R should increase in parallel. One way to increase the magnitude of I_R is by treating R15 neurons with 5-HT. As determined by noise analysis, the increase in I_R caused by 5-HT seems to be due to an increase in the number of active channels rather than an increase in the open probability of individual channels (Gunning, 1987). Hence, in the presence of increasing concentrations of 5-HT, one might expect an increasing amount of I_R inactivation elicited by a constant influx of Ca^{2+}.

The dose-response curve of Fig. 5A shows that increasing concentrations of 5-HT do elicit increasing amounts of I_R, up to a saturating concentration of between 1 and 10 μM. In contrast, the inactivation of I_R elicited by Ca^{2+} influx increases in parallel only up to a 5-HT concentration of 1 μM (Fig. 5B). At higher concentrations of 5-HT the inactivation of I_R decreases dramatically. The decrease in Ca^{2+}-dependent inactivation of I_R in the presence of high levels of 5-HT occurs despite the increase in Ca^{2+} current caused by 5-HT. Hence high levels of 5-HT can prevent the inactivation of I_R, even though there are more I_R channels avilable to be inactivated, and even though the Ca^{2+} current is increased. We have observed a similar supression of the Ca^{2+}-dependent inactivation of I_R using the adenylate cyclase activator forskolin instead of 5-HT to increase the cyclic AMP-dependent activation of I_R. These observations are inconsistent with a direct action of Ca^{2+} on I_R channels. A scenario that would better explain the results is that Ca^{2+} counteracts the cyclic AMP activation of I_R, and that increasing cyclic AMP synthesis by addition of high levels of 5-HT or forskolin overwhelms the effect of Ca^{2+}.

Experiments with 5-HT also suggest that there is an important difference between the mechanism of Ca^{2+}-dependent inactivation of I_R, and the mechanism of Ca^{2+}-dependent inactivation of the subthreshold I_{Ca}. In the presence of a high concentration of 5-HT (50 μM) both I_R and I_{Ca} are

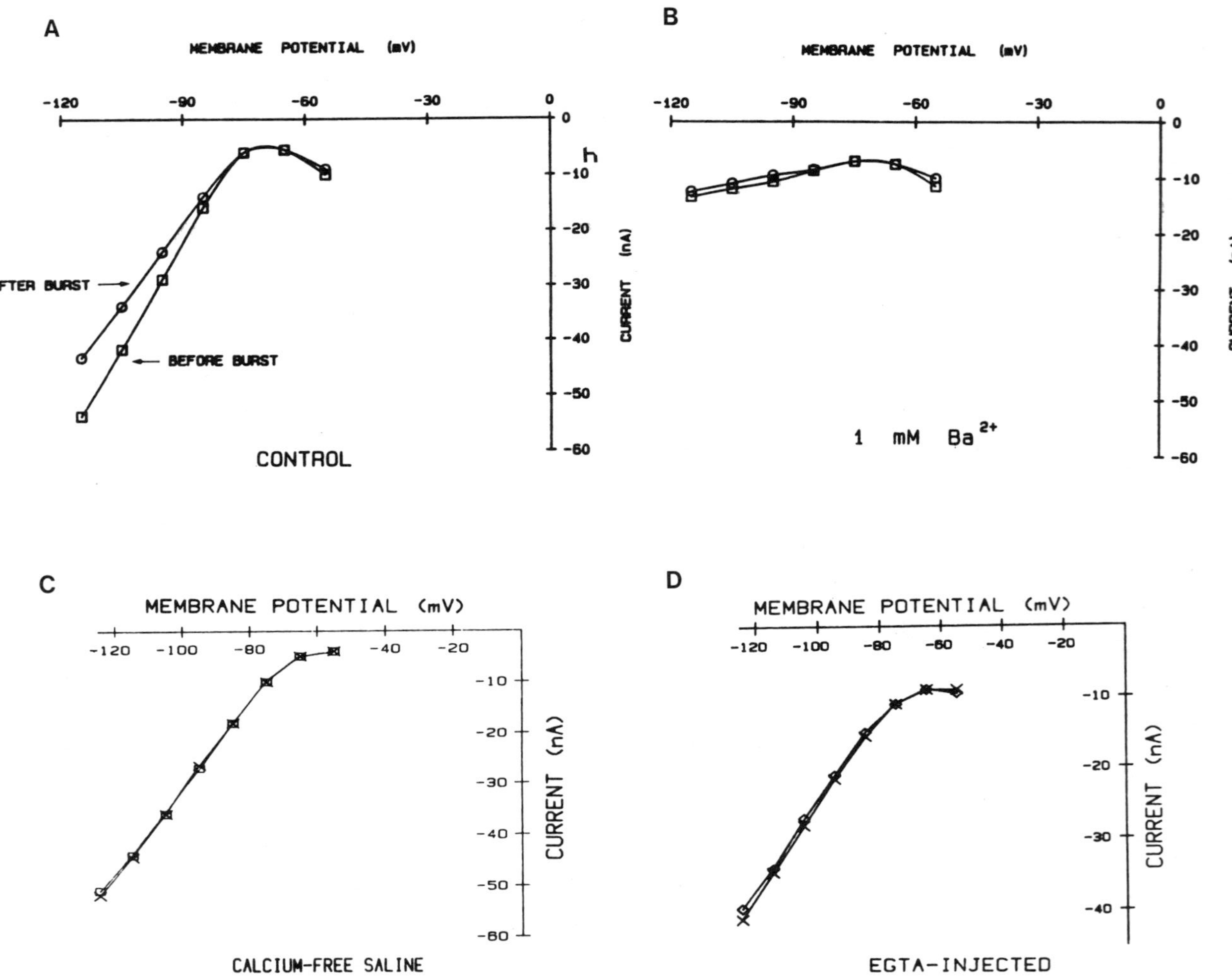

A
MEMBRANE POTENTIAL (mV)
CURRENT (nA)
AFTER BURST
BEFORE BURST
CONTROL

B
MEMBRANE POTENTIAL (mV)
CURRENT (nA)
1 mM Ba 2+

C
MEMBRANE POTENTIAL (mV)
CURRENT (nA)
CALCIUM-FREE SALINE

D
MEMBRANE POTENTIAL (mV)
CURRENT (nA)
EGTA-INJECTED

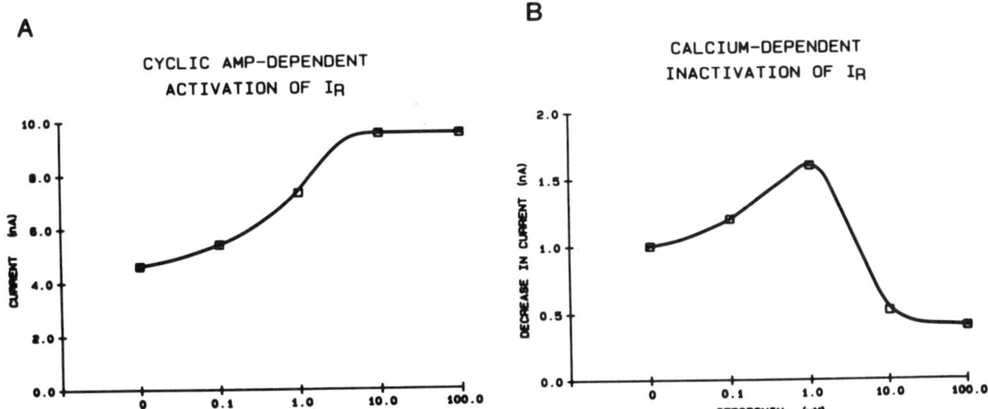

Fig. 5.   Effect of 5-HT on I_R in neuron R15.   **A**. Dose-response curve of the
membrane current activated by 0 to 100 μM 5-HT.   Steady-state
membrane current was measured at the end of 400 msec
hyperpolarizing pulses from -75 to -115 mV.   The change in membrane
current caused by 5-HT application is due entirely to activation of
I_R.   Note the saturation of the activation of I_R between 1 and 10
μM 5-HT.   **B**.   Dose-response curve of the decrease in membrane
current (inactivation of I_R) induced by Ca^{2+} influx in 0 to 100 μM
5-HT.   Bursts of depolarizing pulses were applied to induce Ca^{2+}
influx; each burst consisted of five 200 msec pulses to +15 mV.
The decrease in membrane current is the difference between the
steady-state current measured 90 sec after the burst and the
steady-state current measured before the burst.   Steady-state
membrane current was measured as in **A**.   The decrease in membrane
current induced by the burst is due entirely to Ca^{2+}-dependent
inactivation of I_R.   Note the decline in Ca^{2+}-dependent
inactivation of I_R between 1 and 10 μM 5-HT.

increased.   However, the Ca^{2+}-dependent inactivation of I_R is almost
completely blocked in 50 μM 5-HT (Fig. 6A), while the Ca^{2+}-dependent
inactivation of I_{Ca} is enhanced (Fig. 6B).   This implies that there is a
basic difference in the way Ca^{2+} inactivates the two currents.   Perhaps Ca^{2+}
has a direct effect on I_{Ca} channels, while the effect of Ca^{2+} on I_R is
more indirect, requiring an interaction with cyclic AMP.

Fig. 4.   (Facing page) Effect of a burst of depolarizing voltage clamp
pulses (a simulated burst) on the I-V curve of R15.   **A**. I-V curves
measured before and 90 sec after a burst in normal saline.   Note
the decrease in inward current below -75 mV.   **B**. I-V curves
measured before and 90 sec after the burst in saline containing 1
mM Ba^{2+}, which selectively blocks I_R.   Note that the burst has
little effect on the I-V curve.   **C**.   I-V curves measured before and
90 sec after the burst in saline containing no added Ca^{2+}, and 3 mM
Mn^{2+} to block the residual Ca^{2+} current.   **D**.   I-V curves measured
before and 90 sec after the burst in an R15 cell injected with
EGTA.   Steady-state I-V curves in **A** through **D** were obtained by
applying 400 msec pulses from -75 mV.   The depolarizing burst
consisted of five 200 msec pulses to +20 mV.   Data from **A** and **B** are
from the same R15 cell, data from **C** and **D** are each from different
R15 cells.

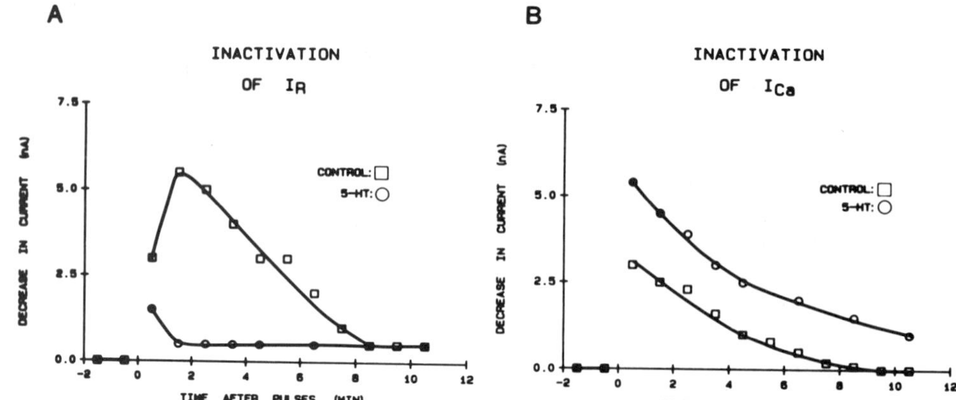

Fig. 6. Ca^{2+}-dependent inactivation of I_R differs from Ca^{2+}-dependent inactivation of the subthreshold I_{Ca}. **A.** Time course of Ca^{2+}-dependent I_R inactivation induced by a burst of depolarizing voltage clamp pulses in normal saline (squares) and in saline containing 50 μM 5-HT (circles). I_R was measured by applying 400 msec pulses from -75 to -115 mV. Note the block of inactivation of I_R after addition of 5-HT. **B.** Time course of Ca^{2+}-dependent inactivation of the subthreshold I_{Ca} induced by a simulated burst in normal saline (squares) and in saline containing 50 μM 5-HT (circles). The subthreshold I_{Ca} was measured by applying 400 msec depolarizing pulses from -75 to -40 mV. Note the increase of I_{Ca} inactivation after addition of 5-HT. In both **A** and **B** the inactivating burst consisted of five 200 msec pulses to +20 mV, and the decrease in current was measured 90 sec after the burst. Data from **A** and **B** were obtained from the same R15 neuron.

Intracellular Ca^{2+} reduces the effect of locally applied 5-HT on I_R

There is another manifestation of the interaction of Ca^{2+} with cyclic AMP in neuron R15. In the experiments discussed above, we examined the effect of a saturating concentration of 5-HT on the Ca^{2+}-dependent inactivation of I_R. In the experiments discussed in this section, we examine the effect of intracellular Ca^{2+} on the modulation of I_R elicited by a non-saturating application of 5-HT. If Ca^+ reduces some step in the cyclic AMP cascade, then Ca^{2+} should reduce the activation of I_R elicited by 5-HT. To test this idea, 5-HT was locally applied to the soma of a voltage-clamped R15 cell that was either continuously voltage clamped to a hyperpolarized membrane potential (-80 mV) to prevent Ca^{2+} influx through voltage-gated Ca^{2+} channels, or periodically depolarized to 0 mV to induce Ca^{2+} influx (Fig. 7). I_R was measured at various times by hyperpolarizing the cell from -80 to -115 mV. Brief (10 sec) applications of 5-HT result in a large, long-lasting increase in I_R when Ca^{2+} influx is prevented by keeping the cell hyperpolarized. In contrast, the response of I_R to the 5-HT puff is about 33% smaller and relaxes much more rapidly if the cell is "loaded" with Ca^{2+} by repeated depolarization of the cell. The response of I_R to locally applied ELH also appears to be reduced as a result of Ca^{2+} loading. Hence, internal Ca^{2+} seems to antagonize the response of R15 to neurotransmitters that operate via cyclic AMP. The antagonism by Ca^{2+} may in fact be a physiological mechanism for terminating a neuron's response to transmitters that operate via cyclic AMP.

<u>Biochemical locus of Ca^{2+} action</u>

At what biochemical step does Ca^{2+} interact with the cyclic AMP system to cause the inactivation of I$_R$? There are at least three biochemical loci at which Ca^{2+}, via the Ca^{2+}-binding protein calmodulin, can potentially regulate the cyclic AMP cascade. First, Ca^{2+}/calmodulin can regulate the synthesis of cyclic AMP by modulating the activity of adenylate cyclase (Brostrom et al., 1978). Second, Ca^{2+}/calmodulin can activate a form of phosphodiesterase, the degradative enzyme of cyclic AMP metabolism (Wolff and Brostrom, 1979). Finally, Ca^{2+}/ calmodulin can activate a phosphatase (calcineurin) that can dephosphorylate substrates phosphorylated by cyclic AMP-dependent protein kinase (Klee et al., 1983).

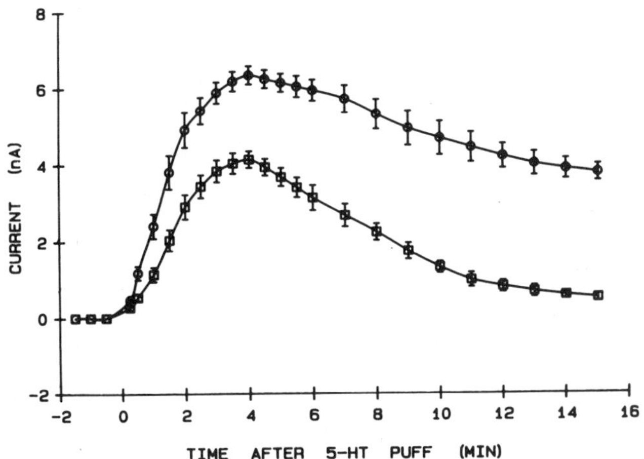

Fig. 7. Effect of Ca^{2+} loading on the response of R15 to locally applied 5-HT. Time course of the average of three responses to 10 sec "puffs" of 5-HT at the soma of R15. 5-HT puffs were applied while the cell was continuosly hyperpolarized under voltage clamp below -80 mV (circles) or during periods when the cell was periodically depolarized (every 5 sec) for 150 msec to 0 mV (squares) to load the cell with Ca^{2+}. The two treatments were alternated. I$_R$ was measured by applying 400 msec hyperpolarizing pulses from -85 to -120 mV. Error bars represent mean $\pm$ standard error of the mean. Note that Ca^{2+} loading both decreases the peak response to 5-HT and speeds the relaxation of the response.

Although there is a great deal of evidence from biochemical studies on cell-free systems that Ca^{2+} regulates these enzymes, there has been little work demonstrating the physiological importance of each of these Ca^{2+}-sensitive steps. Nevertheless, the potential regulation of each of these enzymes by Ca^{2+} does have important implications. For example, regulation of adenylate cyclase activity by intracellular Ca^{2+} has been proposed as a mechanism for long-term changes in responsiveness during associative conditioning in Aplysia (Kandel et al., 1983). Regulation of phosphodiesterase activity by Ca^{2+} may play a role in visual adaptation of photoreceptors (Kawamura and Bownds, 1981). Regulation of phosphatase activity may be responsible for Ca^{2+} channel inactivation by intracellular Ca^{2+} (Eckert and Chad, 1984).

We have not yet determined the exact biochemical mechanism whereby Ca^{2+} interacts with cyclic AMP to cause the inactivation of I_R in neuron R15. However, we have obtained clues that suggest that Ca^{2+} regulation of phosphodiesterase activity is an important locus of action. First, the effect of Ca^{2+} on I_R in R15 is reversibly suppressed if phosphodiesterase inhibitors are added to the saline. Second, we have detected Ca^{2+}/calmodulin-dependent phophodiesterase activity in homogenates of Aplysia ganglia, as well as in extracts from single R15 somata. Finally, we have found that Ca^{2+} influx reduces the content of cyclic AMP in single R15 cells, consistent with the idea that Ca^{2+} activates a phosphodiesterase that reduces the cellular content of cyclic AMP. In a series of radioimmunoassays of pools of isolated R15 somata, we have found that R15 neurons that were hyperpolarized under voltage clamp had 41 $\pm$ 7.3% (mean $\pm$ standard error of mean) more cyclic AMP than R15 neurons that were allowed to burst spontaneously, and hence were presumably loaded with Ca^{2+} (Fig. 8). Hence it is likely that a Ca^{2+}/calmodulin-dependent phosphodiesterase is one site at which Ca^{2+} regulates cyclic AMP in R15. Additional experiments are needed to determine whether other biochemical loci also play a role in mediating the effect of Ca^{2+} on I_R.

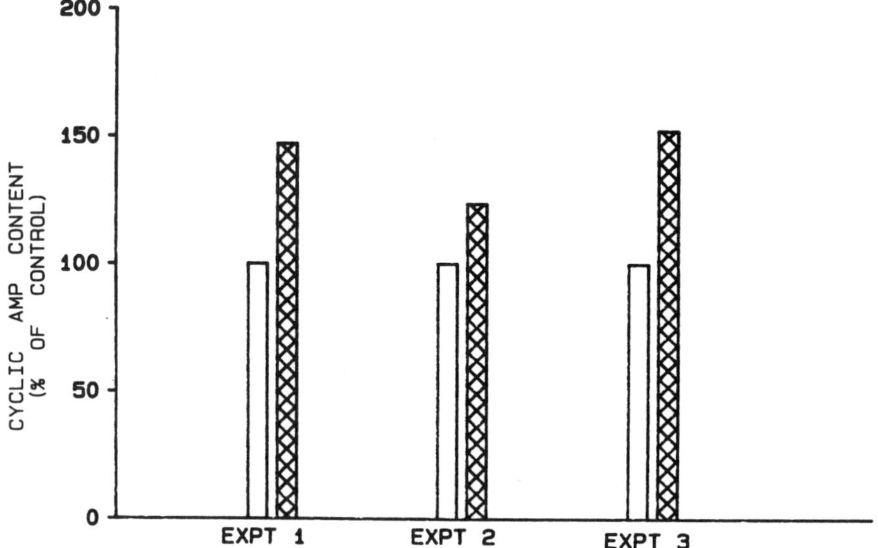

Fig. 8. Cyclic AMP content of R15 cells allowed to burst spontaneously for 20 min (open bars) or continuously hyperpolarized below -75 mV for 20 min (crosshatched bars). Each bar represents data from 5 pooled R15 somata. R15 somata were rapidly frozen after the treatment, were removed frozen from ganglia, and a radioimmunoassay for cyclic AMP was performed as in Levitan and Drummond (1980). Data are normalized with respect to the spontaneously bursting R15 cells.

CONCLUSION

Cellular neurophysiology is rapidly evolving from an examination of
fundamental membrane properties to an understanding of how complex
cytoplasmic regulatory systems modulate those properties. Now that there is
good evidence for the operation of several intracellular messengers in
neurons, it is important to elucidate the physiological interactions between
intracellular messenger systems, and the functional implications of these
interactions. We have begun to make progress in understanding a process
that seems to result from an interaction between Ca^{2+} and cyclic AMP
metabolism in neuron R15. The inactivation of I_R by Ca^{2+} is probably just
one of many consequences of the interplay between Ca^{2+} and cyclic AMP.
Further studies will undoubtedly reveal additional physiological
manifestations of the intricate biochemical machinery that links the cell's
intracellular messenger systems.

ACKNOWLEDGEMENTS

Supported by NIH grant NS17910 to IBL, and NRSA fellowship 1F32 NS07709 to
RHK. We thank Felix Strumwasser for his generous gift of synthetic ELH.

REFERENCES

Adams, W.B., 1985, Slow depolarizing and hyperpolarizing currents which
    mediate bursting in _Aplysia_ neurone R15. _J. Physiol._ 360:51-68.

Adams, W.B. and Benson, J.A., 1985, The generation and modulation of
    endogenous rhythmicity in the _Aplysia_ bursting pacemaker neurone R15.
    Prog. Biophys. molec. Biol. 46:1-49.

Adams, W.B. and Levitan, I.B., 1982, Intracellular injection of protein
    kinase inhibitor blocks the serotonin-induced increase in K^+
    conductance in _Aplysia_ neuron R15. _Proc. Natl. Acad. Sci. USA_ 79:3877-
    3880.

Adams, W.B. and Levitan, I.B., 1985, Voltage and ion dependences of the slow
    currents which mediate bursting in _Aplysia_ neurone R15. _J. Physiol._
    360:69-93.

Benson, J.A. and Levitan, I.B., 1983, Serotonin increases an anomalously
    rectifying K^+ current in the _Aplysia_ neuron R15. _Proc. Natl. Acad.
    Sci. USA_ 80:3522-3525.

Brostrom, M.A., Brostrom, C.O., Breckenridge, B.M., and Wolff, D.J., 1978,
    Calcium-dependent regulation of brain adenylate cyclase. _Adv. Cyclic
    Nucl. Res._ 9:85-99.

Chesnoy-Marchais, D., 1983, Characterization of a chloride conductance
    activated by hyperpolarization in _Aplysia_ neurones. _J. Physiol._
    342:277-308.

Drummond, A.H., Benson, J.A. and Levitan, I.B., 1980, Serotonin-induced
    hyperpolarization of an identified _Aplysia_ neuron is mediated by cyclic
    AMP. Proc. Natl. Acad. Sci. USA 77:5013-5017.

Eckert, R. and Chad, J.E., 1984, Inactivation of Ca channels. _Prog.
    Biophys. molec. Biol._ 44:215-267

Eckert, R. and Tillotson, D., 1981, Calcium-mediated inactivation of the
    calcium conductance in caesium-loaded giant neurones of
    Aplysia. J. Physiol. 314:265-280.

Ewald, D. and Eckert, R., 1983, Cyclic AMP enhances calcium-dependent
    potassium current in Aplysia neurons. Cell. molec. Neurobiol. 3:345-
    353.

Gorman, A.L.F. and Hermann, A., 1979, Internal effects of divalent cations
    on potassium permeability in molluscan neurones. J.Physiol.
    296:393-410.

Gorman, A.L.F. and Thomas, M.V., 1978, Changes in the intracellular
    concentration of free calcium ions in a pace-maker neurone, measured
    with the metallochromic indicator dye arsenazo III. J. Physiol.
    275:357-376.

Gunning, R., 1987, Increased numbers of ion channels promoted by an
    intracellular second messenger. Science 235:80-82.

Kandel, E.R., Abrams, T., Bernier, L., Carew, T.J., Hawkins, R.D., and
    Schwartz, J.H., 1983, Classical conditioning and sensitization share
    aspects of the same molecular cascade in Aplysia. Cold
    Sp. Harb. Symp. Quant. Biol. 48:821-830.

Kawamura, S. and Bownds, M.D., 1981, Light adaptation of the cyclic Gmp
    phosphodiesterase of frog photoreceptor membranes mediated by ATP and
    calcium ions. J. gen. Physiol. 77:571-591.

Klee, C.B., Krinks, M.H., Manalan, A.S., Cohen, P., and Stewart, A.A., 1983,
    Isolation and characterization of bovine brain calcineurin: a
    calmodulin-stimulated protein phosphatase. Methods Enzymol.
    102:227-244.

Kramer, R.H., Levitan, E.S., and Levitan, I.B., 1987, Interaction between
    cyclic AMP-dependent activation and calcium-dependent inactivation of a
    potassium current in Aplysia neuron R15., Submitted for publication.

Kramer, R.H. and Levitan, I.B., 1987, Calcium-dependent inactivation of a
    potassium current in Aplysia neuron R15., Submitted for publication.

Kramer, R.H. and Zucker, R.S., 1985a, Calcium-dependent inward current in
    Aplysia bursting pace-maker neurones. J. Physiol. 362:107-130.

Kramer, R.H. and Zucker, R.S., 1985b, Calcium-induced inactivation of
    calcium current causes the interburst hyperpolarization of Aplysia
    bursting pacemaker neurones. J. Physiol. 362:131-160.

Kupfermann, I. and Kandel, E.R., 1970, Electrophysiological properties and
    functional interconnections of two symmetrical neurosecretory clusters
    (bag cells) in the abdominal ganglion of Aplysia. J. Neurophysiol.
    33:865-876.

Levitan, E.S. and Levitan, I.B., 1987, Serotonin acting via cAMP can produce
    either hyperpolarization or depolarization in the Aplysia bursting
    pacemaker neuron R15. Submitted for publication.

Levitan, E.S., Kramer, R.H. and Levitan, I.B., 1987, Augmentation of
    bursting pacemaker activity in Aplysia neuron R15 by egg-laying hormone
    is mediated by a cyclic AMP-dependent increase in Ca^{2+} and K^+ currents.
    Submitted for publication.

Levitan, I.B. and Drummond, A.H., 1980, Neuronal serotonin receptors and
     cyclic AMP: biochemical, pharmacological and electrophysiological
     analysis. in "Neurotransmitters and their receptors", U.Z. Litteuer et
     al., eds., John Wiley and Sons, London.

Lotshaw, D.P., Levitan, E.S. and Levitan, I.B., 1986, Fine tuning of
     neuronal electrical activity: modulation of several ion channels by
     intracellular messengers in a single identified nerve cell. J. Exp.
     Biol. 124:307-322.

Mayeri, E., Brownell, P., Branton, W.D. and Simon, S.B., 1979, Multiple,
     prolonged actions of neuroendocrine bag cells on neurons in Aplysia. I.
     Effects on bursting pacemaker neurons. J. Neurophysiol. 42:1165-1184.

Mayeri, E. and Rothman, B.S., 1985, Neuropeptides and the control of
     egg-laying behavior in Aplysia. in:"Model Neural Networks and
     Behavior", A.I. Selverston, ed. Plenum, New York.

Mayeri, E., Rothman, B.S., Brownell, P.H., Branton, W.D. and Padgett, L.,
     1985, Nonsynaptic characteristics of neurotransmission mediated by egg-
     laying hormone in the abdominal ganglion of Aplysia. J. Neurosci.
     5:2060-2077.

Meech, R.W., 1972, Intracellular calcium injection causes increased
     potassium conductance in Aplysia nerve cells. Comp. Biochem Physiol.
     42A:493-499.

Moczydlowski, E. and Latorre, R., 1983, Gating kinetics of Ca^{2+}-activated
     potassium channels from rat muscle incorporated into planar lipid
     bilayers: evidence for two voltage-dependent Ca^{2+} binding
     reactions. J. Gen Physiol. 82:511-542.

Wolff, D.A. and Brostrom, C.O., 1979, Properties and function of the
     calcium-dependent regulator protein. Adv. Cyc. Nucleotide Res.
     11:28-88.

FUNCTIONAL IMPLICATIONS OF CALCIUM CHANNEL MODULATION IN EMBRYONIC DORSAL ROOT GANGLION NEURONS

Kathleen Dunlap, Stanley G. Rane, and George G. Holz

Department of Physiology, Tufts University School of Medicine

Boston, Massachusetts 02111

INTRODUCTION

The calcium-dependence of neurosecretion was established more than two decades ago (1). Electrophysiological demonstration of voltage-dependent calcium (Ca) channels in the presynaptic terminal of the squid giant synapse (2) led to the notion that Ca influx through these channels provided the necessary trigger for transmitter release. The relatively recent demonstrations of multiple Ca channel types (3-7) has prompted speculation as to their functional significance; for example, is one type of Ca channel important for regulating threshold and another for controlling exocytosis? If so, these functional differences imply spatial segregation of the various channel subtypes--e.g., those important for secretion would necessarily be located at sites of transmitter release. To address this question, we have begun to investigate the Ca-dependent release of substance P from dorsal root ganglion (DRG) neurons and its control by transmitter receptors and GTP-binding proteins. The results of experiments summarized in this chapter argue that L-type Ca channels (as defined by Tsien and collaborators, see below) play a critical role in exocytosis (8) and their regulation by neurotransmitters and GTP-binding proteins (9) has important functional implications for neurotransmission.

CALCIUM CHANNELS IN DORSAL ROOT GANGLION NEURONS

Channel types. DRG neurons from embryonic chick have been a primary target for investigations of Ca channel function. From the soma membrane of these cells, one can record, both the low (T) and high (L) threshold Ca currents (3,5) commonly seen in many cell types. In addition, there appears to be a third type of Ca current (N) in DRG neurons. These three types of Ca current have been described both macroscopically (5) and at the level of single channels (10). To summarize briefly, T channels (8 pS) are predominantly inactivated at normal resting potentials, exhibit a low activation threshold (-60 mV), and open transiently with steady depolarizations (τ of inactivation is approximately 25 msec at 0 mV). In contrast, L channels (25 pS) are not inactivated at rest, require depolarization to around -20 mV for activation, and produce a relatively long-lasting current with steady depolarization (in EGTA dialyzed cells τ of inactivation is about 1 sec). N channels (13 pS) display some properties of both L and T channnels: like T, they are predominantly inactivated at rest and produce a transient current (with an inactivation τ of 100 msec) but, like L, have a high activation threshold (see chapter by A. Fox, this volume).

Phamacological properties. These channel types differ not only kinetically but also in their susceptibility to pharmacological agents. Cd blocks L and N channels more readily than T (10), omega-conotoxin blocks chick DRG L and N but not T (11), and dihydro-pyridines (DHPs) are selective for L over N and T (12). These findings suggest that

"

pharmacological agents might be used to assay the involvement of the different channel types in cellular processes. This must be done with caution, however. Omega-toxin exhibits substantial inter-species and intraspecific tissue variation. For example, the toxin blocks L and N channels in chick but not in rat DRG, and cardiac muscle L-type Ca channels are also insensitive to the toxin. Use of the DHPs is also somewhat problematic in that their blocking effects are both voltage- and time-dependent in many tissues (13-15).

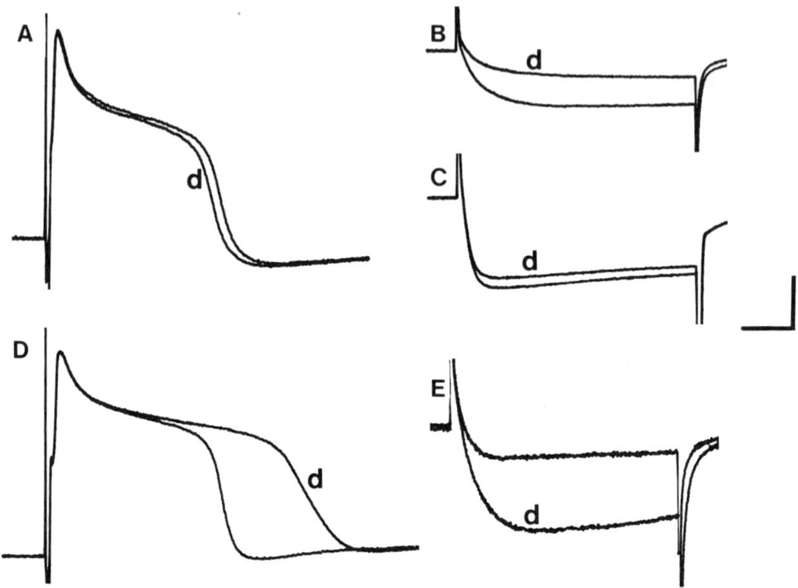

Figure 1: Effect of dihydropyridine antagonist nifedipine (100 nM, A-C) and agonist Bay K 8644 (5 µM, D and E) on dorsal root ganglion neuron action potentials measured with standard intracellular microelectrode recording (A and D) and Ca currents measured under the whole cell recording configuration of the patch clamp method (B,C, and E). The resting potentials of the neurons in A and D were -62 and -66 mV, respectively. The holding potentials were -30 mV (B), -90 mV (C), and -70 mV (E), and the cells were depolarized to 0mV (B and C) and -15 mV (E). Records in B and C are from the same cell. The letter 'd' indicates the traces taken in the presence of 1 µM dihydropyridine. External solutions contained (in mM) 140 NaCl, 2 Ca Cl2, 1 BaCl2, 10 HEPES, and 0.3 µM tetrodotoxin, pH 7.3. The pipette solution for B,C, and E contained (in mM) 150 CsCl, 10 HEPES, 5 MgATP, and 5 BAPTA, pH 7.3. Vertical calibration: 30 mV (A and D), 3 nA (B and C), and 1 nA (E). Horizontal calibration: 10 ms in all traces.

Figure 1 (A-C) illustrates the voltage-dependent effects of the antagonist DHP, nifedipine, on chick DRG soma Ca currents. Nifedipine blocks L-type Ca current (and substance P release, as shown below), but only in cells held at relatively positive holding potentials (15). When brief Ca currents or action potentials are evoked in cells near rest, the DHP antagonists are rendered almost completely ineffective on L channels. The action of nifedipine is also time-dependent; significant reductions in Ca current occur only following prolonged depolarization for periods greater than one second (not shown, see ref. 15). Findings are similar to those previously reported for DHP inhibition of L-type Ca channels in cardiac muscle cells (13,14). The DHP Bay K 8644 also exhibits voltage dependent effects on L-type Ca currents: it is a Ca channel agonist at negative potentials (12 and figure 1E) but exhibits mixed agonist/antagonist effects on Ca current in depolarized cells (16). Consistent with its reported agonist action, Bay K prolonged the Ca-dependent plateau phase of the action potential in chick DRG neurons (figure 1D). These observations underscore the reasons why the term "DHP-insensitivity" should be used with caution, since the efficacy of DHPs depends dramatically upon experimental conditions. This concern is particularly acute in cases in which the state of depolarization cannot be directly measured (in nerve terminals, for example). In short, better tools are needed to allow us to unequivocally demonstrate, using pharmacological means, the involvement of one or another Ca channel type in a particular cellular function.

Ca channel modulation. The function of some types of voltage-dependent Ca channel can be modulated by exogenous application of drugs and neurotransmitters. DRG neuron L-type Ca channels can be inhibited by transmitters such as norepinephrine (NE) and GABA (17). Application of the transmitters in concentrations ranging from 0.1 to 10 μM produces a dose-dependent, reversible decrease in macroscopic L-type Ca current, an effect which is manifest under current clamp as a decrease in the Ca-dependent action potential.

NE and GABA produce their actions on L-type Ca channels via pharmacologically distinct membrane receptors (18,19); the similarity in their actions on the Ca channel, however, suggests that these two transmitters may share a common effector pathway. Indeed, both receptor types are linked to GTP-binding proteins (G-proteins) that transduce the agonist-receptor interation into an effect on Ca channel function (20). Although it is unclear at this time whether the two receptors share a common pool of G-protein, treatment of DRG neurons with pertussis toxin (which inactivates only certain G-proteins, notably G_i and G_o) completely blocks the actions of NE and GABA on L-type Ca channels. These two G-proteins are the predominant substrates ribosylated by pertussis toxin in these neurons (G. Holz, unpublished observations).

Preliminary evidence from our lab indicates that these effects of NE and GABA may be specific for the L-type Ca channel in DRG neurons. The slowly inactivating (non-transient) L-type Ca current activated by depolarization to 0 mV from a holding potential of -50 mV was inhibitied by 10 μM NE with no change in the activation or inactivation time courses (figure 2B). When the cell was held at -90 mV and depolarized to 0 mV, the transient portion of Ca current (T and N) was evoked but not altered by NE (figure 2A and C), suggesting that the modulation is specific for the L-type channel. This suggested, therefore, that the transmitters as well as agents known to interfere with their actions on Ca channel function (such as pertussis toxin), might be useful tools to test for the involvement of L-type Ca channels in exocytosis. Thus, we have begun to test the effects of NE, GABA, and pertussis toxin (PTX), in addition to the DHPs, on the release of substance P (SP) from DRG neurons.

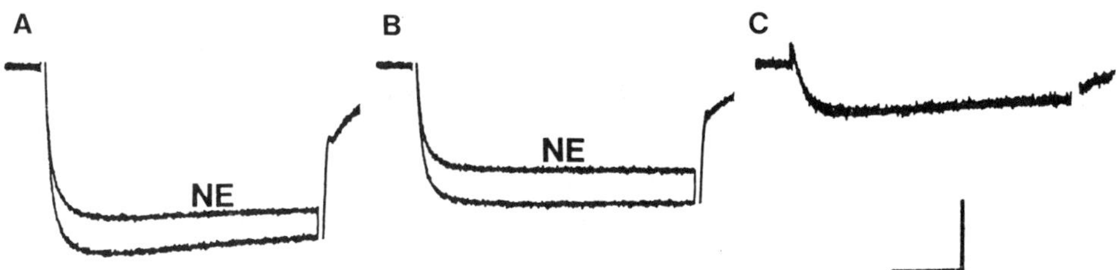

Figure 2: NE effect on long-lasting current. Ca currents in DRG neurons recorded following step depolarization to 0 mV from a holding potential of -90 mV (A) and -50 mV (B), before and during ('NE') application of 10 μM NE. The transient component (N and T) of the current (C) was obtained by subtracting currents in B from those in A. NE does not inhibit this portion of the current, suggesting that the transmitter action is specific for L-type Ca channels. Calibration: 1 nA, 10 ms (A and B) and 500 pA, 10 ms (C).

ELECTRICALLY EVOKED SUBSTANCE P RELEASE

Method. DRG neurons synthesize SP in quantity, both in vivo as well as in dissociated cell culture, and the peptide can be detected with standard radioimmunoassay techniques (21,22). Release of SP can be evoked by depolarizing the membrane, either electrically (8) or with saline solution containing an elevated K concentration (22,23). The former method offers the advantage of studying the release process under conditions that most closely approximate those encountered in vivo; in addition, electrical stimulation enhances the amount of peptide release compared to that evoked by high K.

Electrical field stimulation is achieved by passing current between two extracellular electrodes placed in the culture dish. For the experiments reviewed here, suprathreshold stimuli (3 ms) are applied at 1 Hz for 1.5 to 6 minutes. Using this protocol, several hundred picograms of SP can be released from each dish (of ca. 100,000 cells) per stimulation period. Furthermore, the amount released is reproduceable from trial to trial provided a rest period of about 1.5 hours separates the stimulation periods.

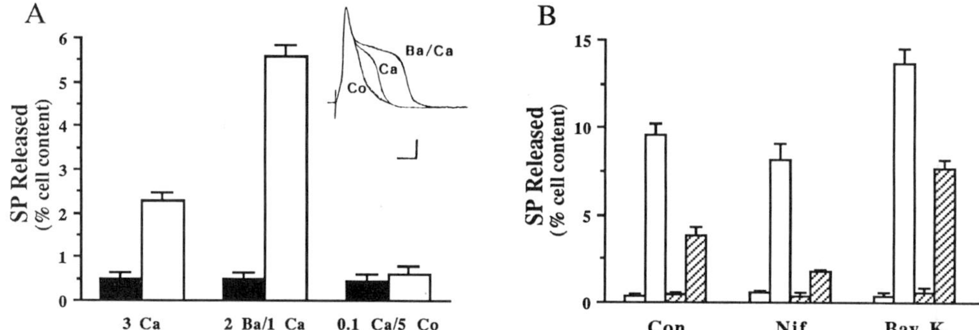

Figure 3: Ca-dependent SP release. A. Electrically-evoked release of SP (plotted as percentage of total cellular content) in three different saline solutions (noted on the abcissa). The filled bars represent baseline and the open bars evoked release (+SEM). The inset illustrates superimposed, representative DRG action potentials recorded in the three different solutions. Calibration bar: 20 mV, 5 ms. B. The effect of 5 µM nifedipine ('nif') and 5 µM Bay K 8644 on SP release evoked by electrical (open bars) or 60 mM KCl depolarization (cross hatched bars). Baseline values for release are plotted to the left of their corresponding experimentals. The total cellular content was 13,250 ± 500 (A) and 14,200 ± 250 (B) pg/culture.

Pharmacology. Electrically-evoked SP release is Ca-dependent; it is blocked when Ca in the saline is replaced by Co or Cd and enhanced with Ba replacement (figure 3A). The latter effect can be explained by the observed prolongation of the action potential duration in Ba-containing solutions which would lead to enhanced Ca entry. DHP agonists and antagonists were employed to investigate the involvement of L-type Ca channels in DRG neurosecretion. The agonist Bay K 8644 potentiated electrically stimulated SP release, but the antagonist nifedipine had no significant effect (figure 3B). One interpretation of the lack of nifedipine action is that SP release is controlled by Ca current not through L-type Ca channels, but rather through DHP-insenstive Ca channels. Considering, however, that the duration of the action potentials evoked by electrical field stimulation are relatively brief (10-20 msec) compared to the above-reported time-dependence of antagonist DHP action (greater than 1 second), it is not surprising that electrically stimulated release of SP is insensitive to nifidipine. In an attempt to circumvent this problem, we studied SP release under conditions of prolonged depolarization, which would allow the DHP antagonists to inhibit L-type Ca channels. High-K stimulated release of the peptide can be readily reduced by nifedipine (figure 4). These results support the notion that L-type Ca channels are important for the control of exocytosis in DRG neurons.

NE and GABA inhibit L-type Ca current without the voltage and time dependent effects characteristic of the DHP antagonists. We have, therefore, tested these transmitters on electrically-evoked SP release from DRG neurons. Both NE (figure 4) and GABA inhibit the release in concentrations known to inhibit L-type Ca current. Furthermore, the pharmacology of the NE and GABA receptors which mediate this action on release is identical to that of the receptors involved in the modulation of soma L-type Ca current (figure 5). This implies not only that the same receptors are used in the regulation of the two processes but encourages the speculation that inhibition of L-type Ca current by the neurotransmitters underlies their inhibition of SP release. One additional line of support for this notion comes from experiments investigating the role of GTP-binding proteins as regulators of SP release. As mentioned above, pertussis toxin (PTX) blocks the GABA and NE receptor-mediated inhibition of L-type Ca current. Following exposure of DRG cells to PTX for a period of time sufficient to eliminate the effects of the transmitters on Ca current, the release process also becomes insensitive to NE (figure 6) and GABA.

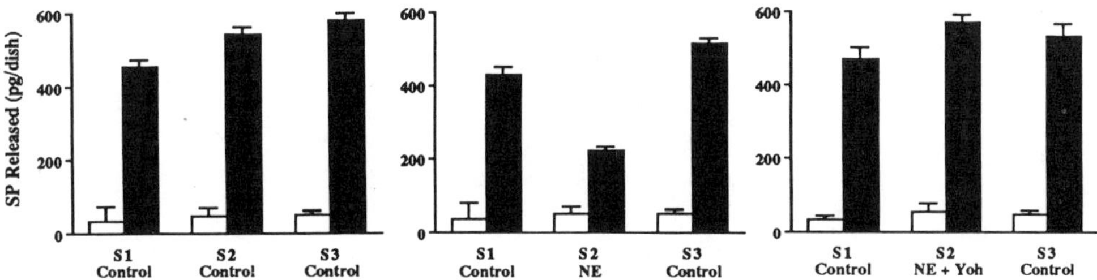

Figure 4: Protocol for electrical field stimulation experiments. Three sets of three cultures each are subjected to three different 90 second stimulation periods (S1-S3) separated by 1.5 hours. During S2, one set of cultures (left) is exposed to control saline only while the other two sets are exposed to experimental saline solutions containing drugs of interest (middle graph, 50 μM NE and right graph, 50 μM NE + 10 μM yohimbine). S1 and S3 are the same for all three sets of cultures and represent stimulation in control solution. The effect of the drugs is assessed by calculating the release ratio (the amount of SP released during S2 divided by the average release during S1 and S3). A release ratio near 1 indicates no effect of the drug. For the three graphs shown in this figure, the release ratios (left to right) are 1.05, 0.4, and 1.02. Total cellular content of SP was 12,700 ± 350 pg/culture.

DISCUSSION

Ca channel types controlling exocytosis. Are L-type Ca channels the only exocytotic Ca channels in DRG neurons? The DHP-insensitive portion of the release process for SP might be explained by the involvement of N and/or T type Ca channels in secretion. Alternatively, the DHP-insensitivity may result from an incomplete block of L-type Ca channels by the antagonist. Both soma L-type Ca current and high-K-stimulated SP release are inhibited approximately 50% by the antagonist. Although this suggests that an involvement of transient Ca currents need not be invoked to explain the partial DHP-insensitivity of the release process in DRG neurons, an involvement of N and/or T channels cannot be ruled out by our present results. However, in contrast to Nowycky et al. (5) who report that N-type Ca currents are a substantial fraction (30-60%) of total Ca current in cultured chick DRG neurons, we see relatively little (less than 10%) transient Ca current (N and T) in the same cells cultured in our lab, raising the somewhat unattractive possibility that the proportions of Ca channel types in these cells may vary, depending upon culture conditions. Direct electrophysiological assay of Ca channel types in nerve terminal membranes remains technically difficult, and a pharmacological assessment of the relative contributions of L, N, and T type channels to exocytosis awaits better, specific antagonists for each of the channel types.

An involvement of L-type Ca channels in exocytosis from DRG neurons is also suggested by the studies of Miller and colleagues on the K-stimulated release of SP from rat DRG neurons (23). Bay K 8644 enhanced while the antagonist DHPs completely inhibited

SP release. The authors suggest that, in rat DRG neurons, L-type Ca channels are the only channels that regulate SP release. In contrast, no effect of the DHP antagonists was observed on NE release from rat superior cervical ganglion (SCG) neurons although Bay K enhanced its release. They conclude, therefore, that different types of Ca channels regulate exocytosis in the different neurons (L-type for DRGs and N-type for SCGs). An alternative possibility, that different types of Ca channels regulate the release of different classes of transmitter substance (e.g., classical transmitters vs. peptides) must also be considered, particularly in light of the demonstration that presynaptic firing patterns regulate the type of neurotransmitter released from sympathetic neurons (24). These interesting possibilities remain to be tested directly.

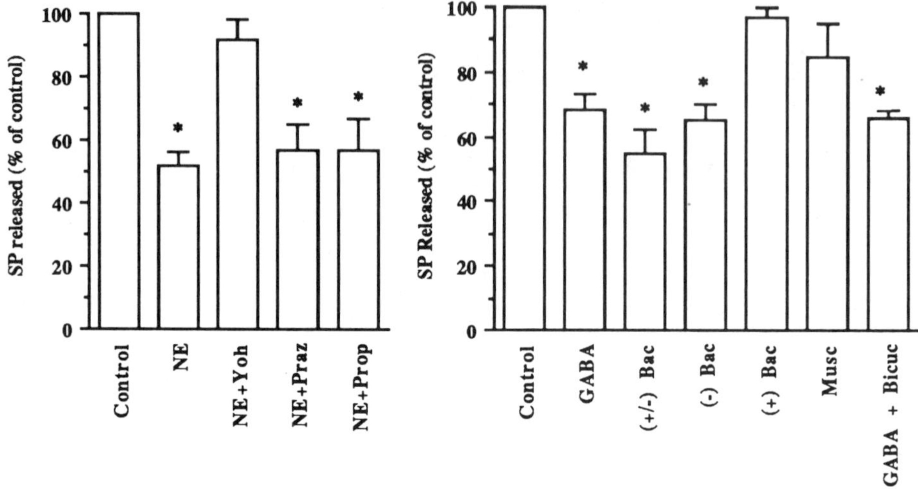

*Figure 5: Summary of the NE (left) and GABA (right) receptor pharmacology. Experiments using the protocol described in figure 4 were performed for each of the drugs illustrated in these summary tables. The NE receptor-induced inhibition of SP release is blocked by 10 μM yohimbine (yoh) but not 10 μM prazosin (praz) or propranolol (prop). The GABA inhibition of SP release was mimicked by 10 μM baclofen (bac) but not 10 μM muscimol (musc) and was insensitive to 50 μM bicuculline (bicuc). The data are plotted relative to the control release in each experiment (+SEM). * indicates values significantly different from control, p ≤ .001.*

Implications for presynaptic inhibition. In the dorsal horn of the spinal cord DRG nerve terminals receive a barrage of synaptic input from spinal cord interneurons and from neurons descending from higher brain centers. It has long been thought that these inputs play a critical role in the regulation of sensory neurotransmission by altering transmitter release from DRG nerve terminals. Although a number of neurotransmitters have been implicated in presynaptic inhibition, the case has been particularly well-developed for GABA. The mechanisms underlying the GABA-mediated inhibition of neurosecretion in DRG neurons, however, are under dispute.

One line of evidence points to changes in resting membrane conductance. GABA can depolarize sensory nerve terminals by activating GABAa receptor-mediated increases in chloride conductance. This was termed primary afferent depolarization (PAD) by Eccles and co-workers (25). It was argued that changes in resting conductance of the nerve terminal membrane would shunt the action potential, thereby decreasing its amplitude and reducing Ca entry. They found that the time course of PAD paralleled that of presynaptic inhibition (25) and that both PAD and presynaptic inhibition were sensitive to the GABAa receptor antagonists bicuculline (26,27). However, a bicuculline-insensitive component to GABA-mediated presynaptic inhibition has also been described. The selective GABAb receptor agonist, baclofen, could substitute for GABA in mediating this component of the inhibition (28). On the basis of the reported action of baclofen on voltage-dependent Ca channel function (18), it was suggested that presynaptic inhibition was, at least in part, due to direct modulation of a Ca channel in the nerve terminal membrane.

The recent description of Ca channels that show substantial inactivation within the normal range of cell resting potentials leads one to speculate about the relationship between PAD and inactivation of the transient Ca currents. In other words, PAD may evoke a decrease in neurotransmitter release not by a non-specific shunt of nerve terminal currents associated with the action potential, but by a relative reduction in the number of functional, transient Ca channels available to open during the action potential. Using this hypothetical scheme, it would be possible to explain the two components of presynaptic inhibition by the involvement of two, kinetically-distinct populations of voltage-dependent Ca channel in the nerve terminal membrane. As with other experimental approaches to the study of Ca channel function, work in this area will be greatly aided by the development of selective pharmacological tools for the different channel types.

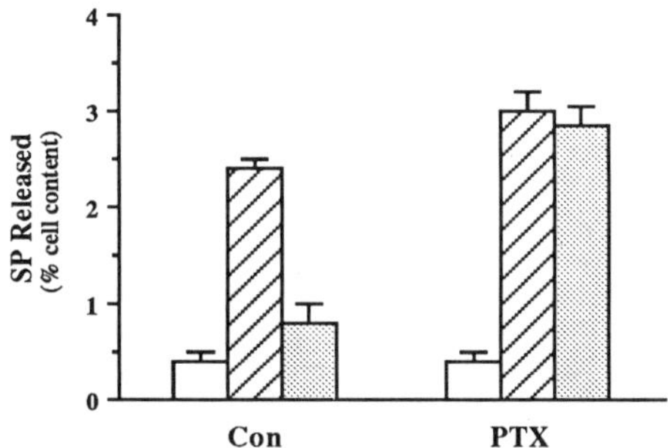

Figure 6: Pertussis toxin sensitivity of the receptor-mediated inhibition of SP release. The amount of peptide released during 6 minutes of electrical field stimulation in 3 mM Ca saline is plottted (as percentage of total cellular content) for control cells (con) and for cells treated 16 hours with 140 ng/ml pertussis toxin (PTX) in the absence (cross hatched bars) or presence (stippled bars) of 50 µM NE. Baseline release is shown by the open bars. All values are shown +SEM. Cellular content of SP was 3000 ± 250 pg/culture in this experiment.

ACKNOWLEDGEMENTS

The authors would like to thank Dr. Richard Kream for discussions and help with the substance P radioimmunoassay. This work is supported by NIH grant 16483 and a Grant In Aid from the American Heart Association (funded, in part, from the Massachusetts Affiliate).

REFERENCES

1. Katz, B. and Miledi, R. (1965) Proc. Roy. Soc. B, 161:496
2. Katz, B. and Miledi, R. (1969) J. Physiol. 203:459
3. Carbone, E. and Lux, H.D. (1984) Nature 310:501
4. Fedulova, S.A., Kostyuk, P.K., and Veselovsdy, N.S. (1985) J. Physiol. 359:431
5. Nowycky, M.C., Fox, A.P., and Tsien, R.W. (1985) Nature 316:440
6. Nilius, B., Hess, P., Lansman, J.B., Tsien, R.W. (1985) Nature 316:443
7. Armstrong, C.M. and Matteson, D.R. Science 227:65
8. Holz, G.G., Dunlap, K., and Kream, R.M. (1987) J. Neurosci. (In Press)
9. Holz, G.G., Kream, R.M., and Dunlap, K. (1987) J. Neurosci. (In Press)
10. Fox, A.P., Nowycky, M.C., and Tsien, R.W. (1987) J. Physiol. (In Press)
11. McClesky, E.W., Fox, A.P., Fledman, D., Cruz, L.J., Olivera, B.M., Tsien, R.W., and Yoshikami, D. (1987) PNAS (In Press)
12. Nowycky, M.C., Fox, A.P., and Tsien, R.W. (1985) PNAS 82:2178
13. Bean, B.P. (1984) PNAS 81:6388
14. Sanguinetti, M.C. and Kass, R.S. (1984) Circ. Res. 55:336

15. Rane, S.G., Holz, G.G., and Dunlap, K. (1987) Pflugers Archiv. (In Press)
16. Sanguinetti, M.C., Krafte, D.S., and Kass, R.S. (1986) J. gen. Physiol. 88:369
17. Dunlap, K. and Fischbach, G.D. (1981) J. Physiol. 317:519
18. Dunlap, (1981) Brit. J. Pharmacol. 74:579
19. Canfield, D.R. and Dunlap, K. (1984) Brit. J. Pharmacol. 82:557
20. Holz, G.G., Rane, S.G., and Dunlap, K. (1986) Nature 319:670
21. Kream, R.M., Schoenfeld, T.A., Mancuso, R., Clancy, A., El-Bermani, W., and Macrides, F. (1985) PNAS, 82:4832
22. Mudge, A.W., Leeman, S.E., and Fischbach, G.D. (1979) PNAS 76:523
23. Perney, T.M., Hirning, L.D., Leeman, S.E., and Miller, R.J. (1986) PNAS 83:6656
24. Ip, N.Y. and Zigmond, R.E. (1984) Nature 311:472
25. Eccles, J.E., Schmidt, R.F., and Willis, W.D. (1962) J. Physiol. 161:282
26. Gallagher, J.P., Higashi, H., and Nishi, S. (1978) J. Physiol. 275:263
27. Curtis, D.R., Duggan, A.W., Felix, D., and Johnston, G.A.R., (1971) Brain Res. 32:69
28. Davies, J. (1981) Brit. J. Pharmacol. 72:377

NEUROTRANSMITTER MODULATION OF CALCIUM CURRENTS IN RAT SENSORY NEURONS

Douglas A. Ewald, Mary W. Walker, Teresa M. Perney, Heinrich J.G. Matthies and Richard J. Miller

Dept. of Pharmacological and Physiological Sciences, University of Chicago, 947 E. 58th Street, Chicago, IL 60637

INTRODUCTION

It is well established that Ca^{2+} acts as an essential second messenger in virtually all cell types (Berridge and Irvine, 1984). Thus, regulation of the concentration of cytoplasmic free Ca^{2+}, $[Ca^{2+}]_i$ is of great importance in the regulation of cell function. One way in which $[Ca^{2+}]_i$ can be readily increased is by opening specific voltage-sensitive Ca^{2+} channels in the cell membrane (Tsien, 1983). In neurons the most important role of this voltage-dependent Ca^{2+} entry is the regulation of neurotransmitter release. Upon depolarization Ca^{2+} channels in the nerve terminal are opened and a transient rise in $[Ca^{2+}]_i$ triggers neurotransmitter release as is discussed elsewhere in this volume (Augustine; Llinas; Smith). This release process can be modulated by other neurotransmitters by at least two types of mechanisms. Modulatory neurotransmitters can change the membrane potential by altering K^+ currents, thereby indirectly increasing or decreasing the entry of Ca^{2+} into the nerve terminal (Miller, 1984; Newberry and Nicoll, 1984; Klein et al., 1980; Williams et al., 1985). Alternatively, modulatory neurotransmitters can change the gating properties of the Ca^{2+} channels themselves and thus directly modulate Ca^{2+} entry [See for example, Cherubini and North, 1985; Deisz and Lux, 1985; Dunlap and Fischbach, 1978, 1981; Dolphin and Scott, 1986; MacDonald and Werz, 1986; Mochida and Kobayashi, 1986; Robertson and Taylor, 1986; Dolphin et al., 1986; Forscher et al., 1986; Heschler et al., 1987; Madison et al., 1987; Marchetti et al., 1986; Tsunoo et al., 1986; Wanke et al., 1987].

MULTIPLE CA^{2+} CHANNEL TYPES AND NEUROTRANSMITTER RELEASE FROM DRG NEURONS

Sensory neurons cultured from the dorsal root ganglia (DRG) of vertebrates have proven to be an excellent system for studying Ca^{2+} currents. In chick DRG neurons there are three distinct types of Ca^{2+} channels which can be distinguished by both pharmacological and biophysical criteria (Nowycky et al., 1985; Fox and Tsien, this volume). T and N channels give rise to transient Ca^{2+} currents evoked from negative holding potentials whereas L channels give rise to a more sustained Ca^{2+} current which can be evoked from relatively depolarized potentials where the transient currents are inactivated. L channels are powerfully modulated by dihydropyridine (DHP) drugs such as the antagonist nitrendipine and the agonist BAY K8644. T and N channels are relatively resistant to these agents (Fox et al., 1987a,b).

Some DRG neurons release the undecapeptide neurotransmitter substance P when depolarized by electrical stimulation or by raising $[K^+]_O$ (Holz et al., 1987a,b; Perney et al., 1986; Rane et al., 1987). This release is clearly dependent on Ca^{2+} entry as it can be blocked by lowering $[Ca^{2+}]_O$ or by divalent cation blockers of Ca^{2+} channels such as Co^{2+} or Cd^{2+}. An obvious question of interest is which types of Ca^{2+} channels are involved in stimulus/secretion coupling? We investigated this question using cultured rat DRG neurons. We observed that release induced by raising $[K^+]_O$ could be completely blocked by DHP antagonists and enhanced by DHP agonists (Perney et al., 1986) indicating that in these neurons L channels are the most important in providing Ca^{2+} for triggering substance P release. Actually this dominant role of L channels is unusual and in many other cases N channels can be shown to play a predominant role (Miller, 1987b; Perney et al., 1986).

INHIBITION OF CA^{2+} CURRENTS BY MODULATORY NEUROTRANSMITTERS

Several different neurotransmitters can inhibit the evoked release of substance P from DRG neurons (Holz et al., 1987b; Mudge et al., 1979). In many instances this inhibition is associated with inhibition of the Ca^{2+} currents in these cells. Presumably this effect is at least partly responsible for the inhibition of release. Primarily from the work of Dunlap and colleagues, we know that norepinephrine (NE) and γ-amino butyric acid (GABA) inhibit the sustained Ca^{2+} current in DRG neurons. Other neurotransmitters such as adenosine and κ-specific opioids have been found to selectively inhibit the transient Ca^{2+} currents. Table 1 summarizes the presently known inhibitory modulators of Ca^{2+} currents in these neurons. The most recently described modulator is the 36 amino acid peptide neuropeptide Y (NPY). This neuropeptide is found to be very widely distributed in a variety of central and peripheral neurons (Emson and De Quidt, 1984; Sundler et al., 1987).

Table 1. Neurotransmitters that inhibit the DRG Ca^{2+} Current

Norepinephrine (α_2)	
GABA (B)	Dunlap and Fishbach, 1981
Serotonin	
Opioid (κ)	Werz and MacDonald, 1984
Dopamine	Marchetti, Carbonne and Lux, 1986
Adenosine	Dolphin, Forda and Scott, 1986
Neuropeptide Y	present study

Receptors for ^{125}I-NPY are found in high concentrations on cultured rat DRG neurons (Fig. 1). Scatchard analysis demonstrates that two classes of binding sites are present, both of rather high affinity. NPY also effectively inhibits the depolarization induced release of substance P from DRG cells (Perney and Miller, unpublished observations).

To study the effects of NPY on Ca^{2+} currents, we used the whole-cell patch clamp configuration with ionic conditions to optimize isolation of Ca^{2+} currents (see Fig. 2 legend). We alternately evoked the sustained or L current at 0 mV from -40 mV holding potential and, additionally, the transient (N and T) currents at 0 mV from -80 mV holding potential.

The left panel of Fig. 2 shows these currents before and at the end of a 3 min exposure to 100 nM NPY. Both transient and sustained currents are substantially inhibited by exposure to NPY. The total and sustained Ca^{2+} currents were plotted as a function of time (top), the time courses before exposure were extrapolated as a single exponential, and the % reduction at the end of exposure was calculated

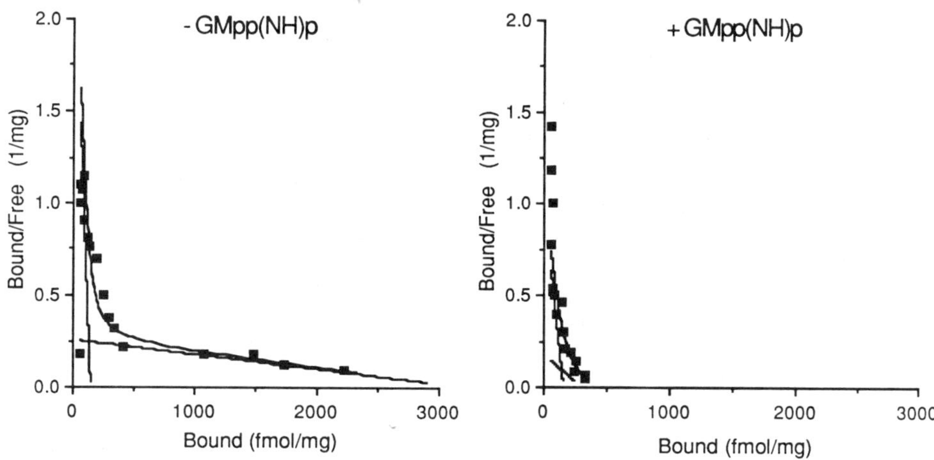

Fig. 1. Binding of ^{125}I-NPY to receptors on membranes from rat DRG neurons. Cells exhibit two types of binding sites with affinities K_{D1}= 60 pM (B_{max}= 8.3 fmoles/mg) and K_{D2}=13 nM (B_{max}=2.7 pmoles/mg). The right panel demonstrates that the stable GTP analogue Gpp(NH)p reduces ^{125}I-NPY binding by reducing the number of low affinity sites.

compared to these extrapolations for the sustained and the transient (total minus sustained) currents. At this concentration the inhibition of both transient and sustained currents ranged from 40 to 70%. Inhibition by 1 nM NPY is, again, approximately equal for transient and sustained currents and is approximately 30% of the extrapolated baselines (Fig. 3, left panel). In this case the traces shown are 1 min after beginning exposure. Plotting the % inhibition of total current and of the separated transient and sustained currents as a function of concentration (Fig. 4, filled symbols), we find that 1 nM is the half-maximally effective concentration in all cases.

We also tested NE at a single concentration (50 μM) and found that the transient currents were inhibited to about the same extent as the sustained. Inhibition ranged from 30 to 60% using the extrapolated baseline technique. The time course of onset of inhibition was much faster than with NPY, often being maximal in less than 20 sec.

MECHANISM OF CALCIUM CURRENT INHIBITION

The mechanism by which inhibitory neurotransmitters block Ca^{2+} currents in DRG neurons remain to be completely elucidated. Moreover, it is not clear whether all inhibitory neurotransmitters work in exactly the same way or indeed whether the mechanisms underlying sustained and transient current inhibition are identical. It is quite clear however that G-proteins are involved in coupling the inhibitory receptors to the Ca^{2+} channel. Thus, Fig. 1 illustrates the effect of a nonhydrolyzable analogue of GTP [Gpp(NH)p] on the binding of ^{125}I-NPY to receptors in DRG cell membranes. Binding is reduced primarily due to the disappearance of low affinity binding sites. This indicates that NPY receptors are regulated in some way by G-proteins. In another type of study it has been shown that injection of GTP or stable analogues into DRG cells inhibits the transient phase of the Ca^{2+}-current and also enhances the inhibitory effect of neurotransmitters on the sustained phase (Scott and Dolphin, 1986; Holz et al., 1986). Furthermore, pertussis toxin clearly blocks the inhibitory effects of the various neurotransmitters on

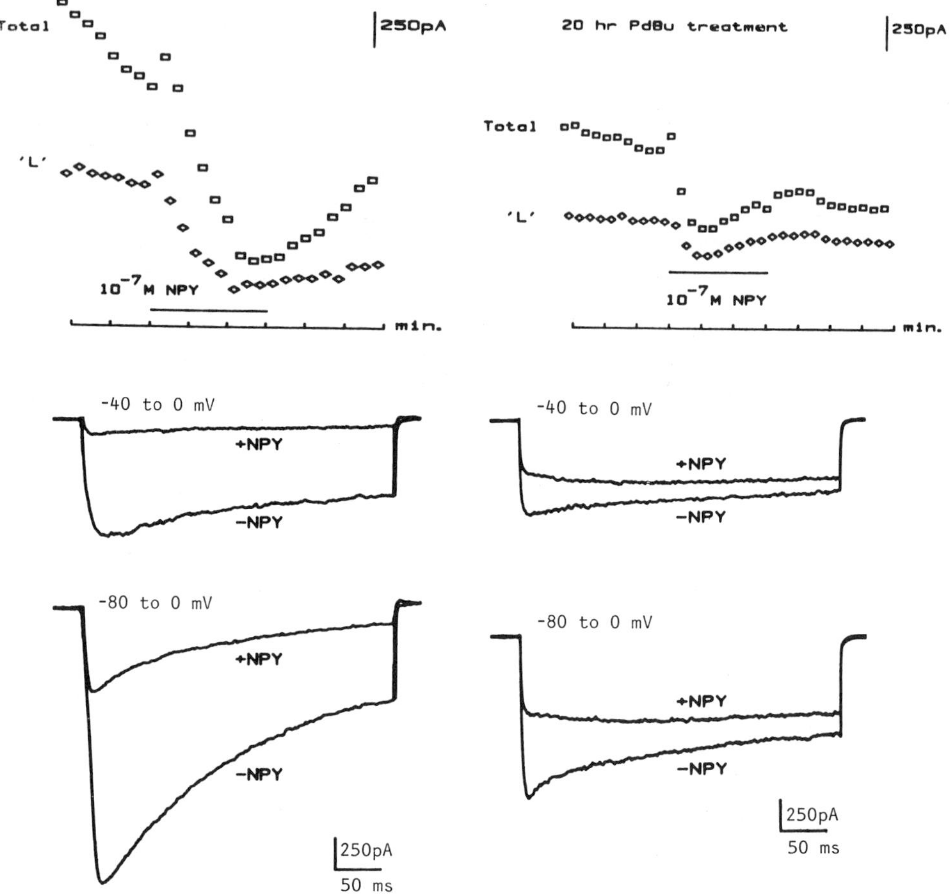

Fig. 2. Inhibition of Ca^{2+} currents by 100 nM NPY. DRG cells were voltage clamped in the whole-cell mode. External solution contained (in mM) 10 CaCl$_2$ and 140 TEACl. Internal solution contained 140 CsCl$_2$, 10 EGTA, 2 ATP and an ATP regenerating system. Voltage clamp pulses to 0 mV (350 msec) were alternately given from holding potentials of -80 mV and -40 mV every 10 sec. The sustained current (L) was evoked from the -40 mV holding potential and the transient currents (N and T) were additionally evoked from the -80 mV holding potential. Traces marked '+NPY' are at the end of a 3 min exposure to 100 nM NPY. Time course of NPY effect on total (sustained plus transient) current and sustained current are plotted at top. Neuron shown in left panel was not treated with phorbol ester. NPY greatly inhibited both transient and sustained currents with a time course of minutes. Neuron shown in right panel was down-regulated for PKC by 20 hr pretreatment with 1 µM phorbol dibutyrate (PdBu) and then rinsed with 1% fatty acid free albumin for 30 min before beginning recording. Inhibition of sustained current was substantially blocked whereas inhibition of transient currents was not.

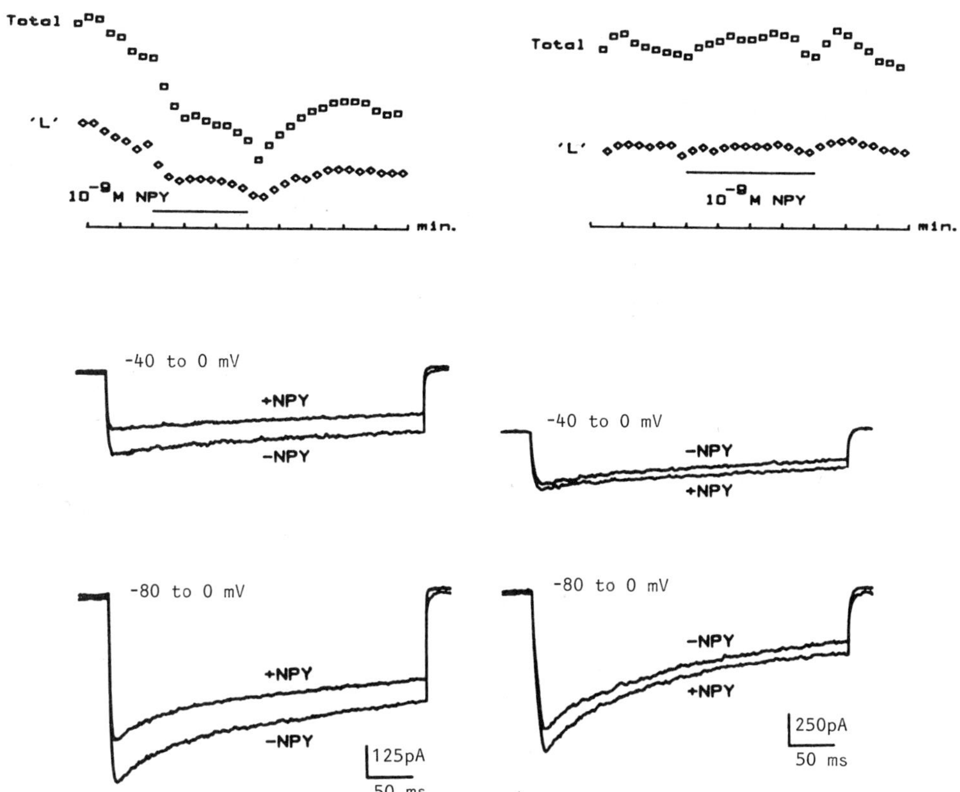

Fig. 3. Inhibition of Ca^{2+} currents by 1 nM NPY. Same format as Fig. 2 for neuron not treated with phorbol ester (left panel) and one down-regulated for PKC with PdBu for 18 hr (right panel). Inhibition of transient and sustained currents caused by 1 nM NPY is completely blocked in the down-regulated neuron. Traces marked '+NPY' are 1 min after beginning exposure.

DRG Ca^{2+} currents (Holz et al., 1986). We have also found that inhibition by NPY (100 nM) is completely blocked by pertussis toxin. The blocking effect of pertussis toxin implies that a G-protein of the G_i or G_o type is involved in the transduction mechanism. There is some controversy at this time as to whether Ca^{2+} channels are "tightly coupled" to the receptor/G-protein complex as with K^+ channels and muscarinic receptors in the heart (Logothetis et al., 1987) or whether a diffusible second messenger is involved. Forscher et al. (1986), reported that ensembles of Ca^{2+} channel activity measured in on-cell patches of DRG cells were unaffected by NE added outside the patch pipette. Moreover, Marchetti et al. (1986), observed inhibition of Ca^{2+} channels in outside-out patches from these neurons. Such observations would indicate tight-coupling of the receptor and the channel. However we have observed that Ca^{2+} channel activity in on-cell patches can be reduced by NE added outside the patch pipette under appropriate conditions (Ewald and Miller, 1987).

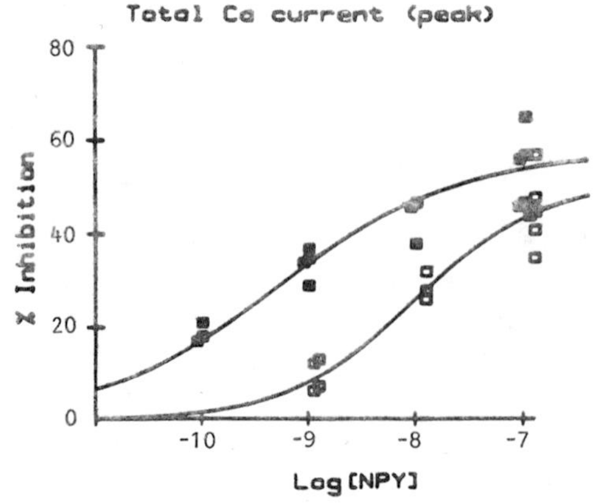

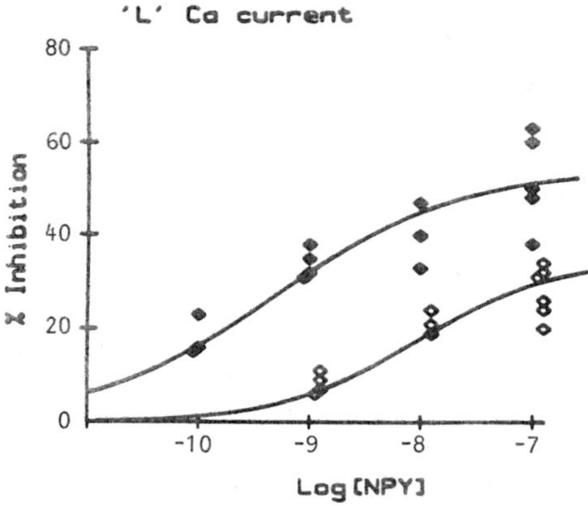

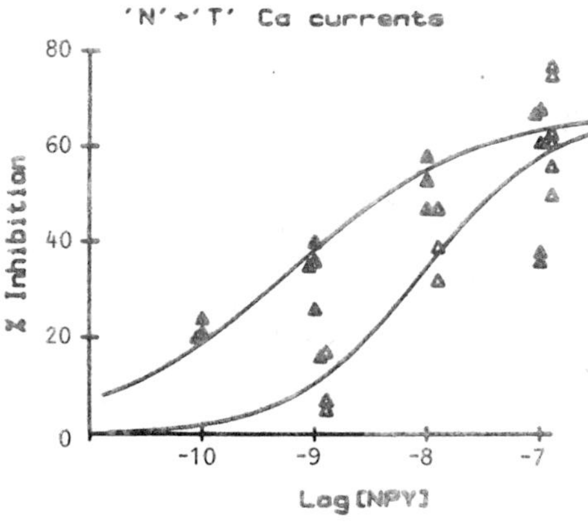

Fig. 4. Dose/response curves for inhibition of total, sustained ('L') and transient ('N' + 'T') Ca^{2+} currents in DRG cells by NPY. Control responses (solid symbols) and responses following down-regulation of protein kinase C by incubation of cells with phorbol dibutyrate (open symbols). Method for calculating % inhibition of various currents is described in the text. Without down-regulation the half-maximal inhibition is 1 nM for both transient and sustained currents. On down-regulated neurons inhibition is blocked at 1 nM and a qualitatively different inhibition is apparent at higher concentrations which is predominantly on the transient currents.

It has been shown that phorbol esters and analogues of diacylglycerol (DAG) can inhibit Ca^{2+} currents in both DRG cells (Rane and Dunlap, 1986) and PC12 pheochromocytoma cells (Harris et al., 1986). We have examined the possibility that DAG and protein kinase C (PKC) may mediate the inhibitory effects of neurotransmitters on DRG Ca^{2+} currents. We found that both NE and NPY stimulated the synthesis of DAG by rat DRG neurons. In both cases two peaks of DAG production could be observed. The first of these was transient and the second was more sustained. Preliminary studies indicate that synthesis of the first peak of DAG can be blocked by pertussis toxin but not the second. Interestingly rather little inositoltrisphosphate (IP_3) is produced indicating that much of the DAG may not be derived from phosphatidylinositol bisphosphate (PIP_2) but possibly from phosphatidylinositol (PI) (Griendling et al., 1986) or even another phospholipid such as phosphatidylcholine (Besterman et al., 1986). Furthermore, little arachidonic acid is released. Thus, these effects can be sharply contrasted with those of bradykinin on DRG cells for example, which stimulates the release of a great deal of IP_3 and arachidonic acid in addition to DAG (Miller, 1987a). It should be noted that in a recent study Worley et al. (1987), mapped the relative distribution of IP_3 and phorbol ester receptors in the nervous system. They found that in the spinal cord rather few IP_3 receptors were observed but an abundance of phorbol ester binding sites were identified.

A further indication that PKC may be involved is that NE appears to produce translocation of the enzyme in DRG cells. Protein kinase C is normally localized in the cell cytoplasm. When activated by a phorbol ester, it moves to the cell membrane fraction to which it binds very tightly (Miller, 1986). Endogenously produced DAG also causes translocation of the enzyme but in this case binding to the cell membrane is not as tight. We found that NE causes the amount of PKC in the DRG cytoplasm to fall suggesting movement to another cellular compartment. We have not yet been able to demonstrate increased association of PKC with the cell membrane. This may be due to rapid dissociation of the enzyme during membrane preparation. At any rate, these biochemical data are consistent with activation of PKC in DRG cells by inhibitory neurotransmitters such as NE and NPY.

A definitive way of implicating PKC would be to inhibit the enzyme pharmacologically and to show that this abolished neurotransmitter induced inhibition of DRG Ca^{2+} currents. Unfortunately drugs are not available which specifically block PKC while leaving other potential second messenger systems unaltered. We attempted to circumvent this problem by preparing DRG cells which lacked the enzyme. This can be achieved by employing the process of "down-regulation" (Matthies et al., 1987). If phorbol esters are added to cultured DRG cells PKC first translocates to the cell membrane. However, if stimulation with the phorbol ester is continued, enzyme activity in the membrane fraction begins to fall. After several hours PKC activity in both membrane and cytoplasmic fractions has fallen to near zero and the cells are essentially enzyme deficient (Fig. 5). This down-regulation of PKC activity is associated with a decrease in enzyme immunoreactivity (Stabel et al., 1987). It appears that upon activation of the enzyme, a protease is also activated which begins to degrade it (Ballester and Rosen, 1985). Down-regulation of PKC can also be demonstrated in cultured sympathetic neurons and in PC12 cells, as well as in many non-neuronal cells (Matthies et al., 1987).

Following down-regulation of the enzyme and removal of the phorbol ester, we tested the effects of NE and NPY on Ca^{2+} currents in the DRG neurons. The range of absolute magnitudes of the Ca^{2+}-currents in down-regulated neurons was not significantly different from those in normal neurons. Furthermore, in preliminary experiments NPY binding to down-regulated DRG neurons was not altered. Thus any changes in the effects of NE and NPY cannot be attributed to loss of modulatable channels or receptors. We found that inhibition by NE was substantially blocked at the single concentration we tested (50 µM). Inhibition of total current was reduced froma mean of 50% to 20% by down-regulation and the blockade was slightly more effective on the sustained current.

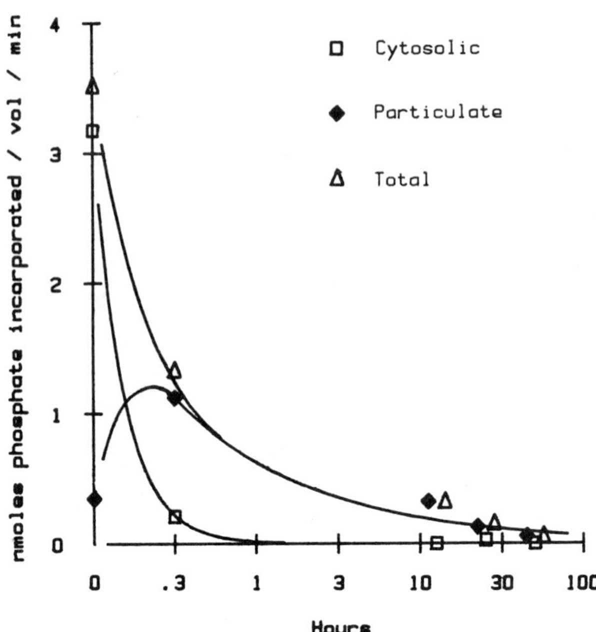

Fig. 5. Time course of translocation and down-regulation of PKC in DRG
neurons. After pretreatment with phorbol ester (1 µM) for times
shown, cultured DRG neurons were homogenized and separated into
cytosolic and particulate fractions by standard ultracentrifugation
techniques. PKC activity was assessed as the difference in phosphorylation of exogenous histone in the presence and absence of
Ca^{2+} (Palfrey and Waseem, 1985). Assay medium contained 50 mM
Tris HCl (pH 7.4), 5 mM $MgSO_4$, 1 mM EGTA or 0.5 mM Ca^{2+},
50 µg/ml phosphatidylserine, 0.2 mg/ml histone H1, and 50 µM ATP
(100-500 mCi/mmol). Time courses of total activity and cytosolic
activity are fitted with single exponential curves and the curve for
particulate activity is the difference between these two exponen-
tials.

The effect of down-regulation on NPY inhibition of Ca^{2+} currents was
distinctly different at 100 nM compared to 1 nM. At the high concentration NPY
clearly inhibited most of the transient current in all experiments on down-regulated
neurons (Fig. 2, right panel). However the remaining sustained current was much
less inhibited compared to non-down-regulated. The remaining sustained current
also consistently exhibited a gradually increasing time course during the pulses. At
lower NPY concentrations, on the other hand, down-regulation fully blocked the inhibitory effect in all experiments (Fig. 3, right panel). The dose-response curves of
Fig. 4 summarize all NPY experiments on down-regulated cells (open symbols).
Down-regulation apparently unmasks an inhibition which is not mediated by PKC via
a receptor with a K_D of approximately 10^{-8} M. This inhibitory mechanism appears
predominantly coupled to the transient currents. This inhibition of Ca^{2+} current

may be very similar to that seen with muscarinic agonists on sympathetic neurons
(Wanke et al., 1987), with adenosine on hippocampal neurons (Madison et al., 1987),
with dopamine on DRG neurons (Marchetti et al., 1986) and with enkephalin on neuroblastoma cells (Tsunoo et al., 1986). In addition to this non-PKC inhibitory mechanism there is another which is apparently mediated by PKC via a receptor of higher
affinity. The physiological significance of such a dual inhibition by a single neurotransmitter remains to be elucidated.

SUMMARY AND CONCLUSIONS

A variety of modulatory neurotransmitters endogenous to the vertebrate
spinal cord can inhibit the release of substance P from DRG neurons. In some, but
not all cases, these neuromodulators inhibit Ca^{2+} currents, an effect which presumably contributes to their ability to inhibit release. Neuropeptide Y is the most
recently described member of this family of neuromodulators. In cultured DRG
neurons we find that it strongly inhibits both the sustained and transient Ca^{2+} currents with a half-maximal effect at 1 nM. Two high affinity binding sites are found
for NPY on these neurons. Binding to the lower affinity site is blocked by the non-
hydrolyzable GTP analog Gpp(NH)p suggesting that a G-protein is linked to this
receptor. We have investigated the possibility that a second messenger, DAG, is
produced by coupling of this G-protein to phospholipase C and that DAG-activated
PKC inhibits Ca^{2+} channel activity. It is known that synthetic analogs of DAG will
inhibit Ca^{2+} currents and we find that NPY stimulates DAG production. We "down-
regulated" DRG neurons for PKC by long-term treatment with phorbol esters and
found that the inhibitory effects of 1 nM NPY were completely blocked. This con-
firms that PKC does, in part, mediate the inhibition of Ca^{2+} currents by a high
affinity receptor. At higher concentrations of NPY on down-regulated neurons,
inhibition of Ca^{2+} currents was clearly evident but qualitatively different from inhibition on non-down-regulated neurons. The effect was predominantly on the transient currents. This suggests the existence of an NPY receptor of lower affinity
which is either directly coupled to Ca^{2+} channels or produces its effect via a different second messenger.

ACKNOWLEDGEMENTS

Supported by PHS grants DA-02121, DA-02575 and MH-40165 and grants
from The Univ. of Chicago Brain Research Inst. and Miles Pharmaceutical Inc.
M.W.W., T.M.P. and H.J.G.M. were supported by PHS training grant GM-07151.

REFERENCES

Ballester, R. and Rosen, O.M., 1985, Fate of immunoprecipitable protein kinase C
    in GH3 cells treated with phorbol-12-myristate-13-acetate, J. Biol. Chem.,
    260:15194-15199.
Berridge, M.J. and Irvine, R.F., 1984, Inositol trisphosphate, a novel second mes-
    senger in cellular signal transduction. Nature, 312:315-321.
Besterman, J.M., Duronio, V. and Cuatrecasas, P., 1986, Rapid formation of diacyl-
    glycerol from phosphatidyl choline: A pathway for generation of a second
    messenger, Proc. Natl. Acad. Sci. (USA), 83:6785-6789.
Cherubini, F. and North, R.A., 1985, µ and κ opiates inhibit transmitter released by
    different mechanisms, Proc. Natl. Acad. Sci. (USA), 82:1860-1863.

Deisz, R.A. and Lux, H.D., 1985, -aminobutyric acid depression of calcium currents of chick sensory neurones, Neurosci. Lett., 56:205-210.

Dolphin, A.C. and Scott, R.H., 1986, Inhibition of calcium currents in cultured rat dorsal root ganglion neurones by (-)-baclofen, Brit. J. Pharmacol., 88:213-220.

Dolphin, A.C., Forda, S.R. and Scott, R.H., 1986, Calcium dependent currents in cultured rat dorsal root ganglion neurones are inhibited by an adenosine analogue, J. Physiol., 373:47-61.

Dunlap, K. and Fischbach, G.D., 1978, Neurotransmitters decrease the calcium component of sensory neurone action potentials, Nature, 276:837-839.

Dunlap, K. and Fischbach, G.D., 1981, Neurotransmitters decrease the Ca conductance activated by depolarization of embryonic chick sensory neurons. J. Physiol., 267:281-298.

Emson, P. and De Quidt, M.E., 1984, NPY - a new member of the pancreatic polypeptide family, Trends in Neurosci., 7:31-35.

Ewald, D.A. and Miller, R.J., 1987, Norepinephrine (NE) decreases the activity of Ca channels and increases the activity of a K-channel in cultured rat dorsal root ganglion (DRG) neurons, Biophys. J., 51:429a.

Forscher, P., Oxford, G.S. and Schulz, D., 1986, Noradrenaline modulates calcium channels in avian dorsal root ganglion cells through tight receptor channel coupling, J. Physiol., 379:131-144.

Fox, A.P., Nowycky, M.C. and Tsien, R.W., 1987a, Kinetic and pharmacological properties distinguishing three types of calcium currents in chick sensory neurones, J. Physiol., (in press).

Gerschenfeld, H.M., Hammond, C. and Paupardin-Tritsch, D., 1986, Modulation of the calcium current of molluscan neurones by neurotransmitters, J. Exp. Biol., 124:73-91.

Griendling, K.K., Rittenhouse, S.E., Brock, T.A., Ekstein, L.S., Gimbrone, M.A. and Alexander, R.W., 1986, Sustained diacylglycerol formation from inositol phospholipids in angiotensin II stimulated vascular smooth muscle cells, J. Biol. Chem., 261:5901-5906.

Harris, K.M., Kongsamut, S. and Miller, R.J., 1986, Protein kinase C mediated regulation of calcium channels in PC-12 pheochromocytoma cells. Biochem. Biophys. Res. Comm., 134:1298-1305.

Heschler, J., Rosenthal, W., Trautwein, W. and Schultz, G., 1987, The GTP binding protein G_O regulates neuronal calcium channels, Nature, 325:445-447.

Holz, G.G., Rane, S.G. and Dunlap, K., 1986, GTP binding proteins mediate transmitter inhibition of voltage dependent calcium channels, Nature, 319:670-672.

Holz, G.G., Dunlap, K. and Kream, R.M., 1987a, Characterization of the electrically evoked release of substance P from dorsal root ganglion neurones: methods and dihydropyridine sensitivity. J. Neurosci., (in press).

Holz, G.G., Kream, R.M. and Dunlap, K., 1987b, Norepinephrine and gamma aminobutyric acid inhibit electrically evoked release of substance P from dorsal root ganglion neurones, J. Neurosci., (in press).

Klein, M., Shapiro, E. and Kandel, E.R., 1980, Synaptic plasticity and modulation of the Ca^{2+} current, J. Exp. Biol., 89:117-157.

Logothetis, D.E., Kurachi, Y., Galper, J., Neer, E.J. and Clapham, D.E., 1987, The subunits of the GTP binding proteins activate the muscarinic K^+ channel in the heart, Nature, 325:321-326.

MacDonald, R. and Werz, M.A., 1986, Dynorphin A decreases voltage dependent calcium conductance of mouse dorsal root ganglion neurones, J. Physiol., 377:237-249.

Madison, D.V., Fox, A.P. and Tsien, R.W., 1987, Adenosine reduces an inactivating component of calcium current in hippocampal CA3 neurones, Biophys. J., 51:30a.

Marchetti, C., Carbone, E. and Lux, H.D., 1986, Effects of dopamine and noradrenaline on Ca channels of cultured sensory and sympathetic neurons of chick, Pflug. Arch., 606:104-111.

Matthies, H.J.G., Palfrey, H.C., Hirning, L.D. and Miller, R.J., 1987, Down regulation of protein kinase C in neuronal cells: Effects on neurotransmitter release, J. Neurosci., (in press).

Miller, R.J., 1984, How do opiates act? Trends in Neurosci., 7:184-185.

Miller, R.J., 1986, Protein kinase C: A key regulator of neuronal excitability, Trends in Neurosci., 9:538-541.

Miller, R.J., 1987a, Bradykinin highlights the role of phospholipid metabolism in the control of nerve excitability, Trends in Neurosci., (in press).

Miller, R.J., 1987b, Multiple calcium channels and neuronal function, Science, 235: 46-52.

Mochida, S. and Kobayashi, H., 1986, Activation of M_2-muscarinic receptors causes an alteration of action potentials by modulation of Ca^{2+} entry in isolated sympathetic neurones of rabbits, Neurosci. Lett.,

Mudge, A.W., Leeman, S.E. and Fischbach, G.D., 1979, Enkephalin inhibits release of substance P from sensory neurones in culture and decreases action potential duration, Proc. Natl. Acad. Sci. (USA), 76:527-532.

Newberry, N.R.and Nicoll, R.A., 1984, Direct hyperpolarizing action of baclofen on hippocampal pyramidal cells, Nature, 308:450-452.

Nowycky, M.C., Fox, A.P. and Tsien, R.Y., 1985, Three types of neuronal calcium channel with different calcium agonist sensitivity, Nature, 316:440-443.

Palfrey, H.C. and Waseem, A., 1985, Protein kinase C in the human erythrocyte, J. Biol. Chem., 260:16021-16029.

Perney, T.M., Hirning, L.D., Leeman, S.E. and Miller, R.J., 1986, Multiple calcium channels mediate transmitter release from peripheral neurones, Proc. Natl. Acad. Sci. (USA), 83:6656-6659.

Rane, S.G. and Dunlap, K., 1986, Kinase C activator 1,2-oleylacetylglycerol attenuates voltage dependent calcium current in sensory neurones, Proc. Natl. Acad. Sci. (USA), 83:184-188.

Robertson, B. and Taylor, W.R., 1986, Effects of γ -aminobutyric acid and (-)-baclofen on calcium and potassium currents in cat dorsal root ganglion neurones in vitro, Brit. J. Pharmacol., 89:661-672.

Scott, R.H. and Dolphin, A.C., 1986, Regulation of Ca^{2+} currents by a GTP analogue: potentiation of (-)-baclofen mediated inhibition, Neurosci. Lett., 69:59-64.

Stabel, S., Rodriguez-Pena, A., Young, S., Rozengurt, E. and Parker, P.J., 1987, Quantitation of protein kinase C by immunoblot-expression in different cell lines and response to phorbol esters. J. Cell. Physiol., 130:111-117.

Sundler, F., Hakanson,R.,Ekblad, E., Uddman, M. and Wahlestadt, C., 1987, Neuropeptide Y in peripheral adrenergic and enteric nervous systems, Ann. Rev. Cytol., (in press).

Tsien, R.W., 1983, Calcium channels in excitable cell membranes, Ann. Rev. Physiol., 45:341-358.

Tsunoo, A., Yoshii, M. and Narahashi, T., 1986, Block of calcium channels by enkephalin and somatostatin in neuroblastoma x glioma hybrid NG108-15 cells, Proc. Natl. Acad. Sci. (USA), 83:9832-9836.

Wanke, E., Ferroni, A., Malgaroli, A., Ambrosini, A., Pozzan, T. and Meldolesi, J., 1987, A novel type of inhibition of voltage gated Ca^{2+} channels via muscarinic receptors in mammalian sympathetic neurones, Proc. Natl. Acad. Sci. (USA), (in press).

Werz, M.A. and MacDonald, R.L. (1984), Dynorphin reduces calcium-dependent action potential duration by decreasing voltage-dependent calcium conductance, Neurosci. Lett., 46:185-190.

Williams, J.T., Henderson, G. and North, R.A., 1985, Characterization of α_2-adrenoceptors which increase potassium conductance in rat locus coeruleus neurons, Neurosci., 14:95-101.

Worley, P.F., Baraban, J.M., Colvin, J.S. and Snyder, S.H., 1987, Inositol trisphosphate receptor localization in brain: Variable stoichiometry with protein kinase C, Nature, 325:159-161.

MODULATION OF POTASSIUM AND CALCIUM CURRENTS BY FMRFamide

IN *APLYSIA* NEURONS: A MECHANISM OF PRESYNAPTIC INHIBITION?

V. Březina and C. Erxleben[*]

Department of Biology, University of California
Los Angeles, CA 90024, U.S.A.

PRESYNAPTIC FACILITATION AND INHIBITION

The search for cellular mechanisms of learning and memory has been
greatly influenced by the analysis, initiated by Eric Kandel and his colleagues, of synaptic plasticity in simple defensive withdrawal reflexes of
the marine snail *Aplysia californica*. Best understood in this system are
events connected with presynaptic facilitation, in which the amount of neurotransmitter released from sensory onto motor neurons is increased following activation of modulatory neurons. The neuromodulatory agent released by
these neurons has not been identified, but its effects can be mimicked by
serotonin and the small cardioactive peptides SCP_A and SCP_B. On the molecular level, presynaptic facilitation appears to result mainly from cAMP-
mediated closure by these agents of a class of background K channels, the
'S' channels, leading to delay in action potential repolarization, enhanced
influx of Ca^{2+} into the presynaptic terminal and greater release of transmitter. In turn, presynaptic facilitation is thought to underlie behavioral
sensitization and classical conditioning (reviewed by Kandel and Schwartz,
1982; Kandel *et al.* 1983; Byrne *et al.* 1986; Siegelbaum, 1987).

Presynaptic inhibition, the converse of presynaptic facilitation, is
not as well understood in *Aplysia*. In the only example analyzed, Kretz *et
al.* (1984, 1986) have shown that presynaptic inhibition of transmitter release from terminals of abdominal ganglion neuron L10 is due to synergistic
activation of a K current and suppression of the Ca current by the modulatory agent, which in this case is probably histamine. Together with Roger
Eckert, we have studied modulation of ion currents in other identified
Aplysia neurons by the endogenous neuropeptide FMRFamide. As described
in this article, we have found that the effects of FMRFamide in these cells
are very similar to those of histamine in neuron L10 and, furthermore, that
the K current activated by FMRFamide is, in at least some of the cells we
have studied, the same 'S' current that is suppressed during presynaptic
facilitation (Březina *et al.* 1987*a,b*). These results suggest a role for
FMRFamide in counteracting presynaptic facilitation, or as an agent of presynaptic inhibition, in *Aplysia*.

---

* Present address: Fakultät für Biologie, Universität Konstanz,
  Postfach 5560, D-7750 Konstanz, Federal Republic of Germany.

Evidence is accumulating that FMRFamide (Phe-Met-Arg-Phe-NH$_2$) acts as a neurotransmitter or neurohormone in *Aplysia*. A gene expressed throughout the nervous system encodes multiple copies of FMRFamide, which appears to be a principal peptide product of identified abdominal ganglion neurons and is also present in neuronal processes and varicosities (Schaefer *et al.* 1985). FMRFamide activates or modulates the activity of a number of *Aplysia* muscles (see, for example, Austin *et al.* 1983; Weiss *et al.* 1984) and central neurons. Using two-electrode voltage clamp and brief (usually 1 s) puffs or occasionally bath application of 1-50 μM peptide (see figure legends), we have found (Březina *et al.* 1987*a*) that FMRFamide elicits an outward current in abdominal ganglion neurons L2-L6 and R2 held at the normal resting potential, usually between -30 and -50 mV (Fig. 1*A*). The FMRFamide-induced current activates slowly, reaching its peak 15-30 s after the peptide puff (at 20°C). The current reverses between -70 and -80 mV, in the range of the potassium equilibrium potential, E_K; the reversal potential shifts with altered extracellular K$^+$ concentration, [K$^+$]$_o$ (but not with [Na$^+$]$_o$, [Ca^{2+}]$_o$ or [Cl$^-$]$_o$), as predicted by the Nernst equation for a current flowing through K-selective channels. The moderate outward rectification seen in current-voltage (*I-V*) plots for the FMRFamide-induced

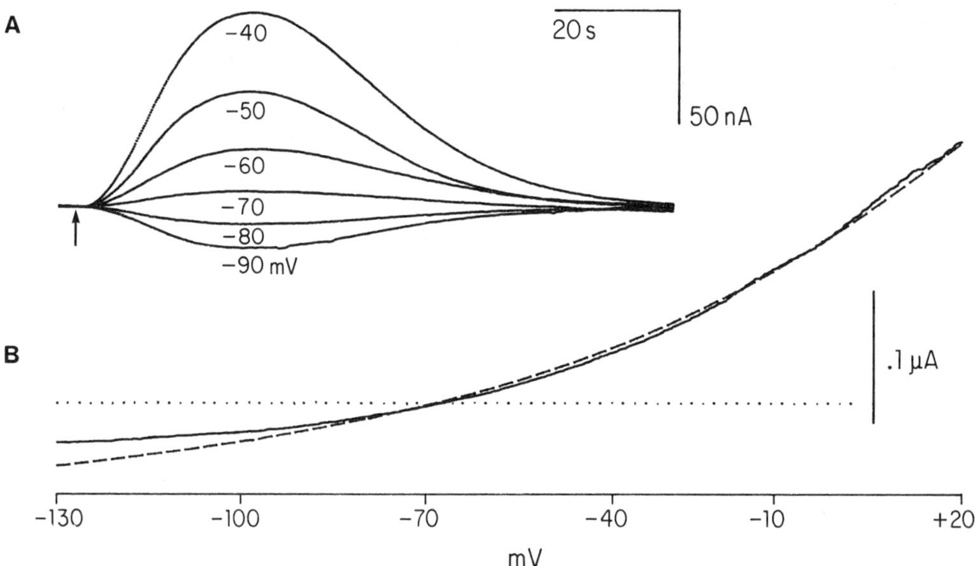

Fig. 1. FMRFamide-induced K current. *A*, currents elicited by 1 s puffs of 50 μM FMRFamide (arrow) in cell R2, clamped to the potentials indicated (mV), in artificial sea water (ASW, containing (mM) 468 NaCl, 10 KCl, 45 MgCl$_2$, 20 CaCl$_2$, 10 glucose, 25 Tris-Cl, pH 7.7). Beginning of traces aligned. *B*, quasi-stationary *I-V* relation of net FMRFamide-induced current, obtained by subtraction of currents elicited using slow (5-10 mV/s) voltage ramps from -130 to +15 mV before and during a long puff of 50 μM FMRFamide, in cell R2 in Ca-free ASW containing 20 mM Co^{2+}. The dashed line is calculated from the constant-field equation assuming an intracellular K$^+$ concentration of 160 mM. Dotted line indicates zero current. (From Březina *et al.* 1987*a*.)

current is well described by the constant-field equation for a K-selective,
voltage-independent conductance (Fig. 1B), suggesting that the outward
rectification is purely a consequence of the asymmetry in K^+ concentration
on either side of the membrane, while gating of the channels opened by
FMRFamide is voltage-independent. The FMRFamide-induced current is also Ca-
independent: it is not suppressed by injection into the cell of the Ca-
chelating agent EGTA, and it can be recorded in Ca-free solution containing
20 mM Co^{2+} (the trace in Fig. 1B, for example, was obtained in this solution in order to eliminate a second effect of FMRFamide, the suppression of
Ca and Ca-dependent K currents described below). The FMRFamide-induced
current is much less sensitive to block by extracellular tetraethylammonium
(TEA; apparent $K_D \approx 75$ mM) or 4-aminopyridine (4-AP; apparent $K_D \approx 6$ mM)
than are other K currents described in $Aplysia$ neurons, such as the early
transient 'A' current, the voltage-dependent K current, or the Ca-dependent
K current. The FMRFamide-induced K current is, however, suppressed when the

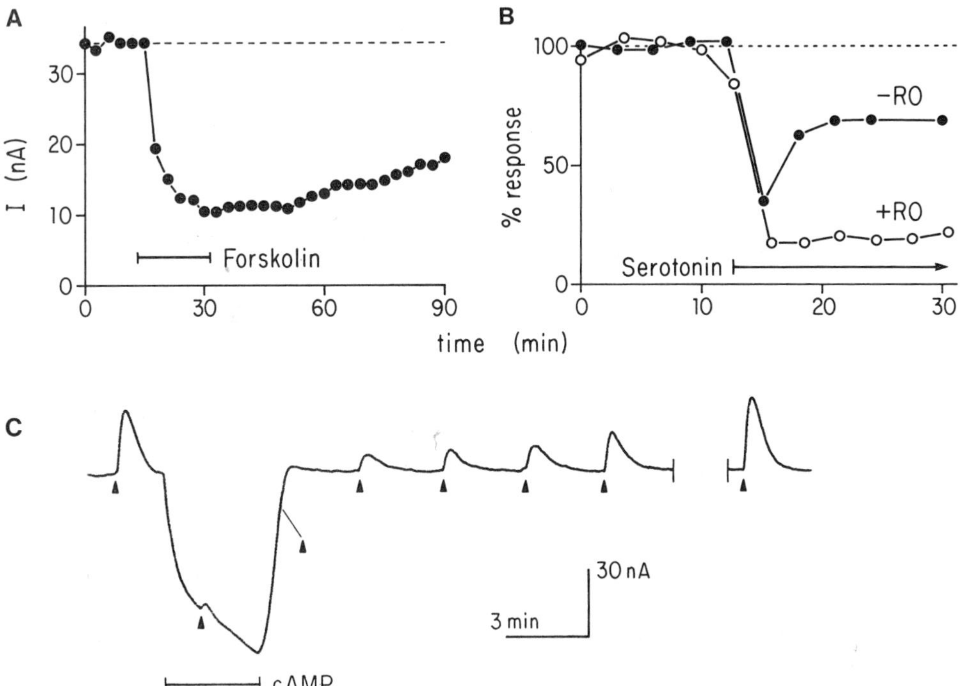

Fig. 2. FMRFamide-induced K current is suppressed by elevation of intracellular cAMP concentration. A, plot of peak currents elicited in
cell R2, held at -40 mV in ASW, by 1 s puffs of 50 μM FMRFamide
every 180 s. 25 μM forskolin was superfused during the time shown.
B, same as A but values are expressed as percentage of base line
(mean of first four measurements). 100 μM serotonin was superfused
as shown. The experiment was carried out either without (filled
circles) or with 100 μM RO 20-1724 added to both control and sero-
tonin-containing solutions (open circles). C, effect of cAMP injection (1 μA iontophoretic current) on currents elicited in cell R2,
held at -40 mV in ASW, by 1 s puffs of 50 μM FMRFamide (arrows).
The large cAMP-induced inward current is carried by Na^+ and is
unrelated to the FMRFamide-induced K current. (From Březina et $al.$
1987a.)

intracellular cAMP concentration is elevated by means of superfusion of forskolin (an activator of adenylate cyclase; Fig. 2*A*), RO 20-1724 (a blocker of the phosphodiesterase that breaks down cAMP), serotonin (a transmitter that increases cAMP levels in some of these cells; Fig. 2*B*), or by direct injection of cAMP into the cell (Fig. 2*C*).

All of these properties of the FMRFamide-induced K current markedly resemble those of the 'S' current, a voltage- and Ca-independent K current suppressed by serotonin and cAMP-dependent phosphorylation in *Aplysia* sensory neurons (Klein *et al.* 1982; Pollock *et al.* 1985; Siegelbaum, 1987). Unequivocal identification of the FMRFamide-induced current as the 'S' current can, however, be made only by determining whether FMRFamide has any effect on identified single 'S' channels in patch-clamp recordings. Due probably to extensive glial-cell covering, we could not routinely obtain gigaohm seals on the large abdominal ganglion neurons in which we recorded the macroscopic FMRFamide-induced current. We therefore turned to the smaller sensory neurons, present in a cluster in each pleural ganglion, that subserve the tail-withdrawal reflex, one of the defensive reflexes subject, as already mentioned, to presynaptic facilitation, sensitization and classical conditioning (Byrne *et al.* 1986). The 'S' current has been

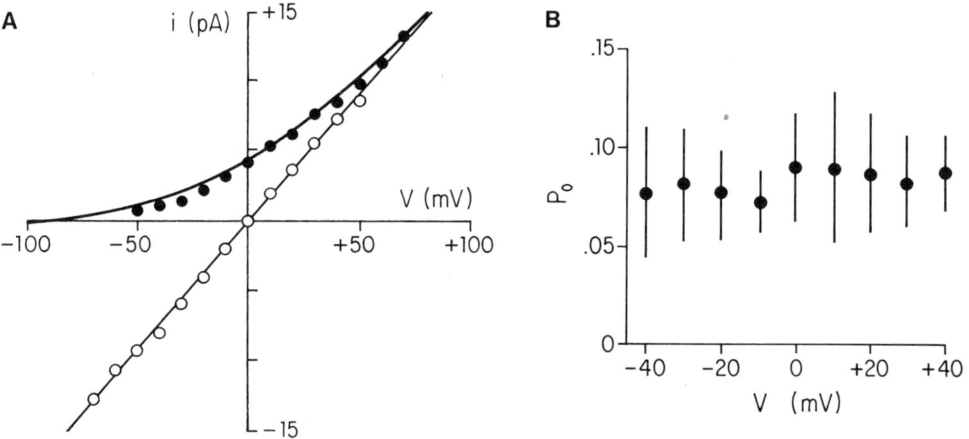

Fig. 3. Characteristics of 'S' channels in pleural sensory neurons. *A, i-V* plots of unitary 'S'-channel current measured in inside-out patches. In one experiment (filled circles) pipette contained ASW ($[K^+]_o$ = 10 mM) and bath contained high-K^+ solution: (mM) 345 KCl, 292 sucrose, 2 $MgCl_2$, 1 $CaCl_2$, 1.5 EGTA, 50 HEPES, pH adjusted to 7.4 with KOH ($[K^+]_i$ ≈ 360 mM). In another experiment (open circles) both pipette and bath contained modified (no $MgCl_2$ or $CaCl_2$, and EDTA instead of EGTA) high-K^+ solution. Theoretical curve through the filled circles is that predicted by the constant-field equation with channel permeability equal to 1.31 x 10^{-13} cm^3/s. Line through the open circles corresponds to a conductance of 183 pS. *B*, probability of being open for the single 'S' channel, P_o, measured in an inside-out patch, as function of membrane potential. Total 'S'-channel current was integrated over records 1.024 s long, and divided by the record length, by the unitary current, and by four, the number of channels in the patch. The plot shows, for each potential, the mean ± standard deviation of 14-41 such determinations. (*B* and part of *A* from Březina *et al.* 1987*a*.)

described in these cells (Pollock *et al.* 1985); indeed, in patches depolarized to inactivate most Na, Ca, early transient and voltage-dependent K channels, the most frequent channel observed has the properties reported for the 'S' channel (Siegelbaum, 1987). In particular, current through the channel reverses at E_K, and the constant-field equation for a K-selective conductance accurately predicts the rectification seen in single-channel current-voltage (*i-V*) plots under conditions of asymmetric K^+ concentration, a feature that, as expected, disappears when the K^+ concentration on the two sides of the channel is made equal (Fig. 3*A*). In contrast, the channel's probability of being open is virtually independent of voltage (Fig. 3*B*), as well as of the Ca^{2+} concentration on the intracellular side of the membrane. 'S' channels in a cell-attached patch close when serotonin is superfused over the rest of the cell (Fig. 4*A*); similarly, 'S' channels in a cell-free patch close when the intracellular surface of the membrane is exposed to the catalytic subunit of cAMP-dependent protein kinase (Fig. 4*B*).

.   While serotonin closes the 'S' channels, FMRFamide, when applied in the same way, increases their activity (Fig. 5), a result recently obtained in these cells also by Belardetti *et al.* (1986*a*, 1987). Like serotonin, bath-applied FMRFamide is able to modulate channels in a cell-attached patch not directly accessible to it, implying mediation of the effect by a diffusible intracellular agent. The apparent symmetry between the opposite actions of serotonin and FMRFamide breaks down, however, upon identification of the microscopic parameters of 'S' channel activity that are affected by each transmitter. In the case of serotonin, the intracellular signal is an increase in cAMP concentration, leading to phosphorylation and long-term closure of some 'S' channels, leaving gating of unphosphorylated channels unaffected (Fig. 4*A*; Siegelbaum, 1987). While FMRFamide is able to reopen (presumably by dephosphorylation) 'S' channels closed by serotonin or cAMP in the sensory neurons (Belardetti *et al.* 1987), although apparently not in the abdominal ganglion neurons in which we recorded the macroscopic FMRFamide-induced 'S'-like current (see Fig. 2), it does this without changing cAMP levels (Ocorr and Byrne, 1985), and Belardetti *et al.* (1987) suggest that the peptide may be stimulating a phosphatase through a cAMP-independent second-messenger system. Of perhaps more significance, however, because it affects even unphosphorylated, active 'S' channels, is another effect of FMRFamide that we (Březina *et al.* 1987*a*) and Belardetti *et al.* (1987) have observed in the sensory neurons, and which may underlie the macroscopic 'S'-like current in the abdominal ganglion neurons. This effect is a FMRFamide-induced increase in the fraction of time which already active 'S' channels spend in the open state, resulting either from a reduction in the rate of closing of the channels, or from an increase in their rate of opening.

FMRFamide SUPPRESSES Ca AND Ca-DEPENDENT K CURRENTS

In addition to activating the 'S'-like K current, FMRFamide has a second effect in abdominal ganglion neurons L2-L6 and R2 (and also in cell R15): it partially suppresses the Ca and Ca-dependent K currents activated on depolarization (in our experiments, usually by means of 100 ms voltage steps from a holding potential of -40 mV to +10 mV; see Březina *et al.* 1987*b*). The Ca current, isolated in Na-free solution containing 200 mM TEA and 5 mM 4-AP to block virtually all contaminating K currents, is maximally suppressed 10-25 s after a puff of FMRFamide. 50 μM FMRFamide typically blocks 10-25% of the Ca current (Fig. 6*A*), but even saturating concentrations (1 mM) of the peptide block only 30-50%, and never 100%, of the current. This, as well as the apparent inability of FMRFamide to block effectively the Ca current activated by voltage steps to potentials below +10 mV (Fig. 6*C*), may perhaps result from the presence of more than one class of Ca channels, only one of which is sensitive to FMRFamide. The

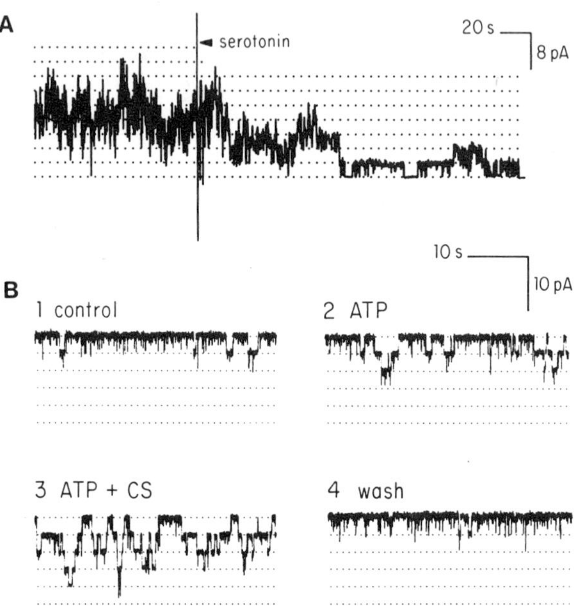

Fig. 4. Closure of 'S' channels in pleural sensory neurons by serotonin and cAMP-dependent protein kinase. *A*, recording from a cell-attached patch with at least nine channels, clamped to about 0 mV by high-K$^+$ bath solution; pipette contained ASW. 25 μM serotonin was superfused in the bath from the time indicated until the end of the experiment. Filtered at 100 Hz. *B*, recording from an inside-out patch with at least four channels, held at 0 mV. Pipette contained ASW; the intracellular surface of the membrane was exposed successively to control high-K$^+$ solution (trace segment 1), high-K$^+$ solution containing 350 μM ATP (segment 2) and ATP plus 100 μg protein/ml of the catalytic subunit of cAMP-dependent protein kinase (segment 3), and control high-K$^+$ solution again (segment 4). Filtered at 45 Hz. (*A* from Březina *et al.* 1987*a*.)

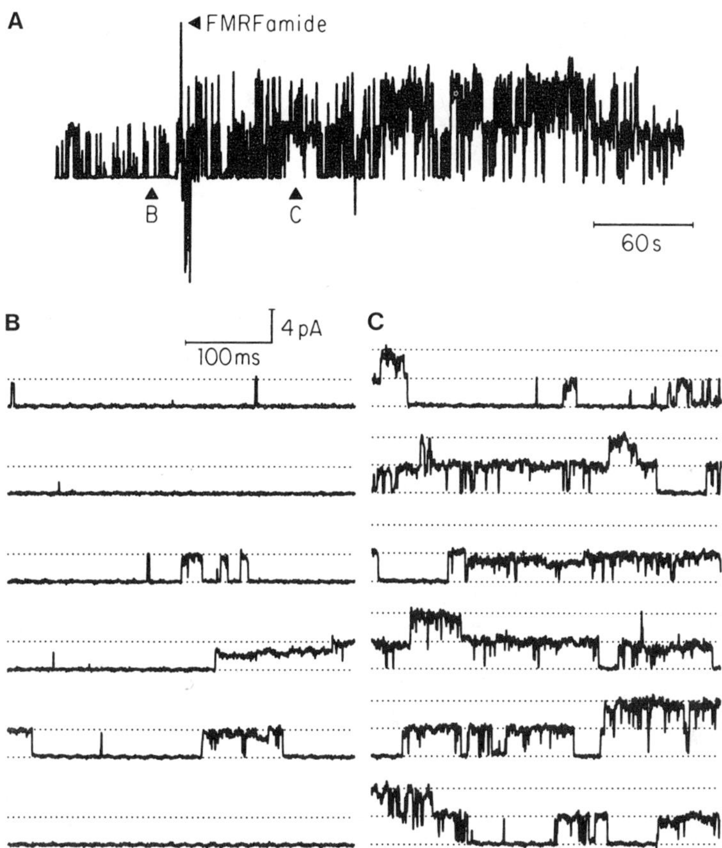

Fig. 5. 'S' channel activity in pleural sensory neurons is enhanced by FMRFamide. *A*, recording from a cell-attached patch clamped to about 0 mV by high-K^+ bath solution; pipette contained ASW. 20 μM FMRF-amide was superfused from the time indicated until the end of the experiment. Filtered at 100 Hz. *B,C*, expansion of sections indicated in *A*, showing 'S'-channel behavior before (*B*) and after (*C*) superfusion of FMRFamide. Filtered at 1 kHz. (From Březina *et al.* 1987*a*.)

peptide eliminates a constant fraction of the Ca current at all potentials
above +10 mV, and has no direct effect on the activation or inactivation of
the remaining current, behavior consistent with a reduction in the number
of functional Ca channels by FMRFamide.

Since these cells also exhibit the 'S'-like K current, whose outward
rectification makes it especially prominent at the depolarized potentials
at which the Ca current is activated, it might be argued that enhancement
by FMRFamide of the 'S'-like current, or of another contaminating outward
current, could be the real cause of the reduction in inward current seen in
experiments such as that in Fig. 6A. Several lines of evidence (Březina *et
al.* 1987*b*) show, however, that this is not the case when, as in the experiments in Fig. 6, high concentrations of TEA and 4-AP are used to block the
'S'-like current, and that the reduction in inward current is the result of
a true decrease in Ca current. For example, Ca tail currents recorded on
repolarization to E_K, thus nulling any possible contaminating K currents,
are suppressed by FMRFamide just as effectively as the Ca current recorded
during the depolarizing step itself (Fig. 6*B*); similar considerations rule
out contamination by currents carried by H^+ or Cl^-. Furthermore, FMRFamide
is able to suppress current carried through the Ca channels by Ba^{2+}, and is

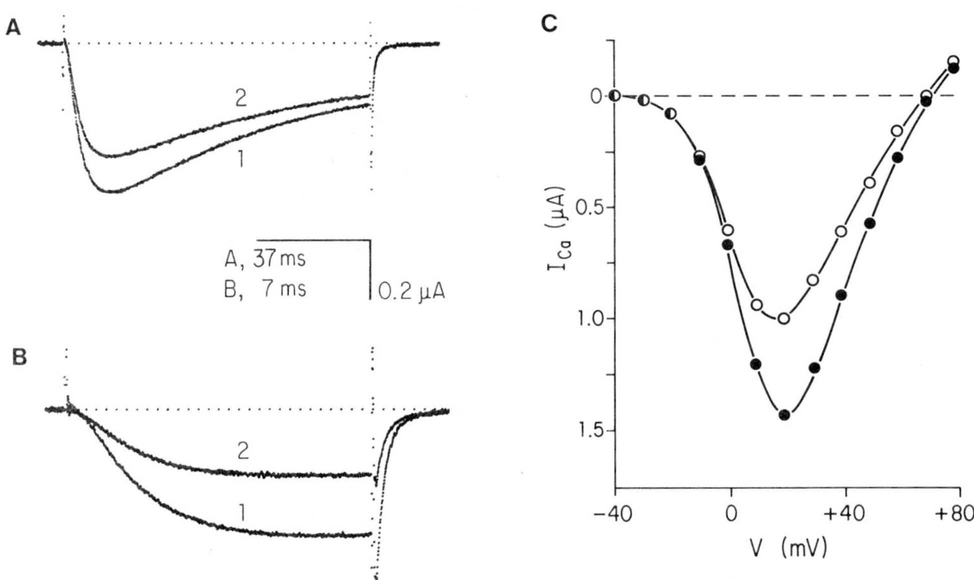

Fig. 6. FMRFamide suppresses Ca current. *A*, Ca current, elicited by voltage
steps from -40 to +10 mV in Na-free ASW containing 200 mM TEA and 5
mM 4-AP (trace 1), is suppressed 20 s after 1 s puff of 50 µM FMRF-
amide (trace 2). Records corrected for leakage current. *B*, Ca
current, elicited as in *A* in cell L6 unusually responsive to FMRF-
amide, and Ca tail current (trace 1) are both suppressed following
superfusion of 10 µM FMRFamide (trace 2). The Na-free, TEA- and
4-AP-containing bath solution contained in addition an elevated (35
mM) K^+ concentration to bring E_K to about -40 mV, the potential to
which the membrane was repolarized for measurement of the tail
currents. *C*, *I-V* relation for peak Ca current, with (open circles)
and without (filled circles) repeated puffed application of 50 µM
FMRFamide. (From Březina *et al.* 1987*b*.)

also effective in cells injected with Cs$^+$, TEA$^+$ or EGTA: each of these treatments might be expected to block a variety of possible contaminating currents, including any remaining 'S'-like current. Perhaps the best evidence, however, for a true suppression of the Ca current by FMRFamide is the finding that the peptide also secondarily suppresses processes normally activated by the influx of Ca^{2+} through the Ca channels; the most prominent such process is the opening of Ca-dependent K channels. Fig. 7C shows that

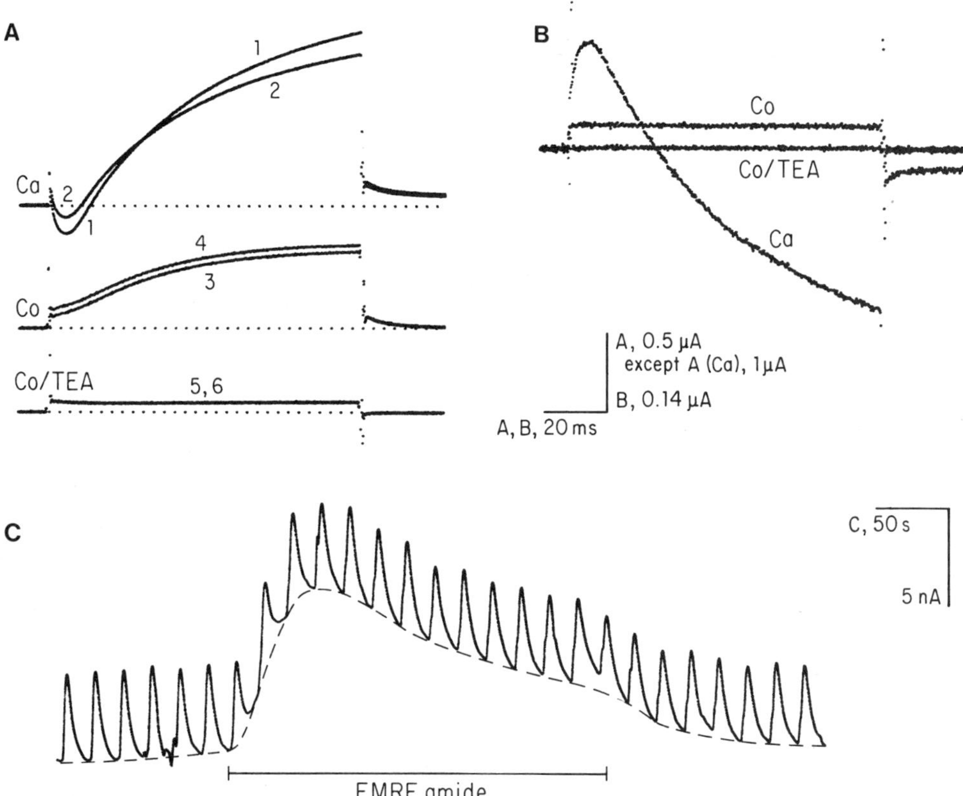

Fig. 7. FMRFamide suppresses Ca-dependent K current. *A*, early inward (Ca) and late outward (mainly K) currents, elicited by voltage steps from -40 to +10 mV in cell L3 bathed in Na-free ASW (trace 1), are both suppressed 20 s after 1 s puff of 50 µM FMRFamide (trace 2). Superfusion of Na- and Ca-free ASW containing 20 mM Co^{2+} eliminates Ca and Ca-dependent currents (trace 3); application of FMRFamide now produces only a simple increase in outward leakage current through the peptide-activated 'S'-like conductance (trace 4), an effect that is blocked in Na- and Ca-free ASW containing 20 mM Co^{2+}, 200 mM TEA and 5 mM 4-AP (traces 5,6). Records not corrected for leakage current. *B*, difference currents obtained from pairs of traces in *A*. *C*, Ca-dependent K current transients, elicited by iontophoretic injections (1 µA iontophoretic current for 300 ms) of Ca^{2+} in cell L4 held at -40 mV (results at more depolarized potentials were similar) in ASW, are not affected by superfusion of 1 µM FMRFamide (larger concentrations of FMRFamide had no effect either). (*A,B* from Březina *et al.* 1987*b*, *C* from Březina *et al.* 1987*a*.)

FMRFamide has no direct effect on Ca-dependent K channels, since it does not suppress the Ca-dependent K current elicited by direct injection of Ca^{2+} into the cell. When, on the other hand, the same current is activated by an influx of Ca^{2+} into the cell following depolarization and opening of Ca channels, FMRFamide effectively reduces its amplitude (Fig. 7A), and we have shown (Březina et al. 1987b) that the effect of the peptide on Ca-dependent K current can be fully accounted for by its effect on the Ca current. In experiments such as that in Fig. 7A, carried out in the absence of K-current blockers, FMRFamide does, of course, activate the 'S'-like current at the same time as it suppresses the Ca and Ca-dependent K currents, producing a larger total suppression of the early inward current, and smaller suppression of the late outward current, than might be expected

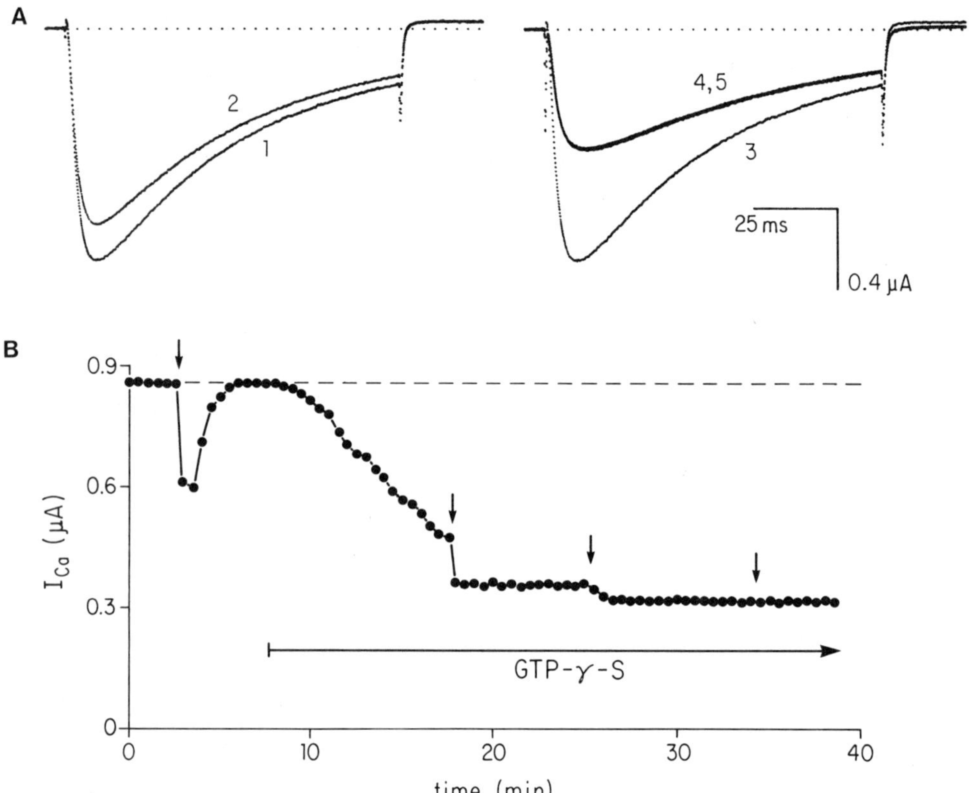

Fig. 8. Injection of GTP-γ-S mimics the effect of FMRFamide on Ca current. A, Ca current, elicited by voltage steps from -40 to +10 mV in cell L3 in Na-free ASW containing 200 mM TEA and 5 mM 4-AP (traces 1,3), is transiently suppressed 20 s after 3 s puff of 50 µM FMRFamide (trace 2) and permanently suppressed after injection of GTP-γ-S (0.1 µA iontophoretic current for 10 min; trace 4). Once GTP-γ-S has been injected, FMRFamide has no further effect (trace 5). Records corrected for leakage current. B, plot of peak amplitude of Ca currents (not corrected for leakage current) elicited as in A. 50 µM FMRFamide was applied in 5 s puffs (arrows); GTP-γ-S was injected (0.5 µA iontophoretic current) during the time shown. (From Březina et al. 1987b.)

from the effect on the Ca and Ca-dependent K currents alone. Nevertheless,
by virtue of the fact that the suppression of Ca and Ca-dependent K
currents can be eliminated in Ca-free solution, which has no significant
effect on the activation of 'S'-like current, the two effects of FMRFamide
can be separated (Fig. 7A) and their magnitudes compared (Fig. 7B).

A GTP-BINDING PROTEIN MAY MEDIATE THE EFFECTS OF FMRFamide

We have found (Březina *et al.* 1987*b*; Březina, 1987) that both the
activation of 'S'-like current and the suppression of Ca and Ca-dependent K
currents by FMRFamide are mimicked by iontophoretic or pressure injection
into the cell of the non-hydrolyzable GTP analog, guanosine 5'-O-(3-thio-
triphosphate) (GTP-γ-S). Thus, injection of GTP-γ-S suppresses the isolated
Ca current by 50-70% (but never completely), and in the absence of
K-current blockers also suppresses the associated Ca-dependent K current.
Suppression of the Ca current by GTP-γ-S, unlike the parallel effect of
FMRFamide, is slow to develop (5-30 min) and irreversible, and is observed
even without prior exposure of the cell to FMRFamide, although, when the
peptide is puffed onto the cell before the suppression of Ca current by
GTP-γ-S is fully developed, its effect also becomes irreversible. Once
GTP-γ-S has fully suppressed the Ca current, FMRFamide has no further
effect (Fig. 8). Similarly, at the holding potential of -40 mV in the
absence of K-current blockers, injection of GTP-γ-S triggers the slow and
irreversible development of an outward current, and at first renders the
activation by FMRFamide of 'S'-like current irreversible and later, once
the GTP-γ-S-induced current has fully developed, prevents activation of any
further outward current by the peptide (Fig. 9). However, FMRFamide is

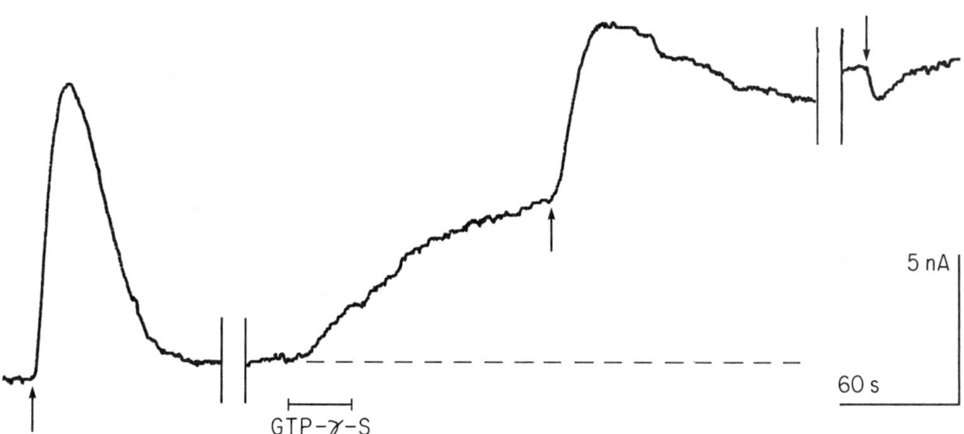

Fig. 9. GTP-γ-S activates 'S'-like current in one of the cells L2-L6 held
at -40 mV in ASW. GTP-γ-S (injected into the cell by brief pressure
pulses during the time shown) triggers the slow development of an
outward current and renders the activation of outward ('S'-like)
current by FMRFamide (applied in 1 s puffs of 50 μM peptide at the
arrows) irreversible. Eventually, FMRFamide is not able to elicit
any further outward current, although it is still able to elicit an
inward current carried by Na$^+$ (which was masked earlier by the out-
ward-current component of the response).

still able to elicit yet another current, not mentioned so far in this article, that it normally induces in some of these cells, an inward current carried by Na$^+$ (Fig. 9; Ruben *et al.* 1986). Further investigation of the GTP-γ-S-induced outward current (Březina, 1987) reveals that in all important properties (selectivity for K$^+$, Ca- and voltage-independence, relative insensitivity to extracellular TEA and 4-AP, and suppression by an elevation of intracellular cAMP) it is identical to, and thus most likely the same current as, the 'S'-like current activated by FMRFamide. Since GTP-γ-S is known to be an irreversible activator of GTP-binding proteins, these results suggest that the suppression of Ca and Ca-dependent K currents and the activation of 'S'-like current (but not activation of the Na-dependent inward current) by FMRFamide in these cells may normally be mediated by such a protein.

FMRFamide AS AN AGENT OF PRESYNAPTIC INHIBITION

Fig. 10 presents a graphical summary of the simplest likely sequence of events by which FMRFamide might activate the 'S'-like current and suppress the Ca (and secondarily the Ca-dependent K) current in the abdominal

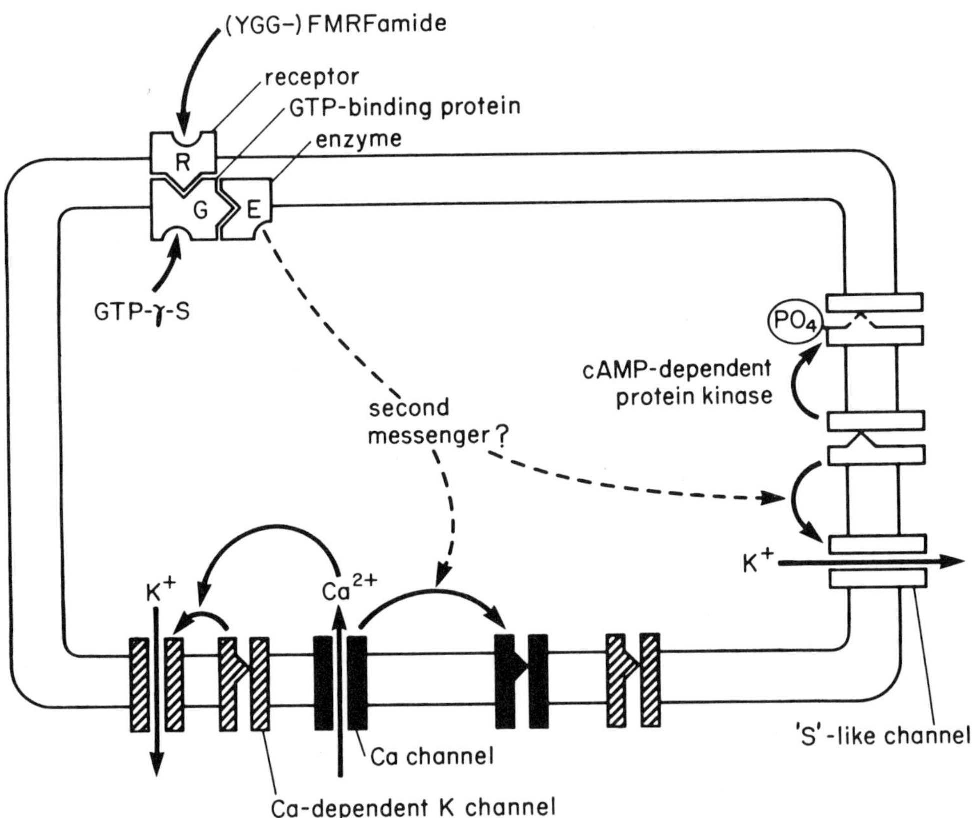

Fig. 10. Probable mechanism of the major effects of FMRFamide in abdominal ganglion neurons L2-L6 and R2. The Na-dependent inward current sometimes activated by the peptide (Ruben *et al.* 1986) is not shown. For further explanation see text. (From Březina *et al.* 1987*b*.)

ganglion neurons. Since both of the primary effects are elicited equally
well (apparent $K_D \approx 10$ μM) by FMRFamide and by its derivative YGG-FMRFamide
(Tyr-Gly-Gly-Phe-Met-Arg-Phe-NH$_2$), but not at all by FMRF (the non-amidated
analog of FMRFamide), both proceed with a similar time course following a
puff of peptide, show similar temperature dependence, and are mimicked by
injection of GTP-γ-S into the cell (Březina et al. 1987a,b), it is likely
that they are both mediated by a single type of FMRFamide receptor coupled
to a GTP-binding protein. The study of 'S' channel modulation in pleural
sensory neurons discussed earlier also suggests participation of a second-
messenger system, which has not yet been identified; alternatively, a
mobile GTP-binding protein subunit might serve as its own second messenger.
In the simplest case, the diffusible agent might then mediate both the
activation of 'S'-like current and the suppression of Ca current, and thus
decreased intracellular accumulation of Ca^{2+} and decreased activation of
Ca-dependent K current.

How might these modulatory changes affect cellular function? The activation of 'S'-like current by FMRFamide leads to hyperpolarization of the
cell (see Fig. 11A_1), suppression of spontaneous spiking (Březina et al.

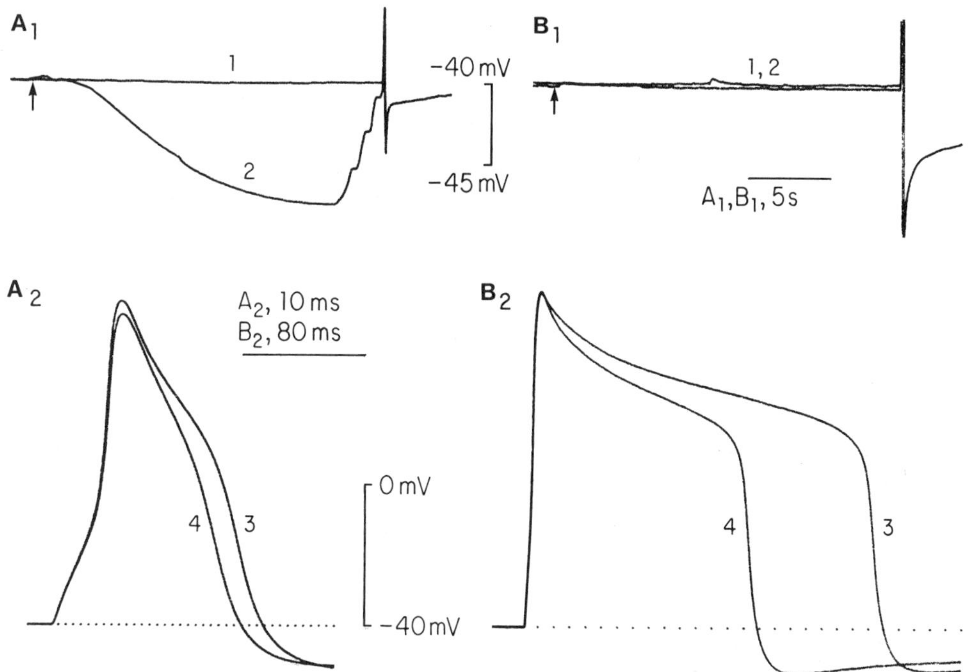

Fig. 11. FMRFamide shortens action potentials. A_1 and B_1 show voltage
traces recorded in cell R2 before (trace 1) and following 3 s puff
of 50 μM FMRFamide (arrow, trace 2). 4 ms current pulse was used
to evoke an action potential (not resolved in A_1 and B_1, expanded
in A_2 and B_2) 20 s following the puff of peptide (trace 4) or at
the corresponding time in the control record (trace 3). Membrane
potential was brought by current injection to -40 mV before each
trace was recorded; when, as in A_1, FMRFamide elicited a hyperpolarization, the potential was returned to -40 mV just before the
action potential. A, ASW. B, ASW containing 200 mM TEA and 5 mM
4-AP. (From Březina et al. 1987b.)

1987*a*) and, by analogy with the opposite effects of serotonin (Klein *et al.* 1986), presumably also to such manifestations of decreased excitability as shunting of and increased accommodation to depolarizing stimuli. Furthermore, when action potentials are nevertheless generated, and even when any direct effect on their shape of the FMRFamide-induced hyperpolarization is experimentally prevented (see Fig. 11), they are decreased in amplitude and shortened by FMRFamide, an effect that presumably results partly from the activation of 'S'-like current by the peptide, since serotonin lengthens action potentials by suppressing 'S' current (Kandel and Schwartz, 1982). However, in the case of FMRFamide, it is clear that the suppression of Ca current can also contribute to the effect, since spikes are shortened even when TEA and 4-AP are used to block the 'S'-like current and its associated FMRFamide-induced hyperpolarization (Fig. 11). Several of these effects of FMRFamide will probably combine to suppress transmitter release. The Ca current necessary for release will be directly reduced in amplitude and, since spikes are shorter, will flow for a shorter time (compare Hochner *et al.* 1986*a*). Moreover, the FMRFamide-induced hyperpolarization will potentiate both of these effects, since Kretz *et al.* (1984) have found that, in *Aplysia*, spikes arising from hyperpolarized resting potentials elicit markedly less transmitter release than those arising from more depolarized potentials, due to the operation of the same two mechanisms: the hyperpolarization turns off a Ca current that is persistently activated at depolarized potentials, and also relieves inhibition of voltage-dependent K currents which are then able to shorten action potentials. In this discussion we have been assuming, of course, that the effects of FMRFamide in the presynaptic terminal resemble the effects that we have described in the cell body, an assumption strengthened by the result that serotonin closes 'S' channels in growth cones, immature presynaptic terminals, as well as in the cell body (Belardetti *et al.* 1986*b*), and further by the finding that FMRFamide does, indeed, suppress transmitter release from *Aplysia* sensory neurons (Belardetti *et al.* 1986*a*). All of these considerations suggest a role for FMRFamide in counteracting the presynaptic facilitation due to serotonin or a transmitter with similar effects, or, like histamine in cell L10 (Kretz *et al.* 1984, 1986), as an agent of presynaptic inhibition in *Aplysia*. In several vertebrate preparations a number of neurotransmitters and peptides have similarly been found to inhibit transmitter release or hormone secretion either by stimulating a K current, and thus shortening action potentials, or by directly suppressing the Ca current or even, as we have proposed here for FMRFamide, by both mechanisms acting synergistically (see, for example, Dunlap and Fischbach, 1981; Cherubini and North, 1985; Luini *et al.* 1986; Dolphin *et al.* 1986; Macdonald and Werz, 1986; Ewald *et al.*, this volume).

The finding that, in the *Aplysia* neurons, both the activation of 'S'-like current and the suppression of Ca current probably contribute synergistically to the action of FMRFamide suggests that the opposite action of serotonin might also result from such a synergistic combination of modulatory effects, by strict analogy from an increase in Ca current in addition to the already described suppression of 'S' current. In *Aplysia* sensory neurons, serotonin does indeed have effects, such as enhancement of the transient increase in intracellular Ca^{2+} concentration elicited by depolarization (Boyle *et al.* 1984), that appear to be independent of the suppression of 'S' current, and that may contribute to presynaptic facilitation (Hochner *et al.* 1986*b*). However, while serotonin does increase the Ca current in some *Aplysia* neurons (see, for example, Lotshaw *et al.* 1986), it apparently does not increase it in the sensory neurons (Klein and Kandel, 1980). Thus, although FMRFamide and serotonin may influence neuronal function in broadly symmetrical, opposite ways, FMRFamide causing presynaptic inhibition and serotonin presynaptic facilitation, the two agents may do this by modulating different currents or perhaps intracellular processes, or, as in the case of the 'S' current, by modulating different microscopic parameters of the same current.

ACKNOWLEDGMENTS

Work in our laboratory described in this article was supported by NSF grant BNS 83-16417, USPHS grant RO1 NS08364 (both to Dr R. Eckert), and by USPHS NRSA GM07185.

REFERENCES

Austin, T., Weiss, S. and Lukowiak, K., 1983, FMRFamide effects on spontaneous and induced contractions of the anterior gizzard in *Aplysia*, *Can. J. Physiol. Pharmacol. 61*:949.

Belardetti, F., Abrams, T.W., Castellucci, V.F., Kandel, E.R. and Siegelbaum, S.A., 1986a, FMRFamide depresses transmitter release and increases the opening probability of the S K^+ channel in *Aplysia* sensory neurons, *Soc. Neurosci. Abstr. 12*:766.

Belardetti, F., Schacher, S., Kandel, E.R. and Siegelbaum, S.A., 1986b, The growth cones of *Aplysia* sensory neurons: modulation by serotonin of action potential duration and single potassium channel currents, *Proc. Natl Acad. Sci. USA 83*:7094.

Belardetti, F., Kandel, E.R. and Siegelbaum, S.A., 1987, Neuronal inhibition by the peptide FMRFamide involves opening of S K^+ channels, *Nature 325*:153.

Boyle, M.B., Klein, M., Smith, S.J. and Kandel, E.R., 1984, Serotonin increases intracellular Ca^{2+} transients in voltage-clamped sensory neurons of *Aplysia californica*, *Proc. Natl Acad. Sci. USA 81*:7642.

Březina, V., 1987, Non-hydrolyzable analog of GTP mimics modulation of K and Ca currents by FMRF-amide in *Aplysia* neurons, *Biophys. J. 51*:251a.

Březina, V., Eckert, R. and Erxleben, C., 1987a, Modulation of potassium conductances by an endogenous neuropeptide in neurones of *Aplysia californica*, *J. Physiol. 382*:267.

Březina, V., Eckert, R. and Erxleben, C., 1987b, Suppression of calcium current by an endogenous neuropeptide in neurones of *Aplysia californica*, *J. Physiol. 388*, in press.

Byrne, J.H., Ocorr, K.A., Walsh, J.P. and Walters, E.T., 1986, Analysis of associative and nonassociative neuronal modifications in *Aplysia* sensory neurons, *in*: "Neural Mechanisms of Conditioning", D.L. Alkon and C.D. Woody, eds, Plenum, New York.

Cherubini, E. and North, R.A., 1985, μ and κ opioids inhibit transmitter release by different mechanisms, *Proc. Natl Acad. Sci. USA 82*:1860.

Dolphin, A.C., Forda, S.R. and Scott, R.H., 1986, Calcium-dependent currents in cultured rat dorsal root ganglion neurones are inhibited by an adenosine analogue, *J. Physiol. 373*:47.

Dunlap, K. and Fischbach, G.D., 1981, Neurotransmitters decrease the calcium conductance activated by depolarization of embryonic chick sensory neurones, *J. Physiol. 317*:519.

Hochner, B., Klein, M., Schacher, S. and Kandel, E.R., 1986a, Action-potential duration and the modulation of transmitter release from the sensory neurons of *Aplysia* in presynaptic facilitation and behavioral sensitization, *Proc. Natl Acad. Sci. USA 83*:8410.

Hochner, B., Klein, M., Schacher, S. and Kandel, E.R., 1986b, Additional component in the cellular mechanism of presynaptic facilitation contributes to behavioral dishabituation in *Aplysia*, *Proc. Natl Acad. Sci. USA 83*:8794.

Kandel, E.R. and Schwartz, J.H., 1982, Molecular biology of learning: modulation of transmitter release, *Science 218*:433.

Kandel, E.R., Abrams, T., Bernier, L., Carew, T.J., Hawkins, R.D. and Schwartz, J.H., 1983, Classical conditioning and sensitization share aspects of the same molecular cascade in *Aplysia*, *Cold Spring Harbor Symp. Quant. Biol. 48*:821.

Klein, M. and Kandel, E.R., 1980, Mechanism of calcium current modulation underlying presynaptic facilitation and behavioral sensitization in *Aplysia*, *Proc. Natl. Acad. Sci. USA 77*:6912.

Klein, M., Camardo, J. and Kandel, E.R., 1982, Serotonin modulates a specific potassium current in the sensory neurons that show presynaptic facilitation in *Aplysia, Proc. Natl Acad. Sci. USA 79*:5713.

Klein, M., Hochner, B. and Kandel, E.R., 1986, Facilitatory transmitters and cAMP can modulate accommodation as well as transmitter release in *Aplysia* sensory neurons: evidence for parallel processing in a single cell, *Proc. Natl Acad. Sci. USA 83*:7994.

Kretz, R., Shapiro, E., Connor, J. and Kandel, E.R., 1984, Posttetanic potentiation, presynaptic inhibition, and the modulation of the free Ca^{2+} level in the presynaptic terminals, *Exp. Brain Res. Suppl. 9*:240.

Kretz, R., Shapiro, E., Bailey, C.H., Chen, M. and Kandel, E.R., 1986, Presynaptic inhibition produced by an identified presynaptic inhibitory neuron. II. Presynaptic conductance changes caused by histamine, *J. Neurophysiol. 55*:131.

Lotshaw, D.P., Levitan, E.S. and Levitan, I.B., 1986, Fine tuning of neuronal electrical activity: modulation of several ion channels by intracellular messengers in a single identified nerve cell, *J. Exp. Biol. 124*:307.

Luini, A., Lewis, D., Guild, S., Schofield, G. and Weight, F., 1986, Somatostatin, an inhibitor of ACTH secretion, decreases cytosolic free calcium and voltage-dependent calcium current in a pituitary cell line, *J. Neurosci. 6*:3128.

Macdonald, R.L. and Werz, M.A., 1986, Dynorphin A decreases voltage-dependent calcium conductance of mouse dorsal root ganglion neurones, *J. Physiol. 377*:237.

Ocorr, K.A. and Byrne, J.H., 1985, Membrane responses and changes in cAMP levels in *Aplysia* sensory neurons produced by serotonin, tryptamine, FMRF-amide and small cardioactive peptide$_B$ (SCP$_B$), *Neurosci. Lett. 55*:113.

Pollock, J.D., Bernier, L. and Camardo, J.S., 1985, Serotonin and cyclic adenosine 3':5'-monophosphate modulate the potassium current in tail sensory neurons in the pleural ganglion of *Aplysia, J. Neurosci. 5*:1862.

Ruben, P., Johnson, J.W. and Thompson, S., 1986, Analysis of FMRF-amide effects on *Aplysia* bursting neurons, *J. Neurosci. 6*:252.

Schaefer, M., Picciotto, M.R., Kreiner, T., Kaldany, R., Taussig, R. and Scheller, R., 1985, *Aplysia* neurons express a gene encoding multiple FMRFamide neuropeptides, *Cell 41*:457.

Siegelbaum, S.A., 1987, The S-current: a background potassium current, *in*: "Neuromodulation", L.K. Kaczmarek and I.B. Levitan, eds, Oxford University Press, New York.

Weiss, S., Goldberg, J.I., Chohan, K.S., Stell, W.K., Drummond, G.I. and Lukowiak, K., 1984, Evidence for FMRF-amide as a neurotransmitter in the gill of *Aplysia californica, J. Neurosci. 4*:1994.

CYTOPLASMIC MODULATION OF TRANSMITTER GATED

K CHANNELS IN CULTURED MAMMALIAN CENTRAL NEURONS

Jerrel L. Yakel, Laurence O. Trussell, and Meyer B. Jackson

Department of Biology
University of California, Los Angeles
Los Angeles CA 90024

INTRODUCTION

Recent years have seen considerable progress in our understanding of slow inhibitory synaptic processes in the mammalian brain. In many studies a surprisingly similar mechanism of inhibition has emerged, though these studies were of different neurotransmitters and in diverse brain regions. In these cases inhibition arises from the activation of a potassium conductance with two unique features. These features are 1) inward rectification and 2) mediation by a GTP-binding protein (G-protein).

The inward rectification of these responses is manifest in their current-voltage relationships as a decreasing response at potentials depolarized beyond a characteristic point. Activation of an inwardly rectifying potassium conductance has been observed with baclofen in hippocampal slices (Newberry & Nicoll, 1985; Gahwiler & Brown, 1985), opiates in locus coeruleus (North & Williams, 1985), and with adenosine and serotonin in cultured hippocampus and striatum (Trussell & Jackson, 1985 & 1987a; Yakel et al, 1986).

The G-proteins that mediate the activation of this potassium channel are members of a family of GTP-dependent regulatory proteins. Some G-proteins have well established and important roles in coupling ligand occupation of a receptor binding site to the modulation of adenylate cyclase (Gilman, 1984). Other types of G-proteins have been found in the mammalian brain (Sternweis & Robishaw, 1984; Neer et al, 1984); the functions of these other G-proteins is still obscure, but at least one study implicates such a protein in the mediation of a neurotransmitter response (Hescheler et al, 1987). The involvement of a G-protein in transmitter-activated inwardly rectifying potassium conductances in the mammalian central nervous system is inferred from experiments with GTP and GTP analogues for baclofen and serotonin (Andrade et al, 1986), as well as adenosine (Trussell & Jackson, 1987a). These studies and an additional study of opiate and α_2-adrenergic receptors (Aghajanian & Wang, 1986) also made use of pertussis toxin, a bacterial toxin that catalyzes the ADP-ribosylation of some G-proteins. The effectiveness of pertussis toxin in blocking these responses allows an inference to be drawn as to the identity of the G-protein involved. Pertussis toxin catalyzes the ADP-ribosylation of G_i, a G-protein involved in inhibition of adenylate cyclase, of G_o, another G-protein found in brain (Sternweis & Robishaw, 1984; Neer et al, 1984), and of transducin. Thus, it is likely that one of these G-proteins mediates the activation of these inwardly rectifying

potassium channels. G-proteins have also been implicated in another
neurotransmitter response involving calcium current modulation (Holtz et
al, 1986; Dunlap, this volume; Hescheler et al, 1987).

Mechanisms of inhibition similar to that described above in mammalian
brain are found in other systems as well. In heart muscle, an inwardly
rectifying potassium conductance is activated by muscarinic agonists, and
this response is dependent on GTP (Pfaffinger et al, 1985), sensitive to
GTP analogues (Breitwieser & Szabo, 1985), and blocked by pertussis toxin
(Pfaffinger et al, 1985). More recently, GTP analogues (Kurachi et al,
1986) and purified G_i subunits (Logothetis et al, 1987; Yatani et al,
1987), have been applied directly to the cytoplasmic face of heart muscle
membrane to activate potassium channels. These studies provide a direct
demonstration of G_i mediation of the response in heart muscle.
Experiments in Aplysia neurons suggest a G-protein involvement in a slow
inhibitory response to carbachol, histamine, and dopamine (Sasaki & Sato,
1987).

The present paper summarizes some recent work on inhibitory responses
of cultured mouse striatal and hippocampal neurons to adenosine and serotonin, using tight-seal whole-cell voltage clamp techniques (Hamill et al,
1981). Responses to these putative neurotransmitters are shown to possess
the two features described above. Some additional features of these
responses are also presented concerning the biophysical properties of the
channel that is activated and the involvement of well known cytoplasmic
signaling systems. Whereever direct comparisons are made, the responses to
adenosine and serotonin are found to be remarkably similar.

ADENOSINE AND SEROTONIN RESPONSES IN STRIATAL AND HIPPOCAMPAL NEURONS

When adenosine or serotonin are applied to voltage clamped neurons in
culture, a current is often observed, the sign and magnitude of which
depend on the voltage relative to the potassium Nernst potential (Fig.
1A). The time course of the potassium currents activated by adenosine and
serotonin are very similar. Pressure ejection of adenosine or serotonin
with a 500 msec pulse produces a response that generally peaks in 1 sec and
ends in about 20 secs. This is the only adenosine response that can be
detected at a constant holding potential (Trussell & Jackson, 1985 &
1987a). Adenosine has an additional effect on a voltage-activated calcium
current in hippocampal neurons (Madison et al, 1987). In striatal
cultures, the adenosine-activated potassium conductance has pharmacological
characteristics that are very similar to the A1 adenosine receptor
(Trussell & Jackson, 1987b).

The serotonin-activated potassium conductance is only one of 3
serotonin responses exhibited by cultured hippocampus and striatum (Yakel
et al, 1986). Pharmacological studies suggest that the serotonin-activated
potassium conductance in cultured neurons (Yakel et al, 1987) is mediated
by the S3 postsynaptic inhibitory receptor (Aghajanian, 1981). Other
responses include a rapidly activated inward current having much in common
with the classical nicotinic response of skeletal muscle, and a slow inward
current which is probably mediated by cAMP. The cAMP mediated effect is
associated with a conductance increase and may result from activation of
one of the charybdotoxin-sensitive cation-permeable channels described by
Kramer & Levitan (1987) in this volume.

POTASSIUM SELECTIVITY AND INWARD RECTIFICATION

Plots of the current versus holding potential, for both adenosine and
serotonin responses, are linear through E_K (the Nernst potential for
potassium), but curve downward at more positive membrane potentials (Fig.
1B). At sufficiently depolarized potentials responses are very small,

disappearing altogether in some instances. The inward rectification seen
here is different from the rectification exhibited by the classical inward
rectifier channel in that adenosine and serotonin can activate moderately
sized outward currents; classical inward rectifier channels pass very
little outward current (Hille, 1984). The muscarinic response of heart
muscle is a classical inward rectifier in this respect. The voltage
dependence of the inward rectifier varies with E_K. This is not explained
by the usual concepts of voltage-gated membrane channels, but can be
explained by blocking models (Hille & Schwarz, 1978), and in heart muscle
actually results from block by intracellular magnesium ions (Vandenberg,
1987; Matsuda et al, 1987).

Current-voltage relations for the adenosine response exhibit similar
behavior when E_K is varied (Fig. 2A). With 4 mM potassium no response to
adenosine is seen at -40 mV, while with 40 mM potassium a large inward
current is seen at the same potential. Thus, this rectification may not
arise from a voltage-induced conformational change in a membrane protein.
Further studies currently underway may show that this rectification results
from a blocking mechanism.

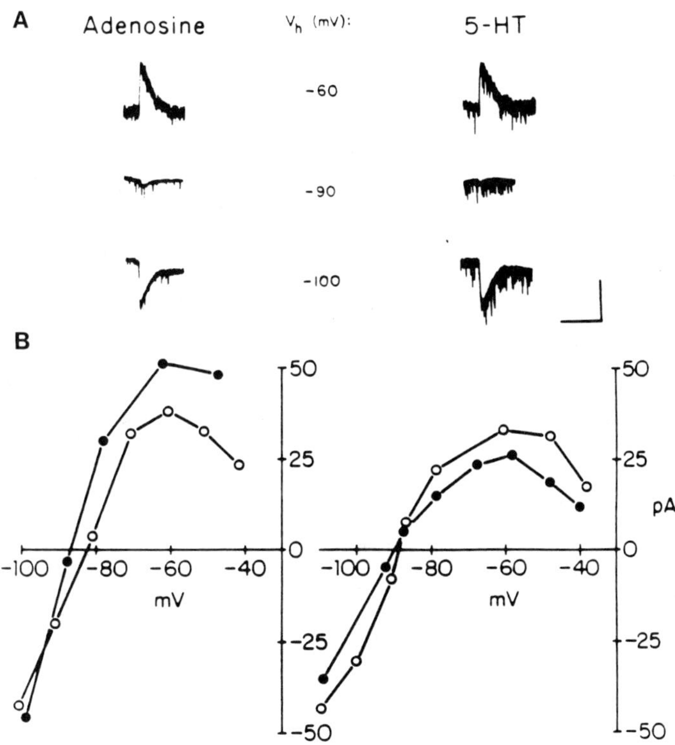

Figure 1. A. Voltage clamped currents in hippocampal neurons during
application of adenosine (50 μM) or serotonin (20 μM). The voltage
is varied to show reversal of the response near the potassium Nernst
potential (-90 mV). The calibration bars are 40 sec and 50 pA. B.
Current-voltage curves for adenosine and serotonin responses in hippocampal
(filled circles) and striatal (closed circles) neurons. For details of
experimental procedures see Trussell & Jackson, (1985 & 1987a).

Most potassium channels are blocked by cesium ions (Miller & Latorre, 1983). While the potassium channels activated by adenosine are not very permeable to cesium, the external application of 1 mM cesium fails to block these responses. Fig. 2B shows that when the interior of a cell is perfused with a solution in which potassium is replaced by cesium, large adenosine-activated inward currents can still be evoked, though outward currents, which would be carried predominantly by cesium, are quite small.

Tetraethylammonium (TEA) is another widely used potassium channel blocker. At 10 mM it does not block adenosine or serotonin responses in cultured neurons. This is a high enough concentration to block delayed rectifier channels and some calcium-activated potassium channels quite effectively (Miller & Latorre, 1983). On the other hand, barium blocks adenosine- (Trussell & Jackson, 1987a) and serotonin-activated potassium currents very well. Our studies of 4-aminopyridine (4-AP) have been limited to the serotonin response, and show no blockade. The calcium channel blockers cobalt and cadmium do not block adenosine (Trussell & Jackson, 1985) or serotonin (Yakel et al, 1987) responses. The similar

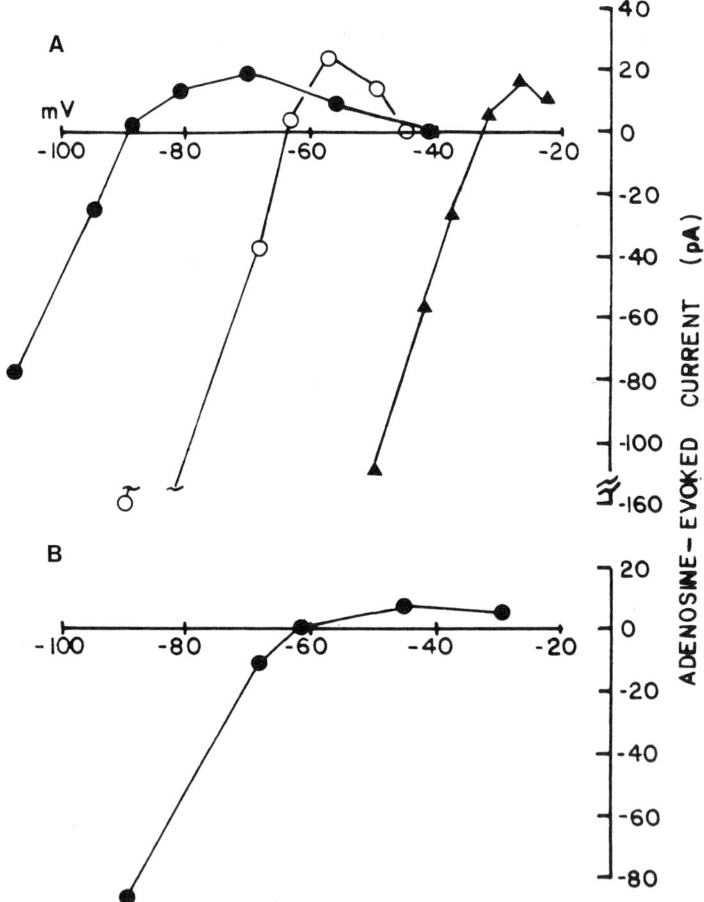

Figure 2. A. Current-voltage curves for adenosine responses in bathing solutions with 3 different potassium concentrations (filled circles - 4 mM, open circles - 10 mM, filled triangles - 40 mM). B. A current-voltage curve for the adenosine response when the recording electrode contained 140 mM CsCl. Data is presented from 4 different cells.

pattern of sensitivity to channel blockers provides another useful basis
for comparison of the responses to adenosine and serotonin. This is
summarized in tabular form below.

TABLE I

Blockade of responses by potassium channel blockers

	10 mM TEA	1 mM barium	1 mM cesium	0.5 mM 4-AP
Adenosine	0/6	5/5	0/5	
Serotonin	0/2	2/2		0/3

Table I. The number of responses successfully blocked/number of responsive
cells tested.

    Studies of other responses or in other preparations show several
similarities with regard to the action of potassium channel blockers. The
inwardly rectifying potassium channel activated by opiates in locus
coeruleus is blocked by barium, but not by 4-AP or TEA (North & Williams,
1985). TEA does not block serotonin responses (Segal, 1980), while 4-AP
does not block the adenosine responses (Segal, 1982; Greene & Haas, 1985),
both in hippocampal slices.

    While adenosine and serotonin activate channels that are identical
with respect to the biophysical characteristics examined so far, a question

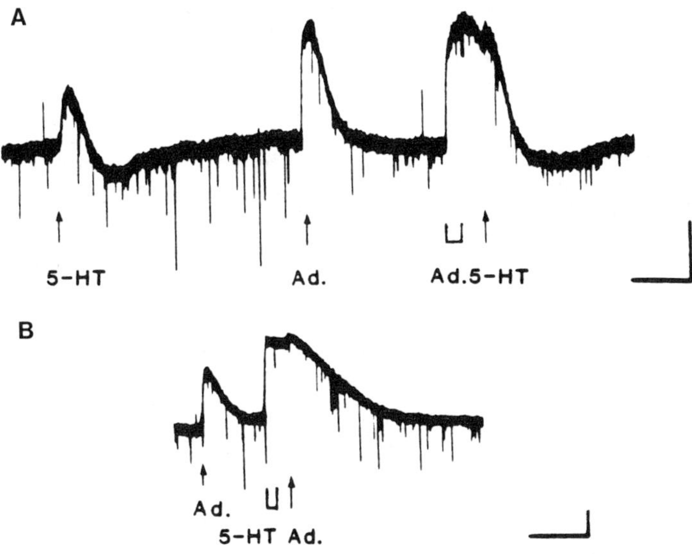

Figure 3. Hippocampal neurons were voltage clamped at -70 mV. Serotonin
(5-HT) and adenosine (Ad.) (both 20 μM in A and 50 μM in B) were
applied with pressure from nearby micropipettes. A. The first response is
to a 0.5 sec pulse of serotonin and the second is to an identical pulse of
adenosine. In the third response adenosine is applied for 5 sec to produce
a prolonged response, then a 0.5 sec pulse of serotonin is again applied,
producing only a small additional outward current. B. The first response
is to a 0.5 sec pulse of adenosine. In the second response serotonin is
applied for 3 sec to produce a prolonged response, then a 0.5 sec pulse of
adenosine is again applied, producing only a small additional outward
current. The calibration bars are 20 sec and 25 pA.

remains as to whether the receptors activate the same physical channel. The receptors could activate different sets of very similar channels through completely separate coupling proteins. Alternatively, receptors could compete for a limited number of coupling proteins or for a limited number of channels. This kind of question can be addressed by examining whether currents elicited by two different ligands are additive. Andrade et al (1986) have shown that voltage changes elicited by serotonin and baclofen are not additive. North and Williams (1985) and Andrade and Aghajanian (1985) obtained similar results with opiate and α_2-adrenergic receptors. A cell challenged with a saturating concentration of one ligand gives little or no response to the other. The studies imply that serotonin and baclofen responses, and opiate and α_2-adrenergic responses, have a common saturable step at some point after receptor activation.

Fig. 3 shows current in a voltage clamped cell presented with both serotonin and adenosine. The concentrations of adenosine (Trussell & Jackson, 1987b) and serotonin (Yakel & Jackson, unpublished) used in this experiment are saturating. The cell is voltage clamped at -70 mV so that shifts in potential during one response, which would compromise the validity of an additivity experiment, do not occur. It is clear that during a saturating response to adenosine the serotonin response is much lower (Fig. 3A), and that during a saturating response to serotonin the adenosine response is much lower (Fig. 3B). Thus, these two receptors converge either on a common channel or on common coupling components.

MEDIATION BY A G-PROTEIN

The whole-cell tight-seal patch clamp allows internal perfusion of a cell (Hamill et al, 1981). Thus, with this technique membrane currents that depend on cytoplasmic factors are lost. This was first observed with calcium currents, which washout or disappear after internal perfusion of a neuron is initiated (Kostyuk et al, 1981; Byerly & Hagiwara, 1982). Trussell and Jackson (1985) demonstrated washout of an adenosine-activated potassium conductance in cultured striatal neurons. The response disappears in approximately 10 to 20 minutes after the establishment of an intracellular recording with a large-tipped patch electrode (Fig. 4A). The loss of this response was due to washout rather than cell deterioration, since other chemically-activated responses examined concurrently do not diminish with time (Trussell & Jackson, 1985; Yakel et al, 1986). The possibility of receptor desensitization was ruled out by using small-tipped

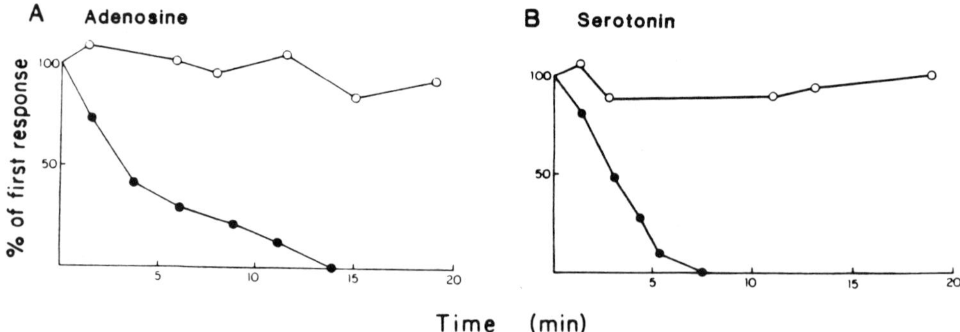

Figure 4. The responses to adenosine (A) and serotonin (B) are plotted versus time after a recording was established. Cells were held at -70 mV. Responses to adenosine and serotonin washout in recordings made with large-tipped (series resistances between 8-15 MΩ) patch electrodes (filled circles). Small-tipped (series resistances between 75-100 MΩ) patch electrodes prevent washout (open circles).

patch electrodes to prevent washout. The serotonin-activated potassium
current is similar in this respect as well: it washes out with a similar
time course, and washout is prevented with a high series resistance patch
electrode (Fig. 4B).

The washout of these responses suggests that they depend on a
cytoplasmic factor. In studies of the adenosine response, this cytoplasmic
factor was subsequently identified as guanosine triphosphate (GTP)
(Trussell & Jackson, 1987a). The addition of 100 μM GTP to the patch
electrode filling solution blocked washout, maintaining the response at
nearly constant levels for more than 30 minutes (Fig. 5). With 50 μM
GTP the response fell to a plateau at about 50% of its initial value (Fig.
5). ATP failed to prevent washout (Fig. 5).

The action of GTP in preventing washout suggests that these responses
are mediated by a G-protein. A nonhydrolyzable analogue of GTP, γ-thio
GTP (GTPγS), is known to bind to G-proteins and activate them
irreversibly (Gilman, 1984). Thus, when GTPγS is added to the patch
electrode filling solution, a large outward current develops rapidly after
the intracellular recording is established (Fig. 6B). Washout of the
adenosine response is accelerated by GTPγS, and is concomitant with the
appearance of this outward current (Fig. 6B). The two processes have been
shown to have a remarkably similar time course (Trussell & Jackson,
1987a). The interpretation of these results is that as GTPγS diffuses
into the cell, it binds to a G-protein, resulting in an enduring activation
of the potassium channels activated by adenosine. The channels that are
then opened in the absence of adenosine can no longer respond to adenosine.

Further evidence for this interpretation was obtained by examining the

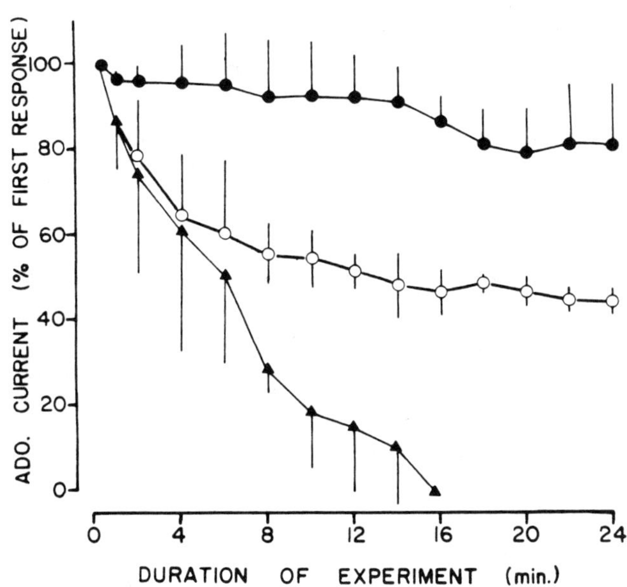

Figure 5. Addition of 100 μM GTP to a large-tipped patch electrode
prevents washout of adenosine (50 μM) responses (filled circles). 50
μM GTP in the electrode allows the response to drop to 50% of its
initial value (open circles). 1 mM ATP in the electrode does not prevent
washout (filled triangles). Experiments from 3 to 6 cells were pooled
together for each plot. Vertical bars show S.D.

effects of acutely applied barium (1 mM). As noted above, barium blocks
adenosine- and serotonin-activated potassium channels. This is illustrated
in Fig. 6A. Barium produces a small inward current as some open channels
that allow outward current to pass are blocked. The development of the
large outward current is concomitant with an increase in the inward current
elicited by barium (Fig. 6B). This is consistent with the idea that
GTPγS opens the same channels activated by adenosine.

SECOND MESSENGERS

Adenosine can both activate and inhibit adenylate cyclase (Phillis &
Wu, 1981). Serotonin can also both activate (Nestler & Greengard, 1984)
and inhibit (DeVivo and Maayani, 1986) adenylate cyclase. For this reason
we investigated the involvement of second messengers in the inhibitory
responses that we were studying. In the experiments with GTP described
above, ATP and cAMP levels should decline during a patch clamp recording.
The fact that the addition of GTP alone is sufficient to maintain the
adenosine-activated potassium current suggests that a second messenger is
not required for the response.

We carried out two different experiments to increase cAMP levels. In
the first experiment cAMP levels were increased by addition of forskolin,
which directly activates adenylate cyclase. Application of forskolin
produces a large long lasting inward current similar to one of the other
serotonin responses described above (Yakel et al, 1986). The use of high
resistance patch electrodes prevented the washout of cytoplasmic factors
essential to adenosine and forskolin responses. A brief pulse of adenosine
was applied after the forskolin-activated inward current had reached a
plateau. It is clear from Fig. 7 that forskolin has no effect on the
adenosine response. The average response in the presence of forskolin was
1.02 ± .11 times as large as the average response before forskolin in 6
similar experiments (Trussell & Jackson, 1987a).

A similar result was obtained in experiments using a low resistance
patch electrode that was filled with 1 mM 8-bromo cAMP. As the cAMP
analogue diffused from the patch electrode into the cell, an inward current
developed similar to that elicited by forskolin. The presence of 100
μM GTP maintained the adenosine response and 1 mM ATP supplied protein

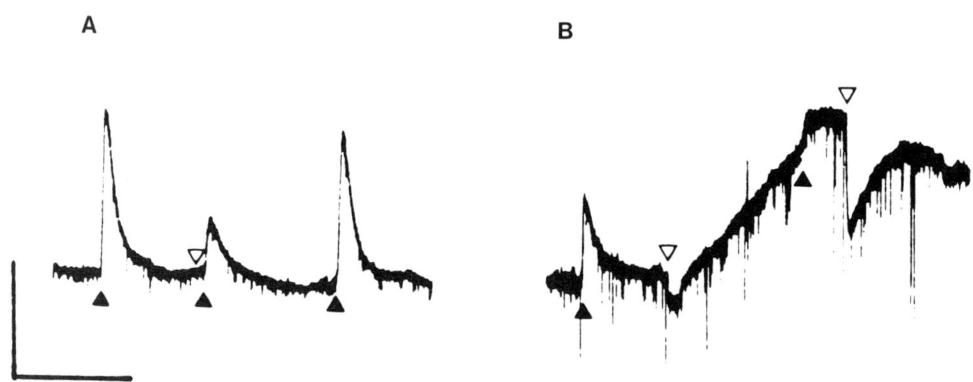

Figure 6. Actions of barium on outward currents in hippocampal neurons.
A. 1 mM barium, applied at the open triangle, blocks an adenosine (50
μM; filled triangles) response by approximately 75%. The calibration
bars are 1 min and 30 pA. B. With a patch electrode filled with 20 μM
GTPγS, 100 μM GTP, and 1 mM ATP, an outward current develops.
Adenosine responses (at the filled triangles) disappear, while inward
currents evoked by application of 1 mM barium (open triangles) become much
larger (from Trussell & Jackson, 1987a).

kinase with the necessary phosphate donor. The adenosine response did not
change significantly over the course of this experiment (Trussell &
Jackson, 1987a), again supporting the hypothesis that the activation of
this outward current is independent of cAMP. Andrade et al (1986) obtained
similar results, showing that 8-bromo cAMP (1 mM) has no effect on
serotonin and baclofen responses in hippocampal slices. However, the
opiate and α_2-adrenergic responses in locus coeruleus are different
in this respect, in that they are inhibited by cAMP analogues (Andrade &
Aghajanian, 1985).

Another second messenger that is receiving increased attention is
diacylglycerol, which, in concert with calcium, activates protein kinase C
(Nishizuka, 1986). Phorbol esters mimic diacylglycerol, and have been
shown to modulate ion currents in hippocampal slices (Baraban, et al, 1985)
and to block the response of hippocampal slices to serotonin and baclofen
(Andrade et al, 1986), and adenosine (Worley et al, 1986). We have found
that serotonin and adenosine responses in cultured neurons are also blocked
by phorbol esters. With high resistance patch electrodes to prevent
washout, adenosine and serotonin normally endure for a prolonged period of
time. These responses are reduced by about 60% after 4 or 5 minutes of
continuous application of 1 μM phorbol 12,13-dibutyrate, and are
reduced by about 80-90% after 10 minutes (Fig. 8). In some instances, the
recording was sufficiently stable to allow partial recovery of the
responses following removal of the phorbol ester. 1 μM 4-β-phorbol
did not block responses in 4 out of 4 experiments. Phorbol 12,13-diacetate
also reduces adenosine and serotonin responses, but less effectively than
phorbol 12,13-dibutyrate. The effects of phorbol ester increase with
concentration. These results are summarized in Table II.

The blocking action of phorbol esters suggests that an essential
cellular component of the response is a subtrate of C-kinase, and that
phosphorylation of that component inactivates it. Since many
neurotransmitters activate C-kinase through stimulation of phospholipase C,
the enzyme that catalyzes the production of diacylglycerol, these other
neurotransmitters could then modulate the inhibitory serotonin and
adenosine responses.

DISCUSSION

The experiments presented and summarized here, along with many other
published studies, suggest that the activation of an inwardly rectifying

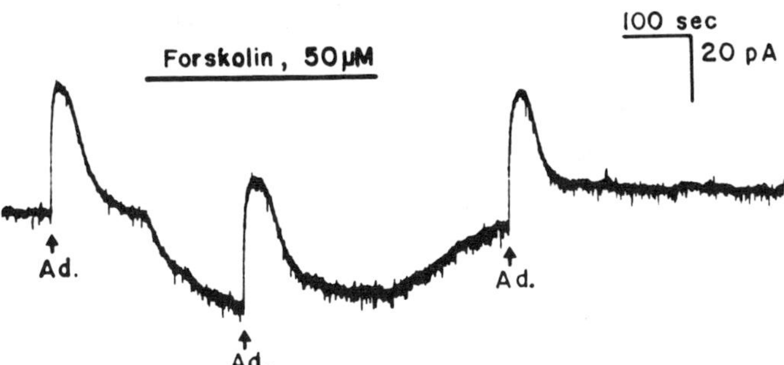

Figure 7. The adenosine (50 μM) response before, during, and after
application of 50 μM forskolin. Forskolin alone produces an inward
current upon which the adenosine response is superimposed (from Trussell &
Jackson, 1987a).

TABLE II

Effect of phorbol esters on adenosine and serotonin responses

	phorbol 12,13-dibutyrate				phorbol 12,13-diacetate			
	1 μM		10 μM		1 μM		10 μM	
5-HT	61±12%	n=5	100%	n=1	24%	n=1	56%	n=1
Ads.	66±18%	n=7	80±0%	n=2	25%	n=1	78±3%	n=2

Table II. Phorbol esters at the indicated concentrations were perfused into the bathing solution as described in Fig. 8. The maximal block calculated as the percent reduction of the response before phorbol ester application (usually achieved after 10-15 minutes) as indicated (± S.D.).

potassium conductance, with coupling between receptor and channel mediated by a G-protein, is an important and ubiquitous mechanism of inhibition in the mammalian brain, as well as in other systems. In most instances this inhibition appears to be independent of cAMP; the response can be blocked by phorbol esters.

A variety of comparisons of serotonin and adenosine responses have been carried out. The time course, voltage dependence, channel blocker sensitivity, washout, and sensitivity to phorbol esters are very similar. These various similarities, together with the nonadditivity of the two responses, make a strong case for the proposal that the receptors for adenosine and serotonin converge on or share the same channel.

The next stage of investigation of this response has already begun. Purified G-proteins as well as combinations of various G-protein subunits have been applied to the cytoplasmic face of heart muscle (Logothetis et al, 1987; Yatani et al, 1987). This approach should ultimately lead to a precise description of the mechanism of action of G-proteins in these

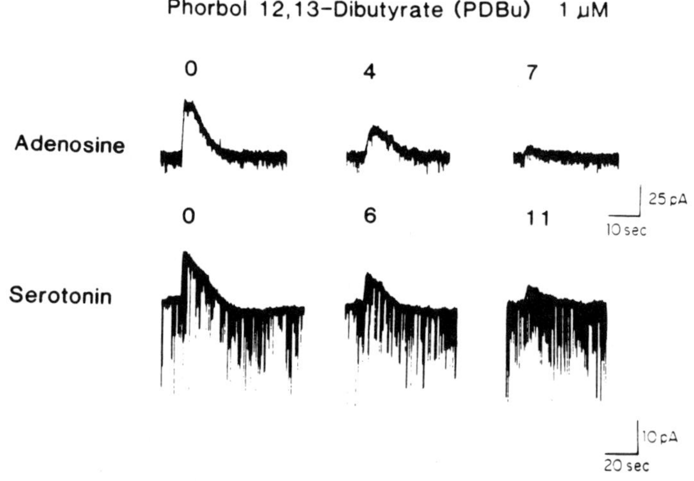

Figure 8. 1 μM phorbol 12,13-dibutyrate was perfused into the bathing solution and adenosine and serotonin (both at 20 μM) responses were monitored. Times after perfusion of phorbol ester in minutes are noted above each response.

processes. This approach should be as useful in the study of neuronal
inhibition as it appears to be in the heart.

REFERENCES

Aghajanian, G.K., 1981, The modulatory role of serotonin at multiple
    receptors of the brain. In: Serotonin neurotransmission and behavior.
    (Eds. B.L. Jacobs and A. Gelperin) MIT Press, Cambridge MA, pp.
    156-185.
Aghajanian, G.K. and Wang, Y.-Y., 1986, Pertussis toxin blocks the outward
    currents evoked by opiate and α_2-agonists in locus coeruleus
    neurons. Brain Res. 371:390.
Andrade, R. and Aghajanian, G.K., 1985, Opiate- and α_2-adrenoceptor-
    induced hyperpolarizations of locus ceruleus neurons in brain slices:
    reversal by cyclic adenosine 3':5'-monophosphate analogues. J.
    Neurosci. 5:2359.
Andrade, R., Malenka, R.C., and Nicoll, R.A., 1986, A G-protein couples
    serotonin and $GABA_B$ receptors to the same channels in hippocampus.
    Science 243:1261.
Baraban, J.M., Snyder, S.H., and Alger, B.E., 1985, Protein kinase C
    regulates ionic conductance in hippocampal pyramidal neurons:
    electrophysiological effects of phorbol esters. Proc. Natl. Acad. Sci.
    82:2538.
Breitwieser, G.E. and Szabo, G., 1985, Uncoupling of muscarinic and
    β-adrenergic receptors from ion channels by a guanine nucleotide
    analogue. Nature 317:538.
Byerly, L. and Hagiwara, S., 1982, Calcium currents in internally perfused
    nerve cell bodies of Limnea stagnalis. J. Physiol. 322:513.
DeVivo, M. and Maayani, S., 1986, Characterization of the 5-hydroxytryp-
    tamine$_{1A}$ receptor mediated inhibition of forskolin-stimulated
    adenylate cyclase activity in guinea pig and rat hippocampal
    membranes, J. Pharmacol. Exp. Ther. 238:248.
Dunlap, K., 1987, This volume.
Gahwiler, B.H. and Brown, D.A., 1985, $GABA_B$-receptor-activated
    K^+-current in voltage-clamped CA3 pyramidal cells in hippocampal
    cultures. Proc. Natl. Acad. Sci. 82:1558.
Gilman, A.G., 1984, G-proteins and dual control of adenylate cyclase. Cell
    36:577.
Greene, R.W. and Haas, H.L., 1985, Adenosine action on CA1 pyramidal
    neurones in rat hippocampal slices. J. Physiol. 366:119
Hamill, O.P, Marty, A., Neher, E., Sakmann, B., and Sigworth, F., 1981,
    Improved patch-clamp techniques for high resolution current recording
    from cells and cell-free membrane patches. Pflugers Arch. 391:85.
Hille, B., 1984, Ionic Channels of Excitable Membranes. Sinauer
    Associates, Sunderland, Mass. pp. 109.
Hille, B. and Schwarz, W., 1978, K channels as multi-ion single file
    pores. J. Gen. Physiol. 72:409.
Hescheler, J., Rosenthal, W., Trautwein, W., and Schultz, G., 1987, The
    GTP-binding protein, G_o, regulates neuronal calcium channels. Nature
    325:445.
Holtz, G.G., Rane, S.G., Dunlap, K., 1986, GTP-binding proteins mediate
    transmitter inhibition of voltage-dependent calcium channels. Nature
    319: 670
Kostyuk, P.G., Krishtal, O.A., Pidoplichko, V.I., 1975, Effect of internal
    fluoride and phosphate on membrane currents during intracellular
    dialysis of nerve cells. Nature 257:691
Kramer, Levitan, 1987, This volume.
Kurachi, Y., Nakajima, T., Sugimoto, T., 1986, On the mechanism of
    activation of muscarinic K^+ channels by adenosine in isolated atrial
    cells: involvement of GTP-binding proteins. Pflugers Arch. 407:264.
Latorre, R. and Miller, C., 1983, Conduction and selectivity in potassium
    channels. J. Membrane Biol. 71:11.

Logothetis, D.E., Kurachi, Y., Galper, J., Neer, E., Clapham, D.E., 1987,
    The $\beta\gamma$ subunits of GTP-binding proteins activate the muscarinic
    K^+ channel in heart. Nature 325:321.
Madison, D.V., Fox, A.P., and Tsien, R., 1987, Adenosine reduces an
    inactivating component of calcium current in hippocampal CA3 neurons.
    Biophys. J. 51:30G.
Matsuda, H., Saigusa, A., and Irisawa, H., 1987, Ohmic conductance through
    inwardly rectifying K channel and blocking by internal Mg^{2+}. Nature
    325:156.
Nestler, E.J. and Greengard, P., 1984, Protein Phosphorylation in the
    Nervous System. Wiley & Sons, New York NY pp. 20.
Neer, E.J., Lok, J.M., and Wolf, L.G., 1984, Purification and properties
    of the inhibitory guanine nucleotide regulatory unit of brain
    adenylate cyclase. J. Biol. Chem. 259:14222.
Newberry, N.R. and Nicoll, R.A., 1985, Comparison of the action of
    baclofen with γ-aminobutyric acid on rat hippocampal pyramidal
    cells in vitro. J. Physiol. 360:161.
Nishizuka, Y., 1986, Studies and perspectives of protein kinase C.
    Science 233:305.
North, R.A. and Williams, J.T., 1985, On the potassium conductance
    increased by opioids in rat locus coeruleus neurons. J. Physiol.
    364:265.
Pfaffinger, P.J., Martin, J.M., Hunter, D.D., Nathenson, N.M. Hille, B.,
    1985, GTP-binding proteins couple cardiac muscarinic receptors to a
    potassium channel. Nature 317:536
Phillis, J.W. and Wu, P.H., 1981, The role of adenosine and its nucleo-
    tides in central synaptic transmission. Prog. Neurobiol. 16:187.
Sasaki, K. and Sato, M., 1987, A single GTP-binding protein regulates
    K^+-channels coupled with dopamine, histamine, and acetylcholine.
    Nature 325:259.
Segal, M., 1980, The action of serotonin in the rat hippocampal slice
    preparation. J. Physiol. 303:423.
Segal, M., 1982, Intracellular analysis of a postsynaptic action of
    adenosine in the rat hippocampus. Eur. J. Pharm. 79:193.
Sternweis, P.C. and Robishaw, J.D., Isolation of two proteins with high
    affinity for guanine nucleotides from membranes of bovine brain. J.
    Biol. Chem. 259:13806.
Trussell, L.O. and Jackson, M.B., 1985, Adenosine-activated potassium
    conductance in cultured striatal neurons. Proc. Natl. Acad. Natl.
    82:4857.
Trussell, L.O. and Jackson, M.B., 1987a, Dependence of an adenosine
    activated potassium current on a GTP-binding protein in mammalian
    central neurons. J. Neurosci. (in press).
Trussell, L.O. and Jackson, M.B., 1987b, Pharmacological characteristics
    of an adenosine activated K^+-conductance. In preparation.
Vandenberg, C.A., 1987, Inward rectification of a potassium channel in
    cardiac ventricular cells depends on internal magnesium ions. Proc.
    Natl. Acad. Sci., 84: (in press).
Worley, P.F., Baraban, J.M., McCarren, M., Snyder, S.H., and Alger,
    B.E., 1986, Muscarinic stimulated phosphoinositide turnover blocks
    inhibitory transmitters in the rat hippocampus. Soc. Neurosci.
    Abstr. 12:735.
Yakel, J.L., Trussell, L.O., and Jackson, M.B., 1986, Serotonergic
    responses in cultured mouse striatal neurons. Soc. Neurosci.
    Abstr. 12:726.
Yakel, J.L., Trussell, L.O., and Jackson, M.B., 1987, Three serotonin
    responses in cultured mouse hippocampal and striatal neurons.
    J. Neurosci. (in press).
Yatani, A., Codina, J., Brown, A.M., and Birnbaumer, L., 1987, Direct
    activation of mammalian atrial muscarinic potassium channels by GTP
    regulatory protein G_K. Science 235:207.

SECTION 4

ION CHANNELS AS CAUSES AND CONSEQUENCES OF DEVELOPMENT

REGULATION OF CORTICAL VESICLE EXOCYTOSIS IN SEA URCHIN EGGS

Laurinda A. Jaffe

Department of Physiology
University of Connecticut Health Center
Farmington, CT 06032

Exocytosis in many cells appears to be regulated by intracellular
free calcium (see Augustine et al., 1987). The exocytosis of cortical
vesicles in the sea urchin egg at fertilization has provided an excellent
example for studying both the processes leading to a rise in intracellular
calcium and the steps by which the rise in calcium causes exocytosis.

The unfertilized sea urchin egg is about 100µm in diameter; underlying the plasma membrane is a monolayer of about 20,000 cortical vesicles,
each about 1µm in diameter (Schroeder, 1979). The vesicles are in close
contact with the plasma membrane, and are surrounded by a membrane reticulum
(Chandler and Heuser, 1979). Exocytosis is initiated by fertilization;
it starts at the point of sperm entry and propagates around the egg
surface. Structural proteins and enzymes released from the cortical
vesicles cause the detachment, elevation, and modification of a glycocalyx
surrounding the egg, termed the vitelline envelope. The elevated envelope
is termed the fertilization envelope; it no longer binds sperm and provides
a barrier through which sperm cannot pass. Thus the fertilization envelope
provides a block to polyspermy (see Jaffe and Gould, 1985), as well
as a protective shell for the developing embryo.

The first response of the sea urchin egg to fertilization is
a depolarization of the egg membrane from about -70mV to $+20$mV (Jaffe,
1976). This fertilization potential provides a fast block to polyspermy,
which protects the egg until the fertilization envelope is fully elevated
about one minute later (Jaffe, 1976). About 30 seconds after the rise
of the fertilization potential, exocytosis of cortical vesicles begins;
exocytosis is complete in another 30 seconds (Jaffe et al., 1978).

Fertilization initiates a rise in intracellular free calcium
(Steinhardt et al., 1977), from a resting level of 0.1µM to a peak of
2µM (Poenie et al., 1985). The calcium rise is first detected at about
the same time the exocytosis begins, about 30 seconds after the rise
of the fertilization potential (Eisen et al., 1984).

The rise in calcium propagates across the egg surface, at about
the same rate as the wave of exocytosis, presumably starting at the
position of the fertilizing sperm (Eisen et al., 1984; Yoshimoto et al.,
1986; Swann and Whitaker, 1986). The source of the calcium is intracellular
(Steinhardt et al., 1977; Schmidt et al., 1982).

That the rise in intracellular calcium is necessary and sufficient
to cause exocytosis has been demonstrated by injection of calcium buffers
into the egg. Buffer injections that maintain intracellular calcium
at less than 0.1µM prevent exocytosis in response to sperm (Zucker and
Steinhardt, 1978; Hamaguchi and Hiramoto, 1981). Buffer injections
that raise intracellular free calcium to greater than 0.3-3.0µM cause
exocytosis in the absence of sperm; estimates of the calcium concentration
required are uncertain due to uncertainties about intracellular pH and
other intracellular conditions (Hamaguchi and Hiramoto, 1981; Turner
et al., 1986; Swann and Whitaker, 1986).

The rest of this article will be concerned with two questions:
1) How does fertilization lead to a release of intracellular calcium,
and 2) how does a rise in intracellular calcium lead to exocytosis?

HOW DOES FERTILIZATION LEAD TO A RELEASE OF INTRACELLULAR CALCIUM?

This section presents evidence that sperm-egg interaction initiates
the activation of a G-protein and production of inositol trisphosphate
($InsP_3$), leading to the release of intracellular calcium (Fig. 1).
This sequence closely resembles signal transduction pathways coupling
neurotransmitter or hormone interaction to calcium release in other
cells (Berridge, 1986).

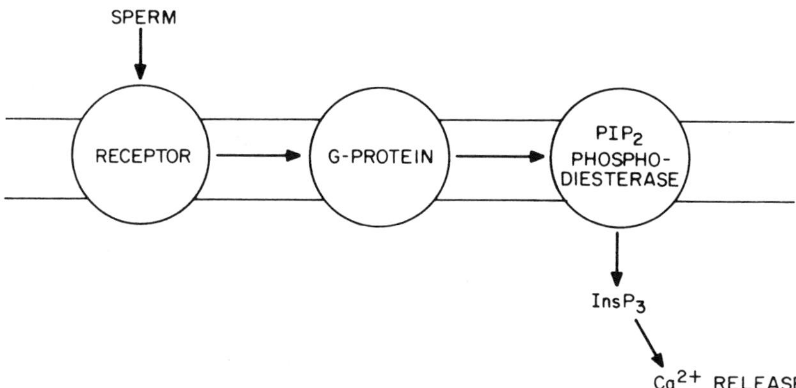

Fig. 1. Proposed sequence by which sperm contact leads to the release
of intracellular calcium in the sea urchin egg. The receptor,
G-protein, and PIP_2 phosphodiesterase are shown located in the
egg membrane. Sperm interaction is proposed to act by way of
this sequence of membrane proteins to cause the release of $InsP_3$
into the egg cytoplasm, leading to release of calcium from intracellular stores.

The first evidence that such a sequence might be involved came
from measurements of a 40% increase in the amount of phosphatidyl inositol
4,5 bisphosphate (PIP_2) in sea urchin eggs at 15 seconds post-fertilization (Turner et al., 1984). This lipid is the precursor of $InsP_3$, and
hence a change in PIP_2 suggests the involvement of $InsP_3$. Soon thereafter,
it was shown that microinjection of $InsP_3$ into sea urchin eggs causes
exocytosis (Whitaker and Irvine, 1984). An increase in $InsP_3$ at 10 sec
post-fertilization was subsequently demonstrated (Ciapa and Whitaker, 1986).

That InsP$_3$ causes exocytosis by releasing intracellular calcium
was tested by preinjecting eggs with an EGTA/CaEGTA buffer to maintain
Ca at 0.1 µM (1.6 mM EGTA). Such a preinjection prevented exocytosis
in response to subsequent injection of InsP$_3$ (28nM) (Turner et al.,
1986; Swann and Whitaker, 1986). EGTA did not prevent exocytosis by
damaging the egg, since control eggs that were preinjected with an EGTA/
CaEGTA buffer at 0.1µM and then injected with a buffer to raise Ca to
0.5µM underwent normal exocytosis (Turner et al., 1986). These and
other injection experiments to be described below were made using a
quantitative pressure injection system originally described by Hiramoto
(1962) and described in more detail by Kiehart (1982). Further evidence
that InsP$_3$ releases calcium from intracellular compartments in sea urchin
eggs came from fura-2 studies with intact eggs (Swann and Whitaker,
1986), and from ^{45}Ca studies with isolated intracellular membranes (Clapper
and Lee, 1985; Oberdorf et al., 1986).

To examine whether a G-protein was involved in the pathway coupling
sperm-egg interaction to InsP$_3$ production, eggs were injected with guanine
nucleotide analogs known to activate or inactivate G-proteins (Turner
et al., 1986; Swann et al., 1987). G-proteins are inactive when GDP
is bound and active when GTP is bound. During stimulation, the G-protein
is activated by contact with an activated receptor; this contact catalyzes
the exchange of GDP for GTP (Stryer and Bourne, 1986). The exchange
reaction proceeds at a slow rate in the resting cell, but does not result
in a significant accumulation of G-proteins in the GTP-bound state,
because G-proteins have GTPase activity that converts GTP back to GDP
at a rate that exceeds the slow exchange. However, if a hydrolysis-
resistant analog of GTP, such as guanosine-5'-0-(3-thiotriphosphate)
(GTP-γ-S) is present, resting exchange results in the accumulation of
G-proteins in the GTP-γ-S bound, and hence active, state. Thus introduction
of GTP-γ-S can activate a G-protein in the absence of receptor interaction.
Conversely, introduction of guanosine-5'-0-(2-thiodiphosphate) (GDP-β-S),
a metabolically stable analog of GDP, can compete with GTP for binding
to the G-protein during stimulation. If sufficient GDP-β-S is present,
the G-protein will not bind GTP and will not be activated by exposure
to an activated receptor.

Eggs injected with GTP-γ-S (28µM) underwent exocytosis (Turner
et al., 1986). Preinjection with an EGTA/CaEGTA buffer (0.1µM Ca^{2+},
1.6 mM EGTA) prevented exocytosis in response to GTP-γ-S (Turner et
al., 1986), and fura-2 measurements showed that GTP-γ-S injection caused
a rise in intracellular calcium (Swann et al., 1987). Eggs injected
with GDP-β-S (3mM) were inhibited from undergoing exocytosis in response
to sperm (Turner et al., 1986; Swann et al., 1987), but not in response
to subsequent injection of InsP$_3$ (28nM) (Turner et al., 1986). These
results support the conclusion that a G-protein is involved in the pathway
coupling sperm-egg interaction to InsP$_3$ production, leading to calcium
release and exocytosis.

Evidence that G-proteins are present in sea urchin eggs came
from studies using the G-protein specific toxins cholera toxin (CTX)
and pertussis toxin (PTX) (Oinuma et al., 1986; Turner et al., 1987).
These toxins catalyze the ADP-ribosylation of α-subunits of specific
G-proteins, and in the presence of ^{32}P-NAD, result in the radioactive
labelling of G-proteins. Particular G-proteins are labelled by either
CTX or PTX or both. Toxin experiments with sea urchin eggs identified
a 47 kD polypeptide labelled by CTX and a 40 kD polypeptide labelled
by PTX, indicating the presence of two separate G-proteins.

CTX- and PTX-catalyzed ADP-ribosylation also modifies the function of G-proteins. CTX results in G-protein activation, while PTX results in G-protein inactivation; particular G-proteins are affected by particular toxins (see Stryer and Bourne, 1986). External application of CTX and PTX had little effect on sea urchin eggs, but microinjection of CTX (30µg/ml) caused exocytosis (Turner et al., 1987). Injection of boiled or non-activated CTX had no effect. Injection of the A-subunit of CTX, the subunit which catalyzes ADP-ribosylation, also caused exocytosis. Exocytosis in response to CTX injection was blocked by preinjection of an EGTA/CaEGTA buffer to maintain Ca at 0.1µM.

As a control that CTX was not acting by stimulating a G-protein that activated adenylate cyclase, eggs were injected with cAMP or with a hydrolysis-resistant cAMP analog, cAMP-S (Turner et al., 1987). These injections, which raised intracellular cAMP to 10,000 times its level in the unfertilized egg (0.1µM), did not cause exocytosis. These results indicate that CTX causes exocytosis by a cAMP-independent, Ca-dependent pathway, and indicate the involvement of a CTX-sensitive G-protein in the pathway coupling sperm-egg interaction to IP_3 production, Ca release and exocytosis.

PTX injection did not cause exocytosis, but it is uncertain how much PTX was injected, due to incomplete solubility of the PTX (Turner et al., 1987). The question of the role of the PTX-sensitive G-protein in the sea urchin egg remains to be examined.

In summary, these results support the conclusion that sperm inter-action with the sea urchin egg activates a G-protein, which activates $InsP_3$ production, which causes Ca release and exocytosis. This conclusion probably also applies to the frog egg (Busa et al., 1985). Other second messengers have also been examined: diacylglycerol (Swann and Whitaker, 1985; Slack et al., 1986; Lau et al., 1986; Shen and Burgart, 1986) and cAMP (Turner et al., 1987) have no effect on exocytosis. Injection of cGMP (10-100µM) causes exocytosis as well as a rise in intracellular calcium (Swann et al., 1987), but whether cGMP in the sea urchin egg changes during fertilization is unknown.

Many other questions remain unanswered: Does the egg plasma membrane contain a sperm receptor, analogous to a neurotransmitter or hormone receptor, for which sperm would serve as an agonist? The presence of receptors which activate other known G-proteins (see Stryer and Bourne, 1986) supports the idea that a sperm receptor may be present. Also unanswered is the question of how activation of a G-protein causes $InsP_3$ production. Possibly the G-protein activates PIP_2 phosphodiesterase, the enzyme which produces $InsP_3$ from PIP_2. Alternatively, or additionally, the G-protein might activate PIP kinase, the enzyme that produces PIP_2 from its precursor lipid PIP. An increase in PIP_2 would increase the production of $InsP_3$. Closely related to the question of how a G-protein initiates exocytosis is the question of how exocytosis propagates around the egg surface. A likely hypothesis is that Ca release at one position activates PIP_2 phosphodiesterase at an adjacent position, leading to Ca release and exocytosis at the adjacent position (Whitaker and Irvine, 1984).

HOW DOES A RISE IN INTRACELLULAR CALCIUM LEAD TO EXOCYTOSIS?

Although this question has not yet been answered for sea urchin eggs or for other cells, studies of sea urchin eggs have provided useful information relevant to this general problem of cell physiology. Much of this information has been obtained by studies of isolated cell surface complexes from eggs. These preparations, often termed cortices, consist

of the egg's plasma membrane with associated cortical vesicles, vitelline
envelope, and membrane reticulum (Vacquier, 1975; Zimmerberg et al.,
1985). They are prepared either by allowing the egg surface to adhere
to a polylysine coated surface and shearing away the egg cytoplasm with
a jet of solution (Vacquier, 1975), or by gently homogenizing the eggs
and separating the cortices from the cytoplasm by centrifugation (Detering
et al., 1977). Cortices prepared by either method can be induced to
undergo exocytosis by addition of micromolar concentrations of Ca (Moy
et al., 1983; Sasaki and Epel, 1983; Whitaker and Baker, 1983; Jackson
et al., 1985). Exocytosis is monitored by observing the disappearance
of cortical vesicles, or by measuring the release of cortical vesicle
enzymes into the medium, or by measuring a change in light-scattering
that accompanies exocytosis.

Like cells perfused with patch pipets, cortices show time-dependent
"washout" of physiological function, in this case, the ability to undergo
exocytosis in response to added Ca (Moy et al., 1983; Sasaki and Epel,
1983; Jackson et al., 1985). The washout phenomenon has, however, led
to an interesting conclusion about the mechanism of exocytosis. The
washout can be prevented by use of a low ionic strength medium, containing
glycine to maintain osmotic strength; when the material extracted by
a high ionic strength medium is added back to cortices in low ionic
strength medium, competence to undergo exocytosis is restored (Sasaki,
1984). The material that restores exocytotic activity is a protein
with a molecular weight in the range of 100,000 (Sasaki, 1984). Inhibition
of exocytosis after treating cortices with trypsin or with N-ethyl-male-
imide, a protein modifying agent (Jackson et al., 1985), provides further
evidence for the involvement of a protein in coupling a rise in Ca to
exocytosis.

In addition to this evidence for involvement of a high molecular
weight protein, calmodulin (17 kD) also appears to be involved in the
sequence coupling a rise in calcium to exocytosis. An antibody against
calmodulin applied to sea urchin egg cortices inhibited exocytosis,
but not if the antibody was preincubated with excess calmodulin (Steinhardt
and Alderton, 1982). These observations suggested the involvement of
a calcium/calmodulin dependent enzyme in the sequence leading to exocytosis.

Sea urchin eggs contain a calcium/calmodulin-dependent kinase
(Chou and Rebhun, 1986), and a calcium/calmodulin-dependent phosphatase
(Iwasa and Ishiguro, 1986). This phosphatase appears to be calcineurin,
since immunoblotting of sea urchin egg proteins with an antibody against
mammalian brain calcineurin identified polypeptides of molecular weights
similar to those of the A and B subunits of calcineurin (M.H. Krinks,
L.A. Jaffe, and C.B. Klee, unpublished). Calcineurin functions in neurons
as a regulator of Ca channel inactivation (Chad and Eckert, 1986), an
observation which led us to examine the possibility that calcineurin
might also function as a regulator of exocytosis.

Injection of the antibody against calcineurin into sea urchin
eggs (0.5-1.1 mg/ml) inhibited exocytosis in response to insemination
(12 of 16 eggs tested), while injection of a control antibody had no
effect (L.A. Jaffe, and C.B. Klee, unpublished). This observation should
be interpreted cautiously, but it does suggest another possible mechanism
by which calcium might regulate exocytosis (see Augustine et al., 1987,
for a review). By activating calcineurin, calcium might stimulate the
dephosphorylation of a protein of the egg membrane or cortical vesicle
membrane. In _Paramecium_, exocytosis of trichocysts is accompanied by
dephosphorylation of a 65 kD polypeptide (Zieseniss and Plattner, 1985).
Perhaps dephosphorylation could lead to exocytosis by reducing the negative
charge on apposing membranes, thus allowing membrane contact and fusion.

ACKNOWLEDGEMENT

        Work from the author's laboratory was supported by NIH grant
HD14939.

REFERENCES

Augustine, G.J., Charlton, M.P., and Smith, S.J, 1987, Calcium action
    in synaptic transmitter release, Ann. Rev. Neurosci., 10: 633.
Berridge, M.J., 1986, Cell signalling through phospholipid metabolism.
    J. Cell Science, 4: 137.
Busa, W.B., Ferguson, J.E., Joseph, S.K., Williamson, J.R., and Nuccitelli,
    R., 1985, Activation of frog (Xenopus laevis) eggs by inositol trisphos-
    phate. 1. Characterization of Ca^{2+} release from intracellular stores,
    J. Cell Biol., 101: 677.
Chad, J.E., and Eckert, R., 1986, An enzymatic mechanism for calcium
    current inactivation in dialysed Helix neurones, J. Physiol., 378: 31.
Chandler, D.E., and Heuser, J., 1979, Membrane fusion during secretion.
    Cortical granule exocytosis in sea urchin eggs as studied by quick-
    freezing and freeze-fracture, J. Cell Biol., 83: 91.
Chou, Y.-H., and Rebhun, L.I., 1986, Purification and characterization
    of a sea urchin egg Ca^{2+}-calmodulin-dependent kinase with myosin light
    chain phosphorylating activity, J. Biol. Chem., 261: 5389.
Ciapa, B., and Whitaker, M., 1986, Two phases of inositol polyphosphate
    and diacylglycerol production at fertilisation, FEBS Lett: 195, 347.
Clapper, D.L., and Lee, H.C., 1985, Inositol trisphosphate induces calcium
    release from non-mitochondrial stores in sea urchin egg homogenates,
    J. Biol. Chem., 260: 13947.
Detering, N.K., Decker, G.L., Schmell, E.D., and Lennarz, J., 1977,
    Isolation and characterization of plasma membrane-associated cortical
    granules from sea urchin eggs, J. Cell Biol., 75: 899.
Eisen, A., Kiehart, D.P., Wieland, S.J., and Reynolds, G.T., 1984, Temporal
    sequence and spatial distribution of early events of fertilization
    in single sea urchin eggs, J. Cell Biol., 99: 1647.
Hamaguchi, Y., and Hiramoto, Y., 1981, Activation of sea urchin eggs
    by microinjection of calcium buffers, Exp. Cell Res., 134: 171.
Hiramoto, Y., 1962, Microinjection of the live spermatozoa into sea
    urchin eggs, Exp. Cell Res., 27: 416.
Iwasa, F., and Ishiguro, K., 1986, Calmodulin-binding protein (55K +
    17K) of sea urchin eggs has a Ca^{2+}- and calmodulin-dependent phosphopro-
    tein phosphatase activity, J. Biochem., 99: 1353.
Jackson, R.C., Ward, K.K., and Haggerty, J.G., 1985, Mild proteolytic
    digestion restores exocytotic activity to N-ethylmaleimide-inactivated
    cell surface complex from sea urchin eggs, J. Cell Biol., 101: 6.
Jaffe, L.A., 1976, Fast block to polyspermy in sea urchin eggs is electri-
    cally mediated, Nature (London), 261: 68.
Jaffe, L.A., and Gould, M., 1985, Polyspermy-preventing mechanisms,
    in: "Biology of Fertilization," C.B. Metz, and A. Monroy, eds., Academic
    Press, Orlando.
Jaffe, L.A., Hagiwara, S., and Kado, R.T., 1978, The time course of
    cortical vesicle fusion in sea urchin eggs observed as membrane capaci-
    tance changes, Develop. Biol., 67: 243.
Kiehart, D.P., 1982, Microinjection of echinoderm eggs: Apparatus and
    procedures, Methods in Cell Biol., 25: 13.
Lau, A.F., Rayson, T.C., and Humphreys T., 1986, Tumor promoters and
    dialglycerol activate the Na^+/H^+ antiporter of sea urchin eggs, Exp.
    Cell Res., 166: 23.
Moy, G.W., Kopf, G.S., Gache, C., and Vacquier, V.D., 1983, Calcium-
    mediated release of glucanase activity from cortical granules of sea
    urchin eggs, Develop. Biol., 100: 267.

Oberdorf, J.A., Head, J.F., and Kaminer, R., 1986, Calcium uptake and
    release by isolated cortices and microsomes from unfertilized eggs
    of the sea urchin _Strongylocentrotus drobachiensis_, _J. Cell Biol._,
    102: 2205.
Oinuma, M., Katada, T., Yokosawa, H., and Ui, M., 1986, Guanine nucleotide
    binding protein in sea urchin eggs serving as the specific substrate
    of islet-activating protein, pertussis toxin, _FEBS Lett._, 207: 28.
Poenie, M., Alderton, J., Tsien, R.Y., and Steinhardt, R.A., 1985, Changes
    of free calcium levels with stages of the cell division cycle, _Nature
    (London)_, 315: 147.
Sasaki, H., 1984, Modulation of calcium sensitivity by a specific cortical
    protein during sea urchin egg cortical vesicle exocytosis, _Develop.
    Biol._, 101: 125.
Sasaki, H., and Epel, D., 1983, Cortical vesicle exocytosis in isolated
    cortices of sea urchin eggs: description of a turbidometric assay
    and its utilization in studying effects of different media on discharge,
    Develop. Biol., 98: 327.
Schmidt, T., Patton, C., and Epel, D., 1982, Is there a role for the
    Ca^{2+} influx during fertilization of the sea urchin egg? _Develop.
    Biol._, 90: 284.
Schroeder, T.E., 1979, Surface area change at fertilization: resorption
    of the mosaic membrane, _Develop. Biol._, 70: 306.
Shen, S.S., and Burgart, L.J., 1986, 1,2-diacylglycerols mimic phorbol
    12-myristate 13-acetate activation of the sea urchin egg, _J. Cell
    Physiol._, 127: 330.
Slack, B.E., Bell, J.E., and Benos, D.J., 1986, Inositol-1,4,5-trisphos-
    phate injection mimics fertilization potentials in sea urchin eggs,
    Am. J. Physiol., 19: C340.
Steinhardt, R.A., and Alderton, J.M., 1982, Calmodulin confers calcium
    sensitivity on secretory exocytosis, _Nature (London)_, 295: 154.
Steinhardt, R., Zucker, R., and Schatten, G., 1977, Intracellular calcium
    release at fertilization in the sea urchin egg, _Develop. Biol._, 58: 185.
Stryer, L., and Bourne, H.R., 1986, G proteins: a family of signal trans-
    ducers, _Ann. Rev. Cell Biol._, 2: 391.
Swann, K., Ciapa, B., and Whitaker, M., 1987, Cellular messengers and
    sea urchin egg activation, _in_: "Molecular Biology of Invertebrate
    Development," D. O'Connor, ed., Alan R. Liss, New York (in press).
Swann, K., and Whitaker, M., 1985, Stimulation of the Na/H exchanger
    of sea urchin eggs by phorbol ester, _Nature (London)_, 314: 274.
Swann, K., and Whitaker, M., 1986, The part played by inositol trisphos-
    phate and calcium in the propagation of the fertilization wave in
    sea urchin eggs, _J. Cell Biol._, 103: 2333.
Turner, P.R., Jaffe, L.A., and Fein, A., 1986, Regulation of cortical
    vesicle exocytosis in sea urchin eggs by inositol 1,4,5-trisphosphate
    and GTP-binding protein, _J. Cell Biol._, 102: 70.
Turner, P.R., Jaffe, L.A., and Primakoff, P., 1987, A cholera toxin-
    sensitive G-protein stimulates exocytosis in sea urchin eggs, _Develop.
    Biol._, 120: 577.
Turner, P.R., Sheetz, M.P., and Jaffe, L.A., 1984, Fertilization increases
    the polyphosphoinositide content of sea urchin eggs, _Nature (London)_,
    310: 414.
Vacquier, V.D., 1975, The isolation of intact cortical granules from
    sea urchin eggs: calcium ions trigger granule discharge, _Develop.
    Biol._, 43: 62.
Whitaker, M.J., and Baker, P.F., 1983, Calcium-dependent exocytosis
    in an _in vitro_ secretory granule plasma membrane preparation from
    sea urchin eggs and the effects of some inhibitors of cytoskeletal
    function, _Proc. R. Soc. Lond. B_, 218: 397.
Whitaker, M., and Irvine, R.F., 1984, Inositol 1,4,5-trisphosphate micro-
    injection activates sea urchin eggs, _Nature (London)_, 312: 633.

Yoshimoto, Y., Iwamatsu, T., Hirano, K., and Hiramoto, Y., 1986, The wave pattern of free calcium release upon fertilization in medaka and sand dollar eggs, _Develop. Growth and Differ._, 28: 583.

Zieseniss, E., and Plattner, H., 1985, Synchronous exocytosis in _Paramecium_ cells involves very rapid (≤ 1 s), reversible dephosphorylation of a 65-kD phosphoprotein in exocytosis-competent strains, _J. Cell Biol._, 101: 2028.

Zimmerberg, J., Sardet, C., and Epel, D., 1985, Exocytosis of sea urchin egg cortical vesicles in vitro is retarded by hyperosmotic sucrose: kinetics of fusion monitored by quantitative light-scattering microscopy, _J. Cell Biol._, 101: 2398.

Zucker, R.S., and Steinhardt, R.A., 1978, Prevention of the cortical reaction in fertilized sea urchin eggs by injection of calcium-chelating ligands, _Biochem. Biophys. Acta._, 541: 459.

STUDIES ON THE DEVELOPMENT OF VOLTAGE-ACTIVATED CALCIUM CHANNELS IN

VERTEBRATE NEURONS

H.D. Lux

Department of Neurophysiology
Max-Planck-Institute for Psychiatry
8033 Planegg, F.R.G.

Ca-dependent excitability is frequent in neurons and the expression
of different Ca channels adds importantly to the arsenal of voltage-
gated membrane channels that define neuronal types. A study of the development of Ca channels may thus provide some insight into the process by
which cells differentiate to specific neurons. Early stages of a developing neuron are difficult to assess for principal reasons. This is
because morphological cell properties which characterize neuronal differentiation cannot a priori be assumed to develop in collateral to membrane channels that provide the substrate for neuronal excitability. It
is thus necessary to study precursive stages of cells which are known to
differentiate to neurons.

Neuronal and glial precursors are frequently studied on cells of
the neural crest, a well identified structure that arises from the
neural primordium at early embryonic times. The cells are, however,
morphologically pluripotential (see Le Douarin and Teillet, 1974; Noden,
1978; Schweizer et al. 1983). Differentiation of neuronal properties,
which in part depends on the environment during migration (see Furshpan
et al. 1976; Patterson, 1978, for autonomic neurons), is not well predictable from initial stages. On the other hand, it is known that neuroblasts which are possibly committed to develop to specific neurons
(neuronal progenitors) exist in embryonic peripheral ganglia for some
time (Le Lièvre et al. 1980; Rohrer et al. 1985). Such progenitors, if
selectable, should be suited for a developmental study of voltage-activated membrane currents that later characterize neurons. Results from
these cells could eventually be used to evaluate findings made in other
populations of developing neurons.

Another point of interest is the relationship of neuronal growth and
voltage-dependent Ca currents. For technical reasons, it appeared profi-
table in this regard to choose a cell line in which growth is strictly
regulated by a nerve growth factor (NGF) and in which growth cones are
well accessible (Streit and Lux, 1987). Although the chosen preparation
(PC12) from rat phaeochromocytoma cells is non-neuronal, some principal
findings may well apply also to neurons.

Development of Ca and Na currents in progenitors of DRG neurons

Neuronal progenitors (Rohrer et al. 1985) can be received from avian
dorsal root ganglion (DRG) cells even some time after their embryonic
migration from the neuronal crest and confinement to ganglia. The dissociated preparation at embryonic day 6 contains differentiated neurons,
glial elements, hystiocytes and immature nonspecific cells. Neuronal progenitors, if present, should be amongst the latter. The availability of
strongly cytotoxic antibodies to both differentiated neurons (IgM Immuno-
globin Q211, Henke-Fahle, 1983; Rösner et al. 1985) and glia cells of
chick (IgM O4, Sommer and Schachner, 1981; Rohrer and Sommer, 1983) is of
considerable advantage to select, by eliminating all differentiated glial
cells and neurons, a cell population which, in fact, fact, contains
neuronal progenitors (Rohrer et al. 1985). These are majority cells
(apart from histiocytes) amongst the cells that remain after the killing
procedure and are identified by their differentiation to neurons within
one day in culture. The differentiation is characterized by the appearance of morphological neuronal markers such as neurofilaments and binding
of Tetanus toxin. Differentiated progenitor cells also develop neuron-
specific surface antigens which then make them accessible to killing by
the antibody.

The investigation of voltage-dependent membrane currents of these
progenitors revealed the quite early appearance or even pre-existence of
a transient Ca current which is activated only under the provision of a
sufficiently negative holding potential. This fully inactivating inward
current (Fig.1) is in all aspects similar to the low-voltage activated
(LVA) Ca current recently found in developed DRG cells (Carbone and Lux,
1984; 1987; Nowycky et al. 1985; Fedulova et al. 1985; Bossu et al. 1985)
and in other neurons of the CNS (see for review Yaari et al. 1987). The
time course of the expression of the LVA-Ca channel cannot be directly
assessed. However, it is observed that shortly (within 20 min to 2 h)

after 'killing', about 50% of the cells (with a large deviation within
the individual preparations) show this Ca current while it is present in
nearly all cells 6 hrs later. This span of time compares well with the
relatively rapid time course of neuronal differentiation. This is indicated by a five-fold increase within 24 hrs of the number of cells which
are stained with neuronal markers including the antibody and which then
almost comprise the neuronal entity (Rohrer et al. 1985). Similarly the
number of cells devoid of any specific marker including the progenitor

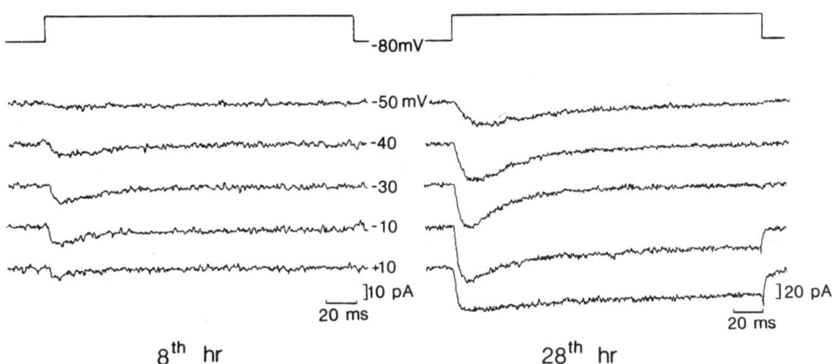

Fig. 1. Ca-current recordings of progenitor cells in the process of differentiation, 8 and 28 hours after elimination of already differentiated
neurons and glia cells. Ca currents at the 8th hour are solely of the
transiently activated type (LVA channels) and were seen to disappear
with reduced holding potential (−60 mV). The Ca currents at the 28th
hour show both LVA and HVA components. The HVA current manifested by its
slow, incomplete inactivation appears with stronger depolarizations, see
traces at −10 and +10 mV. Externally: 5 mM $CaCl_2$, 2 mM MgCl, Choline 120
mM, 0 Na^+_o, TTX 3 μM. Internally: 20 mM TEA, 110 mM N-methyl-D-Glucamin.
pH 7.3.

cells sharply decrease from embryonic age d 6 to d 7 towards barely
detectable levels. All this suggests that the scatter in the age distribution of progenitor cells under the experimental conditions may be not
much larger than half a day which could also be the time needed for the
expression of the LVA Ca channel. The density of Ca channels appears to
be relatively low at early times (within 3 hours after killing) but increases within half a day to values usually found in untreated DRG
neurons of similar preparations (see Carbone and Lux, 1987).

The well known high voltage-activated, slowly and incompletely inactivating Ca channel (HVA channel) needs more time to develop. This
ubiquitous type of Ca channel (see for review Hagiwara and Byerly, 1981;
Eckert and Chad, 1984) was absent during the first 10 hours after 'killing'. However, between 20 to 30 hrs it reached or surpassed the amplitudes of the LVA Ca current at comparable membrane potentials (Fig.1).

It is noteworthy that the delayed appearence of the HVA channel
could be mistaken as a recovery from rundown during the dissociation and
plating procedure or even during the recording period (see Carbone and
Lux, 1987). However, contrary to observations on progenitor cells, a
significant HVA Ca current which was usually larger than its LVA coun-
terpart was regularly observed in differentiated DRG neurons, immediately after plating and in the first 10 min of the recordings. The HVA
current of the progenitor cell thus showed a distinct lag time in
expression with a late rise. A similar observation was made with the
voltage-activated Na channel. Thus, in the first hours of the development of a progenitor cell towards a DRG neuron, electrical excitability
is confined to the LVA Ca channel. Voltage-activated K channels were
observed at most early times but were not especially investigated.

Observations on embryonic hippocampal neurons

Heterogeneity of Ca currents with early presence of the LVA channel
and late appearance of the classical type was recently observed also in
rat hippocampal cells at embryonic age d 18-19 (Yaari et al. 1987). The
cells appear morphologically undifferentiated at this stage, but most of
them are probably committed to become neurons because the contribution
of glia in fresh cultures under the chosen conditions was small. Whole-
cell Ca currents could be activated at relatively low membrane potentials
(-50 to -40 mV) under the conditions of a sufficient negative holding
potential (-90 mV). Activation and inactivation properties of these
early Ca currents were in any respect similar to those of LVA currents
of DRG neurons. The HVA Ca current, if present, was rather small during
the first hours after plating, suggesting some similarity with the
findings on DRG progenitor cells.

After attaching to the substrate, most cells acquired neurite
extensions within a day and intricate sprouting continued during the
first week in culture. This process was associated with the appearance
and gradual increase of the HVA Ca current component. After 24 to 48
hours in culture, HVA-Ca currents were usually larger than the LVA Ca

currents. As with DRG progenitor cells, the Ca current components could
be discriminated on the basis of their time- and voltage-dependent activation and inactivation characteristics. Some results of pharmacological
separation are presented below.

Pharmacological properties

When extracellular Ca was replaced with an equimolar concentration
of Ba, the rat hippocampal LVA currents became reduced, suggesting that
LVA channels are less permeable to Ba^{2+}. In chick DRG cells, the LVA Ca
channel shows similar permeability to Ba and Ca ions (Carbone and Lux,
1987). In contrast, HVA currents in both preparations are enhanced in
this condition. Presumably Ba ions pass through these HVA channels more
easily than Ca ions (Hagiwara and Byerly, 1981, 1983; Eckert and Lux,
1976).

Separation of the neuronal Ca current components by known Ca antagonists proved to be difficult. Both types of Ca currents were strongly
suppressed by 100 µM cadmium. The organic Ca antagonist verapamil reduced LVA and HVA currents of hippocampal neurons to a similar degree at
100 µM, but in DRG neurons the HVA current was preferentially affected.
However, the anticonvulsant drug phenytoin, at a similar dose, more
selectively reduced the LVA Ca current (see Yaari et al. 1987). The
effects of these agents did not appear to depend on age of the cells and
were readily reversible upon washing.

Dihydropyridines (0.05 to 10 µM Nifedipine and Bay K) failed to
excert significant effects which could be separated from run down occurring with the HVA channel. This applies to the neuronal preparations as
well as to the PC12 cells (N.W. Davies, H.D. Lux, M. Morad, J. Streit,
unpublished). Particular use- and voltage-dependent actions were also
not obvious. The substances used were found to be active at even lower
concentrations in similar experiments on heart myocytes. We conclude
that these drugs are not useful for separating neuronal Ca currents.

Ca currents and cell growth induced by nerve growth factor (NGF)

In PC12 cells, a cell line from rat pheochromocytoma, growth is
known to be strongly stimulated by NGF with some reversibility. A con-
ditional role of internal Ca on the effects of NGF (by a cAMP-induced Ca
mobilization) has been suggested (Schubert et al. 1978), but effects of
NGF on transmembrane Ca fluxes were not observed (Landreth et al. 1980).

However, NGF-induced alterations of either properties or distribution of
voltage-dependent Ca channel were recently proposed (Takahashi et al.
1985; Kongsamut and Miller, 1986) in studies of Ca-dependent release of
catecholamines. There is also evidence for the involvement of Ca currents
in the regulation of growth cone sprouting in neuronal cells. An increase
in the area of growth cones is reported to occur with depolarization in
the presence of high Ca_o (Anglister et al. 1982). Furthermore, field
potentials close to active growth cones of cultured retinal ganglion
cells appear to originate from steady Ca-dependent inward currents into
growth cone (Freeman et al. 1985). Evidence for the existence of the
proposed voltage-activated Ca channels in growth cones is thus far provided only by measurements of Ca-dependent action potentials or voltage-
dependent rise of intracellular Ca in or close to growth cones of various cells (Sugimori and Llinas, 1981; Meiri et al. 1981; Grinvald and
Farber, 1981; Anglister et al. 1982; Bolsover and Spector, 1986).
However, no significant Ca currents were found in growth cones of
cultured _Aplysia_ neurons (Belardetti et al. 1986).

In PC12 cells, Streit and Lux (1987) compare the Ca currents of
cell somata and growth cones in an approach to the question of their
relationship with NGF-induced growth of extensions. This study takes
advantage of the reversible effects of nerve growth factor (NGF) on
these cells.

Growth cones

In medium containing serum but no NGF, PC12 cells survive and
divide, but grow only few, short processes. When NGF is added to the
medium, the cells start to grow long processes with relatively large
growth cones at their tips (for a review see Greene and Tischler, 1982).
The appearance of these processes closely resemble the neurites of
neuronal cells and for this reason will be referred to here as neurites.
The first clearly identifiable growth cones could be seen on the second
day of NGF treatment. They varied considerably in size and form (see
O'Lague et al. 1985), but appeared to maintain their configuration
during further neurite growth. To ensure the currents originated from
the growth cones, and to avoid disturbing electrical influences from the
neurites (notch currents), growth cones were mechanically isolated from
the neurites as close as possible (5 μm) to the cone.

Fig. 2 shows Ca currents at different voltage steps as recorded
from an isolated growth cone in the presence of Tetrodotoxin (TTX) and

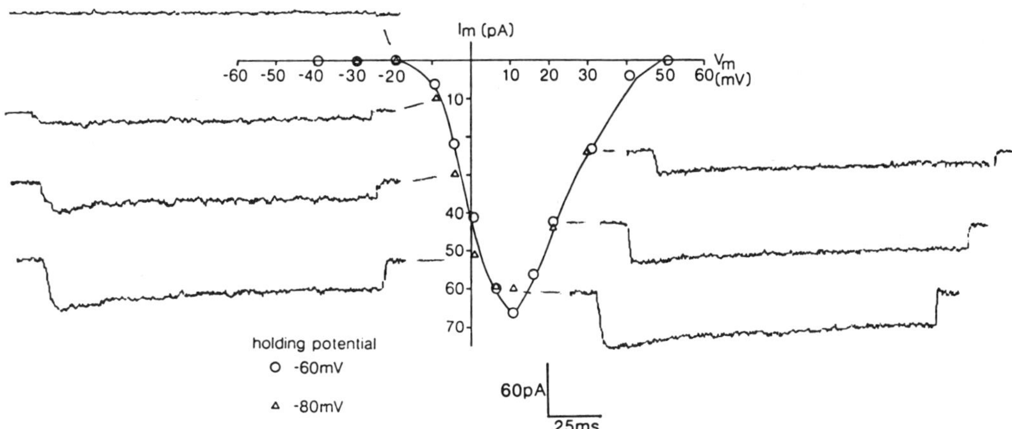

Fig. 2. Voltage-dependent Ca currents in an isolated PC12 growth cone. Original current traces recorded in the presence of TTX and TEA, during depolarizing voltage steps to indicate potentials. Holding potential was -80 mV. Inserted the current voltage relationship from the same growth cone with holding potentials of -60 mV (O) and -80 mV (△). The bath contained 10 mM $CaCl_2$.

TEA. The currents strongly resembled the classical high-voltage-activated Ca currents seen in many neuronal cells (Brown et al. 1982). There was no evidence for the existence of a low-voltage-activated Ca current, either in the growth cones or in the somata of PC12 cells. No difference in activation properties were observed if the holding potentials are varied between -60 and -100 mV and the Ca currents showed no sign of a fast inactivating component such as characteristic for a contribution of the LVA Ca channel.

NGF-induced effects on soma currents

Current recordings including those of growth cones were compared in groups of cells over various lengths of time, either with or without NGF (Streit and Lux, 1987). On the first day after subculturing the cells, the Ca currents are usually small with little time-dependent inactivation. Neither transient nor steady state activation or inactivation were significantly altered by NGF at this time. Because it is known that NGF removal promptly arrests motility and sprouting of growth cones and that these activities are restored within minutes of reapplication of NGF

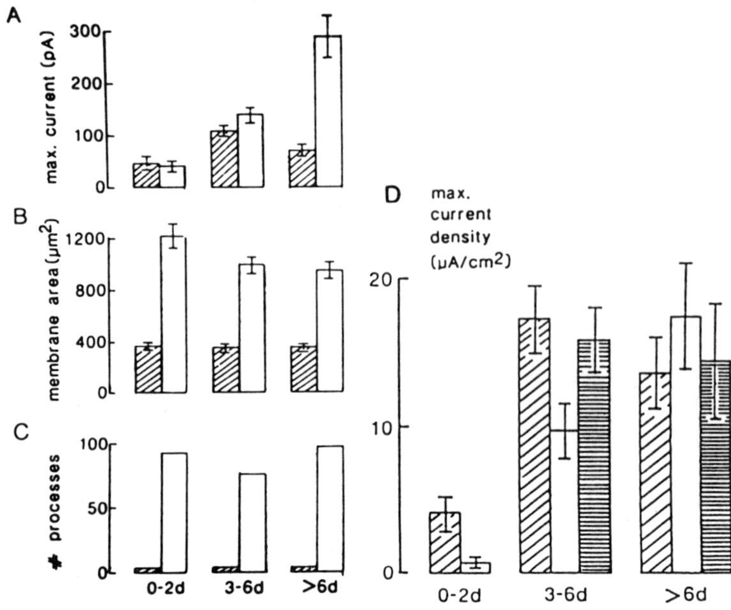

Fig. 3. Comparison between current amplitudes and cell growth either
with (☐) or without (▨) NGF for various times. A) Mean current amplitudes (mean ± SEM) from a total of 8 to 20 experiments per group. A
significant effect of NGF was observed ($p \leq 0.01$, W-test) in the last
group (>6d). B) Mean membrane areas of the cell somata from a total of
29 to 33 cells in each group. Cell processes were not taken into account. The membrane area was calculated from measurements of the cell
diameters assuming an elipsiodal form of the somata. C) The number of
processes, normalized to 100 cells, from a total of 23 to 33 cells in
each group. Only processes longer than 10 µm were considered. The mor-
phometric data were obtained from photographs of culture dishes.
D) Mean current densities of cell bodies and growth cones (mean ± SEM)
from a total of 5 to 12 cells per group. A significant difference
($p \leq 0.05$, W-test) was observed between the first two groups (0-2d); the
differences between the other groups were insignificant. (▨) somata
without NGF; (☐) somata with NGF; (▤) growth cones.

(Greene and Tischler, 1982), NGF was occasionally directly applied onto

the cells but no effects on Ca currents was observed during a period up

to 40 min.

Within two or three days after the last subculturing procedure, the

currents increased independent of NGF treatment. This observation could

be made in each new passage of the cells which suggests a recovery pro-

cess from the subculturing procedure which had decreased the number of

voltage-sensitive Ca channels. To separate such effects from eventual

actions of NGF, a series of experiments was performed whereby NGF was

applied four days after subcultivation of the cells. At this time, the

mean current amplitude had considerably increased but the currents were

still unaffected by NGF during the first two days of NGF application.

The Ca current amplitude of the NGF-treated cells increased for the

duration of the NGF treatment (see Fig.3) reaching a steady maximum at
about the 12th day of passage. Usually the inactivation of Ca currents
of the NGF-treated cells was also more pronounced than those of untreated cells.

Soma growth, as well as outgrowth of neurites, occurs within the
first two days of NGF application. A comparison between the mean Ca
current amplitudes of the cells and the cell growth at various times in
the dishes both in the presence and absence of NGF is shown in Fig.3.
The mean current amplitudes are received for each group from the maximum
current amplitudes of the individual experiments. The mean soma membrane
area values are estimated from measuring the individual cell diameters
assuming an ellipsoidal geometry. The average number of cell processes
longer than 10 μm was determined for the population of cells of each
group. The results suggest that a significant increase of both somatic
area and number of neurites occur only within the first two days of NGF
treatment. However, no differences in current amplitude, time course or
current-voltage relationship were observed within this period. At later
times growth appears to be concentrated in the growth cones where the
neurite elongation is produced.

In summary, no evidence was obtained for the hypothesis that an
alteration of Ca currents is involved in the activation of growth in
PC12 cells by NGF. In addition, the number of Ca channels does not seem
to increase simultaneously with cell growth. The increase in membrane
currents is always observed to occur later than the increase of the
somatic membrane area and the sprouting of neurites. This result suggest
that incorporation of new Ca channels follows with a delay of several
days.

In the growth cone area, the Ca channel density stays fairly constant, however (Fig.3D), and is comparable to that of the soma after 3
days of NGF application. This quantitative comparison of current densities was verified by capacitance measurements of the membrane areas
(Streit and Lux, 1987).

Conclusions

Purified populations of neuronal precursors such as reported can be
used in vitro to analyse the mechanisms involved in differentiation. It
is possible that the time course by which Ca channels develop indicates
the differentiation to specific types of neurons but a general statement
would demand a greater survey of identified neuronal precursors. Undif-

ferentiated cells usually attain morphological neuronal properties
before they become electrically excitable by the development of the Na
system (see Ziller et al. 1983; Bader et al. 1983). A generalization in
this regard does not seem appropriate, however, since a part of the
voltage-dependent Ca system precedes the acquirement of a neuronal
morphology. This holds true also for a parasympathetic preparation (I.
Dietzel, K. Gottmann, H.D. Lux and H. Rohrer, unpublished). Whether
these examples of pre-neuronal populations mark a rule or exceptions
from it remains to be established.

The different time courses of the expression of the two identified
Ca currents strengthen suggestions that these are separate entities.
Whether the earlier expression of the transiently activated, low-voltage
activated Ca channel in the studied neuronal precursors reflects a
commitment to a particular function in developmental processes is not
known.

The incorporation of other Ca channels appears to be delayed also
in respect to growth of both somatic membranes and extensions and they
are less valuable as early markers of neuronal differentiation. From the
results on PC12 cells, an increase in incorporation is concluded to
occur even at times when NGF-stimulated growth comes to arrest. However,
the growth cone may be an exception. If Ca channels in the growth cones
were to appear with delay, one would expect to find active growth cones
of lower current density than that observed in non-growing cell bodies.
However, growing cones showed Ca current densities which were maximum
for the cell soma. It is thus likely that Ca channels are preferentially
incorporated in the frontier of sprout, the growth cone. Whether such
clustering of Ca channels is induced by NGF before sprouting commences,
remains an interesting possibility.

ACKNOWLEDGEMENT

Supported by the Deutsche Forschungsgemeinschaft, SFB 220/A1

REFERENCES

Anglister, L., Farber, I.C., Shahar, A., Grinvald, A., 1982, Localiza-
     tion of voltage-sensitive calcium channels along developing
     neurites: their possible role in regulating neurite elongation,
     Dev. Biol., 94:351.
Bader, C.R., Bertrand, D., Dupin, E., and Kato, A.C., 1983, Development
     of electrical membrane properties in cultured avian neural crest,
     Nature, 305:808.

Belardetti, F., Schacher, S., Siegelbaum, S.A., 1986, Action potentials, macroscopic and single channel currents recorded from growth cones of aplysia neurons in culture, J. Physiol., 374:289.

Bolsover, S.R., and Spector, I., 1986, Measurements of calcium transients in the soma, neurite, and growth cone of single cultured neurons, J. Neurosci., 6:1934.

Bossu, J.L., Feltz, A., and Thomann, i.M., 1985, Depolarization elicits two distinct calcium currents in vertebrate sensory neurons, Pfluegers Arch., 403:360.

Brown, A.M., Camerer, H., Kunze D.L., and Lux, H.D., 1982, Similarity of unitary Ca^{2+}-currents in three different species. Nature, 299:156.

Carbone, E., and Lux, H.D., 1984, A low voltage-activated fully inactivating calcium channel in vertebrate sensory neurons, Nature, 310:501.

Carbone, E., and Lux, H.D., 1984, A low voltage-activated calcium conductance in embryonic chick sensory neurons, Biophys. J., 46:413.

Carbone, E., and Lux, H.D., 1987, Kinetics and selectivity of a low-voltage-activated calcium current in chick and rat sensory neurones, J. Physiol., 386:547.

Carbone, E., and Lux, H.D., 1987, Single low-voltage-activated calcium channels in chick and rat sensory neurones, J. Physiol., 386:571.

Eckert, R., and Chad, J.E., 1984, Inactivation of Ca channels, Prog. Biophys. mol. Biol., 44:215.

Eckert, R., and Lux, H.D., 1976, A voltage-sensitive persistent calcium conductance in neuronal somata of Helix, J. Physiol., 254:129.

Fedulova, S.A., Kostyuk, P.G., and Veselovsky, N.S., 1985, Two types of calcium channels in the somatic membrane of new-born rat dorsal root ganglion neurones, J. Physiol., 359:431.

Freeman, J.A., Manis, P.B., Snipes, G.J., Mayes, B.N., Samson, P.C., Wikswo, J.P., and Freeman, D.B., 1985, Steady growth cone currents revealed by a novel circulary vibrating probe: A possible mechanism underlying neurite growth. J. Neurosci. Res., 13:257.

Furshpan, E.J., MacLeish, P.R., O'Lague, P.H., and Potter, D.D., 1976, Chemical transmission between rat sympathetic neurons and cardiac myocytes developing in microculture: evidence for cholinergic, adrenergic and dual-function neurons, Proc. Natl. Acad. Sci. USA, 73:4225.

Greene, L.A., and Tischler, A.S., 1982, PC12 pheochromocytoma cultures in neurobiological research, Adv. Cell Neurobiol., 3:373.

Grinvald, A., and Farber, I.C., 1981, Optical recordings of calcium action potentials from growth cones of cultured neurons with a laster microbeam. Science, 212:1164.

Hagiwara, S., and Byerly, L., 1981, Calcium channel, Annu. Rev. Neurosci., 4:69.

Hagiwara, S., and Byerly, L., 1983, The calcium channel, Trends Neurosci., 6:189.

Henke-Fahle, S., 1983, Monoclonal antibodies recognize gangliosides in the chick brain, Neurosi. Lett., Suppl.14:S160.

Kongsamut, S., and Miller, R.J., 1986, Nerve growth factor modulates the drug sensitivity of neurotransmitter release from PC12 cells. Proc. Natl. Acad. Sci. USA, 83:2243.

Landreth, G., Cohen, P., and Shooter, E.M., 1980, Ca^{2+} transmembrane fluxes and nerve growth factor action on a clonal cell line of rat pheochromocytoma. Nature, 283:202.

Le Douarin, N.M., and Teillet, A.M., 1974, Experimental analysis of the migration and differentiation of neuroblasts of the autonomic nervous system and of neurectodermal mesenchymal derivatives, using a biological cell marking technique, Dev. Biol., 41:162.

Le Lièvre, C.S., Schweizer, G., Ziller, C.M., and Le Douarin, N.M.,

1980, Restrictions of developmental capabilities in neural crest
derivatives as tested by in vivo transplantation experiments.
Dev. Biol., 77:362.

Meiri, H., Parnas, I., and Spira, M., 1981, Membrane conductance and
action potential of regenerating axonal tip. Science, 211:709.

Noden, D.M., 1978, The control of avian cephalic neural crest cytodif-
ferentiation. Dev.Biol., 67:313.

Nowycky, M.C., Fox, A.P., and Tsien, R.W., 1985, Three types of neuronal
calcium channel with different calcium agonist sensitivity,
Nature, 316:440.

O'Lague, P.H., Huttner, S.L., Vandenberg, C.A., Morrison-Graham, K., and
Horn, R., 1985, Morphological properties and membrane channels of
the growth cones induced in PC12-cells by nerve growth factor.
J. Neurosci. Res., 13:301.

Patterson, P.H., 1978, Environmental determination of autonomic neurotransmitter functions, Ann. Rev. Neurosci., 1:1.

Rösner, H., Al-Aqtum, M., and Henke-Fahle, S., 1985, Developmental
Expression of GD3 and polysialogangliosides in embryonic chicken
nervous tissue reacting with monoclonal antiganglioside antibodies. Dev. Brain Res., 18:85.

Rohrer, H., Henke-Fahle, S., El-Sharkawy, T., Lux, H.D., and Thoenen, H.,
1985, Progenitor cells from embryonic chick dorsal root ganglia
differentiate in vitro to neurons: biochemical and electrophysiological evidence,, EMBO J., 4:1709.

Rohrer, H., and Sommer, I., 1983, Simultaneous expression of neuronal
and glial properties by chick ciliary ganglion cells during
development. J. Neurosci., 3:1683.

Schubert, D., LaCorbiere M., Whitlock, C., and Stallcup, W., 1978, Alterations in the surface properties of cells responsive to nerve
growth factor. Nature, 273:718.

Schweizer, G., Ayer-Le Lièvre, C.S., and Le Douarin, N.M., 1983, Restric-
tion of developmental capacities in the dorsal root ganglia
during the course of development. Cell Differentiation, 13:191.

Sommer, I., and Schachner, M.S., 1981, Monoclonal antibodies (O1 to O4)
to oligodendrocyte cell surfaces: an immunocytological study in
the central nervous system. Dev. Biol., 83:311.

Streit, J., and Lux, H.D., 1987, Voltage dependent calcium currents in
PC12 growth cones and cells during NGF-induced cell growth.
Pflügers Arch., 408:634.

Sugimori, M., and Llinas, R., 1981, Localization of ionic conductances
in soma and denritic regions of Purkinje cells: an in vitro study
in guinea pig cerebellar slices. Neurosci. Abstr., 7:76.

Takahashi, M., Tsukui, H., and Hatanaka, H., 1985, Neuronal differentiation of Ca^{2+} channel by nerve growth factor. Brain Res., 341:381.

Yaari, Y., Hamon, B., and Lux, H.D., 1987, Development of two types of
calcium channels in cultured mammalian hippocampal neurons.
Science, 235:680.

Ziller, C., Dupin, E., Brazeau, P., Paulin, D., and Le Douarin, N.M.,
1983, Early segregation of a neuronal precursor cell line in the
neural crest as revealed by culture in a chemically defined
medium, Cell, 32:627.

GENERATION OF NEURONAL ARCHITECTURE: IONIC REGULATION OF

GROWTH CONE BEHAVIOR

Stanley B. Kater

Program of Neuronal Growth and Development
and Department of Anatomy
Colorado State University
Fort Collins, Colorado   80523

INTRODUCTION

A wealth of knowledge has developed from studies of ion channels on single cells ranging from Paramecium to identified snail neurons. Such studies have given rise to the concept of the single cell as a systems integrator, responsive to environmental change, and capable of precise effector behavior. Recent years have seen the extension of this view to developing systems and we have begun to ask **"To what extent are mechanisms employed in the normal coding processes of adult neurons also used in developing and regenerating neurons to influence differentiation and final connectivity patterns?"** This line of investigation has disclosed a high degree of parallelism between developing neurons and their adult counterparts.

This chapter will review a series of observations on the snail <u>Helisoma</u> which support the view that developing neurons employ mechanisms quite similar to those employed by adult neurons. In development, however, they integrate not for specification of a neural code, but rather for the generation of the neuronal architecture, which will then become the basis of their adult coding capability. Our laboratory has employed embryonic and regenerating neurons <u>in situ</u> and in cell culture and found numerous parallels with established neural coding. For instance: 1) Adult neurons employ neurotransmitters to coordinate the activity among sets of neurons, and these same agents are used developmentally to coordinate the outgrowth of specific neurons in order to establish such circuits (Haydon, *et al.*, 1984 and 1987, Goldberg, 1985 McCobb, *et al.*, 1985, McCobb and Kater, 1986). 2) Action potentials are the conventional means of long distance communications in established circuits, and action potentials play prominent roles in stabilizing neurite outgrowth in developing circuits (Cohan and Kater, 1986). As might be expected, individual ion channels located in specific regions of developing neurons also appear to play prominent roles (Cohan, Haydon, and Kater, 1985) as they also will for adult function. 3) As a final, striking, parallel between developing and established neural circuits, intracellular calcium is now demonstrated to play a most prominent role in the growth status of individual growth cones in a fashion highly reminiscent to the kind of control exerted in synaptic transmission (Cohan, Connor, and Kater, 1986 and 1987).

The aim of this paper is to describe the system of identified neurons in the snail, <u>Helisoma trivolvis</u>, which we have used to make these statements on the use of parallel integrative processes in adult and developing neurons. After summarizing the data leading to our present knowledge of neurotransmitters and action potentials as developmentally relevant signals, this chapter will present our working hypothesis that the behavior of neuronal growth cones is controlled by the intracellular concentration of free calcium.

The investigations described in this paper make use of the ability to identify individual neurons within ganglia of the fresh water snail, <u>Helisoma trivolvis</u>. It has been possible to study the same identified neurons not only from animal to animal with a high degree of consistency, but also under a variety of degrees of neural plasticity. Figure 1 summarizes the conditions under which we study both structural and electrophysiological phenomena. Individual neurons can be identified and studied in the intact ganglia in their stable configurations or during neural plasticity evoked by axotomy. A luxuriant neural outgrowth fills ganglia in response to axotomy. In the context of the complex nervous system, it is difficult, if not impossible, to examine the controls of such outgrowth. We developed methods (Wong, *et al.*, 1981) to study individual identified neurons under the more defined conditions of cell culture. Identified neurons can be excised from intact ganglia and plated as individual, spherical somata in isolated cell culture.

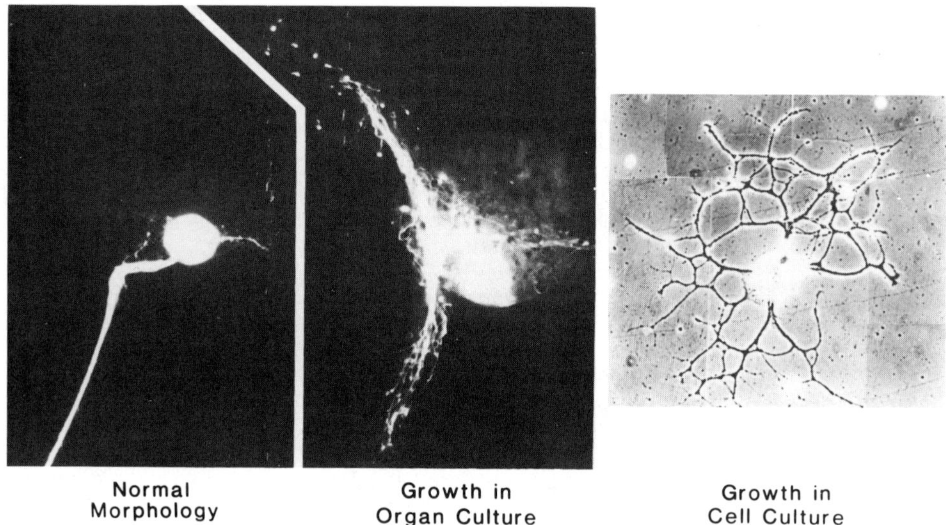

Fig. 1. Buccal ganglia neuron B5 can be examined and manipulated under widely divergent conditions. Our strategy has been to compare individual identified neurons under different environmental conditions. Within normal ganglia, one can examine both physiological and morphological aspects of a normal neuron by standard intracellular recording and dye-injection techniques. Similar methods can be used in axotomized, growing neurons within buccal ganglia either <u>in situ</u> or in organ culture. Such neurons grow profusely (middle panel) and form definable new connections. Any given neuronal cell body can be removed from a normal ganglion and placed in cell culture; (right panel), both physiological and morphological aspects of such neurons can be examined under precisely defined experimental conditions. (From Kater, 1985).

Individual isolated <u>Helisoma</u> neurons remain as spherical somata in defined medium. On the other hand, in the presence of brain conditioned medium (Wong, *et al.*, 1981) growth cones emerge from the spherical cell body and profuse outgrowth is evoked (Fig. 2). Within two hours of plating of individual neurons in cell culture individual neurites begin to emerge. For the next forty-eight hours, outgrowth usually proceeds at relatively constant rates. Growing neurites are tipped with motile growth cones composed of a broad, flattened lamellipodium and numerous filopodia. After about 48 hours of outgrowth, growth cones transform and appear as a phase bright ovoid structure, and elongation ceases. At this point the neurons have achieved a condition that we regard as "stable state".

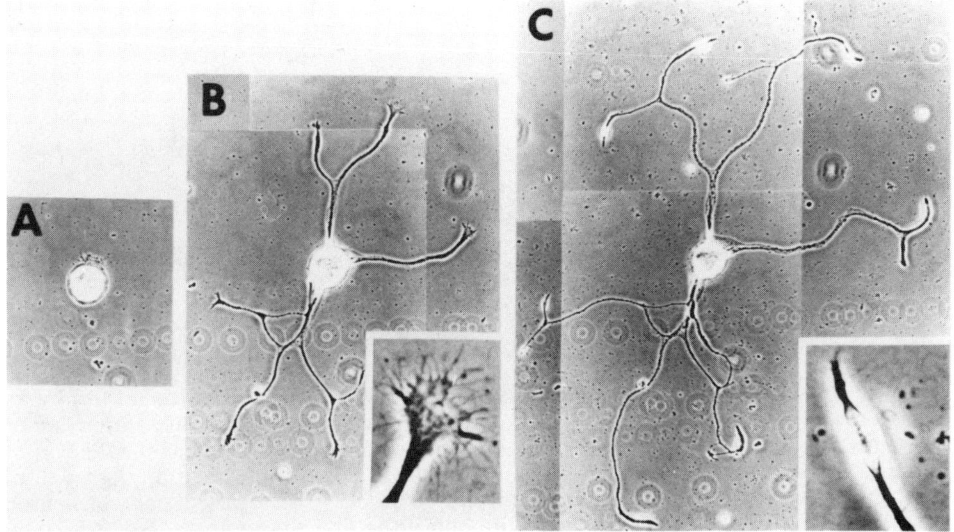

Fig. 2. Characteristic stages of outgrowth of Isolated neuron B5. A) Within two hours of plating veils form around the spherical cell body. B) Active growth occurs after initiation of sprouting. Growth cones (insets) are characteristically large, flat, and phase-dark during active growth. C) A stable morphology is reached after which no more growth occurs. The stable state is characterized by slender, phase bright endings in place of the broad flattened motile growth cones. (From Hadley, Bodnar, and Kater, 1985)

NEUROTRANSMITTER REGULATION OF NEURONAL GROWTH CONES

A set of fortuitous observations in neuronal growth patterns led us to test the role of neurotransmitters in development. Neurons plated in cell culture can display quite predictable morphology.

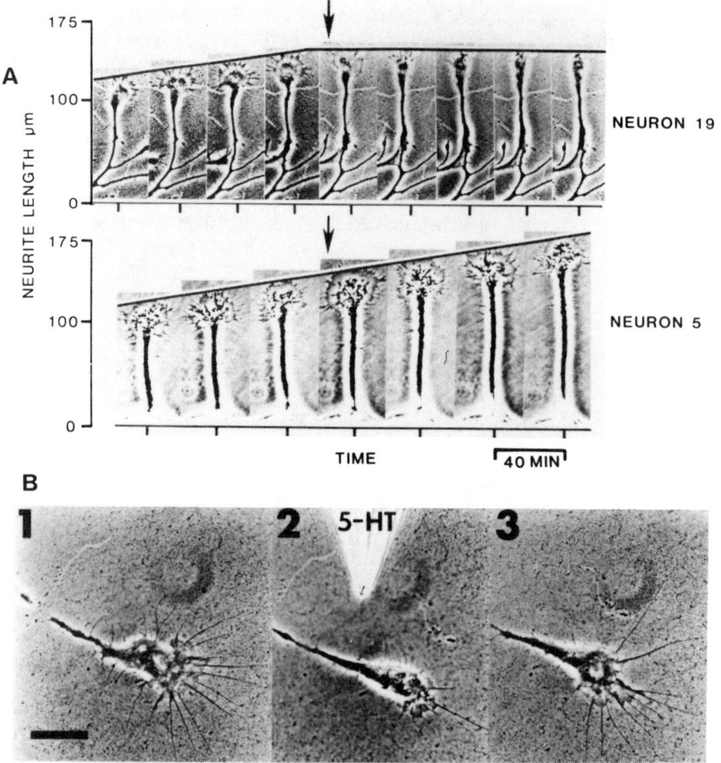

Fig. 3. The responses of intact (A) and isolated (B) growth cones to serotonin. (A) Intact growth cones that are connected to the neuron (not shown) displayed at frame intervals of 40 minutes. Intact growth cones produce neurite outgrowth at a constant rate before serotonin treatment. Application of serotonin (arrows) inhibits the neurite outgrowth of neuron B19 (top) but has no effect on neuron B5 (bottom). (B) A growth cone of neuron B19 isolated by severing the interconnecting neurite with a micropipette (scratch, B1). Isolated growth cones are inhibited by pipette application of serotonin (B2), but after withdrawal of the pipette they resume their characteristic activity (B3). Serotonin's effects on isolated and intact growth cones are virtually indistinguishable, indicating that the growth cones of neuron B19 are directly responsive to serotonin. Calibration bar, 10 µm. (From Haydon et al., 1984). Copyright 1986 by the AAAS.

When pairs of neurons are plated in culture together, there can be significant deviations from expected outgrowth patterns. We tested the possibility that factors released from one neuron could influence the growth pattern of another neuron. In particular, we developed the hypothesis that neurotransmitters might be released from one neuron in culture and affect the growth cones of another. Our first experiments were performed with the neurotransmitter serotonin (Haydon, et. al, 1984 and 1987). These experiments revealed unequivocally that growth cones from specific identified neurons were affected in a predictable fashion by the neurotransmitter serotonin. Growth cones of neuron B5s, for instance, are never affected, even by high concentrations of serotonin. In stark contrast, growth cones of neurons such as neuron B19 immediately ceased outgrowth in response to the presence of serotonin at levels as low as 10^{-7}M (Figure 3). These effects are mediated directly by the neuronal growth cone. Growth cones can be isolated from the cell body simply by cutting the neurites (e.g., Bray 1982). Under these conditions, the isolated growth cones continue their motile behavior and their responsiveness to serotonin. Subsequent to these initial investigations on serotonin, we have found that other neurotransmitters can affect specific matrices of identified neurons. Dopamine can act as an additional regulator of neurite outgrowth on a different but overlapping set of neurons from the buccal ganglia of <u>Helisoma</u> (McCobb, *et al.*, 1985). Additionally, acetylcholine, while it has no direct effects on the outgrowth of the neurons tested can block the serotonin inhibition of growth cone motility (McCobb and Kater, 1986). Taken together, the diversity of effects of different neurotransmitters on different neurons and the combinatorial affects of sets of transmitters on individual neurons are highly reminiscent of the classical forms of neural integration at the level of membrane potential that are used in functional adult circuits. In this case, however, the final output influenced by these neurotransmitters is the growth status of the neuron.

One can envision during development the existence of environmental influences around nerve cells which act upon growing neurons in a fashion not dissimilar from how they will ultimately come to interact in final adult circuitry. However, at such early stages of the life of the organism the neurotransmitters act to *actually shape the circuitry which will ultimately comprise the adult nervous system*. Indeed, work on <u>Helisoma</u> embryos reaffirms the likelihood that embryonic neurons not only respond to neurotransmitters, but actually use such molecules as developmental signals (Goldberg, *et al.*, 1985).

ACTION POTENTIALS AS REGULATORS OF GROWTH STATUS

It has long been suggested that electrical activities of neurons themselves may play a role in shaping the morphology and connectivity of the nervous system both during development and adult neuroplasticity. The literature in the Neurosciences has many examples of how afferents might alter nervous system function. Electrical activity has been shown to regulate many cellular processes ranging from the synthesis of specific neurotransmitters, to the pattern of synaptic connections. We tested the hypothesis that electrical activity affects neurite elongation in neurons of <u>Helisoma</u> by stimulating individual neurons in isolated cell culture (Cohan and Kater, 1985).

Previous results have demonstrated that impalement with a microelectrode could result in the complete cessation of neurite outgrowth. Using extracellular patch clamp electrodes, it was possible to stimulate and monitor the electrical activity in an identified neuron without the method itself altering the growth pattern (Cohan and Kater, 1986). Concurrent with a patch electrode on the cell body it was possible to optically monitor multiple growth cones of an individual neuron as the neurites elongated. During control periods, the presence of microelectrodes in no way affected growth rates. Evoking action potentials, on the other hand, had profound effects on neurite elongation. Within fifteen minutes of the onset of stimulation, growth cone advance ceased (Fig. 4). The reduction in growth rate resulting from stimulation was highly significant. In fact, over half the growth cones studied showed slight retraction during the stimulation period. When the stimulus was turned off, growth cone advance resumed with growth rates returning to their prestimulus values. From results such as these, it became clear that intrinsic spontaneous activity of a neuron or the activity driven in that neuron by a synaptic partner could have profound effects on the elaboration of neurite morphology. This, therefore, provides a second example of a mechanism normally employed in functional adult neural circuitry which can markedly affect the development of such circuitry.

ION CHANNELS OF NEURONAL GROWTH CONES

The neuronal growth cone is thought to be the primary structure responsible for pathfinding and neurite elongation. We have initiated studies on the ion channels included in the membrane of neuronal growth cones; patch clamp methods were used to study the characteristics of different membrane regions of identified neurons (Guthrie, et al., 1986).

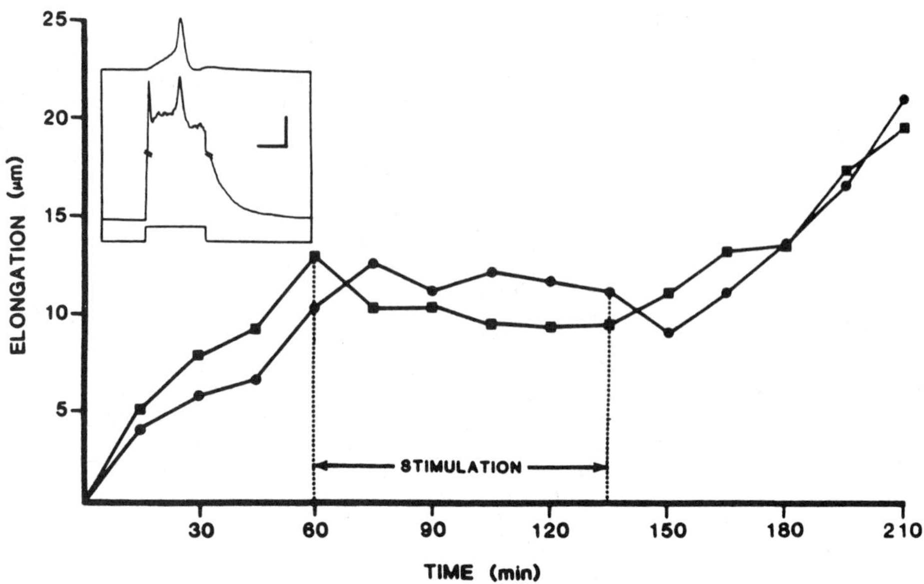

Fig.4. Experimentally evoked action potentials inhibit growth cone advance. Inset: control experiment to determine the stimulus parameters. Stimulus pulses were applied under current clamp conditions through an extracellular patch-pipette sealed to the soma of neuron B19, and an intracellular microelectrode (top trace) was used to monitor the membrane potential. A 20 msec current pulse (bottom trace; amplitude 1.0 nA in this example) through the extracellular pipette (voltage recording on middle trace) caused a small depolarization that evoked an action potential in the soma as recorded through both microelectrodes. Calibration 40 mV, 10 msec. Generation of action potentials (with only the patch pipette present) reversibly suppressed growth cone advance. During the first hour (control), no pulses were passed through the patch-pipette, and neurite elongation proceeded at a constant rate (8.7um per hour). Action potentials evoked by current pulses during the second period stopped neurite elongation. Pulses were turned off during the third period and neurite elongation resumed. (From Cohan and Kater, Copyright 1986 by the AAAS).

We had previously known that the soma of a given identified neuron (both in situ or in cell culture) displayed a characteristic action potential wave form (Cohan, et. al, 1985). Guthrie, et al., (1986) were able to record from isolated growth cones and determined that these structures were also able to generate action potentials. The wave forms of the action potentials of the growth cones of a given neuron were also characteristic of that neuron. One could predict the wave-form of an action potential in a growth cone on the basis of the wave-form of the identified neuronal somata from which it was recorded. While intriguing, the question still remains whether there is a relationship of growth cone ionic currents to growth cone behavior.

In addition to comparing properties of the membrane of different regions of a neuron, we have examined how growth cone properties change as a function of a growth status of a neuron. The morphological changes described in Figure 2 prompted an investigation of physiological changes which might accompany this striking alteration. We initiated an investigation designed to study changes in single channel ionic currents using patch clamp recording techniques (Fig. 5). We have observed at least one ion channel in active growth cones having a conductance of 70 pS (Cohan, et. al, 1985). Recordings from the cell attached patches demonstrated that this channel is normally active in growing growth cones but is not seen in stable growth cones (Fig. 6). This absence is not due to the loss of the channel from the membrane of the growth cones; the channel is present in an inactivatable state. When patches are removed from inactive growth cones, the single channel reappears indicating that conditions of the cytosol in stable neurons may act to mask expression of this individual channel (Fig. 7). While this channel still requires considerable characterization, it is clear from the data presently at hand, that the differences in the state of channels in the neuronal growth cone may provide signals that are responsible for differences in the growth status of the growth cones themselves. A series of experiments now implicates the transmembrane flux of calcium as the key to such a control mechanism. (c. f. Llinas; 1979: MacVicar and Llinas, 1985).

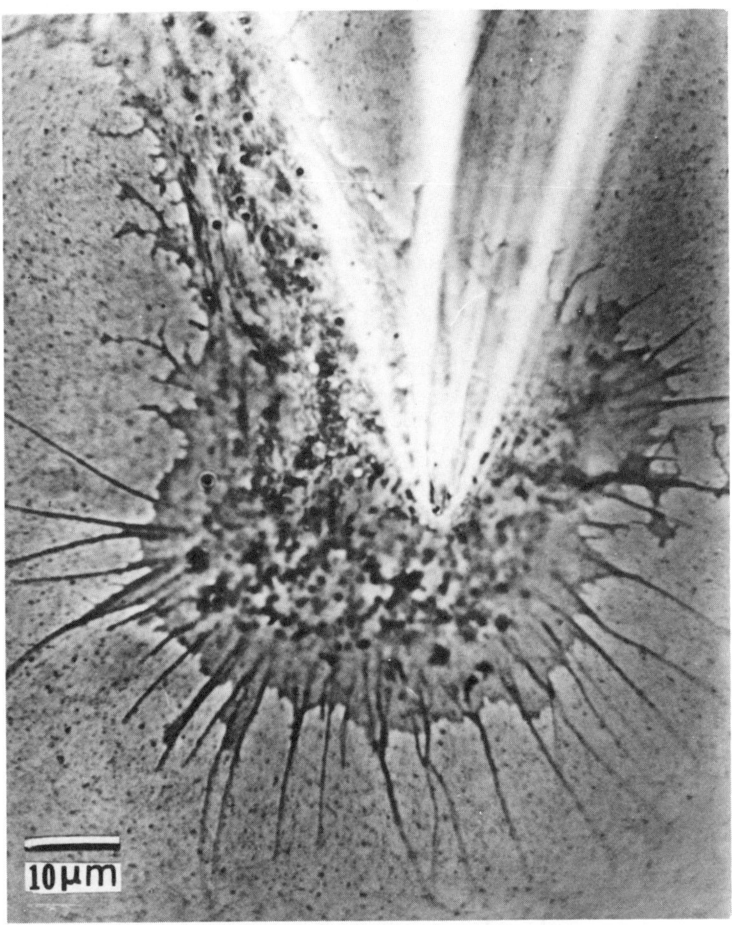

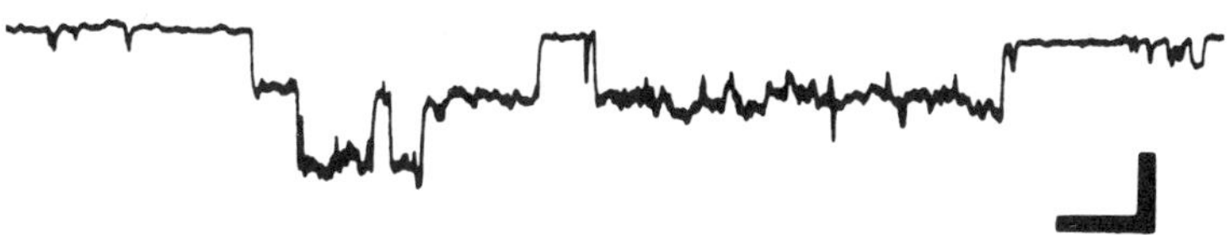

Fig. 5. Patch-clamp recording from a growth cone of _Helisoma_. The upper panel is a photomicrograph of the growth cone of a neuron B5 with a patch-clamp pipette in place. The lower trace shows exemplary records of channel openings and closing as seen in a cell-attached patch-recording configuration. Calibration, 400 msec, 4 pA. (From Kater, 1985).

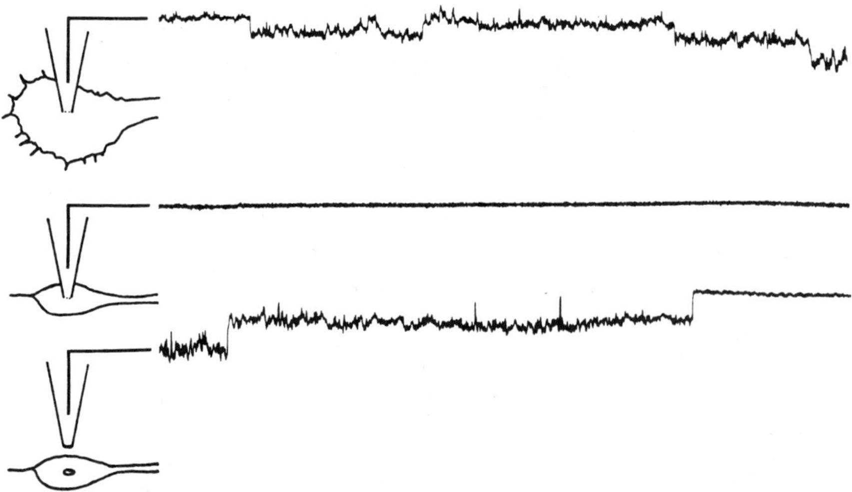

Fig. 6. Differences in single channel activity of growing as compared to stable growth cones. Drawings on the left indicate the type of patch and the growth state of the growth cone. Channel activity that is present in cell-attached patches of growing growth cones (top trace) is absent from cell-attached patches of stable growth cones (middle trace). The onset of channel activity in patches from stable growth cones after the pipette and underlying membrane is removed from the cell (bottom trace) indicates that channels are still present in stable growth cone membrane but were previously silent. Middle and bottom traces are from the same membrane patch. Calibration: 6pA, 600msec. (From Cohan et al., 1985).

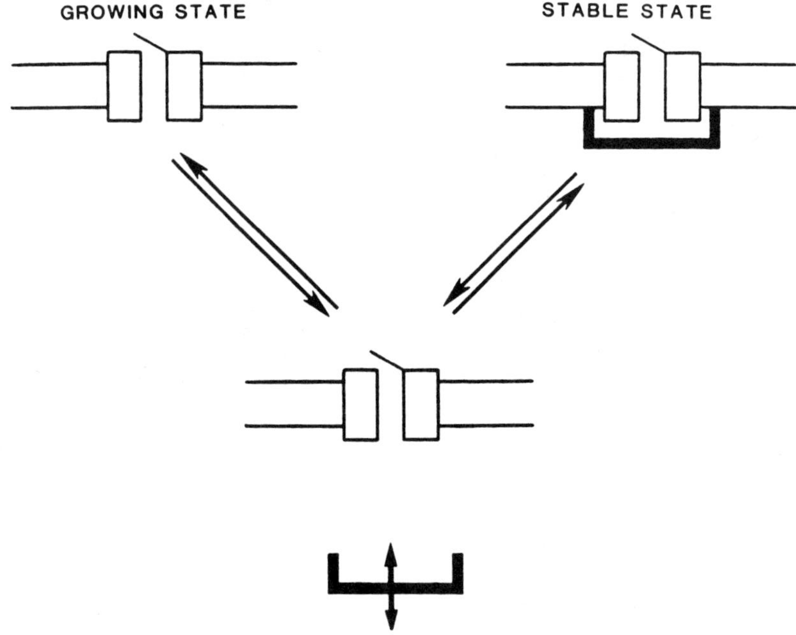

Fig.7. Hypothetical model of the relationship between membrane channels and the growth state of neuronal growth cones. The physiological state of ion channels in the growth cone membrane may be controlled by the binding of an intracellular molecule(s) to channel-associated elements. This hypothesis is based on the fact that silent channels from non-growing growth cones become active when patches are excised from cytoplasmic continuity. (From Cohan et al., 1985).

INTRACELLULAR CALCIUM AND REGULATION OF NEURITE OUTGROWTH

Research during the last year has been directed toward testing the hypothesis that **growth cone behavior and neurite elongation are regulated by the specific levels of intracellular calcium present in the advancing growth cone.** This idea has been reached in collaboration with Drs. C. Cohan, J. Connor, and M. Mattson. It has been possible to test this hypothesis by the combination of techniques employing the intracellular calcium indicator Fura 2 and the use of agents known to alter transmembrane calcium fluxes such as ionophores and inorganic calcium channel blockers. In addition to this combination of techniques, we have had available three separate experimental manipulations for regulating neurite outgrowth including; 1) the spontaneous stabilization of neurite outgrowth that occurs normally after two to three days in cell culture (Fig. 2); 2) neurotransmitter initiated stabilization of growth cones (Fig.3) and 3) electrical activity generated experimentally to inhibit neurite outgrowth (Fig.4). This combination of methods, summarized briefly below, has led to the hypothesis illustrated graphically in the final figure of this paper.

We have used the fluorescent calcium indicator Fura 2 to measure directly intracellular calcium concentrations in growth cones under different experimental conditions. Our findings (Cohan, Connor, and Kater, 1986) demonstrate a significant difference in the rest level of calcium in the growing versus spontaneously stable state growth cones. Growing growth cones display intracellular calcium concentrations of approximatley 130 nM while stable state growth cones at intracellular concentrations of about 50 nM. Such data indicate that there is a decrease in the level of intracellular calcium correlated with the spontaneous cessation of neurite outgrowth. Subsequent experiments showed that increasing calcium concentration can also stabilize growing neurites. Using either electrical stimulation or serotonin on neuron Bl9s to inhibit its growth status, we found that intracellular calcium would rise to nearly 800 nM (Fig. 8) under these conditions. The calcium channel blocker, cobalt, prevented the calcium influx caused by electrical activity. Taken together, these data indicate that inhibition of neurite outgrowth by electrical activity or serotonin results in a rise in intracellular free calcium which may be causally related to the abrupt inhibition of growth cone motility. To test for a causal relationship between changes in calcium influx and neurite outgrowth, Dr. Mark Mattson and I have employed a battery of standard pharmacological agents known for their effects on transmembrane calcium flux.

Addition of the calcium ionophore, A23187, to the bathing medium is an accepted means of increasing intracellular calcium. Under such conditions, we found that all neurons tested immediately ceased outgrowth (Fig. 9). Neurite elongation was reduced to near zero within thirty minutes of treatment with A23187 at 10^{-7} M. This inhibition was reversible. This effect is precisely as would have been predicted from our Fura 2 measurements. A further test of this hypothesis was made by Mattson using inorganic calcium channel blockers. The effects of serotonin are directly related to an influx of calcium from the extracellular medium. Lanthanum in the medium could completely block the suppressive effects of serotonin on the elongation of neurites of neurons l9. This kind of experiment, has led to the general idea that neurons have a particular calcium "set point" and that calcium appears to be a general regulator of neurite outgrowth. The calcium set point has been operationally defined (Kater and Mattson, 1987) as the intracellular concentration of calcium of a neuron under a given set of experimental conditions. On review of the literature, one can see that the outgrowth of some neurons is significantly inhibited by a rise in calcium while the outgrowth of other neuronal cell types is, in direct opposition, augmented. By recognizing that different neurons may indeed have different calcium set points, a general hypothesis of calcium regulation of neurite outgrowth as presented in Figure 10 becomes quite tenable.

CONCLUSION

The goal of this paper has been to illustrate a parallelism between mechanisms used in our standard view of neural coding and those mechanisms used for specification of neuronal architecture in developmental processes. The role of specific levels of intracellular calcium in all of these processes is quite apparent and a fundamental understanding of those mechanisms which regulate intracellular calcium is imperative. It seems reasonable, on

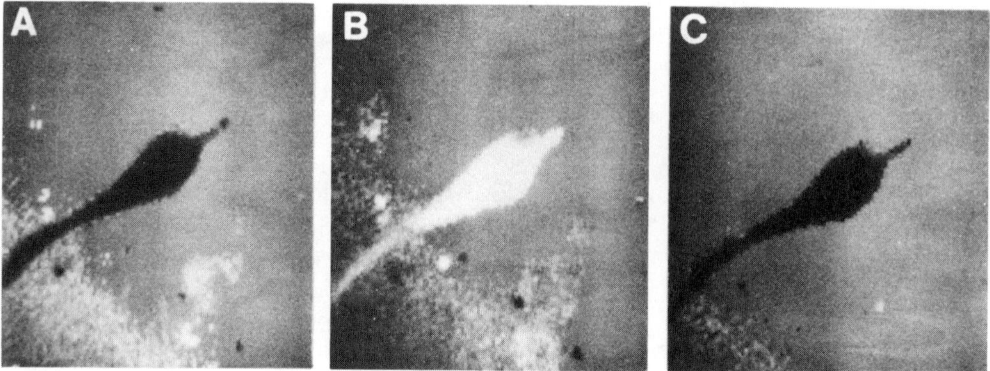

Fig. 8. An example of the analysis of the local intracellular calcium concentrations in a growth cone, as revealed by the calcium fluorescent dye Fura 2. Initially low calcium levels (A) increase while action potentials are evoked in the soma (B). Calcium levels recover to rest levels after action potentials are terminated (C). (Modified from Cohan et al., 1987).

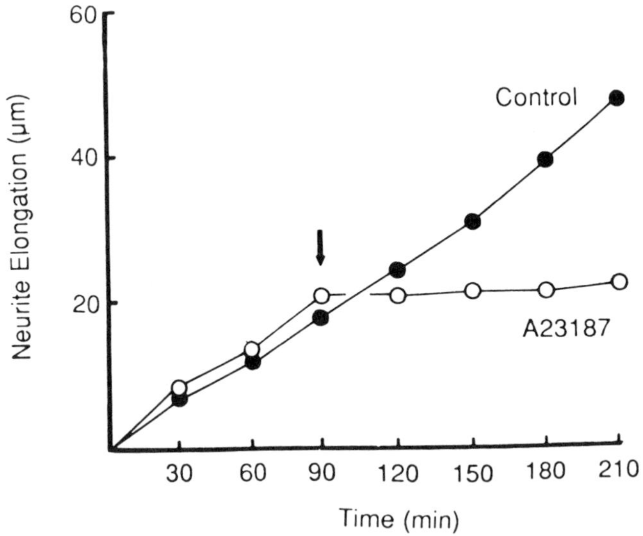

Fig. 9. Time course for inhibition of neurite elongation by A23187. Neurons B5 and B19 were preincubated 90 min followed by a 2 hr incubation in the absence (Control) or presence of 10^{-7} M A23187 (arrow designates addition of A23187); examples shown are representative of all 33 neurites assessed.

examining the ability of neuronal growth cones to regulate their intracellular calcium, that precise levels of intracellular calcium are obtained by a *balance* ; this balance is between the influx of calcium through specific ion channels,and the efflux of calcium through regulatory exchange mechanisms or pumps. Thus, in concert, these two opposing regulatory forces could produce the kind of dynamic balance that determines the stability or growth status of the architecture of neurons throughout the life history of the organism.

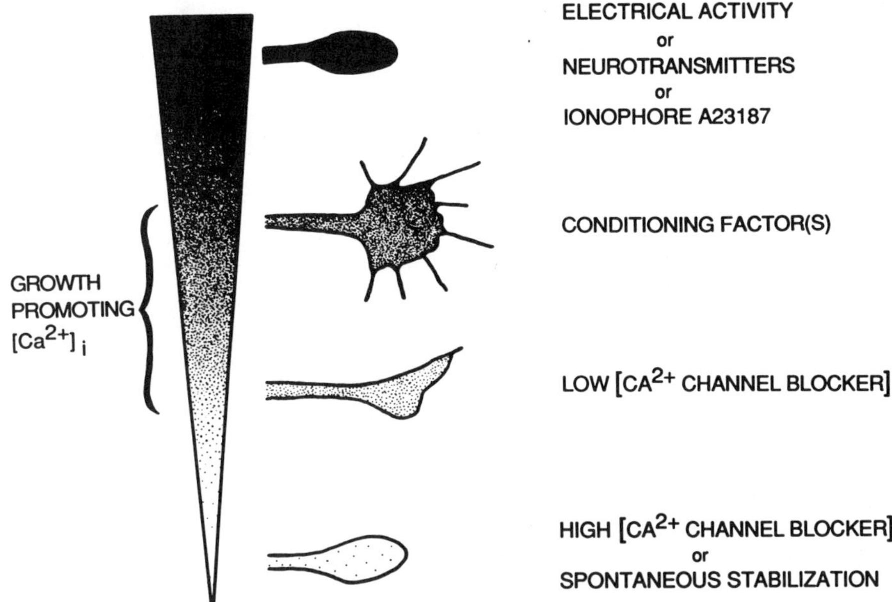

Fig.10. Model for regulation by calcium of growth cone motility and neurite elongation. In the presence of conditioning factor(s), intracellular calcium is maintained within a narrow, optimum range (second drawing from top) which allows for the formation and activity of growth cones. The top drawing depicts the effects on neurite outgrowth of increases in calcium induced by, for example, neurotransmitters, electrical activity or A23187. Such increases inhibit both elongation and growth cone motility. In the third drawing it can be seen that a moderate decrease in calcium such as occurs in response to relatively low concentrations of calcium channel blockers results in decreased growth cone activity but accelerated neurite elongation. Large reductions in internal calcium as induced by high concentrations of calcium channel blockers or in neurites which have attained, spontaneously, a stable growth status leads to cessation of growth cone motility and neurite elongation (bottom drawing).

REFERENCES

Bray D. (1982) Filopodial contraction and growth cone guidance. In Bellairs R., Curtis A., Dunn G. (eds): "Cell Behavior." London/New York: Cambridge University Press, pp. 299-317.

Cohan, C. S., J. A. Connor, and S. B. Kater (1986) Signals which inhibit growth cone motility also increase intracellular calcium levels within the growth cone. Soc. Neurosci. Abstr. 12:370.

Cohan, C. S., P. G. Haydon, and S. B. Kater (1985) Single channel activity differs in growing and nongrowing growth cones of isolated identified neurons of Helisoma. J. Neurosci. Res. 13:285-300.

Cohan, C. S., and S. B. Kater (1986) Suppression of neurite elongation and growth cone dynamics by electrical activity. <u>Science</u> 232:1638-1640.

Cohan, C. S., J. A. Connor, and S. B. Kater (1987) Electrically and chemically mediated increases in intracellular calcium in neuronal growth cones. <u>J. Neurosci.</u> in press.

Goldberg, J. I. and S. B. Kater, (1985). Experimental reduction of serotonin content during embryogenesis alters morphology and connectivity of specific identified <u>Helisoma</u> neurons. <u>Soc. Neurosci.</u> Abstr. 11:158.

Guthrie, P. B., R. E. Lee, and S. B. Kater (1986) Growth cones of identified neurons - unique and shared ultrastructural and electrophysiological characteristics. <u>Soc. Neurosci. Abstr.</u>12:370.

Hadley, R. D., D. A. Bodnar, and S. B. Kater (1986) Formation of electrical synapses between isolated, cultured <u>Helisoma</u> neurons requires mutual neurite elongation. <u>J. Neurosci.</u> 5:3145-3153.

Haydon, P. G., C. S. Cohan, D. P. McCobb, H. R. Miller, and S. B. Kater (1985) Neuron-specific growth cone properties as seen in identified neurons of <u>Helisoma</u>. <u>J. Neurosci. Res.</u> 13:135-147.

Haydon, P.G., D.P. McCobb, and S.B. Kater (1987) The regulation of neurite outgrowth, growth cone motility, and electrical synaptogenesis by serotonin. <u>J. Neurobiol</u>. in press.

Haydon, P. G., D. P. McCobb, and S. B. Kater (1984) Serotonin selectively inhibits growth cone dynamics and synaptogenesis of specific identified neurons. <u>Science</u> 226:561-564.

Kater, S.B. (1985), Dynamic regulators of neuronal form and connectivity in the adult snail <u>Helisoma</u>. In: Selverston, A.I. (ed) "Model Neural Networks and Behavior" Plenum Pub. Corp., New York.

Llinas, R. R. (1975) The role of calcium in neuronal function. In, Schmitt F.O., Worden, F. G. (eds); (The neurosciences: fourth study program). Cambridge, MA: MIT Press, pp. 551-571.

MacVicar, V. A. and Llinas, R. R. (1986) Barium action potentials in regenerating axons of the lamprey spinal cord. <u>J. Neurosci. Res</u>. 13:323-335.

Mattson, M. P., and S. B. Kater. Calcium Regulation of Growth Cone Motility and Neurite Elongation in Identified <u>Helisoma</u> Neurons. Submitted to <u>J. Neurosci</u>.

McCobb, D.P., P. G. Haydon, and S. B. Kater (1985) Dopamine: An additional regulator of neurite outgrowth in <u>Helisoma</u>. <u>Soc. Neurosci. Abstr.</u> 11:761.

McCobb, D. P., and S. B. Kater (1986) Serotonin inhibition of growth cone motility is blocked by acetylcholine <u>Soc. Neurosci. Abstr.</u> 12:1117.

Wong, R. G., R. D. Hadley, S. B. Kater, and G. C. Hauser (1981) Neurite outgrowth in molluscan organ and cell cultures: The role of conditioning factor(s). <u>J. Neurosci.</u> 1:1008-1021.

ACKNOWLEDGEMENTS

The author wishes to acknowledge the constructive comments on this manuscript by Drs. J. Goldberg, P. Guthrie, M. Mattson, and L. Mills. and also the technical assistances of M. Mischke and D. Dehnbostel. This work was supported by NIH grants NS24683, NS24561, NS15350, and a gift from the Monsanto Corporation.

TARGET CELL CONTACT MODULATES SPONTANEOUS QUANTAL AND NON-QUANTAL

ACETYLCHOLINE RELEASE BY <u>XENOPUS</u> SPINAL NEURONS

Ida Chow, Steven H. Young, and Alan D. Grinnell

Department of Physiology, Ahmanson Laboratory of
Neurobiology and the Jerry Lewis Neuromuscular Research
Center, UCLA School of Medicine, Los Angeles, CA 90024

<u>Xenopus</u> nerve-muscle cell culture has proved to be an excellent
preparation for study of the development of neuronal ion channel and
membrane properties (Spitzer, 1976; Spitzer and Lamborghini, 1976;
Willard, 1980) and for studies of synaptogenesis. Much is already known
about the properties of the synapses formed in culture, and the changes
that take place during maturation, particularly post-synaptically. At
synapses formed by growing neurites on muscle cells in 1-2 day old
cultures, there are quantal miniature endplate potentials (mEPPs) at
frequencies ranging from 0.5 - 4 Hz, with skewed amplitude histograms
(Kidokoro et al., 1980; Chow and Poo, 1985). It appears that, at the
time of initial synaptic contact, there is no postsynaptic specialization
at the site of release--the responses are mediated simply by the acetylcholine (ACh) receptors spread diffusely over the muscle cell's surface.
Only after a time course of hours to days, does the sub-synaptic site
develop a high concentration of ACh receptors and other specializations
(Anderson and Cohen, 1977; Anderson et al., 1977; Weldon and Cohen, 1979;
Kidokoro et al., 1980; Cohen and Weldon, 1980; Takahashi et al., 1987;
Chow et al., 1987).

Because functional synapses can be produced within seconds or minutes
simply by mechanically moving a muscle cell into contact with a neurite,
or even the soma of a neuron (Chow and Poo, 1985), <u>Xenopus</u> nerve-muscle
culture is also a favorable preparation for the study of transmitter
release and of factors influencing it, before, during, and after synaptogenesis. We report here a series of experiments describing the release
properties of growth cones and the inductive influence of appropriate
target cells in increasing the levels of release, using intracellular and
patch recording techniques, allowing control of both extra- and intracellular solutions.

ACh release from neuronal growth cones

Even before making contact with a target cell, the growth cones of
cholinergic neurons, in culture, release ACh spontaneously. This has
been shown convincingly in <u>Xenopus</u> by Young and Poo (1983) and in chick
by Hume et al., (1983), using outside-out excised patches of ACh-
sensitive muscle cell membrane as probes. In standard external saline
(125 mM NaCl, 2 mM K Cl, 2 mM $CaCl_2$, 1 mM $MgCl_2$, 20 mM Hepes, pH 7.8),
these responses from <u>Xenopus</u> neurons take a variety of forms, suggesting

different types of release: extremely infrequent, large summating
bursts of synchronous channel openings with a stepwise falling phase
("staircase" or miniature endplate current (mEPC)-like events); bursts of
smaller amplitude consisting of non-synchronous openings lasting up to
several hundred milliseconds ("flat bursts"); and relatively asynchronous
series of low-frequency single channel openings (see Figure 1, also Young
and Poo, 1983). Of these, the large mEPC-like events were seen from only
a few growth cones (15%), and even in these cases they were extremely infrequent (about 0.1 events/minute). It is possible that these represent
quantal release of ACh in a form that, at a synapse, would result in a
typical mEPP, or mEPC when recorded under whole-cell patch clamp mode.
The smaller and less synchronous bursts are of less obvious homology to
anything observed at well-formed junctions. Some of these could simply
be examples of mEPC-like events recorded at a greater distance from the
release site. However, it seems unlikely that this is the case for most
of the bursts, since they characteristically begin abruptly, maintain a
relatively constant frequency of channel openings throughout, and last
much longer than would be expected of ACh diffusing from a distant source
of release of a single quantum of transmitter.

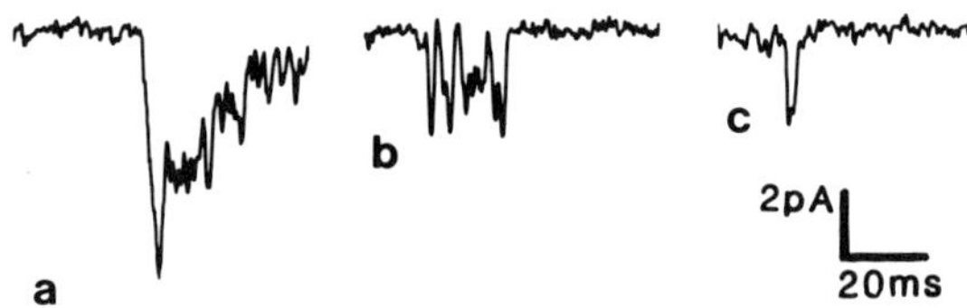

Fig. 1. Tracings of single channel current recordings
showing different types of ACh release from neuronal
growth cones, as detected by outside-out excised patches
of muscle cell membranes. a. Large summating burst of
synchronous channel openings with a stepwise falling
phase--"staircase" or mEPC-like event. b. Non-
synchronous burst of channel openings--"flat" burst.
C. Single channel opening.

When the external saline is replaced by one with elevated Ca^{++}
(10 mM Ca^{++}, 1 mM Mg^{++}), both the fraction of growth cones showing
spontaneous transmitter release and the frequency of release events
decreased. A similar reduction in release was seen when the Ca^{++} was
kept at 2 mM, but Mg^{++} was raised to 9 mM. In low Ca^{++} saline (0 mM
Ca^{++}, 3 mM Mg^{++}), both the fraction of growth cones showing spontaneous
release and the frequency of release events were the same as in standard
saline. Thus spontaneous transmitter release from growth cones of
isolated neurons appears not to be dependent on the presence of extracellular millimolar levels of Ca^{++}, and is suppressed by high external
concentrations of Ca^{++} or Mg^{++}. (These conclusions apply to the non-
synchronous bursts of channel openings. It is not clear that they apply
to the large mEPC-like events as well, since these are normally so
infrequent that a decrease in frequency or probability is difficult to
establish.)

Target cell induction of transmitter release

When a muscle cell is manipulated into contact with the soma or the
neurite of an isolated cholinergic neuron, mEPP-like depolarizations are
recorded from the muscle cell after a short delay (a few seconds to 17
minutes, 4.0 ± 3.9 min (mean ± S.D.) in the 116 cell pairs studied).
This response was detected within 4 min of contact in 64% of the neurite-

muscle cell pairs and in 35% of the soma-muscle pairs. This suggests
that the neurites are more prompt to release ACh upon contact than the
soma. These events are blocked by curare and α-bungarotoxin, hence
are a result of ACh release from the neuron, and their characteristics
(fast rise and slow decay) are very similar to those of mEPPs recorded
from naturally occurring neuromuscular junctions in the same cultures.
However, the average amplitudes are half or less the values found in the
mEPPs at more mature junctions, formed spontaneously on similar muscle
cells in cultures of the same age.

The frequency of these MEPP-like events was normally very low at
first (0.03 - .2/sec in the first minute or two of contact), and lower
than at spontaneously formed junctions (Chow and Poo, 1985). However, in
many cases, within 10 - 15 min there was a sharp increase in frequency,
some examples of which are shown in Figure 2. In other cases, no change
in frequency was observed, for up to one hour contact. (See also Chow
and Poo, 1985).

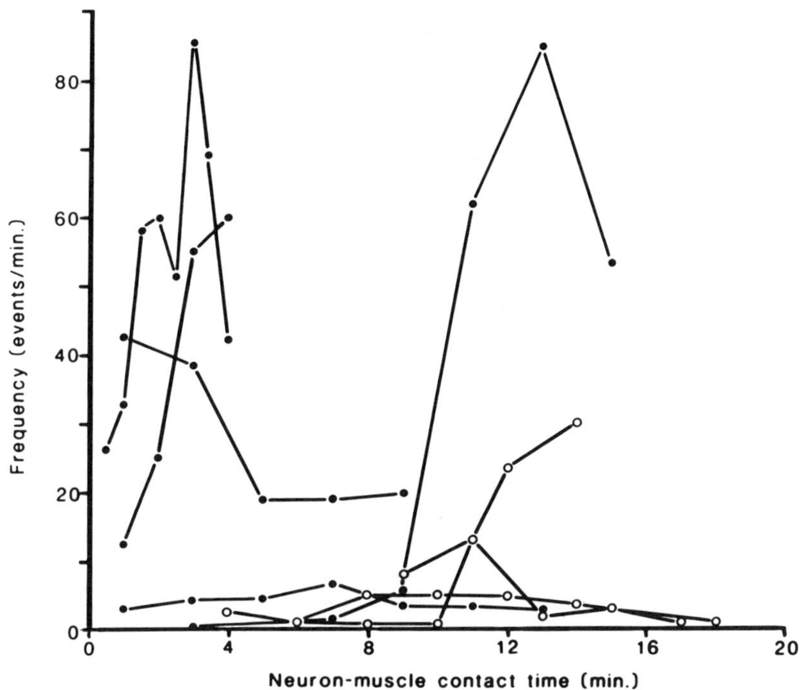

Fig. 2. Changes in the frequency of mEPP-like potentials with
neuron-muscle contact time. Open symbols represent different
soma-muscle cell pairs. Filled symbols represent different
neurite-muscle cell pairs. Note the variability of frequency
changes and the corresponding time course.

The nature of this muscle contact-induced release of ACh is still
unclear; but muscle cell contact has a significantly stronger effect on
the neurotransmitter release than contact with another neuron. Moreover,
preliminary data suggest that cell surface interactions are sufficient
for the induced effect. To determine whether living target cells are
essential, muscle cells were repeatedly impaled with a microelectrode to
cause cell death (resting membrane potential less than -10 mV with most

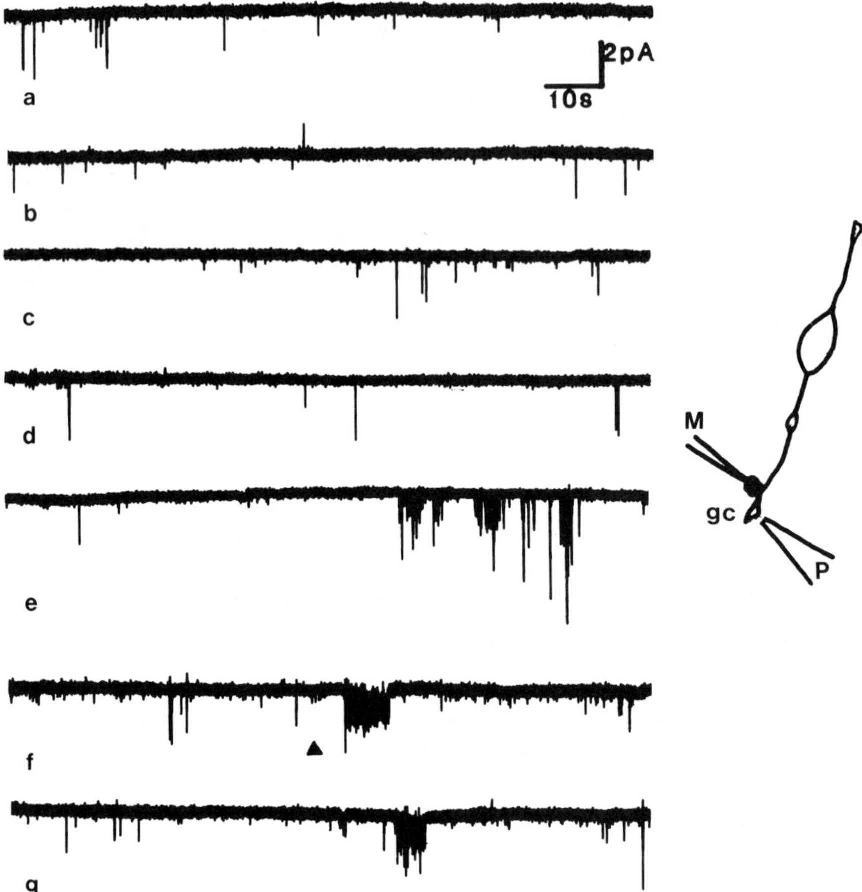

Fig. 3. Tracings of single channel current recordings
from outside-out excised patches of muscle membrane (P)
positioned next to a neuronal (N) growth cone (gc) before
and after contact with lysed cell (neuronal or muscle)
membrane mass (M), from 2 different experiments.
a-e: Experiment 1, holding potential -70 mV. a- before
contact; b- immediately after dead neuronal cell mass;
c- following removal of the neuronal membrane mass after
10 min contact; d- 3 min after muscle membrane mass contact;
e- 18 min after the same muscle mass contact. Note the
appearance of bursts of channel openings.
f-g: Experiment 2, holding potential -60 mV. f- immediately
before and after muscle membrane mass contact (arrowhead),
note the appearance of "flat" burst of channel openings 6 s
after contact; g- 8 min after the same muscle mass contact
showing the presence of occasional single channel events and
a "flat" burst.

of the cell contents extruded). This lysed cell mass consisting
mainly of cell membrane was then brought into contact with the neuron as
the release of ACh was being monitored with an outside-out excised muscle
membrane probe. The same maneuver was done with membrane masses from
other neurons found in the culture and the extent of ACh released
by their contact similarly monitored. The results show that the muscle
membranes induced greatly increased release, while neuron membrane mass
contact sometimes induced some ACh release, but at a much lower frequency.
The neuronal membrane mass had no effect or induced only a few isolated
single channel openings, whereas the muscle mass induced both single
channel openings and bursts of channel openings (Figure 3). Figure 4
shows the plot of frequencies of events before and after contact of
neurites by dead muscle cell membrane, as detected with excised outside-
out patches of muscle cell membrane. Thus it appears that one or more
molecular species built into the muscle cell membrane or the matrix
surrounding it, but absent or in much lower concentration on the membranes of other cells, has a strong inductive influence on motoneurons
to increase their spontaneous transmitter release.

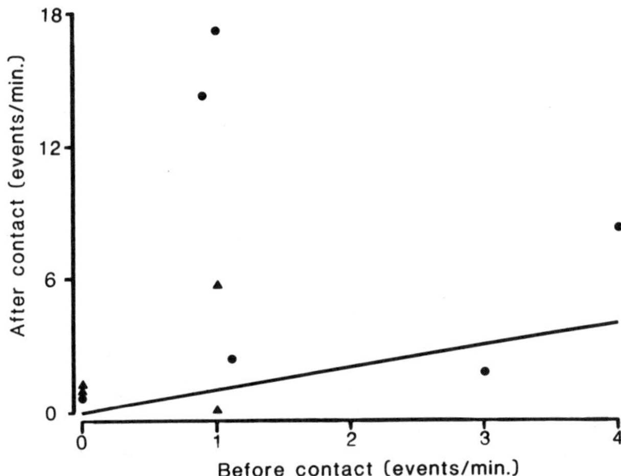

Fig. 4. Plot of frequencies of events before and
after contact of neurites by dead muscle cell masses,
as detected with excised outside-out patches of muscle
cell membrane. The straight line indicates the position
of points if the frequencies before and after were the
same. Circles - frequency of single events. Triangles -
frequency of bursts.

A great deal of research, which we will not attempt to review,
has been done in recent years on possible motoneuron trophic or growth
factors. In most cases the effect of these factors on transmitter
release has not been studied, so we have no basis on which to conclude
that the signal responsible for the induction of increased release in our
system might or might not be the same as one of the other factors
described.

<u>The nature of spontaneous transmitter release at newly formed "synapses"
and the properties of non-quantal "leak"</u>

Much of the increased transmitter release by a neuronal soma after
contact with a muscle cell is in the form of large events that, post-
synaptically, resemble closely the mEPPs seen at established neuromuscular junctions. It seems probable that they are the result of
typical vesicle-fusion and release, although it is possible to calculate
that the ACh content is much less, at early stages, than for the quanta
at mature junctions, both <u>in vivo</u> (Kullberg et al., 1977) and in culture
(Kidokoro et al., 1980). This target-cell induced release is not
affected by an increase in the external calcium concentration from 2 mM
to 10 mM (Chow and Poo, 1985). This suggests that the target-induced
release, seen mostly in the form of mEPPs, is qualitatively different
than the spontaneous release before contact, which is reduced by increased divalent cation concentrations (see above).

The input impedance of muscle cells in culture (ca. 75-200 MΩ)
is sufficiently high that a whole-cell patch clamp electrode is able
to detect small numbers of ACh channel openings; hence it is also
possible to monitor other forms of spontaneous release besides large
full sized quanta. Such recordings from innervated muscle cells in 1-2
day cultures show evidence not only of large mEPCs, but also of much
smaller, less synchronous bursts of openings that can be attributed to
non-quantal "leak". These are surprisingly rare, being seen in only
about 15% of innervated cells in one day cultures, about 40% in 2 day
cultures. Moreover, where they do occur, they are highly intermittent,
an episode of such bursts occurring on the average only about 0.8/minute,
and lasting for an average of about 1.2 sec (Grinnell and Young, 1987).
Although these small events represent very little transmitter release
compared with that producing the large mEPCs, it can be calculated that,
if such release is proportional to synaptic area and if the area is no
more than 0.1-0.2% of the area of a mature neuromuscular junction, the
amount of "leakage" would be compatible with values derived for mature
nerve terminals. Interestingly, in newly formed synapses, the numbers
and temporal properties of the relatively non-synchronous channel
openings are not unlike those of the release events (excluding mEPC-like
events) seen from growth cones before they contact a target cell. Thus
it is tempting to conclude that the growth cone shows mostly non-quantal
spontaneous "leak" of ACh, which continues after synapse formation, but
which has super-imposed on it a greatly increased amount of quantal
release.

<u>Evidence against one form of "vesigate" hypothesis of transmitter
release</u>

Although it is generally accepted that quantal postsynaptic
potentials are the result of synaptic vesicle fusion with the presynaptic
membrane and release of its contents (Katz, 1969; Heuser et al., 1979),
alternatives have been proposed (Israel et al., 1979; Tauc, 1982; Dunant
and Israel, 1985; Dunant, 1986; Israel et al., 1986). One such alternative is the hypothesis that cytoplasmic ACh is released in quantal
amounts through a plasma membrane pore or channel that is briefly gated
by depolarization or calcium influx ("vesigate hypothesis"). While the
properties of such a channel are not specified, it is likely that any
channel capable of passing a molecule as large as ACh would also be
significantly permeable to smaller ions. Using the assumptions that each
quantum represents release of 2000 or more molecules of ACh (since a
large mEPC in one of the cultured muscle cells is the result of opening
of about 1000 ACh receptor channels under voltage clamp conditions (Young,
unpublished data), and that any such channel would pass a significant

amount of Na$^+$ and/or K$^+$ ions, the current through a single channel passing a quantum of ACh should be 2.5 pA or more. We have made simultaneous intracellular recordings of muscle cell membrane potential and whole-cell patch recordings of membrane currents in the neuronal soma immediately before, during and after the occurrence of mEPP-like potentials. Of 6 cell pairs and a total of 822 mEPP-like potentials, we have detected no channels of this size. This argues strongly against the hypothesis that large quanta of ACh are released via specific channels in the neuronal membrane (Young and Chow, 1987).

Conclusions

Xenopus nerve-muscle cultures are proving very useful for investigating the nature and quantity of spontaneous ACh release by the motoneuronal soma and neurite growth cones before and after contact with target muscle cells. We find a low level of spontaneous release, some probably quantal but mostly non-quantal, before contact with a target cell, and much increased release, particularly in the form of quanta of transmitter, after contact with a muscle cell. This inductive effect is relatively specific to muscle cells (as opposed, for example, to other neurons), and is fully simulated by contact with a mass of membranes from killed muscle cells. Non-quantal release is seen to persist and increase after synapse formation, presumably constituting the "leak" reported for mature neuromuscular junctions. Evidence is also presented that there are no large channel openings in the presynaptic membrane before, during or after release of a quantum of transmitter.

Acknowledgements

This work is supported by grants from NSF (BNS 84-19893, BNS 85-10794), NIH (PHS 5R01 NS06232-22) and the Muscular Dystrophy Association.

References

Anderson, M. J., and Cohen, M. W., 1977, Nerve-induced and spontaneous redistribution of acetylcholine receptors on cultured muscle cells, J. Physiol. (Lond.), 268:757.

Anderson, M. J., Cohen, M. W., and Zorychta, E., 1977, Effects of innervation on the distribution of acetylcholine receptors on cultured muscle cells, J. Physiol. (Lond.), 268:731.

Chow, I., and Poo, M-m, 1985, Release of acetylcholine from embryonic neurons upon contact with muscle cell, J. Neurosci., 5:1076.

Chow, I., Young, S. H., Cheng, J., and Grinnell, A. D., 1987, Development of properties of transmitter release during initial stages of synaptogenesis, Abstr. Soc. Neurosci., 13: in press.

Cohen, M. W., and Weldon, P. R. 1980, Localization of acetylcholine receptors at nerve-muscle contacts in culture: Dependence on nerve type, J. Cell Biol., 86:388.

Dunant, Y., 1986, On the mechanism of acetylcholine release, Prog. Neurobiol., 26:55.

Dunant, Y., and Israel, M., 1985, The release of acetylcholine, Scientific American, 252 (4):58.

Grinnell, A. D., and Young, S. H., 1987, Non-quantal release of transmitter at developing neuromuscular junctions of Xenopus in culture, Abstr. Soc. Neurosci., 13: in press.

Heuser, J. E., Reese, T. S., Dennis, M J., Jan, Y., Jan, L., and Evans, L., 1979, Synaptic vesicle exocytosis captured by quick freezing and correlated with quantal transmitter release, J. Cell Biol., 81:275.

Hume, R. I., Role, L. W., and Fischbach, G. D., 1983, Acetylcholine
      release from growth cones detected with patches of acetylcholine
      receptor-rich membranes, Nature 305:632.
Israel, M., Dunant, Y., and Manaranche, R., 1979, The present status of
      the vesicular hypothesis, Prog. Neurobiol., 13:237.
Israel, M., Morel, N., Lesbats, B., Birman, S., and Manaranche, R., 1986,
      Purification of a presynaptic membrane protein that mediates a
      calcium-dependent translocation of acetylcholine, PNAS USA,
      83:9226.
Katz, B., "The Release of Neural Transmitter Substances," Liverpool
      Univ. Press, Liverpool (1969).
Kidokoro, Y., Anderson, M. J., and Gruener, R., 1980, Changes in
      synaptic potential properties during acetylcholine receptor
      accumulation and neurospecific interactions in Xenopus nerve-
      muscle culture, Dev. Biol., 78:464.
Kullberg, R., Lentz, T.L., and Cohen, M. W., 1977, Development of the
      myotomal neuromuscular junction in Xenopus laevis: an electro-
      physiological and fine-structural study, Dev. Biol., 60:101.
Spitzer, N., 1976, The ionic basis of the resting potential and a slow
      depolarizing response in Rohon-Beard neurons of Xenopus tadpoles,
      J. Physiol. (London), 255:105.
Spitzer, N., and Lamborghini, J., 1976, The development of the action
      potential mechanism of amphibian neurons isolated in culture,
      PNAS USA, 73:1641.
Takahashi, T., Nakajima, Y., Hirosawa, K., and Onodera, K., 1987,
      Structure and physiology of developing neuromuscular synapses in
      culture, J. Neurosci., 7:473.
Tauc, L., 1982, Non-vesicular release of neurotransmitter, Phys. Rev.,
      62:857.
Weldon, P. R., and Cohen, M. W., 1979, Development of synaptic ultra-
      structure at neuromuscular contacts in an amphibian cell culture
      system, J. Neurocytol., 8:239.
Willard, A., 1980, Electrical excitability of outgrowing neurites of
      embryonic neurones in cultures of dissociated neural plate of
      Xenopus laevis, J. Physiol., 301:115.
Young, S. H., and Chow, I., 1987, Quantal release of neurotransmitter is
      not associated with opening of large channels on the neuronal
      membrane. Submitted for publication.
Young, S. H., and Poo, M-m, 1983, Spontaneous release of transmitter
      from growth cones of embryonic neurones, Nature, 305:634.

DEVELOPMENT AND REGULATION OF ACETYLCHOLINE RECEPTOR FUNCTION

Paul Brehm, James Lechleiter, and Leslie Henderson

Dept. of Physiology, Tufts Medical School, Boston, MA. 02111

Jesse Owens and Richard Kullberg

Biology Department, University of Alaska, Anchorage, AK. 99508

Nicotinic acetylcholine (ACh) receptors are among the first proteins to be expressed on the membranes of newly differentiated muscle cells. During development, the functional properties of the receptors and their associated channels undergo major changes in many skeletal muscles (Schuetze and Role, 1987). The kind of changes as well as the timetable of the changes depends on the type of muscle being studied (Schuetze, 1980; Kullberg and Owens, 1986). Changes in receptor expression also occur in adult skeletal muscle following denervation (Neher and Sakmann, 1976) and during subsequent reinnervation (Brenner and Sakmann, 1983). Studies on adult muscle have led to the proposition that the development of receptor function depends on the proximity of the receptor to the site of nerve-muscle contact (Cull-Candy et al., 1982; Brenner and Sakmann, 1983). The mechanisms which underlie the regulation of receptor function in developing and adult muscle have been the subject of a large number of studies. In this paper we summarize the results of our studies on the development of ACh receptor function in amphibian muscle and include, as well, the results of complementary studies on both innervated and denervated adult mammalian muscle. We address the issues related to neural regulation of ACh receptor properties by comparing properties of synaptic and nonsynaptic receptors in developing and adult muscle. The possible underlying molecular mechanisms and functional significance of our observations are discussed.

DEVELOPMENT OF ACH RECEPTOR CHANNEL FUNCTION IN XENOPUS MUSCLE

Synaptogenesis takes place with remarkable speed in developing amphibian muscle. In *Xenopus* myotomal muscle, ACh receptors appear within 21 hours of fertilization of the egg (Blackshaw and Warner, 1976; Kullberg et al., 1977; Chow and Cohen, 1983). Just two hours later, the first synaptic activity can be recorded at newly formed neuromuscular junctions (Kullberg et al., 1977). ACh receptors immediately begin to aggregate at the site of nerve-muscle contact, and within another few hours the deposition of acetylcholinesterase at the synapse begins (Kullberg et al., 1980).

During the development of *Xenopus* myotomal muscle, there is a major change in the time course of the synaptic currents (Kullberg et al., 1980). As illustrated in Figure 1, the synaptic currents at newly formed myotomal synapses decay with a time constant of about 7 msec, while at mature synapses, the synaptic current decays in about 1 msec. The explanation for this change in duration lies in the development of acetylcholinesterase and in a reduction in the open time of ACh receptor channels.

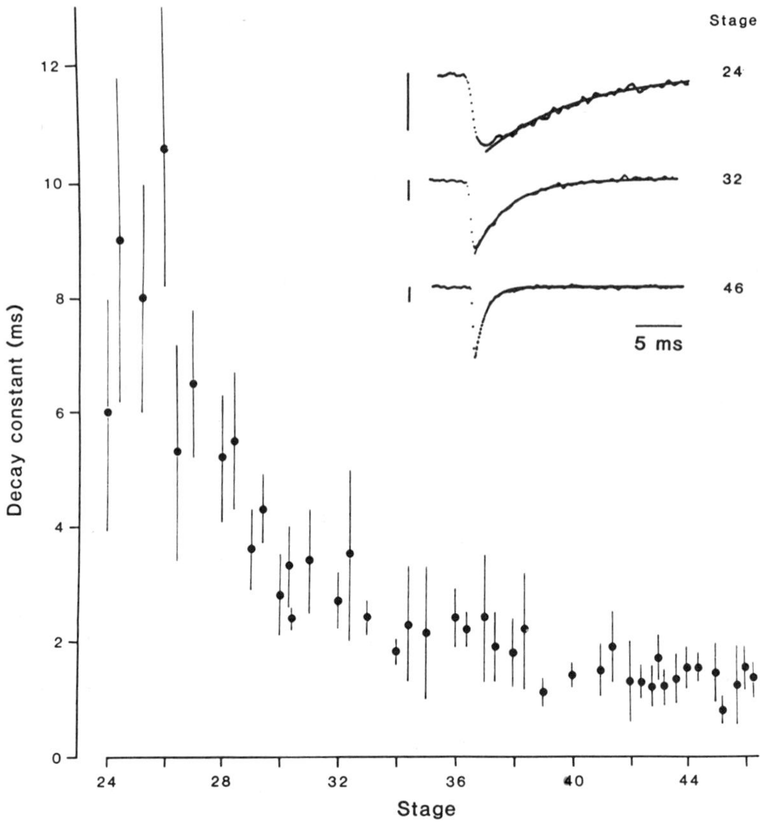

Fig. 1. *The decay constants (mean +/- S.D.) of mepcs at different stages of development. Each data point represents the mean decay constant of mepcs obtained from several recording sites. The average number of recording sites for each data point was 11 (range=3 to 38). The range of stages shown on the abscissa corresponds to 26 hours of age (stage 24) to 106 hours of age (stage 46). Inset: Examples of extracellular miniature endplate currents (mepcs) recorded from* <u>Xenopus</u> *myotomal muscle at different developmental stages. The declining phase of each mepc has been fitted by a single exponential function. The decay constants from top to bottom are 9.2 msec, 3.5 msec, and 1.1 msec. The vertical bar indicates 0.1 mV.*

Changes in the open time of ACh receptors in developing *Xenopus* muscle were first revealed by analysis of ACh-induced current noise (Kullberg et al., 1981; Brehm et al., 1982). When a pipette containing dilute ACh was placed over the synaptic regions, current fluctuations arose from the random opening of ACh receptor channels. Spectral analysis of these fluctuations suggested that the channel open time decreased at least 3 fold during development. Surprisingly, it was found that the open time of ACh receptors in nonsynaptic regions underwent a similar decrease (Kullberg et al., 1981; Brehm et al. 1982). Examination of high resolution spectra from myotomal muscle revealed that two kinetically distinct receptor types were expressed during development (Brehm et·al., 1982; Kullberg and Kasprzak, 1985). Initially, the muscles had a homogeneous population of channels with a mean open time of about 3 to 4 msec. Rather abruptly, at about 45 hours of age, a second class of channels with a mean open time of less than 1 msec appeared. Over the next 4 weeks of development, the faster class of channels became the predominant channel type in the nonsynaptic membrane.

The conclusions based on noise studies have been confirmed by single channel recordings from ACh receptors throughout development (Brehm et al., 1984a,b; Kullberg et al., 1986; Owens and Kullberg, 1986). Single channel recordings from nonsynaptic regions (Fig. 2) revealed a class of low conductance (40 pS), long open time (3 to 4 msec) ACh receptor channels which appeared on the embryonic membrane within 21 hours of

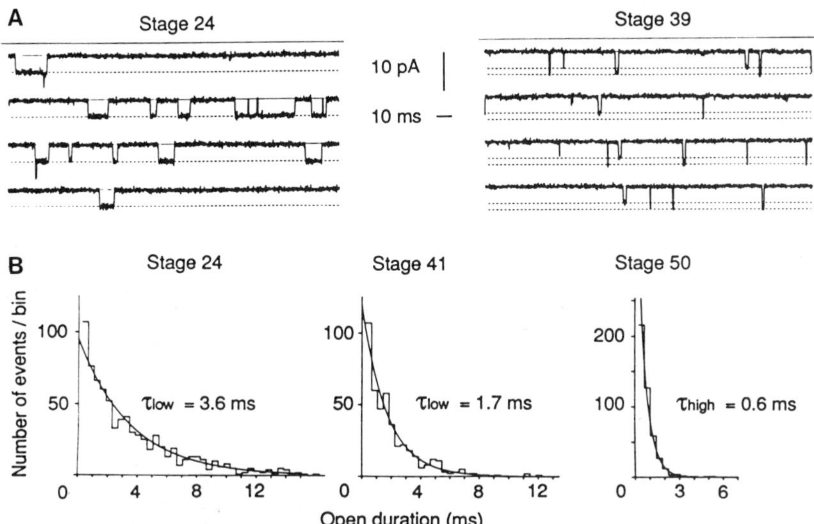

Fig. 2.   *A. Examples of single ACh receptor channel openings recorded from nonsynaptic regions of stage 24 (left panel) and stage 39 (right panel) myotomal muscle. Recordings were made at +40mV applied potential from cell-attached patches (21-24°C). Each panel shows a set of consecutive traces. Only one conductance level (dotted line) was evident in the recordings from the early stage. Two conductance levels were seen in the later stage recordings and the open durations were shorter than those of the earlier stage. B. Open duration histograms of ACh receptor channels. These histograms illustrate the decrease in channel open time which occurs during development of myotomal muscle. The histograms were compiled from single openings which were separated by gaps of 0.2 msec or more from neighboring events. The recordings were made at 0 mV applied potential from nonsynaptic regions of myotomal muscle. Each histogram is fitted by the maximum likelihood estimate of a single exponential function. Note that the histograms at stages 24 and 41 correspond to the low conductance channel openings and the histogram for stage 50 corresponds to openings by the high conductance channel .*

fertilization. At 45 hours of age, a second class of higher conductance (60 pS), brief open time (<1 msec) channels began to be expressed in significant numbers and, over the course of 4 to 5 days, became the most frequently observed channel type on the nonsynaptic membrane. At about the time that the high conductance channel appeared, we also observed a change in the properties of the low conductance channel. Its open time decreased from 3 msec to approximately 1.5 msec and remained at that value for the rest of development.

Are there any differences in the functional properties of ACh receptors at synaptic and nonsynaptic regions of developing muscle? The data from noise studies (Kullberg et al., 1981; Brehm et al., 1982; Kullberg and Kasprzak, 1985), single channel recordings (Brehm et al., 1984b) and endplate current decay (Kullberg and Kasprzak, 1985) all suggest that the development of ACh receptor function proceeds in parallel at synaptic and nonsynaptic regions. To demonstrate conclusively the similarity of channel function in the two regions, it is necessary to compare single channel records at synaptic and nonsynaptic regions. Such comparative studies in developing *Xenopus* muscle are difficult because the synaptic regions, once deprived of the their innervation (which is necessary to access the synaptic receptors), do not show any obvious morphological specialization under the light microscope. To visualize the postsynaptic membrane, we labelled the synaptic receptor aggregates on dissociated myotomal muscle cells with rhodamine-conjugated alpha-bungarotoxin (Fig. 3). The toxin concentration and incubation time was adjusted so that many of the synaptic receptors were not labelled. Following a brief incubation in toxin, the muscle was thoroughly washed free of toxin insuring that binding to additional ACh receptors did not occur. This technique permitted visualization of the synaptic receptor aggregates for the purpose of placement of the electrode while leaving substantial numbers of unbound receptors for physiological recording. The predominant channel revealed by these recordings had a high conductance (60 pS) and brief open time (<1 msec) which were identical to the properties of the predominant class of nonsynaptic channel (Kullberg et al., 1986). A second, less numerous class of channels with a lower conductance (32 pS) was also observed at synaptic regions. Although the conductance of this smaller class of channel was similar to that seen in the nonsynaptic region, the open time was briefer (<1 msec). We do not yet know whether this is a synaptic specialization or an effect of the dissociation procedure on the function of the low conductance channel.

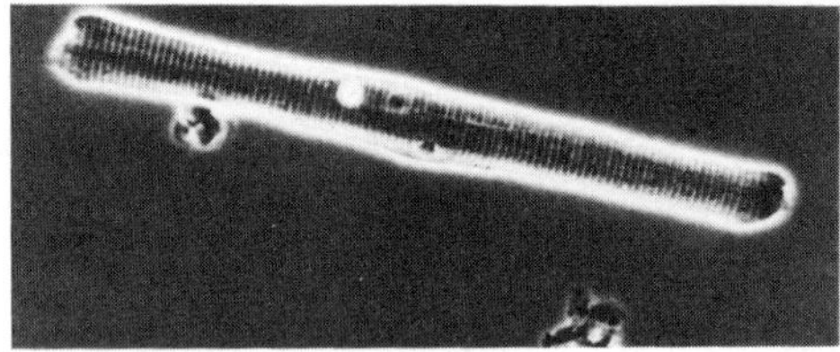

Fig. 3.   *Phase-contrast (left) and fluorescence (right) photographs of a single dissociated myotomal muscle cell from a stage 46 Xenopus tadpole. The postsynaptic aggregates of ACh receptors at either end of the muscle cell are fluorescently labelled with rhodamine-conjugated alpha-bungarotoxin.*

How do different muscles compare in their development of ACh receptor function? ACh receptor development is dependent on the type of muscle in which the receptors are found. For example, chick pectoral muscle differs from *Xenopus* myotomal muscle and rat skeletal muscle in that it exhibits no change in functional properties of ACh receptors during development (Schuetze, 1980). Comparison of *Xenopus* muscle types, differing in the contraction speed and synaptic structure, reveals different patterns in the development of synaptic currents. This implies that differences exist in the development of ACh receptor channel function in those muscles. For example, the interhyoideus, which is a slowly contracting and rhythmically active mouth muscle, has slow synaptic currents throughout development. By contrast, the extraocular muscle, which is a fast twitch muscle, develops fast synaptic currents. Single channel studies of nonsynaptic ACh receptors in these two muscles reveal the presence of ACh receptor channels with kinetic properties consistent with measurements of synaptic current decay (Kullberg and Owens, unpublished observations). Whether it is the muscle or the nerve or both which dictates the properties of developing ACh receptors remains to be determined.

In summary, our results show that the ACh receptor channel function is dependent both on the developmental stage and the type of muscle. However, within a given muscle type there appears to be no topographical segregation of different receptor types. Our findings further indicate that the terms 'embryonic', 'adult', 'junctional', and 'nonjunctional' in reference to the type of channel type must be used with caution. Channels which were once thought to be confined to synaptic sites are present elsewhere on the muscle, and channels which were once thought to be present mainly in embryonic muscle may be the predominant channel in some types of mature muscle.

THE MOLECULAR BASIS OF CHANGE IN ACH RECEPTOR CHANNEL FUNCTION

The underlying causes of the developmental changes in channel function have been a subject of great interest. As described above, two chronologically separable alterations in channel function have been observed in developing *Xenopus* muscle. The earliest alteration which occurs during development is a two fold reduction in the mean channel open duration of the low conductance channel (Leonard et al., 1984; Kullberg et al., 1986; Owens and Kullberg, 1986). Little is known about the factors which govern this particular change in function. The second developmental change that occurs is the appearance of an ACh receptor channel which has both a shorter open duration and a 1.5 fold greater conductance than the channels initially present on this membrane.

Two independent studies on embryonic *Xenopus* muscle have supported the idea that the appearance of the high conductance channel requires synthesis of a new type of channel. In these studies inhibitors of protein synthesis (Carlson et al., 1985; Brehm et al., 1987) and mRNA production (Brehm et al., 1987) were both effective in blocking the developmental appearance of the high conductance channel. More direct evidence for structural differences between the two different types of ACh receptors comes from molecular biological studies. Injection into *Xenopus* oocyte of mRNA coding for the alpha, beta, delta, and gamma subunits of calf ACh receptor results in channels exhibiting the conductance and open duration characteristic of the low conductance type of channel (Mishina et al., 1986). By contrast, when mRNA coding for alpha, beta, delta, and the newly discovered epsilon subunit were expressed in the oocyte, a channel was observed which had the functional properties characteristic of the high conductance type of channel. Northern blot analysis of the mRNA species present in embryonic and adult muscle further support different subunit composition for the high and low conductance ACh receptor channels. Specifically, embryonic calf muscle had high levels of gamma subunit mRNA in contrast to adult muscle which had low levels of mRNA for this subunit. The converse was observed for epsilon subunit mRNA; the levels were low in embryonic muscle when compared to mature calf muscle.

Are we to conclude, on the basis of these findings, that developmental changes in ACh receptor function require synthesis and assembly of a new type of ACh receptor? In contrast to our findings on *Xenopus*, there are several electrophysiological studies on developing rat muscle which support an alternative idea; that post-translational mechanisms are involved in the switch from low conductance to high conductance type channels (Michler and Sakmann,1980; Schuetze and Vicini, 1986). The most direct evidence comes from the observation that the changes in channel function occur more rapidly than the appearance of alpha-bungarotoxin binding sites. The conclusion from these studies on rat muscle was that pre-existing low conductance ACh receptors must have undergone post-translational conversion to the high conductance form in order to account for the rapid functional changes. A second, and less direct argument, stems from the observation that ACh receptor function is thought to depend upon the location of the protein in the muscle membrane (Cull-Candy et al., 1982). Channels at the synapse are hypothesized to be specifically converted to the high conductance type (Brenner and Sakmann, 1983). Localized conversion of assembled membrane receptors is intuitively difficult to explain on the level of transcriptional control. One way in which the differences between the physiology and molecular findings on mammalian muscle could be reconciled is if substitutions between epsilon and gamma subunits could be made within fully assembled channels in the membrane. If this were true, then alterations in function could occur in the absence of an altered number of toxin-binding sites. Such a mechanism could also account for synaptic localization of the high conductance channels if, for instance, synthesis of epsilon subunit were to occur specifically beneath the synapse. We mention these as possible explanations, but point out that there is no direct evidence to support these ideas.

Our physiological findings on *Xenopus* myotomal muscle are consistent with the idea that the functional properties of the ACh receptor are determined prior to insertion into the plasma membrane. In *Xenopus* myotomal muscle, unlike in developing mammalian muscle, the rate of appearance of alpha-bungarotoxin binding sites is consistent with the observed rate of change in channel function (Brehm et al., 1985). Further, our comparisons of ACh receptor channel function at synaptic and nonsynaptic sites indicate that synapse-specific channel function does not exist in developing amphibian muscle (Kullberg et al., 1981; Kullberg and Kasprzak, 1985). These observations in amphibian muscle prompted us to re-examine the question of synapse-specific ACh receptor function in mammalian muscle.

COMPARISON OF SYNAPTIC AND NONSYNAPTIC ACH RECEPTOR CHANNELS IN ADULT MAMMALIAN MUSCLE

To examine the functional properties of ACh receptors on mammalian muscle we used the flexor digitorum brevis (fdb) muscle from the hind feet of adult mice (Brehm and Kullberg, 1987; Henderson et al., 1987). This muscle was a convenient preparation because it could be enzymatically dissociated into individual fibers, each of which retained a single, readily indentifiable postsynaptic structure (Fig. 4). Cell-attached single channel recordings of ACh-activated currents were made from the synapse as well as from sites all along the muscle cell. Due to the high receptor density at the synapse, the event frequency was generally higher than that observed at nonsynaptic locations. Inclusion of either 250 or 500 nM ACh in the patch pipette in most cases elicited sufficient numbers of ACh receptor channel openings for analysis at all locations, but did not cause substantial desensitization.

We first examined freshly dissociated cells in order to determine the properties of ACh receptor channels on fdb muscle prior to the onset of denervation effects. In this 'innervated' muscle approximately one-third of the membrane patches exhibited only one amplitude class of channel currents (Table 1). This class of channels measured 69 pS, indicating that the openings were from the high conductance channel type. In two-thirds of the patches we also observed the infrequent occurrence of a smaller amplitude class of currents, which measured 45 pS. This channel type corresponded to the low conductance channel observed for both embryonic and denervated muscle (see below). The open duration for the high conductance channel on fdb measured approximately 2 msec when the patch was hyperpolarized an additional 60 mV beyond the cell membrane resting potential. Too few events could be collected in freshly dissociated tissue to obtain accurate measurements of open duration for the low conductance channel class. From these observations we conclude that both low and high conductance channel types are represented in innervated muscle, but the vast majority of channel openings were contributed by the high conductance channel type.

The ACh receptor channels were also examined on fdb muscle which had been denervated *in vivo* and maintained in a denervated state for periods up to 3 weeks. Denervations longer than 5 days required that the muscle be repeatedly denervated at 4 day intervals to retard reinnervation. In order to examine newly synthesized ACh receptors, those present at the time of denervation were irreversibly blocked by subcutaneous injections of alpha-bungarotoxin. For the purpose of single channel recordings, the denervated fdb muscle was removed and dissociated immediately prior to recording. Like the innervated muscle, we observed both high and low conductance channel classes at all times following denervation. However, a remarkable difference in the proportions of the low and high conductance channel openings was observed between innervated and denervated muscle. The low conductance channel contributed over 70% of the total events at all times beyond 5 days after denervation. This contrasts to <3% of total events by this channel type in innervated muscle. These single channel observations confirm previous noise analysis studies which indicated that increased density of a lower conductance, longer open time channel accompanied denervation of skeletal muscle (Katz and Miledi, 1972; Neher and Sakmann, 1976) From these observations it is clear that the state of innervation controls, to a large extent, the type of ACh receptor which is present in muscle membrane.

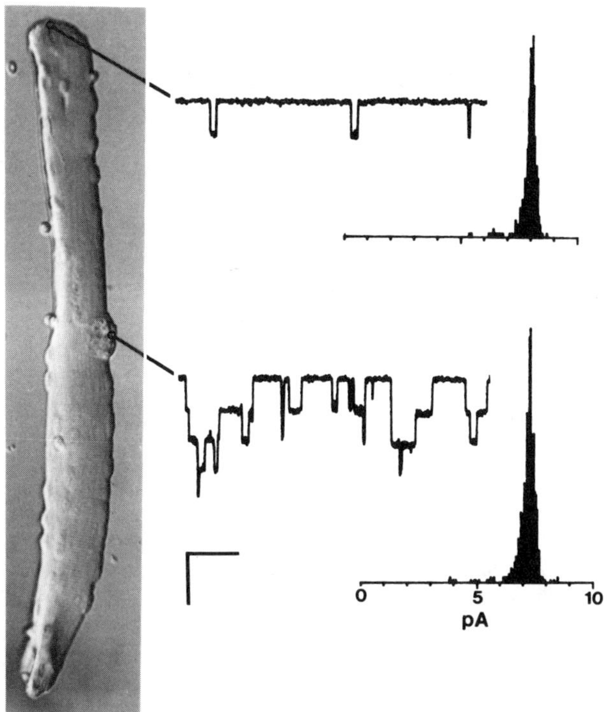

Fig. 4.   *Recordings of ACh-activated currents from the synapse (lower) and the myotendinous endings (upper) of a muscle cell. The cell was dissociated from the flexor digitorum brevis (toe) muscle of an adult mouse. The location of the two recordings are shown on the interference contrast photograph of the muscle cell. The amplitude histogram for the single channel currents is shown for each recording site after multiple level openings were excluded. Calibration bars represent 15 msec and 10 pA (vertical calibration corresponds to 50 μm for the photograph). Data taken from Brehm and Kullberg, 1987.*

Can any evidence be found for synapse-specific channel function on innervated muscle? In both innervated and denervated muscle the channel conductance for both channel types was similar in synaptic and nonsynaptic membrane. In denervated muscle the mean open time of the both channel types was independent of recording location. Thus, on fdb muscle, we found no evidence for functional differences in channel properties on the basis of proximity to the synapse. What about possible differences in the distribution of the two channel types on fdb muscle? Previous studies on mammalian muscle have suggested that the high conductance channel predominates at the synapse while the low conductance channel predominates in nonsynaptic membrane (Cull-Candy et al., 1982; Brenner and Sakmann, 1983). To test this hypothesis, we quantitated the relative number of openings by high conductance channels at 3 different locations on the cell; at the synapse, at the myotendinous ends of the cell, and in nonsynaptic locations between the synapse and the ends of the cell. In innervated muscle we observed no dependence on the relative predominance of either channel type on recording location. The high conductance channel constituted 97% of the total openings at synaptic, nonsynaptic, and myotendinous regions. On the basis of these findings we conclude that, contrary to a widely held viewpoint, the high conductance channel type is not selectively expressed at the synapse in innervated fdb muscle.

Table 1. Properties of the high conductance receptor on innervated fdb muscle

Region	Conductance (pS)*	Open time (msec)*	% High γ channels[†]
Synaptic	68+/-7 (10)	1.9+/-0.3 (10)	97 (2 of 10)
Nonsynaptic	70+/-5 (7)	1.8+/-0.4 (9)	97 (4 of 9)
Myotendinous	69+/-7 (7)	2.0+/-0.5 (8)	96 (6 of 10)

*Mean values (+/-SEM) are shown. Values in parentheses refer to the number of patches tested.

†Mean percentage of openings by high conductance (γ) channels for all patches tested. Values in parentheses indicate the number of patches that showed exclusively high conductance channel openings.
Data taken from Brehm and Kullberg, 1987.

Does the lack of segregation between low and high conductance channel types also hold true for denervated fdb muscle? To address this question we quantitated the proportion of openings by the low conductance channel type at synaptic and myotendinous regions of denervated muscle (Henderson et al., 1987). Examination of these fibers following various periods of denervation demonstrated that, overall, no significant difference could be found between the proportion of low conductance channels present at the synapse and at the ends of the muscle cell (Fig. 5). Moreover, within 5 days after denervation we observed greater than 70% of the channel openings to be of the low conductance channel type. This lack of difference in the proportion of low conductance channels continued to hold true up to the longest times tested (3 weeks). These findings support the view that channels are not specifically modified to assume the high conductance mode at the synapse of denervated muscle. We propose that in both innervated and denervated muscle the distribution of channel types along the muscle membrane is independent of proximity to the synapse.

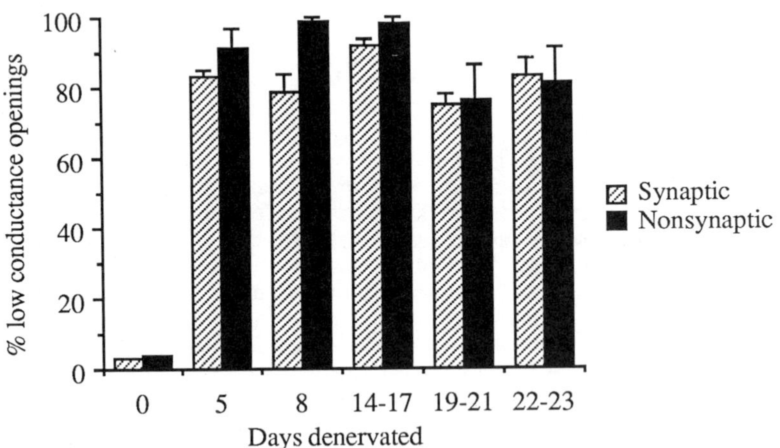

Fig. 5. *Time-dependent changes in the proportion of low conductance ACh receptor channels following denervation of fdb muscle. Receptor channels present at the time of denervation were blocked by injections of alpha-bungarotoxin into the mouse footpads. The proportion of low conductance openings were quantitated from the frequency of single channel events recorded from either the synaptic region or the nonsynaptic region. Taken in part from Henderson et al., 1987.*

FACTORS GOVERNING THE EXPRESSION OF THE DIFFERENT ACH RECEPTOR CHANNELS

There is a large body of literature supporting a role of nerve in the regulation of nonsynaptic ACh receptor density. For example, during development a dramatic decrease in nonsynaptic receptor density occurs subsequent to innervation of both mammalian (Bevan and Steinbach, 1977) and amphibian (Chow and Cohen, 1983) muscle. Additionally, both the density (Axelsson and Thesleff, 1959) and the rate of synthesis (Brockes and Hall, 1975; Merlie et al., 1984) of ACh receptors increase following denervation of adult mammalian muscle. This increase in nonsynaptic receptor density is suppressed by reinnervation (Frank et al., 1975) or by direct stimulation of the muscle cell (Lomo and Rosenthal, 1972) (which presumably mimics the excitatory action played by nerve).

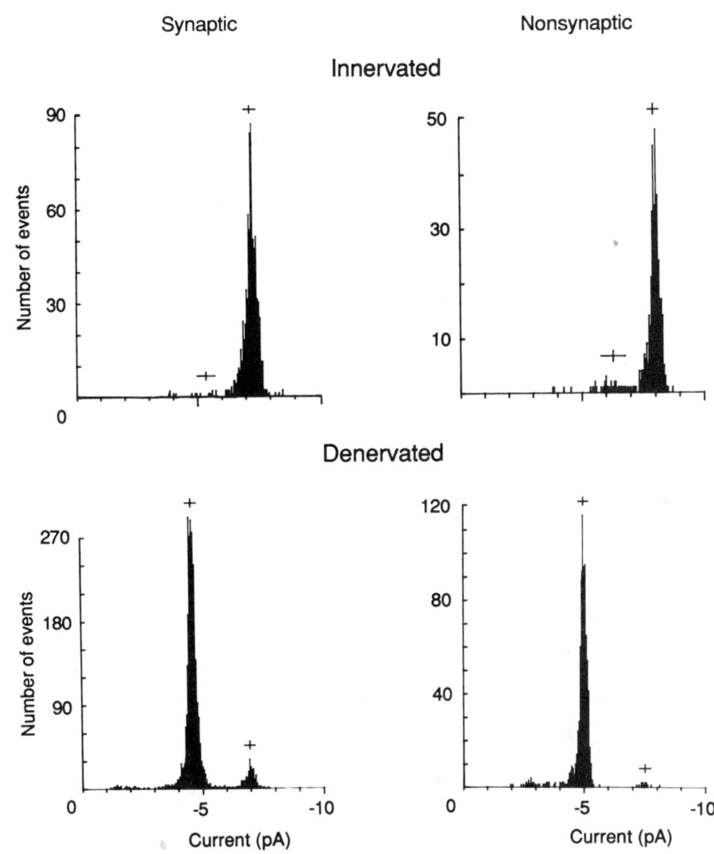

Fig. 6.   *Representative amplitude histograms of ACh-activated single channel currents comparing synaptic and nonsynaptic recordings from both innervated and denervated mouse fdb muscle. These histograms demonstrate that synaptic and nonsynaptic receptors do not differ within muscle cells but do differ between innervated and denervated muscle.*

Does the state of innervation control the type of ACh receptor which is expressed in muscle? Our findings on fdb muscle (see above) as well as many earlier studies (Katz and Miledi, 1972; Neher and Sakmann, 1976) have indicated that denervation supersensitivity is largely the result of an increase in the appearance of the low conductance type of ACh receptor channel (see Fig. 6). On the basis of channel openings, the low conductance channel comprises only 3% of channel population in both synaptic and nonsynaptic membrane of innervated fibers (Fig. 6). However, this channel accounts for more than 70% of either synaptic or nonsynaptic channels in denervated muscle (Fig. 6). The finding that denervation causes increased synthesis of low conductance channels is a compelling argument for neuronal control over this channel type in adult muscle. Similar control by nerve is suggested by the observation that low conductance channels predominate on embryonic muscle prior to establishment of nerve contact (Brehm et al., 1982; Brehm et al., 1984; Siegelbaum et al., 1984). Thus, in some muscles, innervation rather specifically suppresses the synthesis of the low conductance channel type.

In addition to the regulation of synthesis of the low conductance channel, is there direct evidence for neuronal regulation of the high conductance channel? Two separate lines of evidence suggest that the expression of the high conductance channel is much less dependent on innervation than the low conductance channel.

First, our data from denervated adult mammalian muscle indicate that high conductance channels continue to be synthesized at a characteristic low rate, long after removal of nerve. In addition, Brenner et al. (1983) observed that when *transient* points of nerve contact were made on denervated muscle the developmental switch to the high conductance channel type still occurred. Therefore, even though the nerve was no longer contacting the muscle, the high conductance channels continued to be synthesized. Both of these observations are consistent with a programmed developmental commitment in muscle to synthesize the mature form of the ACh receptor.

The second line of evidence supporting the idea that expression of the high conductance channel is independent of nerve comes from observations on dissociated cell cultures of embryonic muscle. In both rat myoblasts (Siegelbaum et al., 1984) and embryonic *Xenopus* myotomal muscle (Brehm et al., 1982; Brehm et al., 1984a) the high conductance channels appear in the absence of nerve. Primary cultures of embryonic rat muscle show very little time dependent increase in the proportion of high conductance channels. However, in aneural *Xenopus* muscle cultures, the developmental increase in the proportion of high conductance channels was comparable to that observed for *in vivo* muscle (Fig. 7). These electrophysiological measurements support the idea that nerve is not required for the synthesis of the high conductance channel. However, these measurements do not provide quantitative estimates of channel number. Therefore, it is possible that innervation is required for the eventual acquisition of the full complement of high conductance channels.

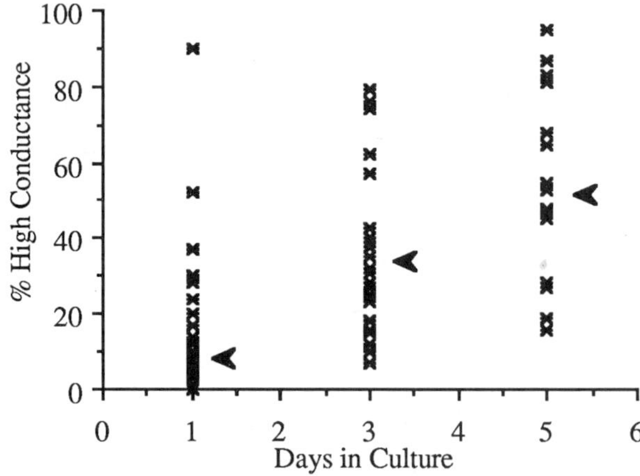

Fig. 7.   Developmental changes in the proportion of openings by the high conductance channel type on aneural Xenopus myotomal muscle in cell culture. The cultures were prepared from stage 17 myotomes and grown for the indicated period (data taken in part from Brehm et al.,1984). The mean percentage of high conductance channels at each time is indicated by the arrow.

How might the developmental changes in receptor function be regulated by the state of innervation? The many observations from embryonic and denervated muscle suggest the following hypothetical model for development of fast twitch skeletal muscle (shown in Fig. 8): ACh receptors appear prior to innervation and these channels are predominantly the low conductance type. Importantly, however, there are some high conductance channels prior to innervation. Contact by nerve provides a signal for clustering of the channels at the synapse (Anderson et al., 1977), but as shown in several studies, the aggregation forces are likely to act equivalently on both channel types (Brehm et al., 1984a; Kullberg and Kasprzak, 1985). Largely as a result of synaptic activity by the nerve, the synthesis of the low conductance channel is specifically suppressed, leaving almost exclusive territorial rights to the high conductance channel. Although the high conductance channel is most highly concentrated at the synapse, it also dominates the nonsynaptic membrane. This highly simplified scheme assumes that the principal effect of innervation is to down regulate the level of synthesis of the low conductance channel. This idea is illustrated in the lower panel of Figure 8 in which denervation of adult muscle elevates the number of low conductance channels without substantial alterations in the number of high conductance channels. One important direction for future studies will be to quantitate the levels of low and high conductance channels under the conditions illustrated in Figure 8.

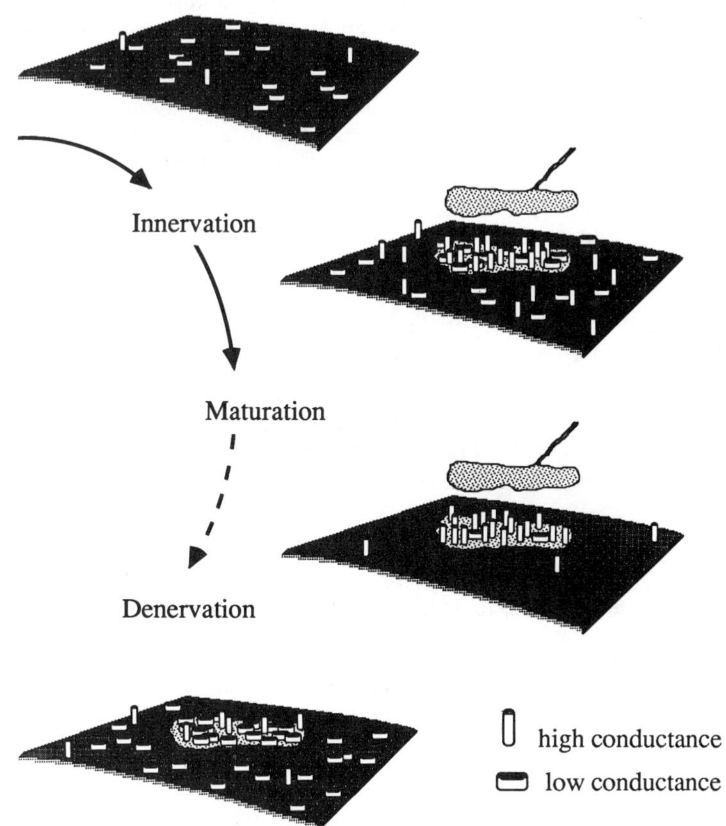

Fig. 8.   Schematic model showing changes in the proportions and distributions of the two different ACh receptor channel types which occur during development and following denervation of skeletal muscle.

WHY ARE TWO DIFFERENT ACH RECEPTOR CHANNEL TYPES EXPRESSED ON SKELETAL MUSCLE?

The ACh receptor channel, like many other types of membrane channels, has multiple functional classes. The significance of expressing multiple channel types is not known and any ideas in that regard remain speculative. With two channel types, a muscle can regulate the duration of its synaptic currents by altering the relative numbers of fast and slow channels. Muscles could tailor the time course of their synaptic currents to match their physiological functions, such as contraction speed. Slowly contracting muscles, such as the classical slow amphibian muscle fibers, which do not generate action potentials and have a widely distributed innervation, would benefit from having channels which pass charge over a prolonged period of time. Such slowly gating channels would, however, be a serious impediment to the function of fast twitch muscles which need to generate rapid, repeated contractions. Thus, we hypothesize that ACh receptor channels are regulated mutually with other muscle proteins, such as myosin ATPase and voltage-gated channels, which give muscles their distinctive slow or fast contractile properties. Under this hypothesis, a developmental change in ACh receptor channel properties could be related to a changed pattern of contractile activity of the muscle throughout development. Indeed, the first appearance of high conductance, fast gated channels in *Xenopus* myotomal muscle is coordinated with hatching and the onset of swimming activity.

However, the differences in functional properties may not be the principal reason for evolving two structurally distinct channel types. As previously discussed, an important difference in channel types resides in the effectiveness by which nerve regulates their rate of synthesis. This difference in the regulation between low and high conductance channels might have important developmental consequences. For instance, on noninnervated (either embryonic or denervated) muscle, a high receptor density would be provided by the synthesis of low conductance receptors, so that any contact by nerve would result in functional synaptic signalling. Following prolonged signalling by nerve, the density of the low conductance channels would be reduced, due to selective suppression of their synthesis. In this way the maintenance of the high density of synaptic receptors, characteristic of adult innervated muscle, would be left to the more slowly synthesized high conductance ACh receptor. To test these ideas will require new approaches. Application of recombinant DNA technology and the development of antibodies which distinguish between the types of receptor channels should provide important information. Such interdisciplinary approaches toward the study of the ACh receptor channel should eventually resolve many of the questions about development and regulation of ACh receptor channel function.

ACKNOWLEDGEMENTS

This work was supported by grants from the NIH (NS 18205 to PB and NS 17875 to RK) and from the NSF (BNS 86-03870 to RK). Further support provided by fellowships from the NIH (LPH and JL) and the NSF (JO).

REFERENCES

Anderson, M.J., Cohen, M.W., Zorychta, E., 1977, Effects of innervation on the distribution of acetylcholine receptors on cultured muscle cells, J. Physiol., 268:731.

Axelsson, J. and Thesleff, F., 1959, A study of supersensitivity in denervated mammalian skeletal muscle, J. Physiol., 149:178.

Bevan, S. and Steinbach, J.H., 1977, The distribution of alpha-bungarotoxin binding sites on mammalian skeletal muscle developing *in vivo*, J. Physiol., 267:195.

Blackshaw, S. and Warner, A., 1976, Onset of acetylcholine sensitivity and endplate activity in developing myotome muscles of *Xenopus*, Nature, 262:217.

Brehm, P., Bates, L., Kream, R. and Moody-Corbett, F., 1985, Inhibition of acetylcholine receptor incorporation blocks developmental changes in channel gating in *Xenopus* muscle, Soc. Neurosci. Abstracts, 11:849.

Brehm, P., Kidokoro, Y., and Moody-Corbett, F., 1984a, Acetylcholine receptor channel properties during development of *Xenopus* muscle cells in culture, J. Physiol., 357:203.

Brehm, P., Kream, R., and Moody-Corbett, F., 1987, Translational and transcriptional requirements for developmental alterations in acetylcholine receptor channel function in *Xenopus* myotomal muscle, Dev. Biol. (in press).

Brehm, P. and Kullberg, R., 1987. Acetylcholine receptor channels on adult mouse skeletal muscle are functionally identical in synaptic and nonsynaptic membrane, Proc. Natl. Acad. Sci. USA, (in press).

Brehm, P., Kullberg, R., and Moody-Corbett, F., 1984b, Properties of nonjunctional acetylcholine receptor channels on innervated muscle of *Xenopus laevis,* J. Physiol., 350:631.

Brehm, P., Steinbach, J.H. and Kidokoro, Y., 1982, Channel open time of acetylcholine receptors on *Xenopus* muscle cells in dissociated cell culture, Dev. Biol., 91:93.

Brenner, H.R., Meier, Th., and Widmer, B., 1983, Early action of nerve determines motor endplate differentiation in rat muscle, Nature, 536.

Brenner, H. and Sakmann, B., 1983, Neurotrophic control of channel properties at neuromuscular synapses of rat muscle, J. Physiol., 337:159.

Brockes, J. and Hall, Z., 1975, Synthesis of acetylcholine receptor by denervated rat diaphragm muscle. Proc. Natl. Acad. Sci. USA, 72:1368.

Carlson, C.G., Leonard, R.J. and Nakajima, S., 1985, The aneural development of the acetylcholine receptor in the presence of agents which block protein synthesis and glycosylation. Soc. Neurosci. Abstracts, 11:156.

Chow, I. and Cohen, M.W., 1983, Developmental changes in the distribution of acetylcholine receptors on the myotomes of *Xenopus laevis*. J. Physiol., 339:553.

Cull-Candy, S.G., Miledi, R. and Uchitel, O.D., 1982, Properties of junctional and extrajunctional acetylcholine-receptor channels in organ cultured human muscle fibres, J. Physiol., 333:251.

Frank, E., Jansen, J.K., Lomo, T., and Westgaard, R.H., 1975, The interaction between foreign and original nerves innervating the soleus muscle of rats, J. Physiol., 247:725.

Henderson, L.P., Lechleiter, J. and Brehm, P. 1987, Single channel properties of newly synthesized acetylcholine receptors following denervation of mammalian skeletal muscle, J. Gen. Physiol., (in press).

Katz, B. and Miledi, R., 1972, The statistical nature of the acetylcholine potential and its molecular components, J. Physiol., 244:703.

Kullberg, R.W., Brehm, P. and Steinbach, J.H., 1981, Nonjunctional acetylcholine receptor channel open time decreases during development of *Xenopus* muscle. Nature, 289:411.

Kullberg, R. and Kasprzak, H., 1985, Gating kinetics of nonjunctional acetylcholine receptor channels in developing *Xenopus* muscle. J. Neurosci., 5:970.

Kullberg, R.W., Lentz, T., and Cohen, M., 1977, Development of myotomal neuromuscular junction in *Xenopus laevis:* an electrophysiological and fine structural study, Dev. Biol., 60:101.

Kullberg, R.W., Mikelberg, F.S. and Cohen, M.W., 1980, Contribution of cholinesterase to developmental decreases in the time course of synaptic potentials at an amphibian neuromuscular junction, Dev. Biol., 75:255.

Kullberg, R. and Owens, J.L., 1986, Comparative development of endplate currents in two muscles of *Xenopus laevis*, J. Physiol., 374:413.

Kullberg, R., Owens, J.L. and Brehm, P., 1986, Development of nicotinic acetylcholine receptor function, Proc. Eighth Annual Conference of IEEE in Medicine and Biology 8:948.

Leonard, R.J., Nakajima, S., Nakajima, Y., and Takahashi, T., 1984, Differential development of two classes of acetylcholine receptors in *Xenopus* muscle in culture, Science, 226:55.

Lomo, T. and Rosenthal, J., 1972, Control of ACh sensitivity by muscle activity in the rat, J. Physiol., 221:493.

Merlie, J.P., Isenberg, K.E., Russell, S.D. and Sanes, J.R., 1984, Denervation supersensitivity in skeletal muscle: analysis with a cloned cDNA probe, J. cell. Biol., 99:332.

Michler, A. and Sakmann, B., 1980, Receptor stability and channel conversion in the subsynaptic membrane of the developing neuromuscular junction, Devel. Biol., 80:1.

Mishina, M., Takai, T., Imoto, K., Noda, M., Takahashi, T., Numa, S., Methfessel, C.

and Sakmann, B., 1986, Molecular distinction between fetal and adult forms of muscle acetylcholine receptor, Nature, 321:406.

Neher, E. and Sakmann, B., 1976, Noise analysis of drug induced voltage clamp currents in denervated frog muscle fibres, J. Physiol., 258:705.

Owens, J.L. and Kullberg, R., 1986, Development of acetylcholine receptor channel function *in vivo*, Soc. Neurosci. Abstr., 12:546.

Schuetze, S.M., 1980, The acetylcholine channel open time in chick muscle is not decreased following innervation, J. Physiol., 303:111.

Schuetze, S.M. and Role, L.W., 1987, Developmental regulation of the nicotinic acetylcholine receptor, Ann. Rev. Neurosci., (in press).

Schuetze, S. and Vicini, S., 1986, Apparent acetylcholine receptor channel conversion at individual rat soleus endplates *in vitro*, J. Physiol., 375:153.

Siegelbaum, S., Trautmann, A. and Koenig, J., 1984, Single acetylcholine-activated channel currents in developing muscle cells, Dev. Biol., 104:366.

STEROIDAL REGULATION OF mRNA CODING FOR POTASSIUM CHANNELS IN UTERINE

SMOOTH MUSCLE

M.B. Boyle and L.K. Kaczmarek

Departments of Pharmacology and Physiology
Yale University School of Medicine
New Haven, CT.

INTRODUCTION

Excitable tissues can produce changes in behavioral and physiological
states of an animal by undergoing prolonged changes in their
excitability. One clear example of such a system is the smooth muscle of
the adult mammalian uterus, whose electrical properties are closely linked
to the action of steroid hormones (Burnstock et al., 1963; Marshall, 1974;
Kuriyama and Suzuki, 1976). During the course of pregnancy, the uterus
undergoes extensive development to a relatively quiescent state at
mid-gestation, and then alters its electrical properties to become highly
excitable shortly before birth occurs. The progressive shifts in
electrical properties follow changes in the hormonal status of the mother,
with progesterone levels relatively high at mid-gestation and estrogen
concentrations showing a sharp rise at the end of pregnancy.
Electrophysiological changes resembling those occurring during pregnancy
can be produced by administration of steroid hormones to ovariectomized
animals. In general, estrogen treatment enhances the uterine smooth
muscle excitability and progesterone reduces it.

These changes in excitability appear relatively slowly compared with
the majority of previously investigated neurotransmitter-induced effects
on cellular excitability. An example of a rapid electrophysiological
effect of estrogen is its action on pituitary tumor-derived GH cells which
begin to fire repetitively after minutes or seconds of exposure to
estrogen or to the peptide TRH (Dufy et al., 1979). Like other
neuromodulatory effects, this action of estrogen probably involves its
interaction with a specific membrane receptor and the rapid generation of
one or more intracellular second messengers which in turn affect the
activities of pre-existing ion channels and other proteins. In contrast,
the enhanced excitability of the uterine smooth muscle appears many hours
or days after treatment of an animal with estrogen. Based on what is
known about the molecular mechanisms of action of estrogen, it seems more
likely that the action of estrogen on uterine excitability involves
interaction with intracellular steroid receptors which can bind to DNA and
increase the transcription of specific genes (Ringold, 1985). As a
result, new proteins are synthesized in the cell. Such an action is
likely to underlie the appearance of new gap junctions in uteri exposed to
estrogen (Dahl et al, 1980, 1981). Accordingly, one possible explanation
for the modification of uterine excitability by steroid hormones is that

they alter the complement of mRNAs coding for voltage-dependent ion
channels, and consequently the synthesis of ion channels themselves.

This hypothesis cannot be tested directly and exhaustively at this
time because molecular probes are available for only a few ion channels
and/or their corresponding mRNAs. However, the work of Barnard and Miledi
and their coworkers, showing that the _Xenopus_ oocyte translation system
(Gurdon et al, 1971) could be used to express mRNA species coding for ion
channels, has provided an alternative approach (Barnard et al., 1982;
Gundersen et al., 1984a,b,c). The membrane of a _Xenopus_ oocyte contains
few endogenous ion channels. _Xenopus_ oocytes, when injected with polyA+
RNA from a variety of excitable sources, however, have been shown to
develop, over a period of days, a number of voltage-dependent ion currents
not normally present in oocytes but which correspond to ion channels found
in the tissue from which the RNA was extracted (Barnard et al., 1982;
Miledi et al., 1982; Miledi and Sumikawa, 1982; Gundersen et al.,
1984a,b,c; Sumikawa et al., 1984; Werner et al., 1985; Dascal et al,
1986). Thus, the _Xenopus_ oocyte may be used as a bioassay system to
detect mRNA species coding for ion channels.

IONIC CURRENTS EXPRESSED IN XENOPUS OOCYTES INJECTED WITH mRNA FROM RAT
BRAIN AND UTERUS

Figure 1 shows a comparison of the currents recorded from oocytes
injected with brain or uterine polyA+ RNA. The left-hand panel illustrates

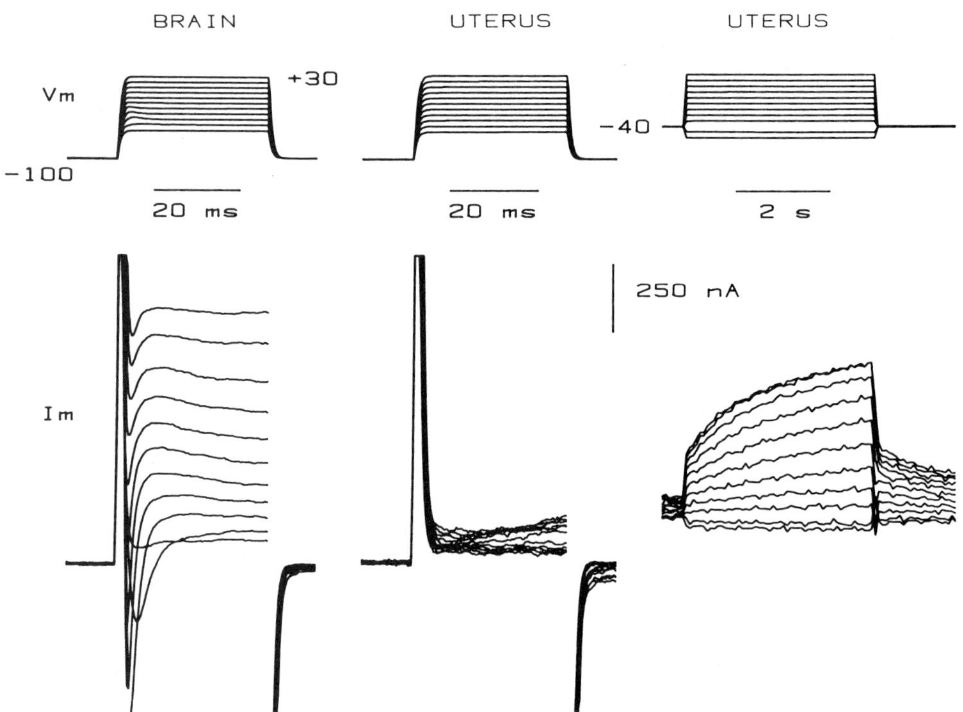

Fig. 1. Voltage-clamp records from oocytes injected with 50 nl aliquots of
brain (2 ug/ul) or uterine (4 ug/ul) RNA. For preparation of the uterine
RNA, female rats were ovariectomized and implanted with estrogen
capsules. After 3 days the animals were sacrificed and RNA was prepared
using a modification of the method of Chirgwin et al. (1979).

the voltage-dependent ionic currents in an oocyte injected with mRNA
extracted from the brain tissue of 10-day-old rats. The induced currents
are similar to those first reported by Gundersen et al. (1984c). When the
oocyte is depolarized for a number of milliseconds from a holding
potential of -100 mV, several voltage-dependent ionic currents are
observed. At potentials more positive than about -50 mV, TTX-sensitive
inward sodium currents appear. Larger depolarizations elicit outward
currents showing a prominent, rapidly activating, transient component
carried by potassium ions.

The middle panel of Figure 1 shows similar brief depolarizations of
an oocyte injected with polyA+ RNA from the estrogen-primed rat uterus.
The currents observed resemble those that would be recorded from
non-injected or water-injected control oocytes. Thus the uterine RNA
appears to contain insufficient amounts of mRNA coding for classical
sodium channels or for rapidly activating voltage-dependent potassium
currents to be readily detected in the oocyte assay. These findings have
been replicated in six independent batches of RNA from estrogen-primed
uterus. The right-hand panel shows, however, that when these RNA-injected
oocytes are subjected to much longer depolarizing pulses, an extremely
slowly activating voltage-dependent outward current is elicited. Upon
repolarization to -40 mV, the pulse-activated conductance decays very
slowly, giving rise to the prolonged tail currents shown. This slowly
activating current has not been observed in any oocytes injected with rat

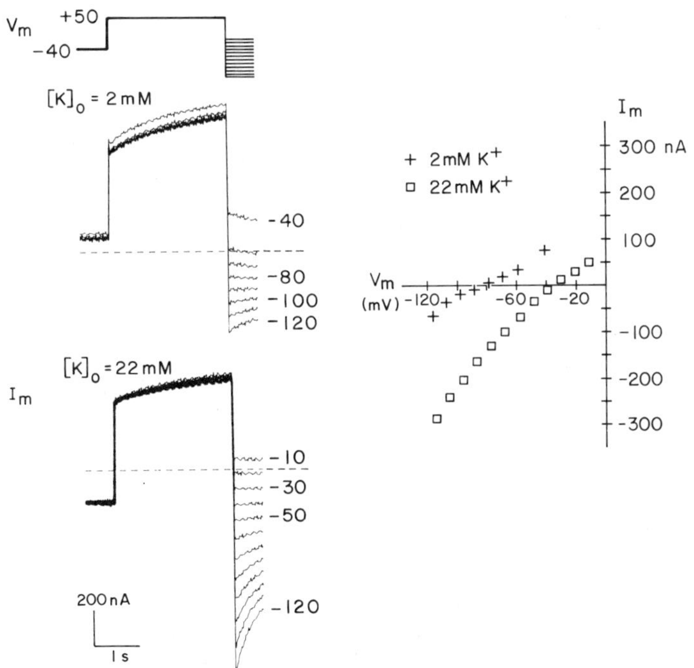

Fig. 2. Sensitivity of the slowly activating outward current to changes in
external potassium concentration. Traces on the left show activation of
the current in 2 mM and 22 mM external K+. The right hand side of the
figure plots the amplitude of the tail currents in these two potassium
concentrations as a function of membrane voltage (from Boyle et al., 1987).

brain RNA or in any water-injected or non-injected control oocytes (Boyle
et al., 1987).

PROPERTIES OF THE RNA-INDUCED SLOWLY ACTIVATING CURRENT

The slowly activating voltage-dependent outward current appears to be
highly selective for potassium ions (Boyle et al., 1987). Figure 2 shows
the activation of the current followed by repolarizations to a series of
more negative potentials in different concentrations of extracellular
potassium ions. The slowly decaying tail currents reverse at about -90 mV
in 2 mM K^+, a potential close to the reversal potential for potassium
ions in _Xenopus_ oocytes. In 22 mM external potassium, reversal of the
tail currents shifts to about -35 mV, in good agreement with the
predictions of the Nernst equation for potassium ions. The slowly
activating current is insensitive to changes in extracellular chloride
ions.

The induced current is not blocked by replacement of the normal,
calcium-containing extracellular medium with a calcium-deficient
EGTA-containing medium (Figure 3). Thus this current differs from many
slowly activating K^+ currents that respond to previous entry and
accumulation of free calcium ions near the inner surface of the channel.

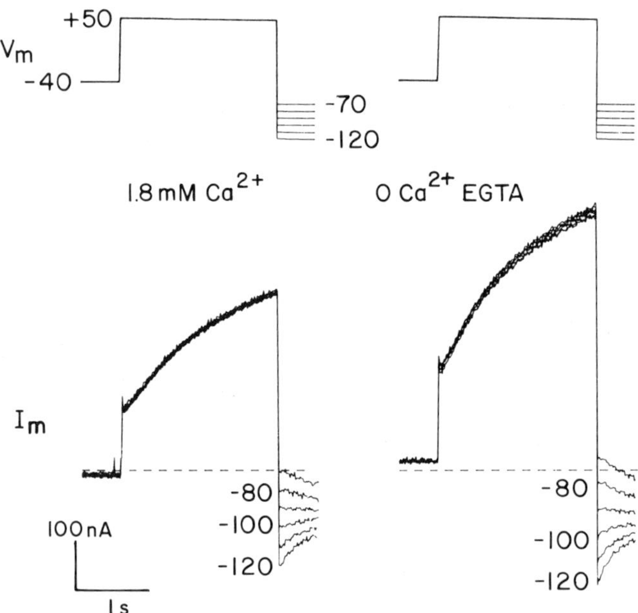

Fig. 3. Insensitivity of the slowly activating potassium current to
removal of extracellular calcium ions (from Boyle et al., 1987).

DEPENDENCE OF POTASSIUM CHANNEL EXPRESSION ON ESTROGEN

When polyA+ RNA from non-estrogen-treated uteri is injected into
Xenopus oocytes no expression of the slowly activating potassium current

can be detected. This finding is illustrated in Figure 4. The left hand
set of traces shows the expression of the slowly activating current in
oocytes injected with RNA from estrogen-primed uterus. In this experiment
the oocytes were stepped to each test potential for 4 seconds and the tail
currents at -40 mV are shown for the first 8 seconds following

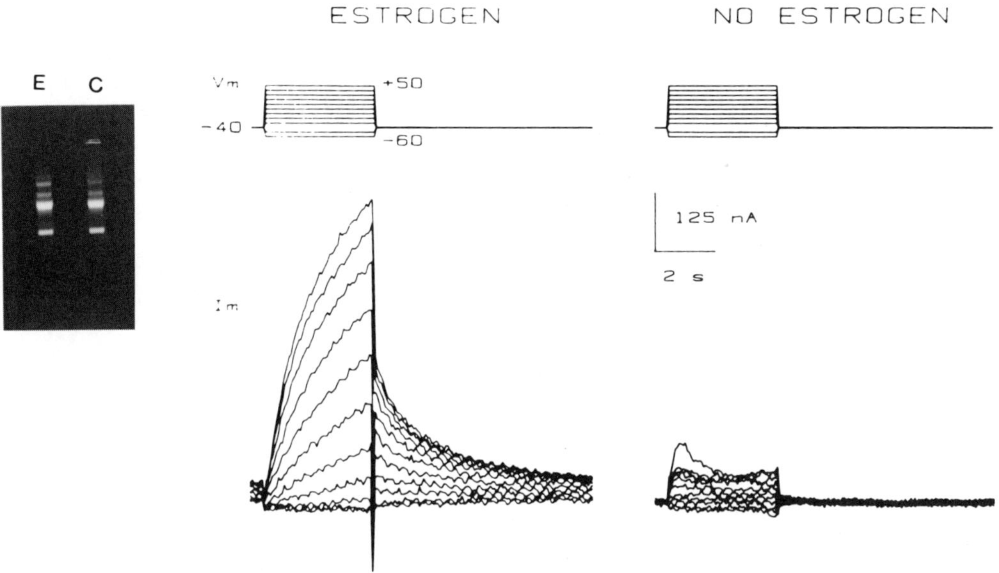

Fig. 4. Left: RNA samples from estrogen-treated (E) and
non-estrogen-treated (C) uteri after electrophoresis on a nondenaturing
agarose gel. Right: Voltage-clamp records from oocytes injected with RNA
extracted from estrogen-treated ("estrogen") and deprived ("no estrogen")
uteri. The rats from which the control ("no estrogen") RNA was obtained
were ovariectomized but not implanted with estrogen-containing silastic
capsules. The same results have been obtained from two independent pairs
of RNA samples from estrogen-primed and non-estrogen-primed rats.

repolarization. The traces on the right show the records obtained when the
same voltage-clamp protocol is applied to an oocyte that has been injected
with the same amount of polyA+ RNA from estrogen-deprived uteri. In
experiments with oocytes from five different frogs showing good expression
of the slow potassium current when injected with RNA from estrogen-primed
uteri, no expression of the slow current has been detected in any oocytes
(n=25) injected with RNA from rat uteri not exposed to estrogen. The
currents shown for the oocyte injected with the RNA from the
non-estrogen-treated uteri resemble those seen in control uninjected
oocytes from the same frog.

The quality (i.e., intactness) of each total RNA preparation has been
tested by electrophoresis on nondenaturing or denaturing agarose gels.
The left side of Figure 4 shows 1 µg of each type of RNA run under
non-denaturing conditions and stained with ethidium bromide. Both the RNA

from estrogen-primed animals (E) and the control RNA from
non-estrogen-treated animals (C) showed sharp ribosomal bands with no
apparent degradation.

EXPRESSION OF THE SLOWLY ACTIVATING POTASSIUM CURRENT INDUCED BY mRNA FROM PREGNANT UTERI

We have also examined the expression of this current in oocytes
injected with RNA from pregnant uteri. Pregnant rats were sacrificed
either at mid-gestation (day 15) or at term (day 21), but before delivery
occurred. The uterus was removed and the fetuses, placentas, and
placental attachment sites separated from the rest of the uterine horn
tissue before preparing RNA. The results described for term-uterine RNA
have been established using two independent samples of RNA. The data
obtained with mid-gestation RNA should be regarded as more preliminary as
they are based on experience with only one sample of RNA.

Figure 5 shows current records from oocytes of a single frog injected
with polyA+ RNA from estrogen-primed, term-pregnant, or mid-gestational
uteri. In each case the oocytes were depolarized for 4 seconds to +70 mV
from a holding potential of -40 mV. The current is shown at high gain so

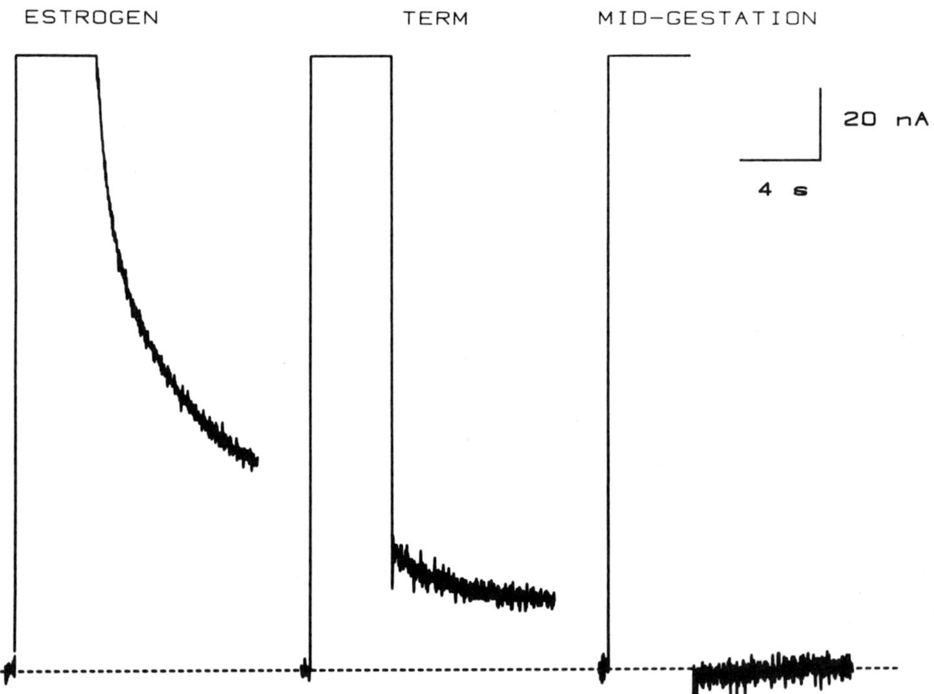

Fig. 5. Current traces showing the relative amplitudes of the slowly
decaying tail currents at -40 mV recorded in oocytes injected with RNA
extracted from estrogen-treated and pregnant uteri (Boyle, MacLusky,
Naftolin, and Kaczmarek, unpublished results). The currents were recorded
in a Na-free, Ca-free bathing solution. Slow tail currents were recorded
in 9 of 14 oocytes (representing 3 frogs) injected with RNA from term
pregnant uteri and in 0 of 5 oocytes (from 2 frogs) injected with RNA from
mid-gestational uteri.

that the relative sizes of the tail currents during the 8 seconds
following the depolarizing pulse can be compared. (The current during the
pulse is off-scale.) Under these ionic conditions control non-injected
oocytes from the same frog showed essentially no slow tail currents, in
either the outward or inward direction, upon repolarization to -40 mV.
Thus the expression of the uterine current could be quantitated accurately
by measuring the very slowly decaying outward tail currents at -40 mV.

The slowly activating potassium current is expressed in oocytes
injected with RNA from term-pregnant uteri, although at a lower level
than in oocytes injected with RNA from estrogen-primed uteri. The
left-hand and middle traces illustrate this difference. On the other
hand, we have seen no detectable expression of the current in oocytes
injected with polyA+ RNA from mid-gestational uteri. The current trace
shown on the right for the oocyte injected with RNA from mid-gestational
uterus closely resembles those in control uninjected oocytes.

The hypothesis that estrogen regulates the concentration of the mRNA
species coding for this ion channel in the uterus predicts that the
mid-gestational uterus, which is progesterone-dominated, should contain
less of this mRNA species than the term uterus, which is
estrogen-dominated. Since the blood estrogen concentrations in the
estrogen-primed animals were designed to mimic those of late pregnancy, it
might seem that similar levels of expression should be obtained from the
estrogen-primed and term-uterine RNA. However, the free estrogen of late
pregnancy is, in fact, lower because of the presence in the blood of the
estrogen-binding protein, alpha-fetoprotein (Nunez et al., 1974). In
addition, progesterone is also present during late pregnancy and would act
to attenuate responses to estrogen through down-regulation of estrogen
receptors (Hsueh et al., 1976; Saito et al., 1985).

CONCLUSIONS

Using the Xenopus oocyte translation system to assay uterine mRNA, we
have found that an unusual potassium channel is expressed with polyA+ RNA
extracted from estrogen-treated, but not estrogen-deprived, uterine
tissue. Since the myometrial smooth muscle accounts for the bulk of the
uterine tissue, it seems most likely that the mRNA species whose
expression results in the slowly activating current originates in the
smooth muscle. The mRNA species coding for this channel appears to occur
in much higher concentrations in uterine tissue than in brain. Of course,
it is possible that some relatively small subset of brain neurons contains
this mRNA species in large amounts but that it cannot be detected because
of dilution effects from the mRNA of other cells.

The hypothesis that estrogen regulates the concentration of the
messenger RNA responsible for the expression of the slowly activating
current is supported by the presence of this RNA species in the
estrogen-dominated term uterus but not in the mid-gestational uterus.
Further experiments will be required to determine the cellular and
molecular targets of estrogen action. However, since the myometrial smooth
muscle cells themselves contain estrogen receptors (Clark et al., 1977),
it seems likely that steroid hormones may act directly on these cells to
control the expression of genes for ion channels and other components that
regulate excitability. The actions of estrogen in increasing the
concentration of an mRNA species could involve increased rates of gene
transcription and/or increased stability of the mRNA species. In order to
examine rigorously the molecular mechanisms underlying the differences in
expression in oocytes, it will eventually be necessary to obtain a nucleic
acid probe for this mRNA species by molecular cloning.

Estrogen-treated and parturient uteri generate intense spontaneous
electrical activity consisting of very slow waves of depolarization that
trigger trains of action potentials, followed by slow interburst
hyperpolarizations (Kuriyama and Suzuki, 1976). A slowly activating
potassium current such as that described here could, in principle,
function in a number of different ways to regulate this electrical
activity. For example, the slow activation of this current during a
maintained burst of action potentials could play a part in terminating the
burst. Prolonged hyperpolarizations between bursts might also serve to
remove voltage-dependent inactivation from other currents and influence
the timing and intensity of the bursts of action potentials. Experiments
to test some of these hypotheses and to compare the properties of the slow
potassium current in the injected oocytes with slow potassium currents in
smooth muscle are in progress.

The electrical individuality of a cell is largely set by the number
and nature of the potassium channels in the plasma membrane. Acute
regulation of the properties of pre-existing potassium channels in a cell
has proved to be a major strategy in the control of excitability by
neurotransmitter substances and hormones (Kaczmarek and Levitan, 1986).
The experiments described here have provided some evidence for the
induction of one class of potassium channel by estrogen in smooth muscle
during a gradual and sustained change in excitability. Changes in the
number or type of potassium channels through changes in gene expression
may prove to be a general mechanism in the responses of a variety of cell
types to steroid hormones and other factors that induce very prolonged
alterations in excitability.

ACKNOWLEDGMENTS

We would like to thank Sharon Cooperman and Gail Mandel for their
generous advice and help in making RNA and for the photograph in Figure 4.

REFERENCES

Barnard, E.A., Parker, I. & Sumikawa, K., 1982, Translation of exogenous
messenger RNA coding for nicotinic acetylcholine receptors produces
functional receptors in Xenopus oocytes, Proc. Roy. Soc. Lond. B
215:241.

Boyle, M.B., Azhderian, E.M., MacLusky, N.J., Naftolin, F., and Kaczmarek,
L.K., 1987, Xenopus oocytes injected with rat uterine RNA express very
slowly activating potassium currents, Science , 235:1221.

Burnstock, G., Holman, M.E., and Prosser, C.L., 1963, Electrophysiology of
smooth muscle, Physiological Reviews 43:482.

Chirgwin, J.M., Przybyla, A.E., Macdonald, R.J. and Rutter, W.J., 1979,
Isolation of biologically active ribonucleic acid from sources enriched in
ribonuclease, Biochem. 18:3294.

Clark, J.H., Hsueh, A., and Peck, E.J., 1977, Biochemical actions of
progesterone and progestins, Ann. N.Y. Acad. Sci. , 286:161.

Dahl, G., Azarnia, R., and Werner, R., 1980, De novo construction of
cell-to-cell channels, In Vitro , 16(12):1068.

Dahl, G., Azarnia, R., and Werner, R., 1981, Induction of cell-cell channel formation by mRNA, Nature , 289:683.

Dascal, N., Snutch, T.P., Lubbert, H., Davison, N. & Lester, H.A., 1986, Expression and modulation of voltage-gated calcium channels after RNA injection in Xenopus oocytes, Science , 231:1147.

Dufy, B., Vincent, J., Fleury, H., DuPasquier, P., Gourdji, D., and Tixier-Vidal, A., 1979, Membrane effects of thyrotropin-releasing hormone and estrogen shown by intracellular recording from pituitary cells, Science , 204:509.

Gundersen, C.B., Miledi, R. & Parker, I., 1984a, Messenger RNA from human brain induces drug- and voltage-operated channels in Xenopus oocytes, Nature , 308:421.

Gundersen, C.B., Miledi, R. & Parker, I., 1984b, Slowly inactivating potassium channels induced in Xenopus oocytes by messenger ribonucleic acid from Torpedo brain, J. Physiol. , 353:231.

Gundersen, C.B., Miledi, R. & Parker, I., 1984c, Voltage-dependent channels induced by foreign messenger RNA in Xenopus oocytes, Proc. Roy. Soc. Lond. B 220:131.

Gurdon, J.B., Lane, C.D., Woodland, H.R., and Marbaix, G., 1971, Use of frog eggs and oocytes for the study of messenger RNA and its translation in living cells, Nature , 333:177.

Hseuh, A.J., Peck, E.J., Clark, J.H., 1976, Control of uterine estrogen receptor levels by progesterone, Endocrinology , 98:438.

Kaczmarek, L.K. and Levitan, I.B., eds., 1986, "Neuromodulation: The biochemical control of neuronal excitability," Oxford University Press, New York.

Kuriyama, H. & Suzuki, H., 1976, Changes in electrical properties of rat myometrium during gestation and following hormonal treatments, J. Physiol. 260:315.

Marshall, J.M., 1974, Effects of neurohypophysial hormones on the myometrium, in: "Handbook of Physiology, Sect. 7: Endocrinology", Volume IV, Part 1, Knobil, E., and Sawyer, W.H., eds., American Physiological Society, Washington, D.C.

Miledi, R., Parker, I., and Sumikawa, K., 1982, Synthesis of chick brain GABA receptors by frog oocytes, Proc. Roy. Soc. Lond. B , 216:509.

Miledi, R., and Sumikawa, K., 1982, Synthesis of cat muscle acetylcholine receptors by Xenopus oocytes, Biomed. Res. , 3(4):390.

Nunez, E., Vallette, G., Benasssayag, G., and Jayle, M-F., 1974, Comparative study on the binding of estrogens by human and rat serum proteins in development, Biochem. Biophys. Res. Comm. , 57:126.

Ringold, G.M., 1985, Steroid hormone regulation of gene expression, Ann. Rev. Pharmacol. Toxicol. , 25:529.

Saito, Y., Sakamoto, H., MacLusky, N.J., and Naftolin, F., 1985, Gap junctions and myometrial steroid hormone receptors in pregnant and postpartum rats: A possible cellular basis for the progesterone withdrawal hypothesis, Am. J. Obstet. Gynecol. , 151:805.

Sumikawa, K., Parker, I., Amano, T., and Miledi, R., 1984, Separate fractions of mRNA from Torpedo electric organ induce chloride channels and acetylcholine receptors in <u>Xenopus</u> oocytes, EMBO J. , 3:2291.

Werner, R., Miller. T., Azarnia, R., and Dahl, G., 1985, Translation and functional expression of cell-cell channel mRNA in <u>Xenopus</u> oocytes, J. Membr. Biol. , 87:253.

SECTION 5

NEW APPROACHES TO ION CHANNEL FUNCTION AND REGULATION

FAST PATCH-PIPETTE INTERNAL PERFUSION

WITH MINIMUM SOLUTION FLOW

E. Neher and R. Eckert

Max-Planck-Institut für biophysikalische Chem.
3400 Göttingen, FRG

INTRODUCTION

Means for changing the filling solution of patch pipettes are highly
desirable whenever the influence of solution composition on cellular
processes is to be studied. This refers both to patch configurations
(e.g. studies of ionic selectivity or drug action on the single channel
level) and to tight-seal whole cell recording. In the latter case it is
particularly interesting to load cells with second messengers, regulating
proteins, antibodies or toxins. Such substances very often interfere with
seal formation, demanding seal formation to be performed while the pipette
tip contains inert solution. A fast exchange of pipette filling solution
at later times allows a timed application of the substance of interest.

To achieve this goal several experimental arrangements have been
described in the literature so far (Cull-Candy et al., 1980; Soejima &
Noma, 1984; Jauch, Petersen & Läuger, 1986; Lapointe & Szabo, 1987). One
feature common to all of these techniques is the application of pressure
or suction for exchange of the solution. This has the disadvantage that
volume flow cannot be accurately controlled. It may change with the
specific configuration at the perfusion pipette tip such as constrictions
by clogging particles or deformations of the very fragile and - in case it
is made from plastic tubing. Thus, relatively large volumes are usually
perfused for safe operation.

Here we describe a device similar to that of Lamb and Matthews (1985)
which is based on controlled displacement of very small volumes, thus
providing the following advantages

1) Parsimony with respect to the solutions, which may be very precious.
2) Ease and convenience of operation, because one filling of the
    application tool may be used for several pipettes.
3) One need not worry about where the waste goes, since the pipette
    internal volume is more than sufficient to take it up.

REQUIREMENTS FOR VOLUME FLOW

We consider the arrangement as depicted in Fig. 1A. A double
barreled plastic tubing is inserted into a patch pipette and approaches

the tip to within approximately 300 μm. The volume included within this
final 300 μm of pipette length is approximately 1 nl and would therefore
require a flow rate of 1 nl/sec in order to be exchanged within 1 second.
If solution leaves one of the barrels at this flow rate during a patch
clamp experiment it will be diverted backwards to slowly fill up the
barrel of the patch pipette. The flow pattern will be distinctly laminar,
since the Reynolds number for aqueous flow at these dimensions approaches
a critical value only above 1000 nl/sec. Thus, the solution in front of
the perfusion tool will be exchanged only over some 50 to 100 μm by bulk
flow. The remaining volume (including the cell interior in a whole cell
recording) has to be exchanged by diffusion. The mean time t required for
a particle to diffuse over a distance of 200 μm is given by the
Einstein-Smoluchowski equation

$$(x^2) = 2Dt$$

where (x^2) is the mean square linear displacement and D the diffusion
constant. With $D = 1.9 \cdot 10^{-5}$ cm^2 sec^{-1} (for KCl) a diffusion time
of approximately 10 seconds is obtained. This means that the flow rate
given above is sufficient to maintain stationary boundary conditions for
this diffusion process. Furthermore, this flow rate will prevent solution
constituents in the remainder of the pipette from diffusing towards the
tip. An argument analogous to the above one, applied to the region
between 500 μm and 150 μm from the pipette tip, shows this.

For safety, we used approximately 3 times higher flow rates (3
nl/sec) for maintaining a certain solution composition at the tip and
approximately 10-30 times higher flow rates (10-30 nl/sec) while changing
solution (i.e. stopping flow in one barrel and starting it in the other
one). During a typical experiment 1 to 4 solution changes are required.
This, together with a recording period under stationary flow conditions of
up to 1/2 hour, leads to a maximal solution requirement of 7 μl per
experiment. Our pipettes, made from relatively wide glass capillaries (2
mm o.d. 0.2 mm wall thickness, Borosilicate from Hilgenberg, Malsburg, W.
Germany) have approximately 50 μl of void inner volume, if filled to
about one third of their length. Thus, the reservoir is more than
sufficient to accommodate the solution injected.

We use a total amount of perfusion fluid of about 150 μl per barrel
(see Fig. 1.). With the thermal expansion coefficient of water
($2.1 \cdot 10^{-4}$ °C^{-1}) this leads in the worst case (no expansion of the
tubing) to 30 nl/°C. A perfusion rate of 3 nl/sec is thus sufficient to
dominate thermal artifacts unless changes in temperature larger than 1°
C/min occur.

FEP (PERFLUORO-ETHYLENE-PROPYLENE) IS AN IDEAL MATERIAL FOR FABRICATING
PERFUSION TOOLS

It is well known that plastic tubings made from polyethylene or
polypropylene can be drawn into fine capillaries under heat. We found
that perfluoro-ethylene propylene (FEP) is much better suited for that
purpose because it can be pulled into finer capillaries and is more rigid
after cooling. FEP tubing was obtained from IFK-Isofluor, Neuss 21,
W.-Germany (An American source is Chemplast Inc. Wayne, N.J.). Double
barrelled perfusion tools as depicted in Fig. 1A can be made by twisting
two lengths of ≈0.6 mm o.d. tubing (thin walled) around each other and
by pulling after local heating with a hot air jet. It sometimes helps to
further constrict the tubing by a second cycle of heating and pulling.
With some practice triple barrelled tools can be obtained, too. FEP-tools
have the advantage over glass tools that they are flexible and not so

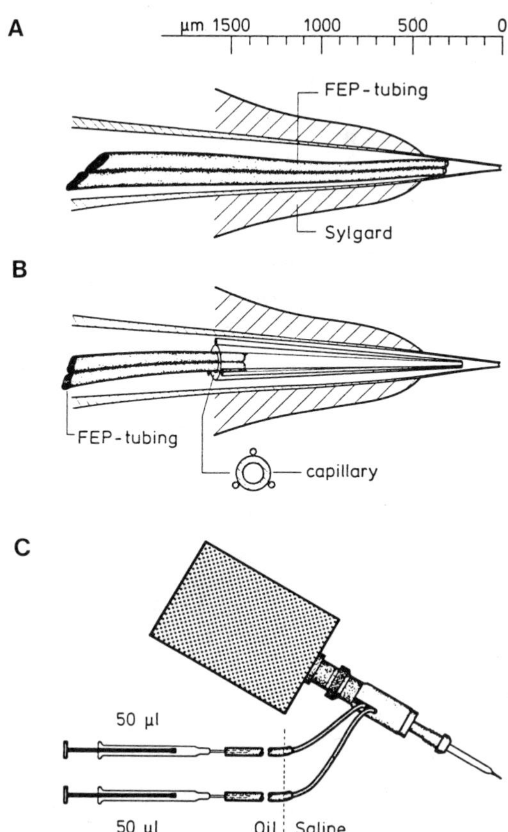

Fig. 1. Schematic drawing of parts of the apparatus. A and B: Scale
drawing of pipettes as used in most of the experiments. A
double-barrelled FEP-tubing is either inserted directly into
the pipette (part A) or glued (using silicon rubber) into
the wide end of a glass tip. The glass tip was pulled from
capillaries with cross section as indicated.
C: Schematic of head stage, pipette holder, tube connections
and microliter syringes. The tubings which insert into a
slot of the adapter piece (see text) extend to the tip of
the pipette (not drawn). Not drawn for clarity is the port
for suction on the pipette holder.

easily damaged when inserted into patch pipettes. Double-barrelled tools,
which have ∞-shaped cross section, can be pushed into the tip of the
pipette until one can observe bending of the remote parts of the tubing.
With some care they will not be deformed to the degree that outflow from
one of the barrels is occluded. With triple-barrelled tools one has to be
very careful not to advance them too far. Otherwise they will form a seal
around the perimeter of the pipette which locks the return path of the
perfusion fluid. In order to avoid some of these problems it is
advantageous to prolong the perfusion tool by a single-barrelled glass
tip, approximately one millimeter in length as shown in Fig. 1B. The tool
has to be glued into the glass tip. The special cross section depicted
in Fig. 1B secures the return path of the perfusion fluid. The single
glass tip also allows closer approach of the perfusion fluid to the
pipette tip, resulting in faster exchange times. Tips as shown can
readily be pulled on microelectrode pullers from capillaries of this

special cross section. Since we do not know a commercial source of such
capillaries, they have to be custom made by the glass blower. It should
be noted that the internal volume of the glass tip is of the order of
magnitude of 10 nl. Its content is therefore exchanged at the specified
flow rates (for solution change) in less than a second.

CONTROLLED VOLUME DISPLACEMENT

     The prime requirement for controlled volume displacement in the
nanoliter range is that the total volume connected to the perfusion tool
should be small and should be filled entirely with a noncompressible
medium. Any air bubble larger than 100 nl will degrade the performance of
the device drastically. The drawing of Fig. 1C shows the two FEP tubings
of the perfusion tool to extend from the pipette holder. A holder,
similar to the List Elektronik, nonshielded type has been supplemented
with a slitted adapter. Inside this adapter the two tubings are fed
through and glued into a 12 mm section of pipette capillary together with
the recording Ag-AgCl-wire. An O-Ring pushed over the capillary section
provides a seal such that suction can be applied to the pipette as usual.

     The ends of the FEP-tubings are pushed into approximately 20 cm
lengths of 1.52 mm o.d., o.5 mm wall thickness Tygon tubing, which connect
to Hamilton microliter syringes. Optionally, a reservoir of about 50 µl
made from a glass capillary can be inserted in between FEP and Tygon
tubing. The total volume of a perfusion barrel is only approximately 45
µl (plus the reservoir and plus the internal volume of the microliter
syringe). The syringes are driven by two independent motorized micrometer
spindles (not shown in Fig. 1) similar to Motor Mike actuators (Oriel
Comp. Stanford, Conn.) The microliter syringes and most part of the Tygon
tubings are filled with low viscosity paraffin oil (Merck). The rest of
the perfusion tool is filled with saline.

PROCEDURES

     At the beginning of an experiment the perfusion tool is mounted onto
the head stage of the patch clamp amplifier. At this time the FEP-tubing
and the Tygon tubing are disconnected, the Tygon tubing (and the reservoir
if there is any) is filled with oil up to the end. The microliter
syringes have their plungers fully inserted. The double-barrelled tip is
dipped into a drop of filtered, pipette-filling solution (pipette
removed). Filtered solution is sucked up through one of the perfusion
barrels by means of a 1 ml, hand-operated syringe. Once the FEP tubing is
filled to the end it is connected to one of the Tygon tubings taking care
that no bubble forms at the junction. The plunger of the microliter
syringe can now be withdrawn to suck more pipette filling solution into
that barrel. This barrel will provide control saline during the
experiment.

     To fill the other barrel the tool is dipped into a drop of test
solution and the filling procedure as described above is repeated. If
test solution is very precious, one can fill the bulk of the tool with
control saline and only the final 5 to 10 µl with test solution. It is
good practice to expel approximately 0.5 µl from each of the barrels
before mounting a pipette, and to watch little droplets form at the tip of
the tool to make sure that solution is being displaced.

     Patch pipettes filled to about 1/3 with either Ringer or pipette
filling solution can be pushed onto the tool and the recording wire while
tool and pipette holder are mounted on the head stage. For locating the

tip of the tool inside the patch pipette a stereo lens as used for
microsurgery, such as the OPMI 99 (Zeiss), or a good magnifying lens is
very helpful. It is best to use bright illumination from the side (e.g.
by a fibre optics bundle) and a bright background for these
manipulations. Painting the tool with a waterproof marking pen improves
visibility. Before the tip is brought to its final position some more
fluid should be expelled from both barrels in order to clear bubbles or
debris from the tip. The pipette can then be pushed back further until
the plastic tubing is seen to deform. From this time on we usually have
the control saline flowing at minimum speed (3 nl/sec) throughout the
experiment, except for episodes when test solution is applied. One
filling is sufficient for several hours of experiments, with several
pipettes used in sequence.

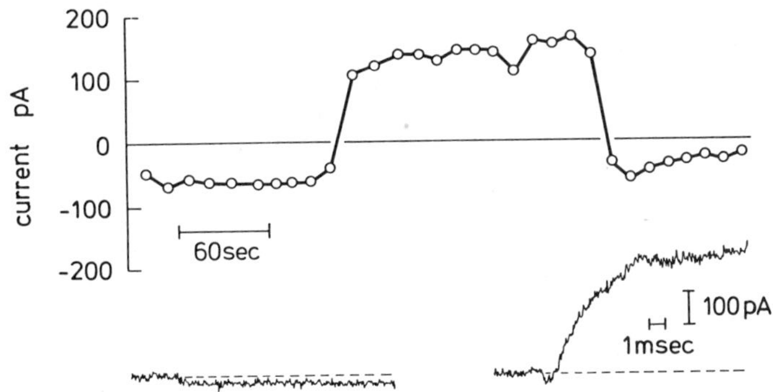

Fig. 2.   Time course of solution changes as indicated by
          voltage-clamp currents in chromaffin cells (see text for
          explanation).

Following an experimental session, the FEP tubing is disconnected
from the Tygon tubing, emptied, rinsed with distilled water, and dried by
aspiration. This way, we were able to use one tool for several weeks.
The useful lifetime of a tool is limited by wearing of material or by
collapse at the tip region. Occasional cleaning with methanol and
ultrasound is recommended to avoid accumulation of dirt which might
interfere with seal formation.

In case there are seal problems it may be helpful to abandon the
procedure of applying positive pressure during the approach of the pipette
towards the cell. If the bath is thoroughly rinsed with filtered saline,
seals form readily, even if there is solution inflow into the pipette
during the approach. Still, it may be helpful to apply positive pressure
while transferring the pipette tip through the air-water-interface.

To test the perfusion system we recorded whole-cell currents from
chromaffin cells of bovine adrenal medulla. Cells were kept in short term

tissue culture as described by Marty & Neher (1984). The control solution
was a Cs-rich solution, similar to that used by Fenwick, Marty & Neher
(1982) to record voltage dependent calcium currents. The test solution
was a K-rich solution as used by Marty & Neher (1984) to study outward
currents.

Na-current were blocked by TTX

     Cells were held at -70 mV and depolarizing test pulses of 80 mV
amplitude were applied at intervals of 13 seconds. A P/4 procedure was
used to correct for leak and linear capacitance. The resulting current
waveforms are shown in Fig. 2 (lower part). On the left side a typical
Ca-current under control conditions (Cs-rich internal solution) is shown.
This particular record was obtained later in the experiment, when
Ca-current had already partly washed out. Thus, its amplitude was very
small. The record on the right side shows a typical K-current as measured
under test conditions. When the current measured at the end of the
depolarizing episodes is plotted against time, Fig. 2 (upper part) is
obtained. Here the control solution flow was stopped after 9 depolarizing
stimuli had been applied. At the same time flow in the test barrel was
started. Depolarizing test pulses were given every 13 seconds. It is
seen that the bulk of the change in current occurred between two
successive stimuli, i.e. within 13 seconds. Similarly, when returning
from test solution to control solution the current returned within one
interpulse interval to near control conditions.

     Thus, the change of solution inside the cell is taking place within
10 to 15 seconds. This time includes both the time required for diffusion
of ions from the tip of the perfusion tool to the tip of the patch pipette
and the diffusional exchange time between pipette and cell. The latter
has been shown by Pusch & Neher (1987) to depend on pipette resistance and
on the molecular weight of the diffusing molecules.

CONCLUSIONS

     A device for internal perfusion of patch pipettes is described which
combines ease of operation with perfusion speed. It differs from
previously published perfusion systems by the use of fluorinated
polyethylene-propylene (FEP) as an ideal material for pulling single and
multi-barrelled perfusion tools, and by the use of controlled volume flow
rather than air pressure to expel fluid from the perfusion barrels. The
solution in the tip of a patch pipette can be changed within 5 to 20
seconds. Several patch pipettes can be used in sequence without refilling
the perfusion system or remounting of the patch pipette holder.

REFERENCES

Cull-Candy, S. G., Miledi, R. and Parker, I., 1980, Single
     glutamate-activated channels recorded from locust muscle fibres with
     perfused patch-clamp electrodes. J. Physiol. 321:195-210.
Fenwick, E. M., Marty, A., and Neher, E., 1982, Sodium and calcium
     channels in bovine chromaffin cells. J. Physiol. 331:599-635.
Jauch, P., Petersen, O. H., Läuger, P., 1986, Cotransporter in pancreatic
     acinar cells: I. Tight-seal whole-cell recordings. J. Membrane Biol.
     94:99-115.
Lamb, T.D. and Matthews, H.R., 1985, A precision microperfusion
     apparatus. J. Physiol. 369:4P.
Lapointe, J.-Y. and Szabo, G., 1987, A novel holder allowing internal
     perfusion of patch-clamp pipettes. In press.

Marty, A. and Neher, E., 1985, Potassium channels in cultured bovine
    adrenal chromaffin cells. <u>J. Physiol</u>. 367:117-141.
Pusch, M. and Neher, E., 1987, Kinetics of loading small cell with various
    compounds by use of patch pipettes. <u>Pflüger's Archiv</u>, Spring Meeting
    of the Physiol Ges.
Soejima, M. and Noma, A., 1984, Model of regulation of the ACh-sensitive
    K-channel by the muscarinic receptor in rabbit atrial cells.
    <u>Pflügers Arch</u>. 400:424-431.

EVIDENCE FOR A BICARBONATE CONDUCTANCE IN NEUROGLIA

Richard K. Orkand

Institute of Neurobiology
University of Puerto Rico MSC
201 Blvd. del Valle
Old San Juan PR 00901 U.S.A.

In his modest little textbook on <u>Animal Physiology</u> (2nd Edition, W.H.
Freeman and Co., New York, \$32.95) Roger Eckert (1983) devotes 12.2 cm of
text to the potassium-buffering function of neuroglia (see glial cells).
Indeed, the selective permeability of glial cells to K^+ and the role of
glial cells in K^+ homeostasis in the neuronal microenvironment has been
reasonably well established during the past 20 years (Kuffler, 1967; Coles,
1985). More recently, evidence has accumulated indicating that glial cells
may express voltage sensitive channels, albeit at low density, to other
ions, including Na^+, Cl^-, and Ca^{++} (for review see Ransom and Carlini,
1986). However, such channels appear to open at membrane potentials far
removed from those naturally occurring in glial cells; there is no evidence
they participate in glial function.

To obtain evidence for channels for ions, other than those to K^+
(Kettenmann et al., 1984), in glial cells in Necturus optic nerve, and to
study their function, Astion and colleagues (Astion et al., 1986) blocked
the K^+ channel with barium. The addition of a few mM of barium depolarizes
the glial cells from about -90 mV to -50 mV.   As shown in Fig. 1B, under
this condition the addition of HCO_3^- and CO_2 (at constant pH,) produces a
prompt hyperpolarization (Astion et al., 1987). An important observation is
that in normal Ringer the effect of HCO_3^- on membrane potential is still
present (Fig. 1A), but reduced in amplitude presumably because of the
shunting effect of the K^+ conductance. If the superfusion with raised HCO_3^-
continues, the membrane potential returns toward the resting level over the
next few minutes. When the HCO_3^- is withdrawn a transient depolarization
ensues.

It is probably possible to devise schemes using a variety of electrogenic transport systems and varying conductances to account for these
effects of HCO_3^- on the glial membrane potential. However, all the observations to date can most readily be explained by supposing that the membrane
is conductive to HCO_3^-, as are some other epithelial membranes such as the
frog choroid plexus (Saito and Wright 1984), and peritubular membranes in
rat kidney (Burckhardt et al., 1984). This simple hypothesis is consistent
with a variety of observations such as the following: 1) There occurs a
transient depolarization after the end of an exposure to HCO_3^- lasting
several minutes (Fig.1). Presumably HCO_3^- accumulates in the cells during
exposure and leaves when it is removed from the bathing solution. This

produces an outward current of a negative ion. 2) Depletion of intracellular Cl^-, by exposure to Cl^- free solutions for up to 4 h, had no apparent effect on the potential changes produced when varying HCO_3^- . Cl^- is required by many types of cells, for recovery from acidification (Astion et al. 1987; Thomas, 1984) and Cl^- movements could produce comparable potential changes. In fact, Cl^- effects on membrane potential, although difficult to resolve due to changes in junction potentials, were apparently less than those to HCO_3^- (Astion and Orkand, unpublished). This makes it unlikely that HCO_3^- crosses the membrane via Cl^- channels. 3) Besides CO_2, another way to change cell pH is to add and remove NH_4^+ from the bathing fluid. Withdrawl of NH_4^+ did not produce a detectable hyperpolarization in Necturus glial cells (unpublished experiments described in Bracho et al., 1975). This suggests that the potential changes observed with HCO_3^- were not simply secondary to pH regulation processes. Furthermore, the primary membrane potential effect of adding high CO_2, with insufficient HCO_3^- to keep pH_o constant, is a reversible depolarization (Tang et al., 1985).

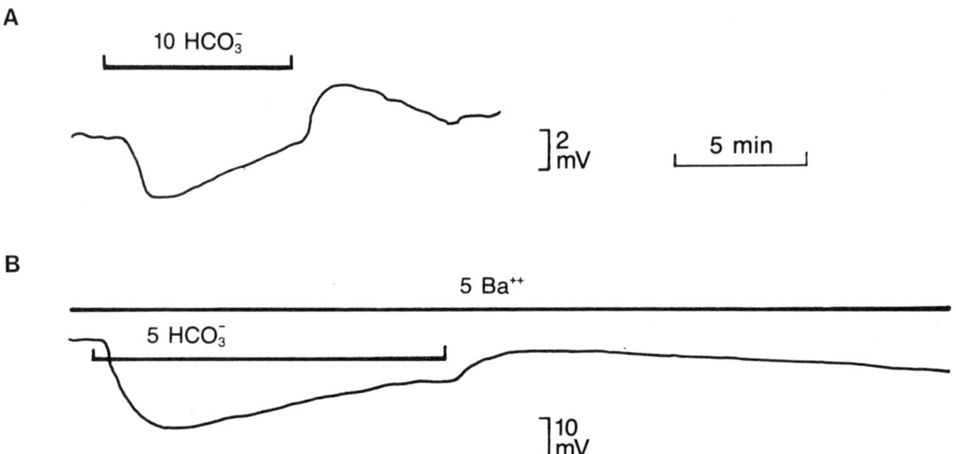

Fig. 1. Effects of HCO_3^- on glial membrane potential. These are tracings of membrane potential recordings from glial cells in the isolated optic nerve of Necturus. A. Resting potential -90 mV. The superfusing Ringer was changed from normal to one containing 10 mM HCO_3^- for the period indicated by the bar. With the addition of HCO_3^- the membrane hyperpolarized and repolarized slowly. With HCO_3^- removal there was a transient depolarization. B. In the presence of 5 mM Ba^{++} the membrane potential was -46 mV. The addition of 5 mM HCO_3^- produces a much larger hyperpolarization (gain reduced five-fold from A) with a slower repolarization. (Modified from Astion et al., 1987)

To answer the following questions we shall assume simply that there is a conductance for HCO_3^- across the glial membrane, although, until now we have not identified which charged species of the buffer complex is actually moving (HCO_3^-, OH^-, CO_3^{2-}, or H^+ in the opposite direction) or if the movement is coupled to another ion like Na^+ (Burckhardt et al., 1984). This assumption does not require the existence of a HCO_3^- channel.

1) What is the relative permeability of glial cells to HCO_3^- and K^+?

If we assume that the permeability to other ions is small, the ratio $P_{HCO_3^-}/P_{K^+}$ can be calculated using the Goldman-Hodgkin-Katz equation and the measured hyperpolarization to a step change in HCO_3^-. Using this method, Astion et al. (1987) determined that average ratio was 2.4. Since the passive ionic currents through the glial syncytium provide for the spatial buffering of K^+, one might wonder what proportion of the current will be carried by HCO_3^-. Recall that the current carried by an ion is a function of its conductance and the difference between the membrane potential and the equilibrium potential for the ion (Eckert, 1983). Conductance is a measure of actual ion flux and depends on both the permeability and the concentration (Bullock et al. 1977). Thus, in those experiments in which there is no added HCO_3^- in the Ringer, the $[HCO_3^-]$ due to atmosphereic CO_2 is about 0.2 mM and the conductance is very low. Under these conditions very little of the spatial buffer current would be carried by HCO_3^- despite the high permeability.

2) What is the HCO_3^- equilibirum potential?

The available data on the intracellular pH of glial cells in the leech indicates that it is more acid than the extracellular bathing solution. At a pH_o of 7.4 Deitmer and Schlue (1987) found pH_i to be 6.9 in HEPES buffer and 7.2 in bicarbonate buffer. At steady state conditions the Henderson-Hasselbalch equation holds, and we can calculate the HCO_3^- ratio from the following relation

$$pH_o - pH_i = \log\,[HCO_3^-]_i / [HCO_3^-]_o$$

If the intracellular pH is 7.1 the ratio of $[HCO_3^-]_i / [HCO_3^-]_o$ will be 2 and the equilibrium potential from the Nernst relation will be about -18 mV. This would explain why the glial membrane depolarizes when the K^+ conductance is decreased. At the same time it raises the question as to how (and why) the HCO_3^- distribution is maintained away from electrochemical equilibrium?

3) What are the functional consequences of a significant glial HCO_3^- conductance?

The CO_2-HCO_3^- buffer system is considered to be the most important buffer in the body because the concentration of its components can be independently regulated. Since CO_2 is quite soluble, easily and rapidly crosses cell membranes and may be readily converted to HCO_3^- (in the presence of carbonic anhydrase) the question arises as to the need for a HCO_3^- conductance. Although CO_2 and HCO_3^- may be converted to one another they are obviously not equivalent. That is, CO_2 is a volatile acid while HCO_3^- is a base and, the addition of one or the other will affect intracellular pH, and thus intracellular activity dependent on pH in different ways. Furthermore, the rapid interconversion requires carbonic anhydrase - which may not be uniformly distributed in nervous tissue.

There is plenty of room for speculation as to the possible role of a HCO_3^- conductance in glial cells. More interesting, however, is that the observation of an effect of HCO_3^- on glial cell membrane potential suggests a whole range of experiments in which one could monitor intracellular and extracellular pH and HCO_3^- with ion-sensitivie electrodes and vary extracellular pH and HCO_3^- independently. Observation of the reactions of glial cells to these challenges is likely to provide some understanding of the function of glial cells in the pH buffering of the neuronal micro-environment.

ACKNOWLEDGEMENTS

Supported by NIH Grant NS-24913. I thank Jonathan Coles and Helmut Kettenmann for helpful discussions.

REFERENCES

Astion, M.L., Obaid, A.L., Orkand, R.K. and Salzberg, B.M.,1986, Barium
block of potassium channels in glial cells of mudpuppy (<u>Necturus maculosus</u>)
optic nerve: evidence from studies with microelectrodes and voltage sensitive dyes. <u>Soc. Neurosci. Abstr.</u> 12:164.

Astion, M.L., Coles, J.A. and Orkand R.K., 1987, Effects of bicarbonate on
glial cell membrane potential in <u>Necturus</u> optic nerve. <u>Neurosci. Lett.</u>
76:47.

Bullock, T.H., Orkand, R. and Grinnell, A., 1977, "Introduction to Nervous
Systems," W. H. Freeman, San Francisco.

Bracho, H., Orkand, P.M. and Orkand, R.K., 1975, A further study of the fine
structure and membrane properties of neuroglia in the optic nerve of
<u>Necturus</u>. <u>J.Neurobiol.</u> 6:395.

Burckhardt, B.Ch., Cassola, A.C. and Fromter, E., 1984, Electrophysiological
analysis of bicarbonate permeation across the peritubular cell membrane of
rat kidney proximal tubule. II. Exclusion of HCO_3^- effects on other ion
permeabilities and of coupled electroneutral HCO_3^- transport. <u>Pflugers Arch.</u>
401:43.

Coles, J.A., 1985, Homeostasis of extracellular fluid in retinas of invertebrates and vertebrates. <u>Prog. Sensory Physiol.</u>, 6:105.

Deitmer, J.W. and Schlue, W-R., 1987, The regulation of intracellular pH by
identified glial cells and neurones in the central nervous system of the
leech. <u>J.Physiol.</u> In Press.

Eckert, R., "Animal Physiology", 2nd Ed. W.H. Freeman, San Francisco (1983).

Kettenmann, H., Orkand, R.K. and Lux, H.D., 1984, Some properties of
potassium channels in cultured oligodendrocytes. <u>Pflugers Arch.</u> 400:215.

Kuffler, S.W., 1967, Neuroglial cells: physiological properties and a
potassium mediated effect of neuronal activity on glial membrane potential.
<u>Proc. Roy. Soc. B.</u>,168:1.

Ransom, B.R. and Carlini, W.G.,1986, Electrophysiological properties of
astrocytes, <u>in</u> "Astrocytes, Vol.2" A. Vernadakis ed., Academic Press, New
York.

Saito,Y. and Wright, E.M., 1984, Regulation of bicarbonate transport across
the brush border membrane of the bull-frog choroid plexus. <u>J.Physiol.</u>
350:327.

Tang. C-M., Orkand, P.M. and Orkand, R.K., 1985, Coupling and uncoupling of
amphibian neuroglia. <u>Neurosci. Lett.</u> 54:237.

Thomas, R.C., 1984, Experimental displacement of intracellular pH and the
mechanism of its subsequent recovery. <u>J.Physiol.</u> 354:3P.

DIVALENT CATIONS AS MODULATORS OF NMDA-RECEPTOR CHANNELS

ON MOUSE CENTRAL NEURONS

Gary L. Westbrook and Mark L. Mayer

Laboratory of Developmental Neurobiology, NICHD
Bldg. 36, Rm. 2A-21 NIH, Bethesda, MD. 20892

INTRODUCTION

The term neuromodulation usually implies some combination of peptide
or catecholamine neurotransmitters and a second messenger such as cAMP
which together produce a response with a rather slow time course (Kaczmarek
and Levitan, 1986). In contrast, the presumptive transmitter for many
central excitatory synapses, L-glutamate, has been considered to function
primarily as a fast-acting relay between neurons, analogous to the action
of acetylcholine at the neuromuscular junction. However recent studies of
conductance mechanisms activated by L-glutamate on mammalian neurons
indicate that their behavior may be substantially more complex. Three
receptor subtypes for excitatory amino acids can be distinguished pharmacologically based on selective activation by N-methyl-D-aspartic (NMDA),
kainic (KA) and quisqualic (QA) acids (Watkins & Evans, 1981). An important corollary of this classification has been the recognition that
L-glutamate acts at more than one receptor type, but with a particularly
high affinity for the NMDA receptor (Olverman et al., 1984). NMDA receptors
are associated with a voltage-dependent conductance whereas kainate/quis-
qualate receptors are primarily linked to a voltage-insensitive cation
conductance (for review see Mayer and Westbrook, 1987a); however whether
these functionally distinct conductance mechanisms represent separate
molecular entities is controversial (Cull-Candy and Usowicz, 1987; Jahr
and Stevens, 1987).

The conductance mechanism activated by NMDA is of particular interest
since several of its properties suggest a role as a neuromodulator. At a
functional level, antagonism of NMDA receptors has been shown to block
induction of longterm potentiation at the Schaffer collateral/commissural
input to CA1 pyramidal cells in the hippocampus, a widely studied model of
synaptic plasticity (Collingridge et al., 1983). NMDA receptors may also
modulate 'motor programs' in the spinal cord as has been demonstrated by
studies of fictive swimming in amphibian and lamprey spinal cord (Grillner
et al., 1987). In addition, the NMDA-receptor channel itself is subject
to regulation by several agents including the divalent cations present in
physiological saline (Nowak et al., 1984; Mayer et al., 1984; Mayer and
Westbrook, 1985a) and the amino acid, glycine (Johnson and Ascher, 1987).
We summarize here evidence obtained from patch clamp studies with cultured
spinal cord and hippocampal neurons which indicate that physiological
concentrations of endogenous divalent cations have three distinct actions

on NMDA receptor channels: (1) Mg binds to a site within the channel and
blocks in a voltage-dependent manner, (2) Ca has a significant permeability
through the channel, and (3) Zn antagonizes NMDA responses at a site
quite distinct from the Mg-binding site. The effects of these interactions
are important not only to an understanding of this rather unique ion
channel, but have significant implications for both short- and longlast-
ing modulation of neuronal excitability in the CNS.

CHANNEL BLOCK BY MAGNESIUM

Under voltage clamp in physiological saline, the inward current
activated by NMDA has a region of negative slope conductance at membrane
potentials hyperpolarized to -35 mV (MacDonald et al., 1982; Mayer and
Westbrook, 1984). Thus depolarization of the neuron, e.g. by barrages of
synaptic activity will increase the availability of NMDA-receptor channels.
This 'regenerative' characteristic is at least part of the explanation
for the increased tendency for neurons to fire action potentials in
bursts during applications of NMDA. For other ion channels such as the
tetrodotoxin-sensitive sodium channel or voltage-dependent calcium channels, the voltage sensitivity of the channel results from a voltage-induced conformational change in the channel protein leading to "gating" of
the channel between the closed and open states. However, the voltage-dependence of the NMDA-activated conductance has a quite different mechanism,
Mg ions in the extracellular fluid enter and block the channel (Nowak et
al., 1984; Mayer et al., 1984). The channel blocking action is voltage-sensitive and increases with membrane hyperpolarization as the positively-
charged blocking ion enters the membrane electric field and senses the
transmembrane voltage across the ion channel protein. As shown in Figure
1A, removal of Mg from the extracellular fluid removes the negative slope
from the (I-V) relationship of the NMDA-activated conductance. Removal of
Mg is also sufficient to change the nature of the depolarizing response
to applications of NMDA such that in Mg-free medium, the same concentration
of NMDA induces a much larger depolarization which shunts the membrane
resistance and thus fails to induce a burst of action potentials (see
e.g. Mayer and Westbrook, 1985a).

The blocking action of Mg is apparent at concentrations much below
the presumptive concentration of Mg in the extracellular space in vivo.
Nowak et al. (1984) demonstrated that with low concentrations (1-10 μM)
of Mg, NMDA-receptor channels show the flickering block (Figure 1B)
characteristic of the fast open-channel block seen with the action of a
variety of charged molecules on acetylcholine receptor channels (Adams,
1976; Neher and Steinbach, 1978). A simple sequential kinetic model with
a blocked state accessible only from the open state has been used as a
first approximation of the interaction of Mg with the NMDA receptor
channel at low Mg concentrations (Nowak et al., 1984); however several
features of NMDA-receptor channels have demonstrated that a more complex
model will be necessary to describe the action of Mg at physiological
concentrations, i.e. 0.8-1.0 mM. For example, an important prediction of
the sequential model is that the total charge transfer during opening of
a single channel should remain constant in the presence of blocker, i.e.
the total time the channel spends in the open state after initially
opening is invariant. At the single channel level this means that the
burst duration should increase as the blocker concentration increases,
but in the presence 100 μM Mg, the burst duration actually decreases
(Nowak et al., 1984). At the level of the macroscopic current, the sequential model also fails to predict the observed negative slope conductance;
this may be a consequence of the fall in the freqency of single channel
opening also observed in the presence of Mg. A kinetic explanation for
this may involve either 'closed' channel block (Adams, 1976; Lingle,
1983) or enhanced desensitization (Clapham and Neher, 1984) as observed
on other ion channels.

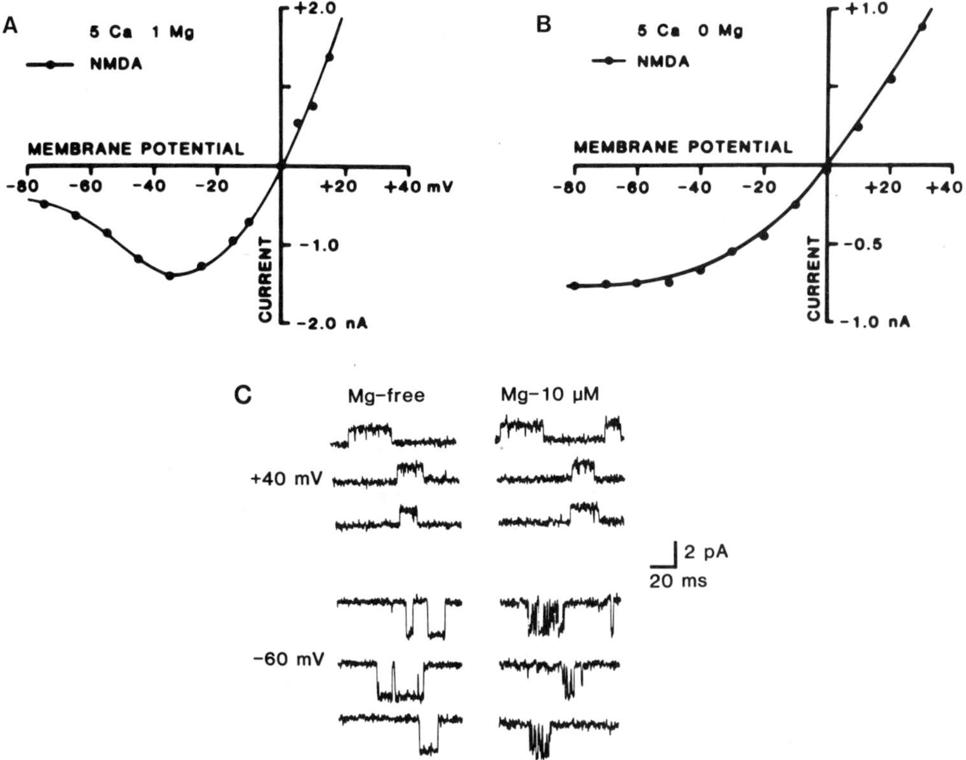

Figure 1. A,B. Current-voltage (I-V) relationship of the NMDA-activated conductance in saline containing 1 mM Mg demonstrates the
negative slope conductance at membrane potentials hyperpolarized to
-30 mV; in "0" Mg, the negative slope is markedly reduced. Two
electrode voltage clamp of cultured spinal cord neurons (from Mayer
and Westbrook, 1985a with permission). C. Low concentrations of Mg
(10 µM) cause flickery block of L-glutamate-activated single channel
currents at -60 mV but not at +40 mV consistent with a fast open
channel block. Outside-out patch of cultured mouse central neuron
(from Nowak et al., 1984 with permission).

Implications of channel block for neuronal excitability

Since the blocking action of Mg underlies the voltage dependence of
the NMDA-activated conductance, factors that alter the degree of channel
block will also influence the effects of NMDA-receptor activation on cell
excitability. As predicted by the sequential channel blocking model these
will include the extracellular [Mg], but whether there are sufficiently
rapid changes in $[Mg]_o$ to in fact alter excitability on this basis is
unclear. The agonist concentration at the synaptic site may also be a
factor since the block by Mg is uncompetitive (Nowak et al., 1984; Mayer
and Westbrook, 1985a), thus a higher agonist concentration will lead to a
greater percentage reduction of agonist-induced current. Finally, the
NMDA-activated conductance will interact with other voltage-dependent
membrane conductances in determining the overall current-voltage (I-V)
relationship of the cell. An example to illustrate this is shown in
Figure 2. In Fig. 2A the I-V relationship of a spinal cord neuron in
response to a ramp voltage stimulus is shown at 'rest' (the neuron is
loaded with Cs to reduce outward current). The downward current deflection
positive to -20 mV reflects activation of conventional voltage-dependent
calcium channels. On application of NMDA in 1 mM $[Mg]_o$ (Fig. 2A, "NMDA"),

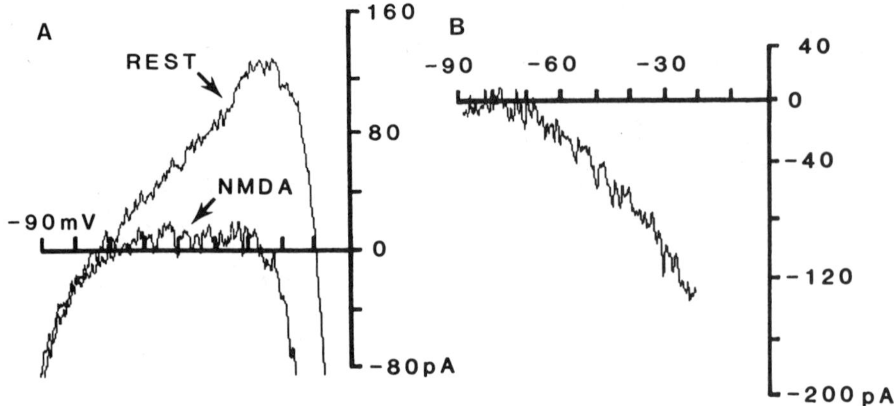

Figure 2. Interaction of the NMDA-activated conductance with other
voltage-dependent membrane conductances in shaping the membrane I-V
relationship. Whole-cell I-V plots obtained at rest and during
application of 50 μM NMDA in response to a ramp voltage clamp from
-100 to 0 at 0.42 mV/ms. Difference I-V relationship shown in B
reflects voltage-dependence of the NMDA-activated conductance. Note
that the threshold for inward current of the whole-cell I-V plot
(determined by activation of voltage-dependent Ca channels) is
shifted to more negative potentials by the activation of NMDA-receptor channels. Patch electrode contained Cs to block outward K
currents and 5 mM Mg-ATP to reduce washout of the Ca current.
Reproduced from Mayer and Westbrook, 1987a with permission.

the slope conductance of the neuronal I-V relationship decreases markedly
and shifts the threshold for net inward current to a more hyperpolarized
voltage (near -30 mV in this example). The threshold for inward current
through NMDA channels is much more hyperpolarized than for the voltage-dependent calcium current activated here (i.e. the L-type channel, Nowycky
et al., 1985) thus the NMDA conductance would serve to increase neuronal
excitability by augmenting "Ca" spikes and enhancing burst firing. This
effect is entirely a result of the region of negative slope of the NMDA-
activated conductance.

Several other divalent cations including Co, Mn and Ni also cause a
voltage-dependent block of NMDA-receptor channels similar to Mg (Mayer
and Westbrook, 1985a; Nowak and Ascher, 1985; Mayer and Westbrook, 1987b).
However Cd has essentially no blocking action on NMDA responses at submillimolar concentrations (Mayer et al., 1984). This relative Cd-resistance
is in marked contrast to N and L-type calcium channels on dorsal root
ganglion neurons (Nowycky et al., 1985). The rate of substitution of
water molecules in the inner hydration shell of hydrated divalent cations
(Diebler et al., 1969) could be a limiting factor in the permeation (or
block) by these ions in the pore of the NMDA channel, since divalents
with slow exchange rates of 10^4-10^5 sec^{-1} (Mg, Ni) are potent blockers,
Mn with a an intermediate exchange rate of 5 X 10^7 sec^{-1} has a measurable
permeability, whereas divalents with faster exchange rates of 10^8-10^9
sec^{-1} (Ca,Ba) have a high permeability (see Figure 3.).

CALCIUM PERMEATES NMDA CHANNELS

<u>Relative permeability</u>

Postsynaptic calcium influx through agonist-gated ion channels could have important functional consequences, particularly if separate calcium domains (Chad and Eckert, 1984; Simon and Llinás, 1985; Fogelson and Zucker, 1985) are determined by the strategic localization of receptors. Early studies using Ca-sensitive electrodes either intra- or extracellularly did suggest that calcium influx accompanies depolarizing responses to L-glutamate or NMDA (Bührle and Sonnhof, 1983; Pumain and Heinemann, 1985). However in physiological saline, the reversal potential of responses to excitatory amino acids is similar and close to 0 mV (Mayer and Westbrook, 1984; Nowak et al., 1984), more consistent with ion channels that have a non-selective permeability to monovalent cations. However from detailed measurement of reversal potentials of agonist-evoked currents under voltage clamp, it is now clear that the NMDA-activated conductance has a significant permeability to Ca (Mayer and Westbrook, 1985b; Mayer and Westbrook, 1987b); calcium's action in the channel is complex in that it both permeates the channel and shows competition with permeant monovalent cations (Na and K) as well as with blocking ions such as Mg (Mayer and Westbrook, 1987b).

The relative permeability of the NMDA channel to calcium as well as monovalent cations was calculated from reversal potential measurements over a wide range of calcium and sodium concentrations. At a constant sodium concentration of 100 mM, increasing $[Ca]_o$ from 0.1 to 50 mM resulted in a progressive shift of the reversal potential from - 7 mV to + 30 mV. As shown in Figure 3, raising $[Ca]_o$ from 0.3 to 3 mM resulted in a 7 mV positive shift of the reversal potential. Recently Ascher and Nowak (1986) have reported similar results for single channel currents activated by NMDA. Using the extended constant field equation and the calculated ionic activities, our results are reasonably well fit by a PCa:PNa of 10.6, assuming PNa = PK = PCs. Based on the shifts in reversal potentials for a divalent cation concentration of 20 mM, the sequence of permeability is

Ca>Ba>Sr>Mn.

No permeability could be demonstrated for Mg, Ni or Co as expected on the basis of their potent blocking action on NMDA-receptor channels. Similar calculations for the reversal potential of responses evoked by kainic acid showed essentially no measurable calcium permeability with a PCa:PNa = 0.14 (Mayer and Westbrook, 1987b). The PCa:PNa ratio of 10.6 for the NMDA channel compares to 0.22 for the acetylcholine receptor channel ($[Ca]_o$ = 20 mM, Adams et al, 1980) while conventional voltage-sensitive calcium channels show much greater selectivity for calcium over sodium and potassium, e.g. PCa:PCs > 1000 for calcium channels in guinea pig heart ventricular cells (Lee and Tsien, 1984). Nonetheless, using the extended constand field current equation, we estimate that approximately 10% of the flux through NMDA receptor channels at -60 mV is carried by Ca.

Activation of NMDA receptors also results in a transient increase in free cytoplasmic calcium as measured by the calcium indicator dye arsenazo III in voltage-clamped spinal cord neurons (MacDermott et al., 1986). Calcium transients in response to NMDA varied with the driving force for inward current through NMDA channels. High concentrations of Mg blocked the calcium transient in parallel with the block of inward currents demonstrating that ion flux through NMDA channels was responsible for the calcium transient. In solutions with no added Mg the peak inward current and peak calcium transient occurred at a membrane potential near -60 mV,

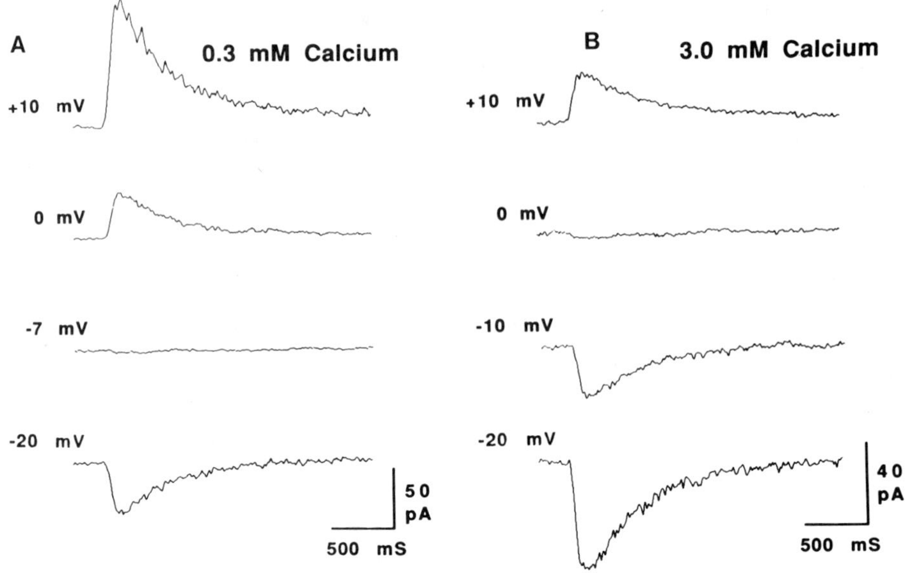

Figure 3. The reversal potential for currents evoked by NMDA is calcium dependent. Under voltage clamp and in the presence of 100 mM Na, raising the extracellular Ca from 0.3 mM to 3 mM shifts the null-current voltage from -7 mV to 0 mV consistent with a significant permeability to Ca. Whole cell recording from 2 cultured hippocampal neurons using a discontinous one-electrode voltage clamp. Reproduced with permission from Mayer and Westbrook, 1987b.

however NMDA-evoked calcium transients were also seen in physiological levels of Mg (Mayer et al., 1987). Since Mg shifts the point of maximal inward current of the NMDA-evoked I-V relationship to more depolarized potentials (near -35 mV in 1 mM Mg), it is likely that under physiological conditions depolarization from -60 to -30 mV would result not only in a larger NMDA-evoked current, but also a larger calcium influx. Therefore, it is conceivable that calcium influx through NMDA channels could be especially significant at membrane potentials subthreshold for L or N-type calcium channels. For example this may explain why conjunctive depolarization of the postsynaptic neuron can trigger longterm potentiation (LTP) in the CA1 region of the hippocampus (Wigström et al., 1986; Malinow and Miller, 1986; Kelso et al., 1986).

<u>Ion interactions</u>

Although the extended constant field equation provides a useful method in this case to compare relative divalent cation permeabilities, it is only a valid description of channel permeation if permeant ions move independently, i.e. there is no interaction between ions or binding of ions within the channel. However several lines of evidence suggest that this is not the case for the NMDA receptor channel. For example, at $[Na]_o < 50$ mM and in the presence of 10 mM $[Ca]_o$, there is an apparent increase in PCa:PNa indicative of competition between the permeating ions (Mayer and Westbrook, 1987). There is also evidence of interaction between divalent cations which primarily act as blockers of the NMDA channel (Mg) and those divalent cations which can permeate the channel (Ca). For example at low Mg concentrations (5-30 μM), increasing $[Ca]_o$ from 0.1 mM to 5.0 mM results in partial relief of the channel block at membrane potentials negative to -80 mV. On the other hand, at depolarized

voltages high [Ca]$_o$ decreased the NMDA-activated conductance consistent
with a competition of Ca with permeant monovalent cations. Of note, the
single channel conductance of NMDA-receptor channels is lower in high
calcium solutions between 0 and -60 mV (Nowak and Ascher, 1985).

ANTAGONISM BY ZINC

Recently we have also examined the effects of Zn on responses to NMDA
on cultured hippocampal neurons (Mayer and Westbrook, 1987). Zn is of particular physiological interest since it is present in high concentrations
in synaptic vesicles of the hippocampus (Perez-Clausell and Danscher, 1984)
and is released from mossy fiber terminals during electrical activity
(Assaf and Chung, 1984; Howell et al., 1984). Our results demonstrate that
Zn at low micromolar concentrations is a potent antagonist of responses
evoked by NMDA but not kainate or quisqualate (Figure 4). The antagonism
by Zn is not competitive, but in contrast to the action of Mg, the effects
of Zn decrease with hyperpolarization. This suggests that the site of
action of zinc is not within the channel, but rather is near the external
surface of the membrane. The effect of zinc was rapidly reversible and
thus differs from the irreversible action of Hg (acting as a sulfhydryl-
reducing agent) on responses to kainic acid (Kiskin et al., 1986). Millimolar concentrations of Cd had an action similar to zinc on responses to
NMDA.

Although more detailed investigations, especially at the single
channel level, will be necessary to develop a complete biophysical description of ion permeation and block of NMDA channels, the action of divalent
cations on NMDA receptor channels can be summarized as follows. Mg enters
the channel and binds to a site deep within the channel thus blocking
flow of permeant ions; this block is increased with membrane hyperpolarization, and except at extreme hyperpolarization (between -150 and -200 mV)
Mg can exit the channel only to the extracellular side. On the other hand
Ca can enter the channel and compete with Mg (not necessarily at the same
binding site, but perhaps by electrostatic repulsion) and displace Mg from
its binding site leading to permeation of both Ca and monovalent cations.
We suggest that there is a second metal-binding site on the external
surface of the receptor-channel complex to account for the action of Zn.

IMPLICATIONS FOR SYNAPTIC TRANSMISSION

Studies with excitatory amino acid receptor antagonists both _in vivo_
and _in vitro_ have generally demonstrated that fast epsps, at pathways
thought to use an excitatory amino acid as the transmitter, are mediated
by a kainate/quisqualate receptor type. In addition the synaptic conductance underlying fast epsps between single Ia afferents and motoneurons
in vivo (Finkel and Redman, 1983) and between cultured spinal cord neurons
(Nelson et al., 1986) is voltage-insensitive, compatible with that seen
with the responses evoked by kainate or quisqualate. However, several
recent studies _in vitro_ have indicated that NMDA receptors are directly
activated at synapses in neocortex, hippocampus and spinal cord (for
review see Mayer and Westbrook, 1987). In most cases these epsps were monosynaptic and contained both an early component due to activation of
kainate/quisqualate receptors and a slower component which was blocked by
the specific NMDA-receptor antagonist, AP5. We have found that monosynaptic
epsps between neurons in both spinal cord and hippocampal cultures involve
both of these conductance mechanisms - a fast component with properties
similar to kainate/quisqualate responses and a slow component, lasting up
to 500 msec with the properties expected for activation of NMDA receptor
channels (Forsythe and Westbrook, 1986). Thus accumulating evidence
suggests that NMDA-receptor channels are functional at a number of central
excitatory pathways.

    The functional significance of the unusual properties of the NMDA-
receptor channel are still to be resolved. The Mg-binding site on the
NMDA-receptor channel has a rather straightforward sequelae, i.e. the resulting voltage-dependence of the NMDA-receptor channel can, in the short-
term, serve to modulate neuronal excitability by augmenting burst firing.
Calcium influx through NMDA channels may further increase the tendency to
rhythmic bursting through activation of calcium-dependent K channels
(Nicoll and Alger, 1981; Grillner and Wallén, 1985). In addition a second
messenger role for the receptor-activated calcium influx could serve to
link a number of phenomena, such as LTP in the CA1 region of the hippocampus, which are both calcium-dependent (Dunwiddie and Lynch, 1979;
Wigström et al., 1979; Turner et al., 1982), and blocked by NMDA receptor

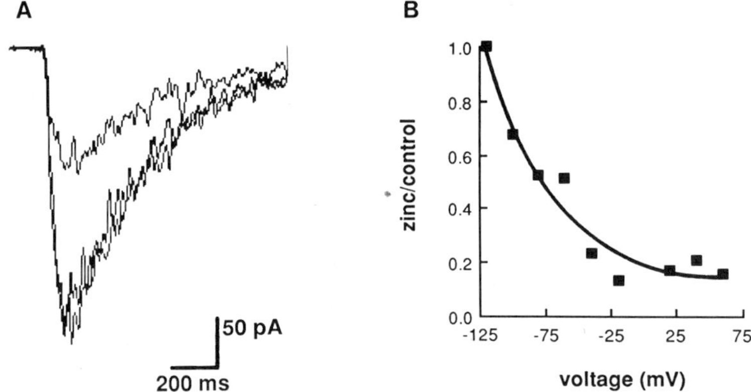

Figure 4. A. Zn (5 μM) reversibly antagonizes the inward current
activated by a brief iontophoretic application of NMDA at a holding
potential of -60 mV. Recovery response is superimposed on the
control response. As shown in B, the antagonism by Zn (expressed as
fractional response in zinc/control) decreased with hyperpolarization. Line drawn by eye. Data from a cultured hippocampal neuron
using whole-cell patch recording with a Cs-contaning electrode.
Recording solution contained 1 mM Ca and no added Mg. Previously
unpublished data.

antagonists (Collingridge et al., 1983; Wigström and Gustaffson, 1984;
Harris et al., 1984). Excitotoxic neuronal damage may also result from
similar sequence of events (Choi, 1985). Whether Zn released from nerve
terminals can also play a role in regulating the available conductance-
through NMDA receptor channels deserves further consideration. Not discussed here but clearly of major importance will be the role of glycine in
potentiating NMDA receptors, possibly through an allosteric site (Johnson
and Ascher, 1987). Hopefully a more quantitative knowledge of these
phenomena will contribute to an understanding of excitatory neurotransmission in the CNS as well as define the role these receptors and their
associated ion channels play in synaptic plasticity and abnormal hyper-
excitability.

REFERENCES

Adams, D.J., Dwyer, T.M. and Hille, B., 1980, The permeability of endplate channels to monovalent and divalent cations, J. Gen. Physiol., 75:493.

Adams, P.R., 1976, Drug blockade of open end-plate channels, J. Physiol. (Lond.), 260:531.

Ascher, P. and Nowak, L., 1986, Calcium permeability of the channels activated by N-methyl-D-aspartate (NMDA) in isolated mouse central neurones, J.Physiol. (Lond.), 377:35P.

Assaf, S.Y. and Chung, S.H., 1984, Release of endogenous Zn^{+2} from brain tissue during activity, Nature, 308:734.

Bürhle, C. P. and Sonnhof, U., 1983, The ionic mechanism of the excitatory action of L-glutamate upon the membranes of motoneurones of the frog, Pflugers Archiv, 396:154.

Chad, J.E. and Eckert, R., 1984, Calcium domains associated with individual channels can account for anomalous voltage relations of Ca-dependent responses, Biophys. J. 45:993.

Choi, D.W. ,1985, Glutamate neurotoxicity in cortical cell culture is calcium dependent, Neurosci. Lett., 58:293.

Clapham, D.E. and Neher, E., 1984, Substance P reduces acetyl-choline-induced currents in isolated bovine chromaffin cells, J.Physiol. (Lond.), 347:255.

Collingridge, G.L., Kehl, S.J. and McLennan, H. ,1983, Excitatory amino acids in synaptic transmission in the Schaffer collateral-commissural pathway of the rat hippocampus, J. Physiol. (Lond.), 334:33.

Cull-Candy, S.G. and Usowicz, M.M., 1987, Multiple-conductance channels activated by excitatory amino acids in cerebellar neurons, Nature, 325:525.

Davies, J.D. and Watkins, J.C. ,1983, Role of excitatory amino acid receptors in mono- and polysynaptic excitation in the cat spinal cord, Exp. Brain Res., 49:280.

Diebler, H., Eigen, M., Ilgenfritz, G., Maas, G. and Winkler, R., 1969, Kinetics and mechanism of reactions of main group metal ions with biological carriers, Pure Appl. Chem., 20:93.

Dunwiddie, T. and Lynch, G ,1979, The relationship between extracellular calcium concentrations and the induction of hippocampal long-term potentiation, Brain Res., 169:103.

Finkel, A.S. and Redman, S.J. ,1983, The synaptic current evoked in cat spinal motoneurones by impulses in single group la axons, J. Physiol. (Lond.), 342:615.

Fogelson, A. and Zucker, R.S., 1985, Presynaptic calcium diffusion from various arrays of single channels: Implications for transmitter release and synaptic facilitation, Biophys. J., 48:1003.

Forsythe, I.D. and Westbrook, G.L., 1986, Monosynaptic activation of NMDA receptors in mouse spinal cord cultures, Soc. Neurosci. Abs., 12:62.

Grillner, S. and Wallen, P., 1985, The ionic mechanisms underlying N-methyl-D-aspartate receptor-induced, tetrodotoxin-resistant membrane potential oscillations in lamprey neurons active during locomotion, Neurosci. Lett., 60:289.

Grillner, S., Wallén, P., Dale, N., Brodin, L., Buchanan, J. and Hill, R., 1987, Transmitters, membrane properties and network circuitry in the control of locomotion in lamprey, TINS, 10:34.

Harris, E.W., Ganong, A.H. and Cotman, C.W. ,1984, Long-term potentiation in the hippocampus involves activation of N-methyl-D-aspartate receptors, Brain Res., 323:132.

Howell, G.A., Welch, M.G. and Frederickson, C.J. ,1984, Stimulation-induced uptake and release of zinc in hippocampal slices, Nature, 308:736.

Jahr, C.E. and Stevens, C.F., 1987, Glutamate activates multiple single channel conductances in hippocampal neurons, Nature, 325:522.

Johnson, J.W. and Ascher, P., 1987, Glycine potentiates the NMDA response in cultured mouse brain neurons, <u>Nature</u>, 325:529.

Kelso, S.R., Ganong, A.H. and Brown, T.H ,1986, Hebbian synapses in hippocampus, <u>Proc. Natl. Acad. Sci. USA</u>, 83:5326.

Kiskin, N.I., Khrishtal, O.A., Tsyndrenko, A.Ya. and Akaike, N., 1986, Are sulfhydryl groups essential for function of the glutamate-operated receptor-ionophore complex?, <u>Neurosci. Lett.</u>, 66:305.

Lee, K.S. and Tsien, R.W., 1984, High selectivity of calcium channels in single dialysed heart cells of the guinea-pig, <u>J. Physiol. (Lond.)</u>, 354:253.

Lingle, C., 1983, Blockade of cholinergic channels by chlorisondamine on a crustacean muscle, <u>J. Physiol. (Lond.)</u>, 339:395.

MacDermott, A.B., Mayer, M.L., Westbrook, G.L. Smith, S.J. and Barker, J.L., 1986, NMDA-receptor activation increases cytoplasmic calcium concentration in cultured spinal cord neurones, <u>Nature</u>, 321:519.

MacDonald. J.F., Porietis, A.V. and Wojtowicz, J.M., 1982, L-aspartic acid induces a region of negative slope conductance in the current voltage relationship of cultured spinal cord neurons, <u>Brain Res.</u>, 237:248.

Malinow, R. and Miller, J.P., 1986, Postsynaptic hyperpolarization during conditioning reversibly blocks induction of long-term potentiation, <u>Nature</u>, 320:529.

Mayer, M.L. and Westbrook, G.L., 1984, Mixed-agonist action of excitatory amino acids on mouse spinal cord neurons under voltage clamp, <u>J. Physiol. (Lond.)</u>, 354:29.

Mayer, M.L. and Westbrook, G.L ,1985a, The action of N-methyl-D-aspartic acid on mouse spinal neurons in culture, <u>J. Physiol. (Lond.)</u> 361, 65-90.

Mayer, M.L. and Westbrook, G.L., 1985b, Divalent cation permeability of N-methyl-D-aspartate channels, <u>Soc. Neurosci. Abs.</u>, 11:785.

Mayer, M.L. and Westbrook, G.L. (1987a) The physiology of excitatory amino acids in the vertebrate central nervous system, <u>Prog. Neurobiol.</u>, 28:197.

Mayer, M.L. and Westbrook, G.L., 1987b, Permeation and block by divalent cations of N-methyl-D-aspartate receptor channels in mouse central neurons. <u>J. Physiol. (Lond.)</u>, submitted.

Mayer, M.L. and Westbrook, G.L. (1987c) Zinc is a potent blocker of the NMDA-activated conductance on hippocampal neurons, <u>Biophys.J.</u>, 51:64a.

Mayer, M.L., MacDermott, A.B., Westbrook, G.L., Smith, S.J. and Barker, J.L., 1987, Agonist- and voltage-gated calcium entry in mouse spinal cord neurons under voltage clamp measured using arsenazo III, <u>J. Neurosci.</u>, submitted.

Mayer, M.L., Westbrook, G.L. and Guthrie, P.B., 1984, Voltage-dependent block by Mg^{2+} of NMDA responses in spinal cord neurones, <u>Nature</u>, 309:261.

Neher, E. and Steinbach, J.H., 1978, Local anaesthetics transiently block currents through single acetylcholine-receptor channels, <u>J. Physiol. (Lond.)</u>, 277:153.

Nelson, P.G., Pun, R.Y.K., and Westbrook, G.L., 1986, Synaptic excitation in cultures of mouse spinal cord neurones: receptor pharmacology and behaviour of synaptic currents, <u>J. Physiol. (Lond.)</u>,372:169.

Nicoll, R.A. and Alger, B.E. ,1981, Synaptic excitation may activate a calcium-dependent potassium conductance in hippocampal pyramidal cells, <u>Science</u>, 212:957.

Nowak, L., Bregestovski, P., Ascher, P., Herbet, A. and Prochiantz, A., 1984, Magnesium gates glutamate-activated channels in mouse central neurones, <u>Nature</u>, 307:462.

Nowak, L. and Ascher, P., 1985, Divalent cation effects on NMDA-activated channels can be described as Mg-like or Ca-like, <u>Soc. Neurosci.Abs.</u>, 11:953.

Nowycky, M.C., Fox, A.P. and Tsien, R.W., 1985, Three types of neuronal calcium channel with diferent calcium agonist sensitivity, <u>Nature</u>, 316:440.

Olverman, H.J., Jones, A.W. and Watkins, J.C. ,1984, L-glutamate has higher affinity than other amino acids for [3H]-D-AP5 binding sites in rat brain membranes, <u>Nature</u>, 307:460.

Perez-Clausell, J. and Danscher, G., 1985, Intravesicular localization of zinc in rat telencephalic boutons. A histochemical study, <u>Brain Res</u>., 337:91.

Pumain, R. and Heinemann. U.,1985, Stimulus- and amino acid-induced calcium and potassium changes in the rat neocortex, <u>J. Neurophysiol</u>., 53:1.

Simon, S.M. and Llinás, R.R., 1985, Compartmentalization of the submembrane calcium activity during calcium flux and its significance to transmitter release, <u>Biophys. J</u>., 48:485.

Turner, R.W., Baimbridge, K.G. and Miller, J.J., 1982, Calcium-induced long-term potentiation in the hippocampus, <u>Neurosci</u>., 7:1411.

Watkins, J.C. and Evans, R.H., 1981, Excitatory amino acid transmitters, <u>Ann. Rev.Pharmacol. Toxicol</u>. 21:165.

Wigström, H. and Gustaffson, B, 1984, A possible correlate of the postsynaptic condition for long-lasting potentiation in the guinea pig hippocampus in vitro, <u>Neurosci. Lett</u>., 44:327.

Wigström, H., Gustaffson, B., Huang, Y.Y. and Abraham, W.C., 1986, Hippocampal long-term potentiation is induced by pairing single afferent volleys with intracellularly injected depolarizing current pulses, <u>Acta. Physiol. Scand</u>., 126:317.

Wigström, H., Swann, J.H. and Andersen, P., 1979, Calcium dependency of synaptic long-lasting potentiation in the hippocampal slice, <u>Acta. Physiol. Scand</u>., 105:126.

FLUORESCENCE IMAGING APPLIED TO THE MEASUREMENT OF Ca^{2+} IN

MAMMALIAN NEURONS

John A. Connor

AT&T Bell Laboratories
Molecular Biophysics Research Department
600 Mountain Avenue
Murray Hill, NJ 07974

INTRODUCTION

Interest in high resolution photometry in conjunction with cytoplasmic indicator molecules has increased markedly in the past few years. The pioneering experiments in the field date back to the late 1960's and early 1970's (c.f. Reynolds, 1972; and for more recent reviews, Ashley and Campbell, 1979; DeWeer and Salzberg, 1986). For the most part these early studies employed the luminescent, Ca binding protein, aequorin, together with intensified target (SIT) TV cameras. Three factors, however, have always limited the use of aequorin: 1) short supply, 2) difficulty in introducing it into cells, 3) kinetics that make calibrated measurements of Ca all but impossible. Synthetic azo dyes with somewhat lower affinities for Ca^{2+} have also been around for a number of years, in use primarily for molluscan neurons and striated muscle (c.f. Brown, et al.,1976; Ahmed and Connor, 1979; Miledi, et al.,1977; Baylor, et al., 1986). More recently such studies have been extended to mammalian neurons and tumor-derived cell lines (Bolsover and Spector, 1986), using arsenazo III, the highest affinity dye of this class. Its large change in extinction coefficient at 660 nm, where tissue absorbance is low, has made it possible to track Ca^{2+} changes during physiological activity in buried processes of molluscan neurons (Graubard and Ross, 1986; Connor, Kretz, and Shapiro, 1986) and in Purkinje cell dendrites in tissue slices (Ross and Werman, 1987). No membrane permeable forms of the azo dyes exist at present, so their use is limited to cells where injections can be made. Both aequorin and arsenazo III continue to be valuable indicators in spite of the inevitable hyperbole that has accompanied the introduction of newer dyes.

Much of the current interest in imaging has been triggered by the development of high efficiency fluorescent indicators that load themselves into cells and subsequently become trapped and activated. This eliminates much of the experimental difficulty associated with the older indicators but introduces other problems, some of them subtle. At present only two classes of indicator have been demonstrated widely, those for Ca^{2+}, quin2 and fura-2 (Tsien, et al., 1982; Grynkiewicz, et al., 1985; Tsien and Poenie, 1986), and others for pH, primarily carboxyfluorescein compounds (Gotoh, et al., 1982; Rink, et al., 1982; Paradiso, et al., 1984).

One hopes that development of indicators for other cellular messengers will be spurred by the present level of activity in the measurement of Ca^{2+} and pH. This chapter deals with some of the basic methodology of using fura-2 to measure intracellular Ca^{2+} in small neurons.

EXPERIMENTAL CONSIDERATIONS

A brief summary of optical principles involved in these measurements is given here with the suggestion that better, more complete treatments may be found elsewhere (Pesce, et al., 1971; Lakowicz, 1983; Udenfriend, 1969). The fluorescence of an indicator, F_{λ_o} (light flux/unit area) is determined by the following factors: the amount of light energy it absorbs at a given wavelength, measured by its molar extinction coefficient, ϵ_λ; 2) the number of fluorescing molecules, measured by concentration, C, and the pathlength through which the excitation beam travels, Δz ; 3) Quantum efficiency ,Q, the percent of the total energy absorbed that is released as light rather than being dissipated as heat or transferred to nearby molecules; 4) the intensity of the excitation light, I_λ. These factors are summarized in the approximation of equation 1.

$$F_{\lambda_o} = I_\lambda \bullet C \bullet \Delta z \bullet Q \bullet \epsilon_\lambda$$

Indicators such as fura-2 or quin2 work, in part, because their extinction coefficients undergo a change when a ligand is bound. The energy absorbed at a given wavelength changes and as a result the fluorescent output undergoes a corresponding change. Additionally there is an increase in quantum efficiency between the bound and unbound states.

Fura-2 and its predecessor, quin2, are somewhat unusual in that the emission wavelength shifts very little between the free and liganded state. The related compound indo-1 (Grynkiewicz, et al., 1985) gives a large emission wavelength change between bound and unbound forms. The excitation and emission spectra for fura-2 are given in Fig. 1A. They are simple spectra in that they do not have a number of maxima such as in perylene. Fluorescence, for a given excitation intensity increases for excitation wavelengths shorter than 360 nm and decreases for wavelengths longer than 360. That juncture is generally called the crossover wavelength, or isostilbic point. Under the carefully controlled conditions in a good spectrofluorimeter, the excitation spectrum is strictly related to Ca concentration $[Ca^{2+}]$.

It should be clear from Equation 1 that the fluorescence of any indicator is dependent on factors other than the concentration of its ligand, the quantity one is interested in. Concentration, and sample thickness can be eliminated (to a first approximation) by taking data points at two different wavelengths and forming the ratio of the fluorescence values. The only uninteresting factor remaining in the ratio is the term involving excitation (I_λ). In principle any two wavelengths may be employed to form the ratio; in practice a number of considerations should be kept in mind.

First, a simplified equation like #1, when all the terms are constant as opposed to being terms under an integral sign, is only meaningful experimentally where the terms are indeed constant or else vary in the same way throughout the sample. The Z-dimension (pathlength) is the one of most concern in the present type of work. One must assume that proportionately the same amount of light at both wavelengths reaches every point in the sample. Absorption and scattering of light are generally wavelength dependent. Therefore, one of the excitation wavelengths in the ratio may undergo more attenuation than the other while passing through the sample. The I_λ terms are then no longer in a fixed ratio. This problem is minimized by making the two excitation wavelengths close together but the cost is a loss in signal for a given Ca^{2+}. Reference to the fura-2 excitation spectra of Fig. 1A shows that a wavelength pair of 340 and 390 nm will just about maximize the ratio for a given change in Ca^{2+}, but closer wavelengths might be desirable under some circumstances.

Absorption and scattering are not serious problems in tissue culture cells since the light path is short and the structures are not particularly refractile even in the 340-400 nm band. In other preparations such as brain slices the problem is much more severe. Many glial cells in culture tend to trap fura-2 very poorly in comparison to neurons (Connor, Tseng, and Hockberger, 1987), and we have been able to show that in some slices, neurons trap dye in contrast to surrounding tissue (unpublished). In this type of preparation there are tens of μm of tissue between the excitation source and the cell of interest making it necessary to determine the optical properties of the intervening tissue with some care. Wavelength dependent absorbance by the indicator itself is a problem in many types of cuvette measurements, but for the 3-10 um pathlengths encountered in tissue monolayers the optical density (O.D.) is $<$ .001 for submillimolar concentrations of indicator and should not introduce significant errors. There are generally physiological complications at concentrations higher than that.

If a linear photodetector is used, intensity variation is not a particular problem in the X-Y directions of a microscope field since illumination is passing through the same set of elements and should have the same spectral characteristics. A ratio will then eliminate this source of unwanted variation in signal.

MEASUREMENT SYSTEM

The most critical elements of the photometer measurement system are a good UV objective and a good quality, low light level camera of some sort. Unfortunately, the latter generally requires a good quality grant, but with the attendant madness that accompanies almost any promising new technique, it has been my impression that such money is not as hard to come by as it would have been a few years ago, especially in medical schools. The makeup of my system is summarized in Fig. 1B. The primary element, a charge coupled device (CCD) camera is represented as a rectangular array of light sensing diodes each with a storage capacitor. In principle, each diode generates electrons as it absorbs light and these are stored on the capacitors. At the end of a capture period, or exposure, the amount of charge on each capacitor (voltage) is read out serially through a high impedance buffer amplifier. It is thus a "still camera" rather than a "video" camera and the preparation is illuminated only during the actual period that light is collected. It is a linear detector, and if a good grade chip is selected, free from non-uniformities in sensitivity across the field. Present day versions of CCD cameras are nearly identical to prototypes made at Bell Laboratories in the 1970's (Sequin and Tompsett, 1975). In order to reduce thermal noise the CCD chip is cooled to approximately -40 degrees C.

The voltage levels from the camera are digitized and stored as arrays in computer memory. It requires from 1 to 1.5 sec to collect and store a pair of images, depending upon their size and the exposure time. These image arrays may then be manipulated in some way or another, as by subtracting background or spatial filtering, or may be written directly to mass storage, a hard disk in this system. The arrays are displayed as grey levels or false color on a video monitor. For the indicator ratioing technique described above, pictures are taken using 340 and 380 nm excitation and then the numerical value of each picture element (pixel) at 340 nm excitation is divided by the value of the corresponding element in the 380 nm picture. This array of ratios is then displayed on a monitor by making higher ratio values brighter or by showing them on a color continuum as has become standard for the various tomography maps.

EXPERIMENTAL ASPECTS OF USING SELF-LOADING INDICATORS

1. Unpredictable processing of the indicator by cells

One highly attractive feature of indicators such as fura-2 and indo-1 /AM is that they load themselves into cells. Lipid soluble groups are attached to the carboxyl groups of the parent compound making the structure very membrane permeable. Once inside the cell, the ester bonds attaching the acetoxymethyl chains to the indicator are cleaved by enzymes present in the cytoplasm (in theory). In practice there are complications that result from this type of loading. For example, while all 4 of the /AM groups must be removed for the molecule to bind Ca with high affinity, the molecule is trapped in the cells when fewer than four are removed. This is easily demonstrated by exposing cells to a loading solution for a reasonable time, say 30 min, at their normal operating temperature, 35-38 degrees C for mammals. If the cells are then washed and held at room temperature, they will hold the indicator for hours but the fluorescence ratio will not change under conditions where companion cells incubated at 37 degrees respond robustly. There is a spectral shift in the emission spectrum as a result of deesterification that can be monitored with the proper equipment but it is difficult to follow in an imaging system and a cuvette of properly functioning neurons is hard to come by. The excitation spectrum for the /AM form of the indicator is very similar to the low Ca spectrum of the active indicator; therefore a mixture of the two species in the cytoplasm will give an underestimate of Ca^{2+} levels. This systematic error can be tolerated in a number of situations where one is just interested in establishing the occurrence of $[Ca^{2+}]$ changes.

Probably the best control is to compare results between /AM loaded cells, and cells where the ionic form is injected directly. This is readily done with molluscan neurons since the soma is a large pincushion and the indicator diffuses rapidly to outlying processes which are of the most interest. It is also most necessary to do this control in molluscan neurons since cells from the marine species such as *Aplysia* hydrolyze fura-2 very poorly. The procedure is more difficult for mammalian cells, but the effort is necessary, at least until a greater body of natural history evolves for the behavior of this class of indicator.

I have observed a wide range of indicator trapping and holding among mammalian CNS cells of various types and ages. Freshly plated ($<$ 3 days in vitro, DIV) cells from embryonic tissue trap fura-2 or quin2 very effectively. Incubation periods of 30 min at nominal indicator concentration of 1-2 μM are sufficient to achieve 100-300 uM levels of trapped indicator. In older cultures, 4-20 DIV, neuronal trapping remains reasonably efficient but glial and fibroblast efficiency is very low in most cases. Small neurons like cerebellar granule cells have proved more efficient than larger cerebellar neurons (among them Purkinje neurons) on the same coverslip. In cultures over 30 DIV, trapping is generally poor even though the cells have all the electrical characteristics of healthy neurons. At all stages, however, it appears that the trapped indicator becomes completely deesterified over the course of a 30-60 min post-load incubation period. Acutely isolated neurons from adult animals load and activate the indicator very rapidly (experiments done with W. Wadman and R. Wong).

2. Intracellular compartmentalization

There is no guarantee that the /AM form of the indicator crosses just one membrane. As a result it may enter into intracellular compartments that do not reflect Ca levels in the cytoplasm at large. These compartments in some instances have enough esterase activity to function at least as traps in which the indicator may or may not be completely activated. Thus, intracellular compartmentalization is also one of the potential costs incurred for the ease of loading, though under some conditions it might be turned to an advantage.

Again, filling a given cell type both ways is a good control where possible. Almers and Neher (1985) have provided a clear example of differences in the signal obtained from dyes loaded by different methods into mast cells. In these cells most of the /AM -loaded indicator wound up in internal vesicles and reported almost none of the [Ca] change that the injected ionic form of the indicator reported. A similar phenomenon occurs in the bag

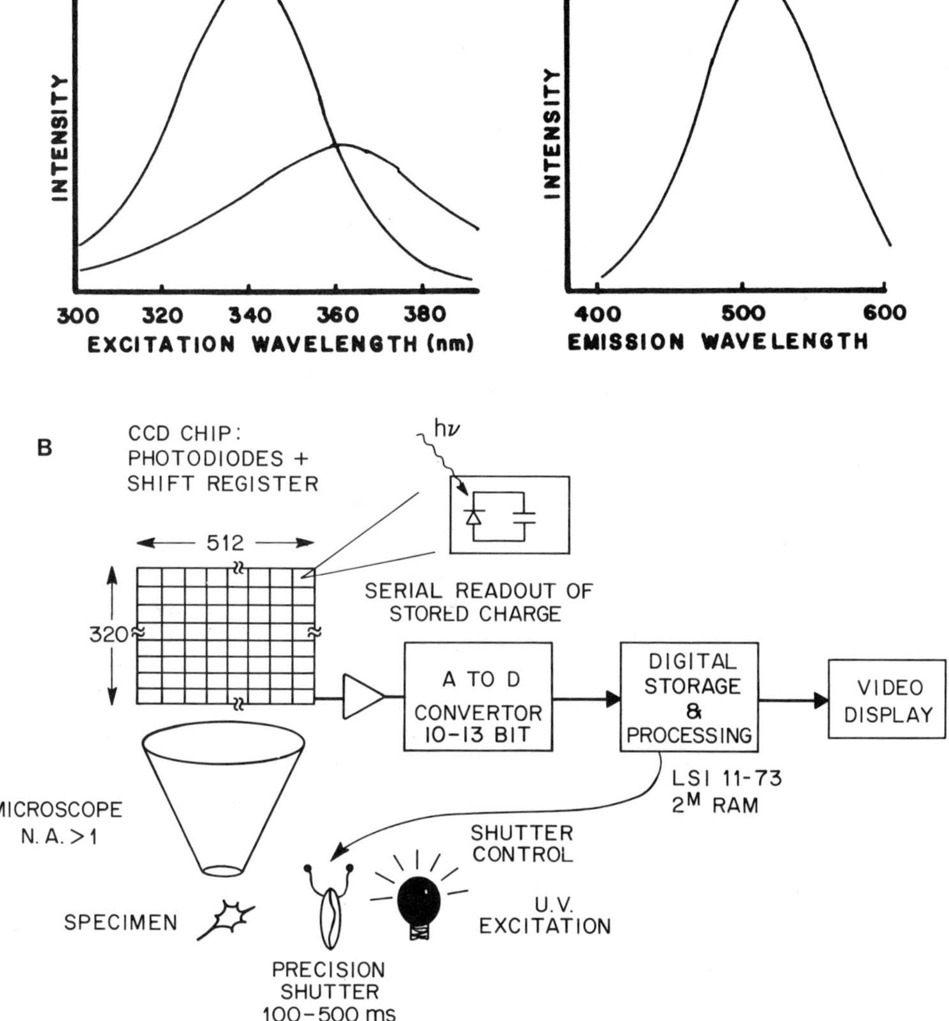

Figure 1. A: Excitation and emission spectra of fura-2. B: Diagram of essential parts of imaging system. Charge coupled device camera is represented as rectangular array of photodiodes and storage capacitors, though in fact both elements are derived from the same continuous substrate. The type used employs a 320x512 array. The values of stored charge on the elements are read out serially, digitized and stored as an array in the computer memory. A DEC LSI-11/73 computer is used in collecting and analyzing data although more modern processors would speed arithmetic and display of data. Changing the wavelength of the UV excitation can be done by manually changing interference filters or any number of quicker methods.

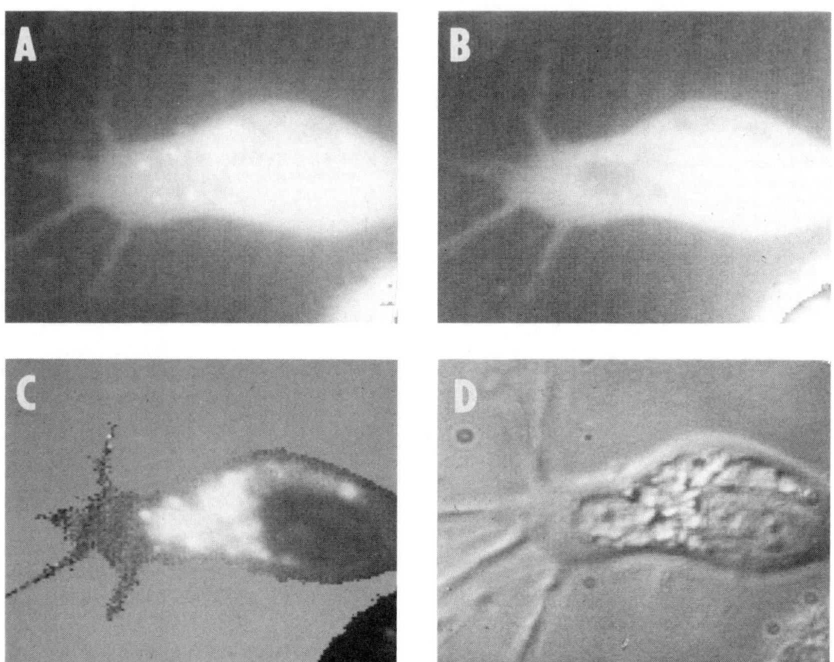

Figure 2. Fluorescence measurements made on a cell from rat embryo (E-18) diencephalon, that had been isolated and plated 6 hours before. The cell was not extending processes at the time of observation. A: Fura-2 fluorescence with 340 nm excitation. B: 380 nm excitation. Note obvious differences in fluorescence intensities between A and B. C: 340/380 ratio image. Regions of high Ca^{2+} have been displayed as brighter than regions of low Ca^{2+}. D: Nomarski picture of cell. Several of the regions of high and low Ca^{2+} can be identified in this picture. Minor axis of cell is approximately 10 μm. Frame size is 140x240 pixels.

neurons of Aplysia (J. Connor, J. Strong, L. Kaczmarek, unpublished). Although there is no direct evidence for it that I know of, it is possible that even the ionic form of the indicator may become partitioned in the cell.

In certain instances, imaging techniques may be used to exploit this behavior. Williams et.al (1985) reported separating what was possibly endoplasmic reticulum levels of Ca from cytoplasm levels. Fig. 2 shows an example of a newly plated cell from rat diencephalon. Ratios are displayed on a grey scale where the brighter the region the higher the ratio. The indicator fluorescence ratio in some, but not all, of the internal compartments is approximately 5X that of the surrounding cytoplasm. It can also be seen that the nuclear region of the cell (surrounded by bright vessicles) is lower than other non-vessicle laden areas.

3. Dye binding in the cytoplasm that alters kinetics of the indicator

This is a potential plague for any type of indicator technique. Diffusion, fluorescence or absorbance polarization, and spectral shift measurements can give an estimate of binding but they say little as to how the Ca binding reaction or the indicator is actually affected. The obvious solution, but a difficult one, is to dialyze the cytoplasm at known $[Ca^{2+}]$ and standardize the dye *in vivo* Almers and Neher (1985) report controlling internal $[Ca^{2+}]$ using a whole cell patch electrode on mast cells. In this type of experiment the patch electrode is filled with buffered Ca saline and a sufficient amount of indicator. The cell interior is assumed to equilibrate rapidly with this mixture upon rupture of the membrane within the electrode. This technique is relatively unreliable for controlling absolute Ca levels in larger neurons.

Williams, et.al. (1985) reported using the Ca ionophore, ionomycin (Behringer) to equilibrate the interior of smooth muscle cells with buffered [Ca] in the external bath. My own experience with this compound on neurons is that it generally fails to control Ca at intermediate levels, but is very useful in establishing maximum and minimum values of the fluorescence ratio in situ and is, therefore, of value in determining whether the indicator is activated and whether the dynamic range is altered by the intracellular environment. For example, incubation of cells in a Ca-free medium with 0.5 mM EGTA reduces most cells that I have worked with to a limiting, low fluorescence ratio within 10-15 min as long as ionomycin (2-4 μM) is present in the media. Conversely, a brief (5-10 sec) exposure to a medium containing 100 μM Ca and ionomycin sends the ratio to a limiting high value.

GROWTH CONES

Fig. 3 illustrates a typical set of observations made on mammalian neurons during the period of outgrowth following dissociation in tissue culture. These cells had been dissociated from rat diencephalon and plated on a poly-D-lysine coverslip about 18 hours before. About 95% of the cells that survive dissociation are simple spheres, devoid of processes (c.f. Ahmed, et al., 1983, 1986). During the first 24 hours in culture they extend processes rapidly but soma neurite associations are still clearly defined. At later times it is difficult to tell from where a given neurite arises.

In Fig. 3B Ca^{2+} levels have been coded on a grey scale, similar to that of Fig. 2. The leftmost cell shows a uniformly high Ca, approximately 250 nM, whereas the somata of the right-hand cells were much lower, 50 to 60 nM. The process arising from the lower cells was active as judged by small shape changes and extension over a 10 min. observation period, and it can be seen that the Ca level is significantly higher in the tip region as compared to more proximal neurite and soma (180 nM vs. 60 nM). I have found that, generally, actively growing cells show internal Ca levels that lie in the range of 200-500 nM, either throughout the cell where outgrowth is near the soma, or in more restricted areas near the tip where substantial neurite has already been laid down

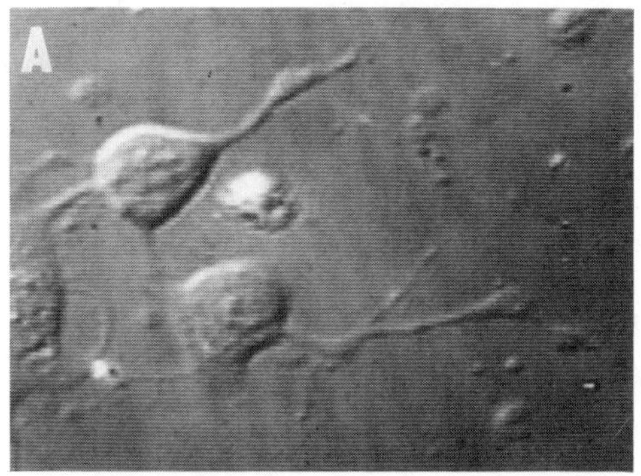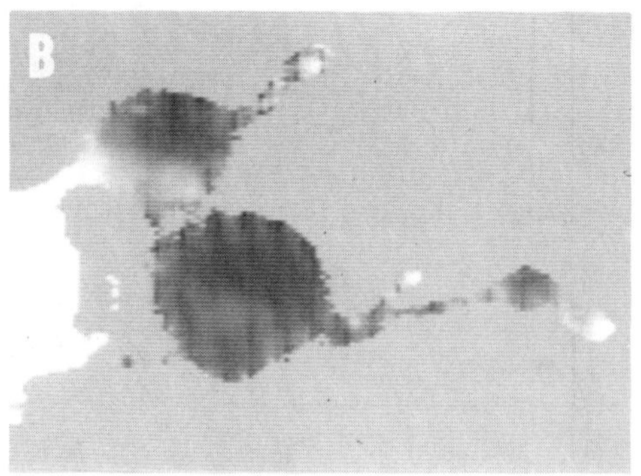

Figure 3. A: Nomarski picture showing 3 rat diencephalon cells that had been growing in culture for 18 hours. The cell on the lower right had extended a long process and was pushing out a proboscis-like structure at the time of recording. The left-hand cell had sent out a process over the top cell. B: Ca^{2+} levels in the three cells, mapped in the same convention as in Fig. 2. Soma of right-hand cell approximately 12 μm.

(Connor, 1986). Examples of both are shown in Fig. 3. These same growth cone-soma differences are also observed in molluscan neurons where growth cones are often a hundred um or more away from the cell bodies and where the indicator can be injected directly into the cell without perturbing growth (Cohan, Connor, and Kater, 1987).

Where neurite extension has stalled (often the case on a microscope stage unless conditions are very well controlled), Ca^{2+} levels are low and much more uniform (Connor, 1986, Cohan, et al., 1987). Occasionally, as illustrated in Fig. 4B, growth cone levels are significantly lower than in the soma. Fig. 4A shows a transmitted light picture of this cell. In these situations it is generally possible to cause a steady increase in internal Ca^{2+} in the cell. For the example shown, the growth cone increased its Ca^{2+} to a level significantly higher than the soma (Fig. 4C). Thus standing gradients in both directions can be demonstrated between soma and growth cones.

In approximately 30% of the cases examined, the K-induced Ca^{2+} increase was accompanied by renewed extension of the cell. It should also be noted that much of CNS tissue culture, our own included, is done using 25 mM K media. It has long been known that high K enhances cell outgrowth and survivability. Cells grown in such media when examined even after 20-30 days in culture show internal Ca^{2+} levels in the 250-350 nM range. When the growth media is replaced (on the microscope stage) by media with normal K the Ca^{2+} levels drop to below 100 nM within 2-3 minutes (Connor, Tseng, and Hockberger, in preparation) Thus high Ca^{2+} levels are maintained for extended periods by high K and from all outward signs are beneficial to the initial stages of cell development.

Ca^{2+} LEVELS IN ELECTRICALLY ACTIVE NEURONS

Fig. 5 shows the reduction in Ca^{2+} level that can be demonstrated when spontaneous firing is inhibited. The cell shown in the micrograph of Fig. 5A is a putative Purkinje neuron grown in culture 15 days. The electrode shown at the bottom of the frame was a patch electrode, containing Krebs saline, that had been sealed to the cell membrane and detected ongoing, irregular, spike activity at 5-10 Hz. Indicator measurements were made before placement (not shown) and immediately after (Fig. 5B). There was no significant perturbation of the Ca^{2+} level resulting from placement of the

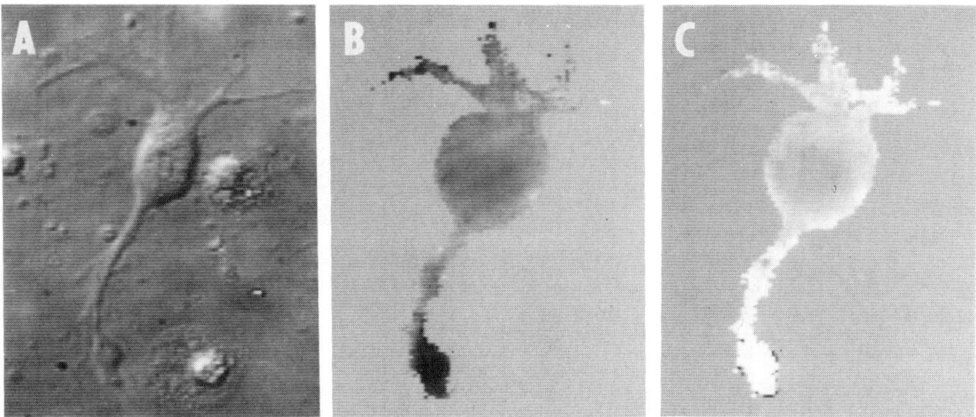

Figure 4. A: Nomarski picture of a diencepahlon cell, 20 hours in culture. Cell extension had ceased at the time of initial observation. B: Ratio image of the cell showing that Ca^{2+} levels in the lower, dormant, growth cone are below those in the soma. C: Ratio image made after the cell was bathed in growth medium with 25 mM K, showing a general elevation of Ca^{2+} and a clear reversal of the gradient. Soma minor axis 8 μm.

electrode. A small negative current applied through the electrode reduced the firing rate to $< 1/s$. The Ca^{2+} levels 20 and 60 s after this reduction are shown in Fig. 5C&D. Soma levels were reduced from 270 nM in B to 170 & 110 nM in C&D. It is to be noted that the dendritic region remains at a higher Ca^{2+} level than the soma after the hyperpolarization. This type of dendrite-soma gradient is common. Attempts at stimulating the cell with such an external electrode have generally resulted in membrane rupture and loss of indicator up into the electrode. It is probably reasonable to assume that negative-going currents can utilize the delayed rectifier channels that are opened by the drop in electric potential across the patch of membrane within the electrode. Thus a considerable amount of steady current can be passed without undue electric field; however in the opposite direction these channels cannot be utilized and an equivalent current requires a much higher field.

CONCLUSION

In spite of the excessive fanfare and initial oversimplification accompanying the introduction of self-loading fluorescent indicators, they have produced some interesting findings and one would expect many more in the future as cell-indicator interactions are explored more thoroughly. Other studies being pursued in this laboratory are those investigating the actions of excitatory and inhibitory neurotransmitters on cerebellar neurons in culture (Connor, Tseng, and Hockberger, 1987; Connor and Hockberger, 1987); angiotensin and high K effects on Ca^{2+} levels that lead to aldosterone secretion in adrenal glomerulosa cells (Connor, Cornwall, and Williams, 1987); hormone mediated effects on Ca^{2+} in fibroblasts (with C. Benjamin, and R. Gorman); neurite extension and Ca regulating factors in molluscan neurons (Cohan, Connor, and Kater, 1987; Fink, Connor, and Kaczmarek, 1987).

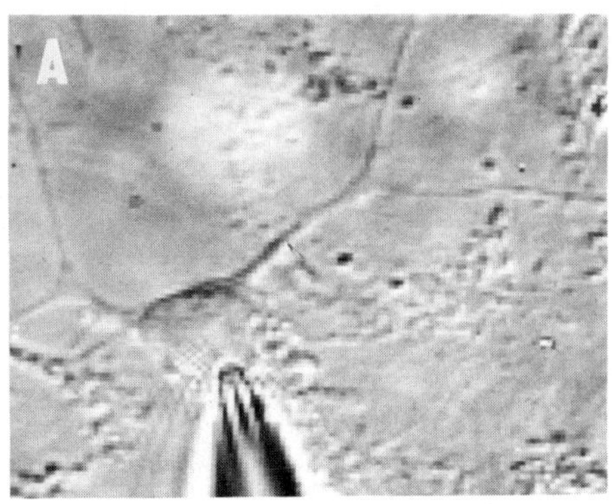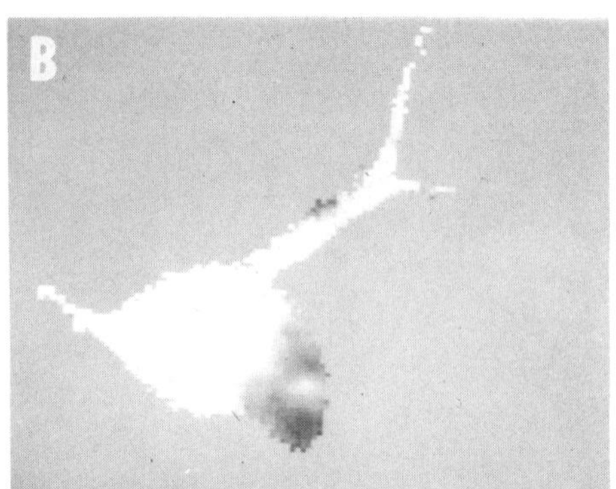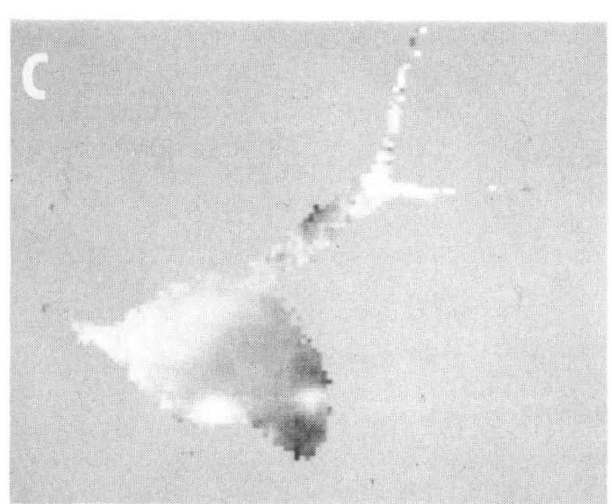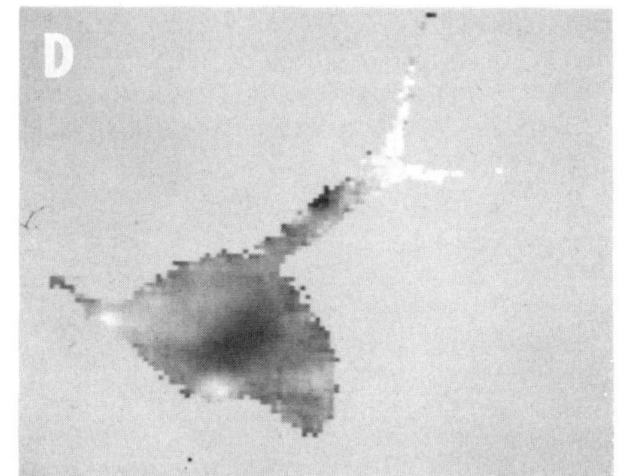

Figure 5. A: Neuron in cerebellar explant culture (15 days) with patch electrode positioned near surface. B,C, and D: Ratio images made during and after cessation of spontaneous firing (see text), showing large reduction in Ca^{2+}. Soma diameter approximately 15 μm. Image size 140x140 pixels.

ACKNOWLEDGMENT

I gratefully acknowledge the support and advice of A.A. Lamola and J.A. Tyson during the initial part of this study, the collaboration of K.L. McMillan in software development, and P.E. Hockberger and H.Y. Tseng in tissue culture experiments. Supported in part by the Air Force Office of Scientific Research: F49620-85-C-0009.

REFERENCES

Ahmed, Z. & Connor, J.A. (1979) Measurement of calcium influx under voltage clamp in molluscan neurons using the metallochromic dye arsenazo III. J. Physiol. (Lond.) 286:61-82.

Ahmed, Z., Connor, J.A., Tank, D.W., and Fellows, R.E. (1986) Expression of membrane currents in rat diencephalic neurons in serum-free culture. Dev. Br. Res., 28:221-231.

Ahmed, Z., Walker, P.S., and Fellows, R.E. (1983) Properties of neurons from dissociated fetal rat brain in serum-free culture. J. Neurosci., 3:2448-2462.

Almers, W. & Neher, E. (1985) The Ca signal from fura-2 loaded mast cells depends strongly on the method of dye loading. Febs. Letters 192:13-18.

Ashley, C.C. & Campbell, A.K. (1979) Detection and Measurement of Free Ca in Cells. Elsevier/North Holland Amsterdam.

Baylor, S.M., Hollingworth, S., Hui, C.S., and Quinta-Ferreira, M.E. (1986) Properties of the metallochromic dyes Arsenazo III, Antipyrylazo III, and Azol in frog skeletal muscle fibers at rest. J. Physiol. (Lond.), 377:89-141.

Bolsover, S.R., and Spector, I. (1986) Measurements of calcium transients in the soma, neurite, and growth cone of single cultured neurons. J. Neuroscience, 6:1934-1940.

Brown, J.E. & Blinks, J.R. (1974) Changes in intracellular free Ca concentration during illumination of invertebrate photoreceptors. Detection with aequorin. J. Gen. Physiol., 64:643-665.

Cohan, C.S., Connor, J.A., and Kater, S.B. (1987) Electrically and chemically mediated increases in intracellular calcium in neuronal growth cones. J. Neurosci. (in press)

Connor, J.A. (1986) Digital imaging of free calcium changes and of spatial gradients in growing processes in single, mammalian central nervous system cells. Proc. Natl. Acad. Sci. USA, 83:6179-6183 De Weer, P. & Salzberg, B.M. (1986) Optical Methods in Cell Physiology. Wiley-Interscience, New York.

Connor, J.A., Cornwall, M.C. and Williams, G.H. (1987) Spatially resolved cytosolic calcium response to angiotensin II and potassium in rat glomerulosa cells measured by digital imaging techniques. J. Biol. Chem., 262:2919-2927.

Connor, J.A. and Hockberger, Phillip E. (1987) Digital imaging of Ca levels in CNS neurons under conditions that induce facilitating increases in Ca levels and sustained Ca elevation. In: Cellular Mechanisms of Conditioning and Behavioral Plasticity. ed. C.M. Woody Plenum (in press).

Connor, J.A., Kretz, R., Shapiro, E. (1986) Calcium levels measured in a presynaptic neuron of *Aplysia* under conditions that modulate transmitter release. J. Physiol., *375*:625-642.

Connor, J.A., Tseng, Hsiu-Yu, and Hockberger, P.E. (1987) Depolarization and transmitter induced changes in intracellular Ca of rat cerebellar granule cells in explant cultures. J. Neurosci., *7*:1384-1400.

Fink, L., Connor, J.A., and Kaczmarek, L. (1987) Ion conductance activated by inositol trisphosphate injection into peptidergic neurons. Soc Neurosci Abstracts (in press)

Gotoh, Y., Sugamura, K., & Hinuma, Y. (1982) Proc. Natl. Acad. Sci. USA. 79:4780-4782.

Graubard, K. and Ross, W.N. (1985) Regional distribution of calcium influx into bursting neurons detected with arsenazo III. Proc. Natl. Acad. Sci. USA. 82:5565-5569.

Grynkiewicz, G., Poenie, M., & Tsien, R.Y. (1985) A new generation of Ca indicators with greatly improved fluorescence properties. J. Biol. Chem. 260:3440-3450.

Lakowicz, J.R. (1983) Principles of Fluorescence Spectroscopy. Plenum, New York.

Levy, S. & Fein A. (1985) Relationship between light sensitivity and intracellular free Ca concentration in Limulus ventral photoreceptors. J. Gen. Physiol., 85:805-841.

Miledi, R., Parker, J., and Schalow, G. (1977) Measurement of calcium transients in frog muscle by the use a Arsenazo III. Proc. Royal Soc. B, 198:201-210.

Moisescu, D.G. & Pusch, H. (1975) A pH-metric method for the determination of the relative concentration of calcium to EGTA. Pflugers Archive. 355:R122.

Paradiso, A.M., Tsien, R.Y., & Machen, T.E. (1984) Na^+-H^+ exchange in gastric glands as measured with a cytoplasmic-trapped, fluorescent pH indicator. Proc. Natl. Acad. Sci. USA. 81:7436-7440.

Pesce, A.J., Rosen, C., & Pasby, T. (1971) Fluorescence Spectroscopy: An introduction for biology and medicine. Marcel Dekker, New York.

Rink, T.J., Tsien, R.Y., & Pozzan, T. (1982) Cytoplasmic pH and free Mg^{2+} in lymphocytes. J. Cell. Biol., 95:189-196.

Reynolds, G.T. (1972) Image intensification applied to biological problems. Quart. Rev. Biophys., 5: 285-347.

Ross, W.M. and Werman, R. (1987) Mapping Ca transients in the dendrites of Purkinje cells from the guinea pig cerebellum in vitro. J. Physiol., (in press).

Sequin, C.A., and Tompsett, F. (1975) Charge transfer devices. Academic Press, New York.

Tsien, R.Y. (1981) A non-disruptive technique for loading calcium buffers and indicators into cells. Nature, 280:527-528.

Tsien, R.Y. and Poenie, M. (1986) Fluorescence ratio imaging: a new window into intracellular ionic signaling. TIBS., 11:450-455.

Tsien, R.Y., Pozzan, T., & Rink, T.J. (1982) Calcium homeostasis in intact lymphocytes; cytoplasmic free calcium monitored with a new, intracellularly trapped fluorescent indicator. J. Cell. Biol., 94:325-334.

Tsien, R.Y., Rink, T.J., & Poenie, M. (1985) Measurement of cytosolic free Ca in individual small cells using fluorescence microscopy with dual excitation wavelengths. Cell Calcium 6:145-157.

Tyson, J.A., Baum, W.A., & Kreidl, T. (1982) Deep CCD images of 3c273. Ap.J. 252, L1.

Tyson, J.A., Boeshaar, P.C. (1983) New limits on the surface density of M-dwarfs from CCD and photographic data. Proc. I.A.U. Colloq. "The nearby stars and the stellar luminosity function", Davis Press, Schenectady, New York.

Udenfriend, S. (1969) Fluorescence Assay in Biology and Medicine Academic Press, New York.

Williams, D.A., Fogarty, K.E., Tsien, R.Y., & Fay, F.S. (1985) Calcium gradients in single smooth muscle cells revealed by the digital imaging microscope using fura-2. Nature 318:558-561.

USE OF FUSED SYNAPTOSOMES OR SYNAPTIC VESICLES TO

STUDY ION CHANNELS INVOLVED IN NEUROTRANSMISSION

S.A. DeRiemer[+*], R. Martin[#], R. Rahamimoff[*%], B. Sakmann[*] and
H. Stadler[*]

+ Dept. of Biological Sciences, Columbia Univ., NY NY 10027
  U.S.A.
* Max-Planck Institute fuer Biophysikalisches Chemie
  Goettingen, F.R.G.
Sektion **Elektronenmikroskopie,** Universitaet Ulm, D7900
  Ulm, F.R.G.
% Dept. of Physiology, Hebrew Univ. Hadassah Medical School
  Jerusalem, Israel.

INTRODUCTION

Ion channel modulation is a process which allows nerve cells to
perform differently under changing requirements. Until recently the
major focus was on the modulation occuring at the surface membrane, in
particular in large cell bodies. Small nerve terminals and intracellular
organelles, such as synaptic vesicles, could not be easily studied.

Our interest in ion channel modulation has brought us to attempt to
develop methods for making small, usually inaccessible structures such as
pre-synaptic nerve terminals available for physiological experimentation.
The approach we would like to present here involves the fusion of
purified pre-synaptic endings (synaptosomes) or purified synaptic
vesicles with dimethylsulfoxide (DMSO) and polyethylene glycol (PEG) to
create structures large enough for recording and analysis with the patch
clamp technique.

While this method depends on the ability to prepare a pure
preparation of sealed membranes, we believe it will be applicable to a
large variety of situations in which this criterion is met. The examples
with which we illustrate the method are synaptosomes and synaptic
vesicles purified from the electric organ of _Torpedo_.

METHODS

The methods for purifying synaptosomes (Gray & Whittaker, 1962;
Israel, et al., 1976; Stadler & Tashiro, 1979) and synaptic vesicles
(Tashiro & Stadler, 1978) have been detailed elsewhere. As an example,
the synaptic vesicle preparation used will be briefly described.

Torpedo _marmorata_ were anesthetized with tricaine methanesulphonate
and the electric organ removed and frozen in liquid nitrogen. The frozen
tissue was crushed and extracted with 0.4 M NaCl/20 mM Tris/0.3 mM

phenylmethylsulphonyl fluoride (PMSF) buffer at pH 7.4 in the presence
absence of 3.5 mM EGTA. The supernatant from a low-speed centrifugation
(10,000 x g x 30 min.) was layered onto a discontinuous sucrose gradient
(0.6 M sucrose/0.1 M NaCl:0.2 M sucrose/0.3 M NaCl) and centrifuged at
67,700 x g x 2.5 hrs. The band at the interface was loaded onto a
continuous isoosmotic sucrose gradient, centrifuged at 171,800 x g x 3
hrs. Fractions containing the peak ATP activity were pooled, diluted
with 0.5 volumes of 400 mM NaCl (or another of the fusion buffers listed
below), and pelleted at 105,000 x g x 3 hr.

Care was paid to remove as much of the supernatant as possible from
the final pellet which was then washed twice by overlaying it with 100 μl
of fusion buffer. After drying the tube, 100 μl of fusion buffer
containing 20% DMSO and 25% PEG 1500 (Boehringer) was added to the pellet
which was then scraped gently from the tube and transferred to a 1.5 ml
microfuge tube. The preparation was incubated for 2 min. at 37°C
followed by addition of 2-3 volumes more fusion buffer (without the DMSO
and PEG) and a further 10 min. incubation at 37°C. Following this, the
preparation was allowed to sit at room temperature (20-25°C) for 2-4
hours during which time there was gradual formation of giant synaptic
vesicles. Once the vesicles had formed, the medium could be diluted
further if an osmotic balance was preserved. Preparations kept on ice
were stable for 1-2 days although proteolysis during this period has not
been ruled out.

Giant vesicles appear as large, grey spheres which are initially
observed at the edges of clumps of unfused vesicles; fusion of
synaptosomes leads to structures which are indistinguishable from the
giant vesicles at the light microscopic level, and an example of giant
synaptosomes is shown in Fig. 1a. Electron micrographs of giant vesicle
preparations show both unilamellar and multilamellar structures; the
structures large enough to correspond to the giant vesicles appeared to
fall primarily into the first category (Fig. 2).

The choice of fusion buffers was determined by the osmolarity of
Torpedo tissues and by the experiments planned. Typical solutions were
(in mM, all pH 7.4) 400 Na-Glutamate, 400 K-Glutamate, 400 N-
methylglucamine chloride for looking at Na, K, and Cl conductances
respectively. Glucose was used to adjust the osmolarity when needed.
During experiments the vesicles were bathed in buffers identical to the
fusion buffer except that a small amount of chloride was added, and there
was of course no PEG or DMSO.

The patch clamp technique was applied to the giant vesicles
primarily in the "cell"-attached configuration (Hamill, et al., 1981).
It was possible to manipulate the ionic composition of both sides of the
membrane even in the "cell"-attached mode by determining the
intravesicular composition during fusion. Pipettes were made of thick-
walled borosilicate and pulled to resistances of 5-10 MΩs when filled
with 400 mM NaCl. Pipette and patch capacitance were cancelled with the
EPC/7 amplifier and signals were collected on video tape after conversion
with a modified Sony PCM701. Pipette potential relative to the bath was
changed manually using the EPC/7 or by applying 1.5 or 5 second ramps
generated by a modified Rockwell AIM microcomputer. Ramps and single
channel data were analyzed using a PDP 11/23 computer and programs
written primarily by Christoph Methfessel.

RESULTS

Ionic conductances and spontaneously occuring single channel

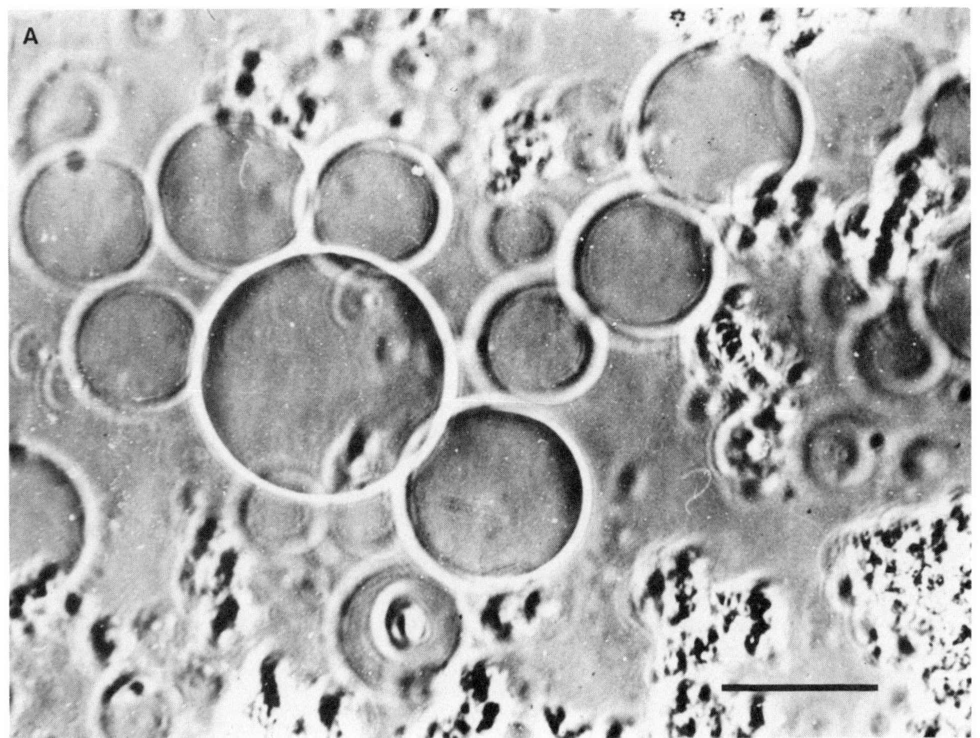

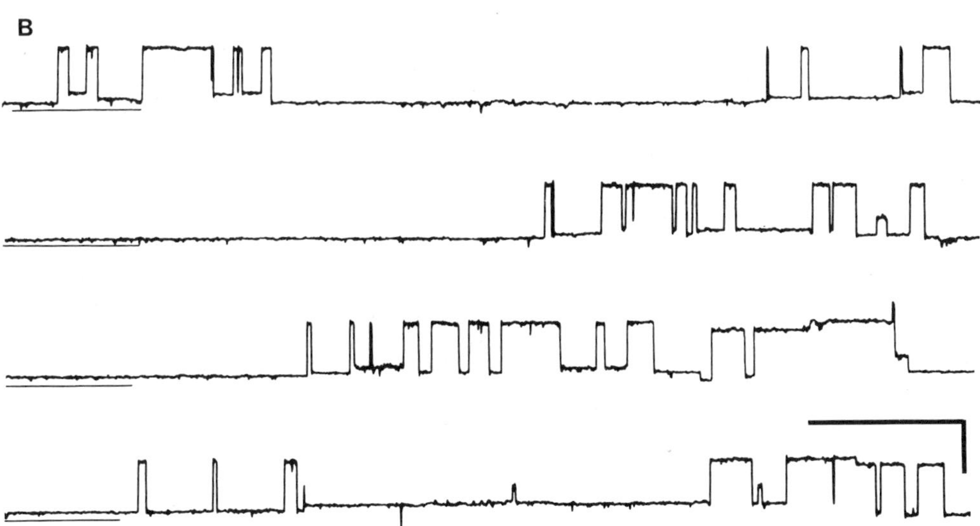

Fig. 1. (A) Light micrograph of giant synaptosomes. Visible are both giant synaptosomes and clumps of unfused synaptosomes. Scale bar: 20 μM.
(B) Channel activity in giant synaptosome. Fusion buffer: 400 mM NaCl/10 mM Tris. Bath solution: 350 mM KCl/10 mM Tris. Pipette solution: 200 mM KCl/100 mM $CaCl_2$/10 mM Tris. Pipette potential: −12 mV. Scale bars: 2 pA, 1 sec. Closed (———).

events were consistently observed in giant structures prepared from either synaptosomes (Fig. 1b) or synaptic vesicles (Rahamimoff et al., 1986; DeRiemer et al., 1987; Figs 3, 4). Indeed, some form of activity was observed in each of the over 135 patches from 29 preparations which we have examined to date. Recordings were stable for periods ranging from minutes to several hours, and it was possible to obtain seals with resistances of up to 200 GΩs. Currents were not observed however when both pipette and fusion media consisted only of impermeant ions (N-methylglucamine glutamate) suggesting that they were not non-selective breakdown events in the membranes.

Because the presence of ionic channels in synaptic vesicles raises the interesting possibility that these channels are involved in aspects of vesicle physiology such as the storage of neurotransmitter or exocytosis and that modulation of ion channels may not be restricted to the plasma membrane, we have focused our attention on the properties of these channels and will discuss these results in more detail.

In recordings from giant vesicles, conductances were observed with reversal potentials within 1-3 mV of zero applied pipette potential when the concentrations of the permeant ions Cl, Na or K were symmetrical in pipette and fusion medium/bath. Substitution of an impermeant ion (glutamate for Cl or N-methylglucamine for the cations) on either side of the membrane, however, produced shifts in the observed reversal potentials for single channel events. The observed shifts were consistent with the presence of conductances for Na, K, and Cl in synaptic vesicle membranes.

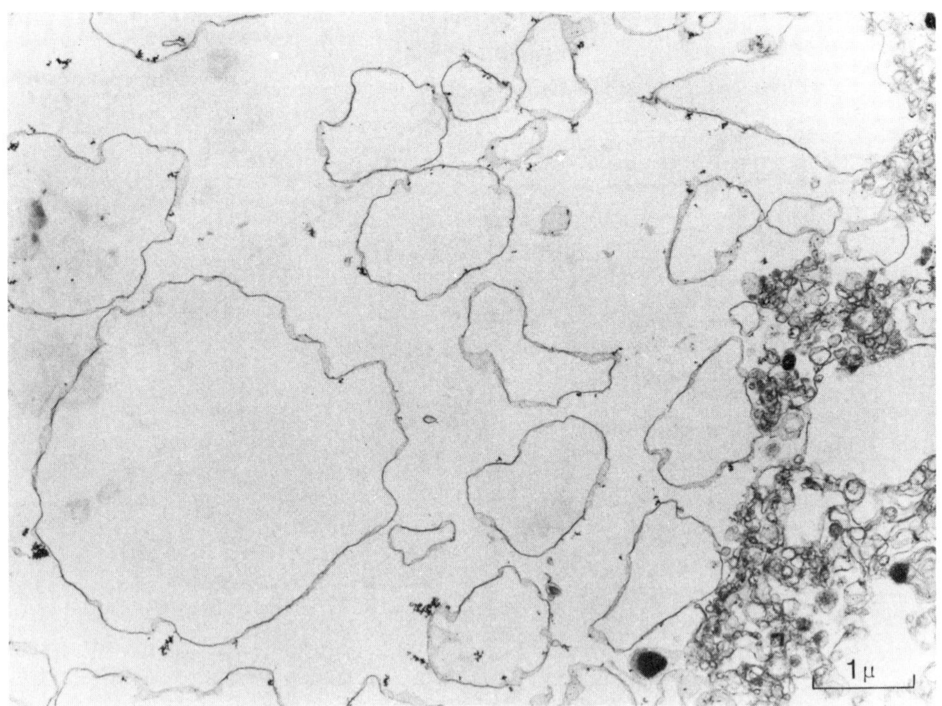

Fig. 2. Electron micrograph of synaptic vesicle preparation after fusion. Synaptic vesicles were fused in 400 mM K-glutamate, fixed in 2% glutaraldehyde in 450 mM Na-cacodylate buffer, pH 7.1, for 24 hrs. post fixed in 2% OsO4 in H2O, then imbeded in Epon and sectioned.

Characterization of single channel events in the presence of
solutions with various compositions suggested that the two major classes
observed represented cation conductances. These two classes have been
distinguished on the basis of their size and relative selectivity for Na
and K ions.

The predominant single channel event seen when potassium is used
as charge carrier is a large channel with multiple conductance levels in
the range 100 to 200 pS (Fig. 3). The reversal potential shifts observed
in patches containing only this channel in response to potassium
concentration gradients suggest that the channel is a cation channel and
that it shows selectivity for potassium over sodium.

A second class of single channel event was also consistently
observed although it was easily masked in the presence of the large
events described above. This channel was also cation conducting; the
reversal potential of 0 mV seen with sodium on one side and potassium on
the other side of the membrane suggests that it shows little selectivity
between these two cations. Examples of activity from this channel and an
amplitude histogram for these events is shown in Fig. 4. The estimated
single channel conductance from these data is approximately 30 pS.

Both types of channels were seen in most of the preparations of
giant vesicles prepared from purified synaptic vesicles. In one series
of 30 patches, the mean number of channels per patch was 5 of which there
were an average of 3 large and 2 small. If one assumes a tip diameter of
1 μM for the patch and the vesicle parameters given above, this leads to
an estimate of 40 vesicles per patch and one channel per 8 vesicles.
There are a number of reasons for arguing that this is an underestimate
of the channel density. Channels may be inactivated during the
purification or fusion, and our experimental conditions (calcium
concentrations, voltage protocols, etc.) may not be optimal for channel
activation.

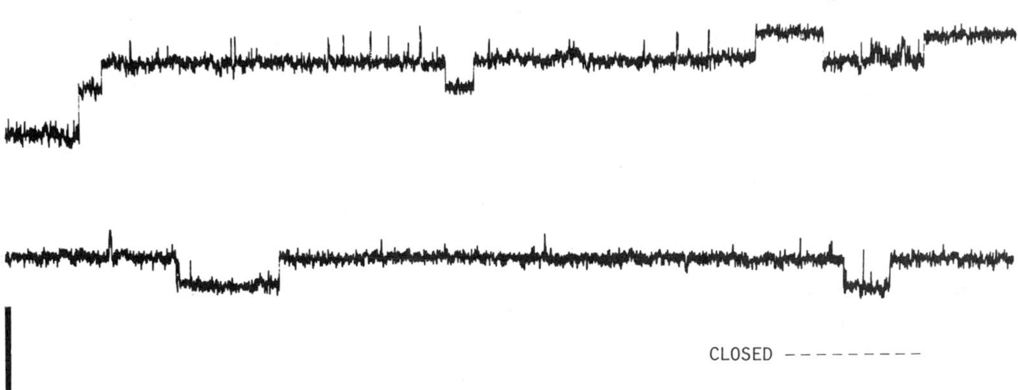

Fig. 3. Large potassium selective channels in giant vesicles. Fusion
        buffer: 400 mM K-glutamate. Bath solution: 400 mM K-glutamate/
        2 mM KCl. Pipette solution: 100 mM K-glutamate/2 mM KCl.
        Pipette potential: 0 mV. Scale bars: 10 pA, 250 msec.
        Continuous trace illustrating several levels of openings.
        Closed (------).

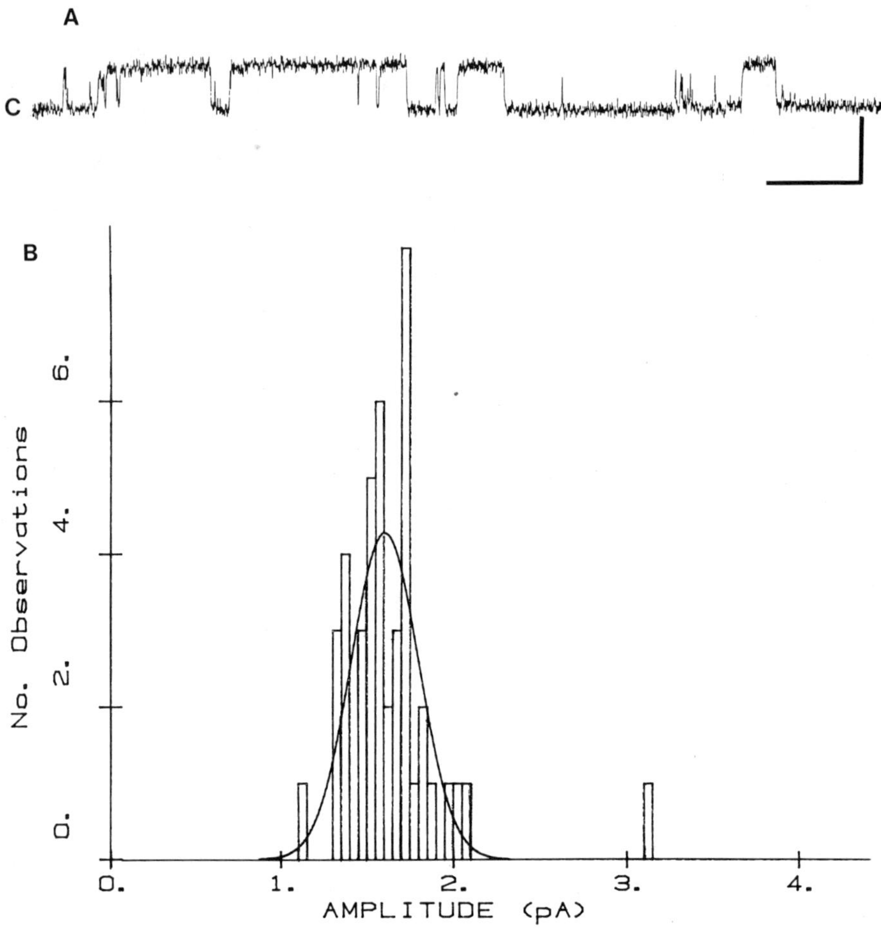

Fig. 4. Small cation-selective channels in giant vesicles. Fusion
buffer: 400 mM K-glutamate. Bath solution: 360 mM K-glutamate/
45 mM KCl/10 mM EGTA/10 mM Tris. Pipette solution: 180 mM
K-glutamate/21 mM KCl/5 mM EGTA/5 mM Tris. Pipette potential:
-20 mV. (A) Record of single channel activity. Scale bars:
2 pA, 100 msec. Closed (c). (B) Amplitude histogram of
activity in the same patch as above. Fit is a Gaussian with
mean = 1.60 pA; s.d. = 0.20; n = 42.

DISCUSSION

In this work we have shown that fusion of subcellular organelles
into giant structures can provide an useful preparation for the study of
channels involved in synaptic transmission. There are several other
methods which have allowed electrical recordings of ion channels in
synaptosomes or secretory vesicles. These include incorporation of
channels into planar lipid bilayers (Stanley, et al., 1986a,b) or
bilayers formed on the tip of a patch pipette (tip dip; Lemos & Nordmann,
1986; Lemos, et al., 1987), fusion of secretory vesicles with liposomes
(Picaud, et al., 1984), and enzyme treatment of synaptosomes to cause
spontaneous formation of "giants" (Umbach, et al., 1984). An additional
method for fusing structures which may be applicable is the electrofusion
technique which has been widely used to fuse nuclei and cells
(Zimmermann, 1986). Each of these techniques has obvious advantages and
disadvantages regarding possible perturbations of membrane proteins and
the frequency of success in obtaining recordings. Probably only a
combination of several methods will give an accurate view of these
channels.

Implications of channels in vesicles

Already, there is a considerable accumulation of evidence supporting
the view that secretory vesicles, including synaptic vesicles, have ion
channels in their membranes. In addition to the data presented here,
there are reports of channels in chromaffin granules (Picaud, et al.,
1984), and in secretory vesicles from the neural lobe of the rat
pituitary (Lemos & Nordmann, 1986; Lemos, et al., 1987; Stanley, et al.,
1986a,b). One is then faced with the question of what roles these ion
channels may play in vesicle physiology, and in the context of this book,
how might their presence influence our views of the role of ion channel
modulation.

The primary function of synaptic vesicles is to store transmitter
and upon activation of the terminal, to release it. Ion channels have
recently been implicated in the uptake and release of compounds stored in
the vacuoles of plant cells, a function which they perform in conjunction
with pumps and carriers also present in the vacuolar membrane (Hedrich,
et al., 1986). An analogous situation can be envisaged for secretory
granules. As for exocytosis, there is strong evidence that fusion of the
vesicular and plasmalemmal membranes requires the presence of an osmotic
gradient (Miller, et al., 1976; Cohen, et al., 1980), and there are a
number of models dealing with this observation which predict the presence
of specific types of ion channels in vesicle membranes (Pazoles &
Pollard, 1978; Grinstein, et al., 1982; Stanley & Ehrenstein, 1985).
With the methods currently available it is possible now to address the
question not only of whether the channels predicted by the models really
are present, but also to use information gained in characterizing their
properties to devise experiments to directly test their possible role in
transmitter release.

Finally, in view of the wide spread occurence of ionic channels and
in view of their capacity for modulation on the one hand and the
plasticity of synaptic transmission on the other hand, the presence of
ion channels on vesicles may indicate a novel site for the regulation of
transmitter release and brain function.

ACKNOWLEDGEMENTS

This work was supported in part by MDA, CTR, EMBO and SFB 236. SAD
was a Hoffmann-La Roche Fellow of LSRF.

REFERENCES

Cohen, R.S., Zimmerberg, J. and Finkestein, A., 1980, Fusion of
    phospholipid vesicles with planar phospholipid bilayer membranes. II.
    Incorporation of a vesicular membrane marker into the planar membrane,
    J. Gen. Physiol. 75:251-270.
DeRiemer, S.A., Rahamimoff, R., Sakmann, B. and Stadler, H., 1987,
    Conductances and channels in fused synaptic vesicles from Torpedo
    electric organ, J. Physiol. (in press).
Gray, E.G. & Whittaker, V.P., 1962, The isolation of nerve endings from
    brain: An electron-microscopic study of cell fragments derived by
    homogenization and centrifugation, J. Anat. 96:79-96.
Grinstein, S., Vander Meulen, J. and Furuya, W., 1982, Possible role of H+-
    alkali cation countertransport in secretory granule swelling during
    exocytosis, FEBS Lett. 148:1-4.
Hamill, O.P., Marty, A., Neher, E., Sakmann, B. and Sigworth, F.J., 1981,
    Improved patch-clamp techniques for high-resolution current recording
    from cells and cell-free membrane patches, Pflugers Arch. 391:85-100.
Hedrich, R., Flugge, U.I. and Fernandez, J.M., 1986, Patch-clamp studies of
    ion transport in isolated plant vacuoles, FEBS Lett. 204:228-232.
Israel, M., Manaranche, R., Mastour-Frachon, P. and Morel, N., 1976,
    Isolation of pure cholinergic nerve endings from the electric organ of
    Torpedo marmorata, Biochem. J. 160:113-115.
Lemos, J.R. & Nordmann, J.J., 1986, Ionic channels and hormone release from
    peptidergic nerve terminals, J. Exp. Biol. 30:271-282.
Lemos, J.R., Ocorr, K.A. & Nordmann, J.J., 1987, Ionic channels in
    neurosecretory granules from rat neural lobe, Biophys. J. 51:64a.
Miller, C., Arvn, P., Telford, J.N. & Racker, E., 1976, Ca++-induced fusion
    of proteoliposomes: Dependence on transmembrane osmotic gradient, J.
    Membrane Biol. 30:271-282.
Pazoles, C.J. and Pollard, H.B., 1978, Evidence for stimulation of anion
    transport in ATP-evoked transmitter release from isolated secretory
    vesicles, J. Biol. Chem. 253:3962-3969.
Picaud, S., Marty, A., Trautmann, A., Grynszpan-Winograd, O. and Henry,
    J.-P., 1984, Incorporation of chromaffin granule membranes into large-
    size vesicles suitable for patch-clamp recording, FEBS Lett.
    178:20-24.
Rahamimoff, R., DeRiemer, S., Ginsburg, S., Sakmann, B., Shapira, R.,
    Silberberg, S.D. and Stadler, H., 1986, Short term and long term
    regulation of transmitter release, Proc. First Congress of the Asian
    and Oceanic Physsiological Societies, p. 15.
Stadler, H. and Tashiro, T., 1979, Isolation of synaptosomal plasma
    membranes from cholinergic nerve terminals and a comparison of their
    proteins with those of synaptic vesicles, Eur. J. Biochem.
    101:171-178.
Stanley, E.F. and Ehrenstein, G., 1985, A model for exocytosis based on the
    opening of calcium-activated potassium channels in vesicles, Life Sci.
    37:1985-1995.
Stanley, E.F., Ehrenstein, G. and Russell, J.T., 1986, Evidence for
    calcium-activated potassium channels in vesicles of pituitary cells,
    Biophys. J. 49:19a.
Stanley, E.F., Ehrenstein, G. and Russell, J.T., 1986, Evidence for anion
    channels in secretory vesicles, Soc. Neurosci. Abstr., Vol. 12,
    Part 2, p. 819.
Tashiro, T., and Stadler, H., 1978, Chemical composition of cholinergic
    synaptic vesicles from Torpedo marmorata based on improved
    purification, Eur. J. Biochem. 90:479-487.
Umbach, J.A., Gundersen, C.B. and Baker, P.F., 1984, Giant synaptosomes,
    Nature 311:474-477.
Zimmermann, U., 1986, Electrical breakdown, electropermabilization and
    electrofusion, Rev. Physiol. Biochem. Pharmacol. 105:175-256.

ION CHANNELS OF THREE MICROBES: <u>PARAMECIUM</u>, YEAST AND <u>ESCHERICHIA</u> <u>COLI</u>

Boris Martinac, Yoshiro Saimi, Michael C. Gustin
and Ching Kung

Laboratory of Molecular Biology
and Department of Genetics
University of Wisconsin
Madison, WI 53706

Although the ion channels of vertebrates, and larger invertebrates have
been the focus of channel research in the past and at the present, some
studies of the channels of microbes have been made. The microbial channels
were found to have many similar and a few different features from those of
the metazoan channels.

Microbes are studied because of

1. interests in their ion channels in their own right,
2. interests in channel evolution and the nature of the primordial
   channel(s),
3. interests in cloning and expressing the genes for metazoan channels
   into microbes and studying them in molecular details.

Microbes offer the following experimental advantages:

1. They have short life cycles and, in some cases, haplophases,
making their genetics easier to study.

2. They can be cultured <u>en</u> <u>masse</u> making their biochemistry easier to
study.

3. Their clonal growth gives populations of cells of identical geno-
and phenotypes. The problem of cellular heterogeneity in a tissue does not
arise.

4. They are unicells. The theoretical complications of cellular
interactions and the practical problems of dissection are obviated.

5. They have been studied for a long time by geneticists, biochemists,
molecular biologists and cell physiologists. There exist a vast and deep
knowledge on many molecular processes in the microbes. Technical advances
in the manipulations of genes and gene products have been largely made with
microbes. The recombinant DNA technology makes possible the cloning and
expressing foreign channel genes in microbes as well as mass producing
native or foreign ion-channel proteins in industrial quantities.

The shortcomings of microbes as subjects of electro-physiological and
biophysical investigations are their small sizes (except protozoa, see

below) and the unexplored nature of their surface. These shortcomings
have recently been overcome by using the patch clamp technique and by
manipulating the cells to generate "patchable" objects of the proper
dimensions (Figs. 1, 5, and 9A). Though all are considered microbes,
the three organisms reviewed here are, in fact, very different in sizes.
Paramecium, yeast and E. coli are approximately 100, 10 and 1 μm in length
respectively. They are respectively a "primitive" animal, a "simple"
plant and a Gram-negative bacterium.

PARAMECIUM TETRAURELIA OR P. CAUDATUM (CILIATED PROTOZOA)

There has been a long tradition in the study of the electrical properties of the membrane of Paramecium since the pioneering work of Kamada and
Kinosita (1940). This was possible because Paramecium is a giant among
microbes and can be examined with conventional electrodes. Kinosita and
coworkers (1964a,b) showed that Paramecium can generate action potentials.
The action potentials were found to cause a reversal of the beat direction
of the cilia. The ionic mechanisms of the action potentials and touch-
receptor potentials were further analyzed in the late 60's and the 70's by
Roger Eckert and Yutaka Naitoh (1968, 1969, 1972), and later, their collaborators. Their work laid much of the foundation for the modern studies
of the ion channels in this ciliated protozoa. The use of the whole-cell
voltage clamp in the late 70's and early 80's allowed us to sort out the
various currents through the plasma membrane of Paramecium (Oertel et al.,
1977, Kung and Saimi, 1982). At least eight different macroscopic currents
are revealed. Depolarizations of the normal (wild-type) membrane induce a
Ca^{++} inward current rapidly (milliseconds) followed by a fast K$^+$ outward
current (the delayed rectifier). Ca^{++}, which carries the fast inward
current and arrives at the cell interior, activates a slow inward Na$^+$
current (tens to hundreds of milliseconds) which is followed by a still
slower Ca^{++}-activated K$^+$ current. We believe that these four currents
constitute the action potential for Paramecium in its natural environment
(Saimi et al., 1983). Hyperpolarization of the Paramecium membrane
activates a separate set of ion channels.

Single-channel recording through patch-clamp electrodes can be made
on blisters dislodged from the surface of Paramecium cells as illustrated
by Fig. 1. We have encountered activities of four types of K$^+$ channels

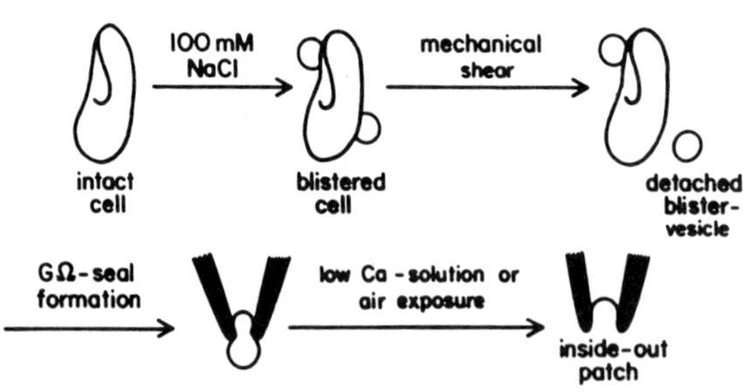

Fig. 1. A cartoon showing a procedure used for single-channel recording in Paramecium. In order to form plasma-membrane blisters the cells were first exposed for 5-10 min to a high-salt solution (100 mM NaCl, 10^{-5} M Ca^{++}, 1 mM EGTA, 5 mM HEPES, pH 7.2). Seals (usually 4 GΩ) were formed on isolated blisters that had been mechanically detached from the cell. The excision of inside-out patches was achieved simply by vigorous perfusion of the bath with a low-calcium solution.

Na$^+$ channel, a Cl$^-$ channel and a cation-nonspecific channel (Table 1).
As an example the current fluctuations at different voltages and Ca^{++}
concentrations of the Ca^{++}-dependent K$^+$ channel with a large conductance
are shown in Fig. 2A. The current-voltage relationship in asymmetric K$^+$
solutions gives a conductance of ca. 150 pS for this channel. In the
presence of Cs$^+$ ions the reversal potential shifts with the calculated
equilibrium potential for potassium (Fig. 2B).

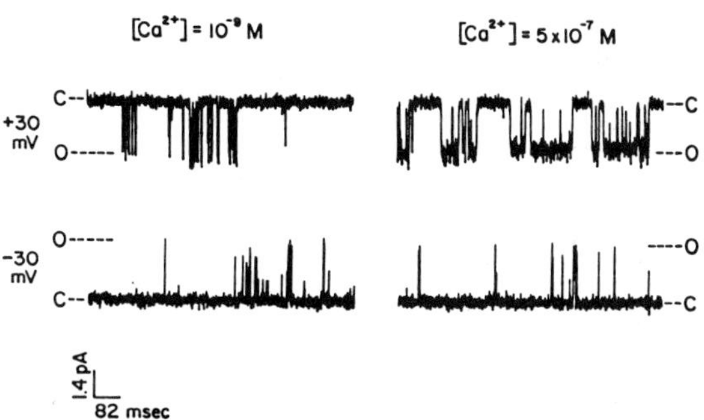

<u>Fig. 2A</u>. Effect of applied membrane potential on single-channel currents
in an excised inside-out blister patch containing one Ca^{++}-dependent K$^+$
channel at two different calcium concentrations. Pipette solution: 100
mM KCl, 3 mM CaCl$_2$, 1 mM MgCl$_2$, 0.1 mM EDTA buffered to pH 7.2 with
HEPES-KOH. Bath solution: 150 mM KCl buffered to pH 7.2 with HEPES-KOH.
The free Ca^{++} was buffered using 1 mM EGTA.

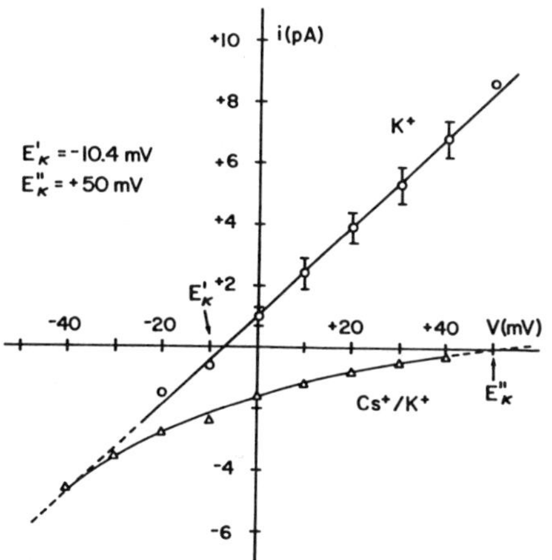

<u>Fig. 2B</u>. Single-channel current as a function of membrane potential in the
presence of 150 mM KCl (circles, means +/- S.D. of three channels except at
voltages of -20, -10, and +50 mV, where every point represents measurements
of one or two channels) and 150 mM CsCl/14 mM KCl (triangles, measurements
from one channel) in the bath solution. From the slope of the current-
voltage relationship in assymetric K$^+$ the channel conductance of 145 pS was
determined. The reversal potential of -6.9 mV was found close to the calculated K$^+$ equilibrium potential value of -10.4 mV. In Cs$^+$/K$^+$ solution
the experimental curve could be extrapolated to the calculated reversal
potential of +50 mV (Martinac et al., 1986).

Table I. Ion Channels in <u>Paramecium</u> Identified to Date

type of channel	known activation factor	approximate conductance (pS)	ion selectivity	ref.
G_K	Ca, depol.	150	$P_K \gg P_{Na}$	a
G_K	Ca, depol.	40	$P_K \gg P_{Na}$	b
G_K	Ca, hyperpol.	70	$P_K > P_{Na}$	b
G_{Na}	Ca	<10	$P_{Na} \gg P_K$	c
G_{Cl}	?	<430	$P_{Cl} \gg P_{Glut}$	b
G_{cation}	ATP, depol.	<150	$P_K \sim P_{Na}$	b
G_K	?	40	$P_K \gg P_{Na}$	a
G_{cation}	?	30	$P_{Ba} \sim P_{Mg} \gg P_K$	d
G_{Ca}	voltage	2	$P_{Ba} \gg P_{Mg}, P_K$	d

Footnotes: All measurements were done in the presence of 50 to 150 mM salts.
a: Martinac <u>et al</u>., (1986);
b: Martinac and Saimi, unpublished
c: Saimi, unpublished
d: Ehrlich <u>et al</u>., (1984).

Table II. Summary of Ion Currents Affected by Mutations in <u>Paramecium</u> (from Hinrichsen et al., 1985)

Type	Complementation Group	Ion-Channel Affected	
Pawn	pwA, pwB, pwC, pwD	I_{Ca}	↓
CNR	cnrA, cnrB, cnrC, cnrD	I_{Ca}	↓
Dancer	Dn	I_{Ca}	↑
pantophobiac	pntA, pntB	I_K^{Ca}	↓
TEA-insensitive	teaA	I_K^{Ca}	↑
fast-2	fna	I_{Na}^{Ca}	↓
Paranoiac	PaA, PaC	I_{Na}^{Ca}	↑

Because the motile behavior of _Paramecium_ is governed by the membrane-potential, induced and selected mutants that behave abnormally (Kung 1971a,b) are often defective in their ion-channel functions (Kung and Eckert, 1972, Saimi and Kung, 1982, Hinrichsen et al., 1985, Ramanathan et al., 1986). When the whole-cell voltage clamp was applied to these mutants, specific current defects were identified (Table 2). Thus, we found mutants in seven complementation groups (_pawns_ and _CNR's_; Haga et al., 1983) with smaller or no Ca^{++} current. Mutants in one complementation group (gene) called _Dancer_ have a stronger Ca^{++} current because the Ca^{++}-channel inactivation is made inefficient by the mutation (Hinrichsen and Saimi, 1984). There are probably more than three complementation groups (_pantophobiacs_), mutations in which turn down the Ca^{++}-dependent K^+ current. There is at least one complementation group (_TEA-insensitive_), mutations in which appear to turn up this current through a more rapid activation. Similarly, there are mutations that turn the Ca^{++}-dependent Na^+ current down (_fast-2_) or up (_Paranoiacs_). Mutants lacking the Ca^{++}-dependent hyperpolarization-activated K^+ current and their supressor mutants have also been isolated and studied (Richard, Saimi and Kung, 1986). These mutational effects are summarized in Table 2 and are periodically reviewed (Kung and Saimi, 1982, Hinrichsen et al., 1985b, Ramanathan et al., 1986).

We have taken a novel approach to find the defective proteins in some of these mutants. Haga and Hiwatashi (1982) found that it is possible to restore the wild-type behavior in one of the _CNR_ mutants by directly injecting it with high-speed supernatant taken from a wild type homogenate. Preparative quantities of wild-type supernatant can then be fractionated through ammonium-sulfate precipitations followed by column chromatography. The presence of the restoration activity is assayed by injecting different fractions into the mutant, and measuring the return of the normal behavior and the normal membrane current. This method, though relatively tedious, ensures that the wild-type substance purified or enriched is functionally active.

pntA, one of the _pantophobiac_ mutants, lacks the Ca^{++}-dependent K^+ current. Because this K^+ outward current participates in the downstroke of the action potential, the _pntA_ mutant tends to have prolonged excitations which result in long backward swimming upon various stimuli (Hinrichsen et al., 1985a). Both the behavioral and electrophysiological defect can be corrected by an injection of wild-type cytoplasm. Fractionation of the wild-type cytoplasm traced the restoration activity to a small, soluble, acidic, heat-stable protein. This protein migrates differently in gel electrophoretic assays depending on whether Ca^{++} is present or not. All these are characteristics of calmodulin, the well-known Ca^{++}-binding protein which is highly conserved among eucaryotes approximately 16,000 daltons in molecular weight. It has four Ca^{++}-binding domains arranged in such a way that a Ca^{++}-binding loop is located between two perpendicular helices (Kretsinger, 1980). The restoration of the Ca^{++}-dependent K^+ current in _pntA_ mutant by microinjection of wild-type calmodulin is shown in Fig. 3. The wild-type calmodulin from _Paramecium_ has been completely sequenced. It consists of 148 amino-acid residues (Schaefer et al., 1987a). The calmodulin of the _pntA_ mutant has also been completely sequenced. The mutant defect was found to be a substitution of the serine at position 101 by a phenylalanine (Fig. 4) (Schaefer et al., 1987b). Serine 101 is conserved from yeast, _Paramecium_ to mammals. Its hydroxyl oxygen is one of the six oxygens in the third Ca^{++}-binding loop which coordinates with the Ca^{++} ion, based on current structural models. It is therefore likely that the third domain of the _pntA_ calmodulin binds Ca^{++} more weakly. The _pntA_ mutant is the first to have an identified molecular defect among all the behavioral mutants with electrophysiological

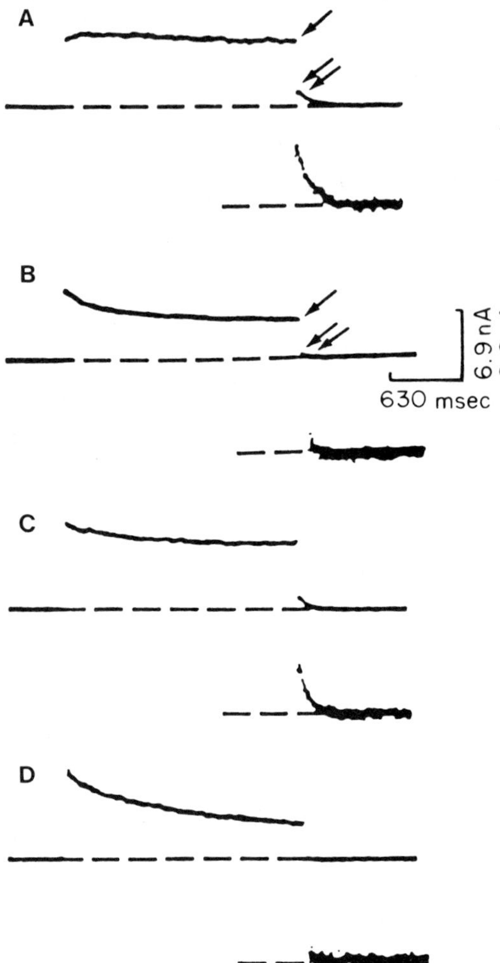

Fig. 3. Whole-cell voltage-clamp analysis of the Ca++-dependent K+ current in _Paramecium_ cells injected with calmodulin or buffer. The cells were micro-injected with either 20 pl of buffer solution or calmodulin (1 pg) 4 hours before the voltage clamp analysis. A depolarization step was given from the resting level of -40 mV to -5 mV and held for 2 seconds before returning to -40 mV. The inset in A to D shows the tail current amplified three times.
A) Wild-type cells injected with 20 pl of buffer solution showed the Ca++-dependent K+ current. The single arrow indicates the late outward current upon depolarization, primarily through the Ca++-dependent K+ channel. The double arrow points to the tail current upon repolarization. This current is shown on the expanded scale in the inset.
B) The _pntA_ mutant, injected with 20 pl buffer solution, virtually devoid of the Ca++-dependent K+ current. The drooping outward current during the depolarization is largely the leakage current and the voltage-dependent K+ current. There is almost no tail current upon repolarization (double arrow and inset).
C) The _pntA_ mutant after the injection of wild-type calmodulin showed a return of both the late outward current and the tail current.
D) In the control experiment the _pntA_ mutant injected with _pntA_ calmodulin showed no restoration of either late outward or tail current (from Hinrichsen et al., 1986).

defects in _Paramecium_ or _Drosophila_. The _pntA_ finding provides the first solid evidence that calmodulin is connected to a Ca++-dependent channel. The nature of that connection is yet to be understood. It is possible that a modified form of calmodulin enters the membrane to become a subunit to mediate the Ca++ dependence of the channel. It is also possible that it modulates the activities of this channel through a covalent modification such as a Ca++-calmodulin dependent phosphorylation. The single-channel recording analysis may give us an answer to these questions.

The method of cytoplasm fractionation and microinjection assay has also been used to partially purify a factor which restores the normal behavior in a Ca++ channel mutant of _P. caudatum_. The factor which restores the normal behavior in _cnrC_ (one of the _CNR_ mutants) has been

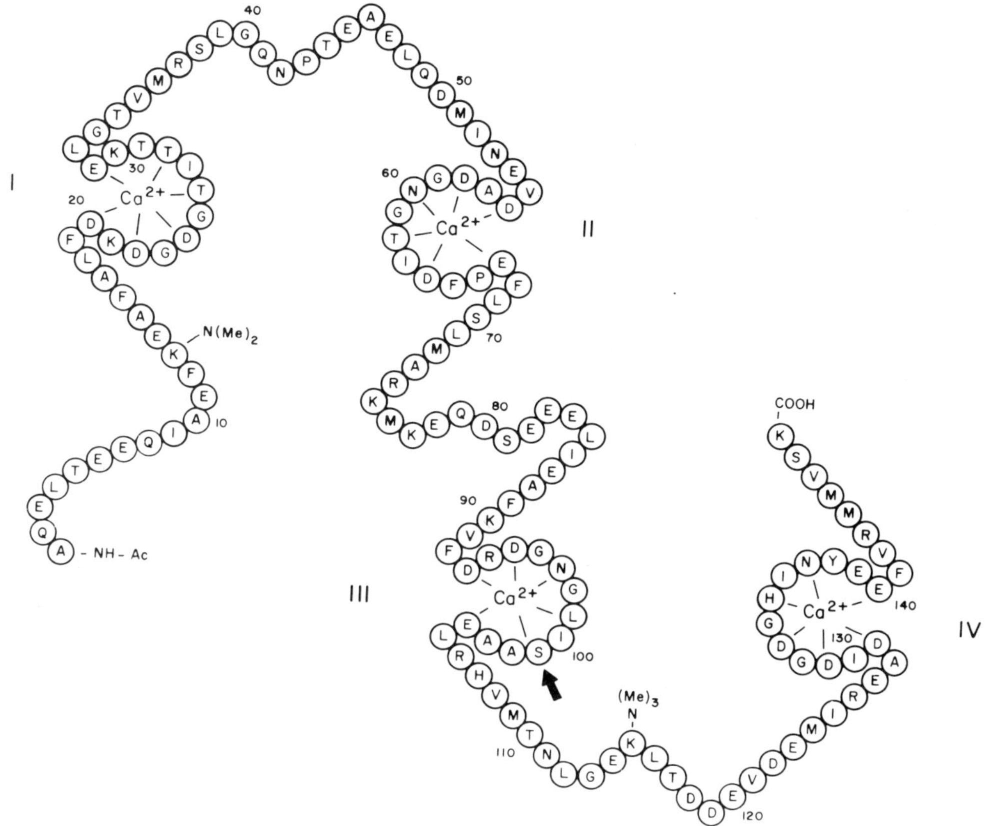

Fig. 4. Scheme of the amino acid-sequence of Paramecium calmodulin molecule. The arrow points to the serine at the 101st location which is substituted with phenylalanine in the pntA calmodulin. The lysine residues at positions 13 and 115 in wild-type Paramecium calmodulin are dimethylated and trimethylated respectively (figure design after Klee et al., 1980, and data from Schaefer et al., 1987b).

enriched almost 600 fold and is most likely a soluble, acidic protein less than 30,000 daltons in molecular weight. It is not calmodulin since it is not heat stable and does not stimulate brain phosphodiesterase (Haga et al., 1984).

SACCHAROMYCES CEREVISIAE (BAKER'S YEAST)

The size of a yeast cell is ideal for patch-clamp experiment. However, it is enclosed by a cell wall. Removal of this wall with zymolyase yields spheroplasts of about 5 to 7 μm in diameter. Consistent gigaohm seals can be formed on the surfaces of these spheroplasts if one delivers strong and sustained suction to the pipette during seal formation (Fig. 5). After the seal is formed, we can then arrive at on-cell, whole-cell, inside-out patch or outside-out patch mode by conventional methods (Hamill et al., 1981).

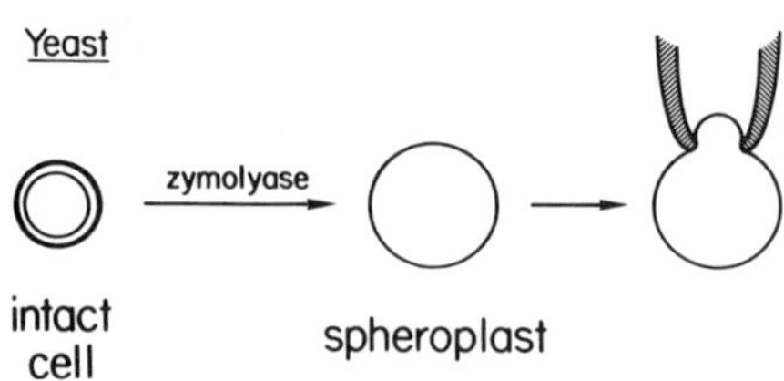

We found two types of channel activities in yeast. The first is the activities of a depolarization-activated K^+-specific channel (Gustin et al, 1986). This channel has a unit conductance of about 20 pS in 100 mM KCl solutions. Analyses of reversal potentials and experiments with biionic solutions showed a strong selectivity for K^+ over Na^+ (Fig. 6C). It is blocked by tetraethylammonium, Ba^{++}, Cs^+ and quinidine; these chemicals block K^+ channels in other systems. Single-channel activities show bursts separated by long (second) interbursts, and rapid (millisecond) flickers between the open and closed states during bursts (Fig. 6A and B). Opening probability and duration increase with membrane

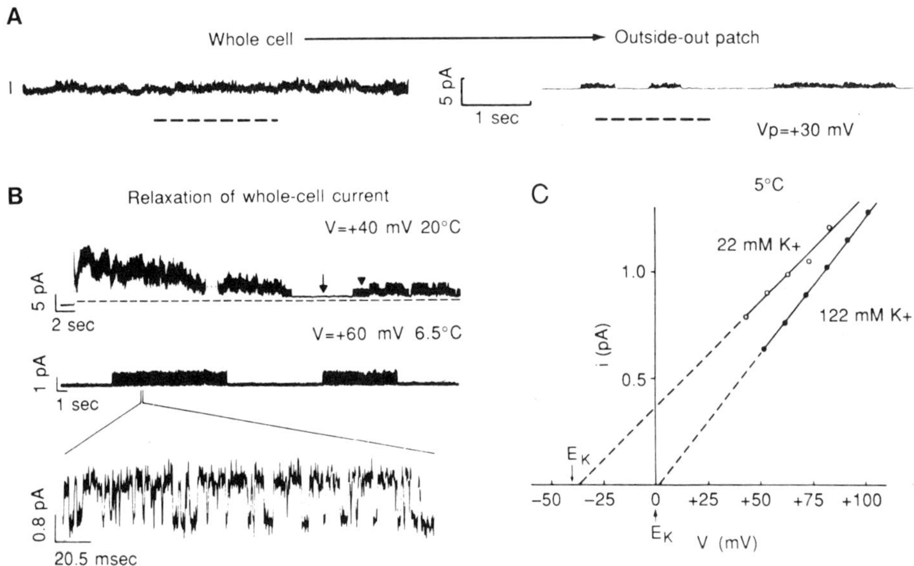

Fig. 6. Single channel K^+ currents in yeast. A) The unitary current underlying the ensemble currents was recorded from an outside-out patch (right) excised from a whole cell (left). After excision of the patch, the current was greatly reduced and showed bursting behavior. Traces are from chart records (dashed line, zero current level). B) Unitary currents were also recorded after relaxation of the whole-cell currents upon a voltage step from −40 to + 40 mV. After the relaxation lasting over tens of seconds (dots) bursts of activities from one or more channels were observed. Data are from chart record (upper trace). Single-channel bursts after whole-cell currents relaxation, examined at higher time resolution, revealed clear open and closed states characteristic of ion-channel behavior (middle and lower trace). C) Current-voltage relationship of unitary current. Calculated K^+ equilibrium potentials (arrows) were 0 and −41 mV, respectively. Changing K^+ concentration in the bath produced shifts in the extrapolated reversal potential (from Gustin et al., 1986).

depolarization. Although we first reported some 10-20 such channels per
yeast cell, improved studies showed up to 100 such channels per cell
(Gustin et. al., unpublished observations).

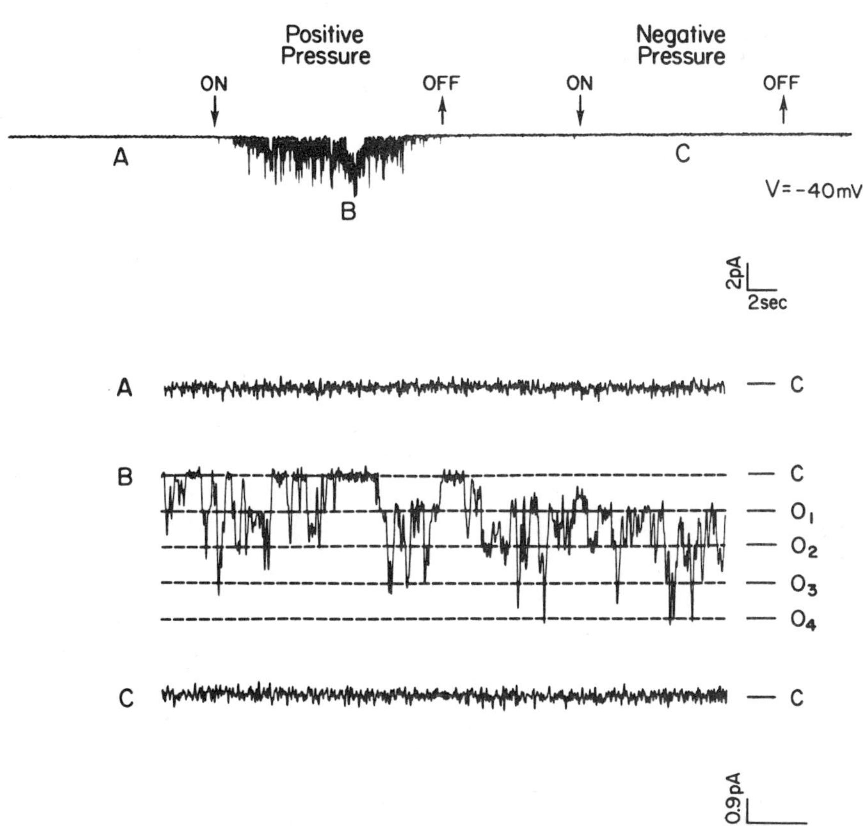

Fig. 7. Pressure-activated channels in yeast. Multiple channel openings
were observed in whole-cell tight-seal recording mode upon application of
positive pressure exerted to the cell through the pipette. The application
of equal magnitude negative pressure was without effect. Current deflections down are channel openings. Note the multiple channel levels and
apparent cooperative kinetics at higher time resolution. Experimental
solutions contained 180 mM KCl, 1 mM CaCl$_2$, 5 mM HEPES/KOH, pH 7.2. The
pipette voltage was -40 mV (Gustin et al., 1987).

    A second type of channel activity appears to be that of a stretch-
activated cation channel (Gustin et al., 1987). A few centimeters mercury
of positive pressure in the whole-cell mode, is enough to open this type
of channels (Fig. 7). The stronger the pressure the more likely the

channels are in their open state. Dilution of the solution bathing the
spheroplasts, creating an outward osmotic pressure, can also activate
these channels in the whole-cell mode. These channels have a unit conductance of 40 pS in hundred millimolar ionic solutions. They pass
cations but not anions. The selectivity among various cations is not
very strong; they can pass alkali ions, TEA$^+$, Ca^{++} and Mg^{++} (Gustin et
al., 1987).

The physiological roles of these two types of ion channels are not
known. Two lines of research are being carried out to find their roles.
First, cells at different stages of cell cycle, at different stages
during mating, or after growth in different osmotic, ionic, pH, aerobic
or other conditions are being examined. Systematic variations of channel
activities in different stages or conditions may suggest channel functions.
Second, mutants defective in channel activities are being sought. Other
phenotypes, such as mating or budding abnormalities, which co-segregate
with the channel defects would also implicate channel functions in the
physiological processes. In the scheme in Fig. 8 these channels are
added to the previously known ion transport pathways in the yeast plasma
membrane.

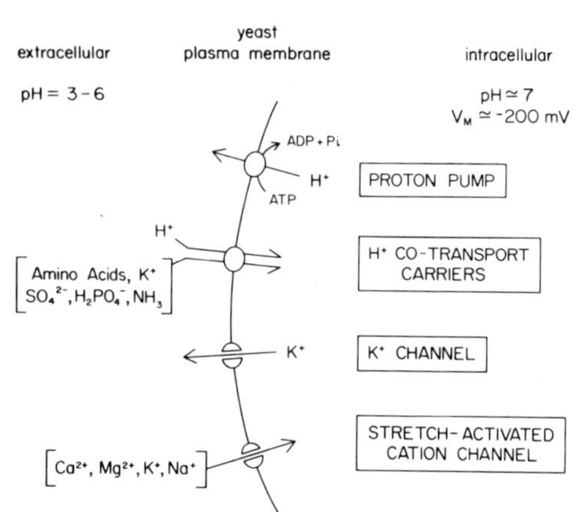

Fig. 8. A scheme summarizing the known ion transport pathways in the plasma membrane of yeast.

S. cerevisiae is most amenable to molecular-genetic manipulations
among eucaryotes. The capability to record from its plasma membrane
coupled with the ability to manipulate its genome should allow us to
examine the functional expression of foreign channel genes after they
have been cloned into the yeast genome. Fujita et al. (1986, 1987)
have succeeded in cloning the genes coding for the alpha and delta
subunits of the nicotinic acetylcholine receptor from Torpedo into
yeast. They showed that the subunit proteins are made and are incorporated into the yeast plasma membrane. It should be very interesting
to see whether a complete, pentameric, functional channel can be
assembled when all the subunit genes are added to the yeast genome. If
this is successful, yeast can be an alternative to frog oocyte as an
expression system in which to further study the molecular structures and
functions of vertebrate channels. The yeast system has added advantages. Unlike the mRNA-injected frog oocytes which are terminal, the
DNA-transformed yeasts are permanant constructions and can be regarded
as new strains of yeasts ready for repeated examination and further
modifications. Yeast can also be grown in fermenters. This fact
coupled with the possible over-expression of the gene products may allow
us to generate enough channel proteins for studies such as X-ray
crystallography.

ESCHERICHIA <u>COLI</u> (A GRAM-NEGATIVE BACTERIUM)

<u>E. coli</u> is only 0.2 μm in diameter and 2 μm in length. It is therefore too small for a direct application of the patch-clamp pipette. However, Ruthe and Adler (1985) showed that giant spheroplasts of the bacterium can be generated. <u>E. coli</u> cells grown in the presence of cephalexin, a penicillin analog, fail to form septa when they try to divide and therefore become long filaments 50 to 150 μm in length. After a treatment with EDTA and lysozyme to digest the peptidoglycan layer (the cell wall), these filaments collapse into giant spheroplasts about 5 to 10 um in diameter, suitable for patch-clamp experiments (Fig. 9A).

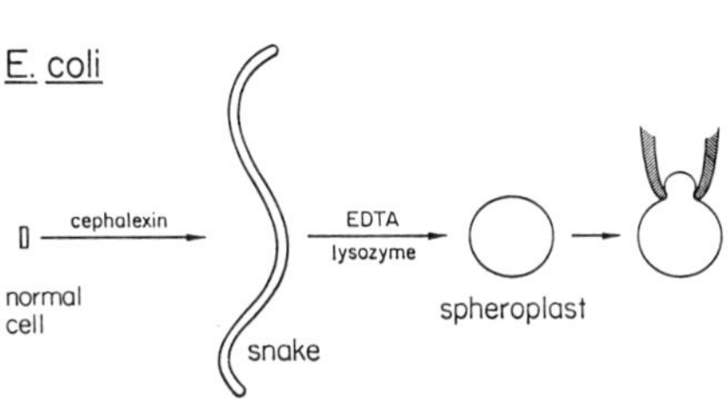

Fig. 9A. A cartoon showing the formation of a giant spheroplast by growing <u>E. coli</u> in the presence of cephalexin into a long filament (a "snake" is usually 50 to 150 μm long) and treating the filament with EDTA and lysozyme. Such spheroplasts are about 5 to 10 μm in diameter and can be used for patch-clamp recording.

As with the yeast spheroplasts, gigaohm seals can be formed consistently on <u>E. coli</u> spheroplasts provided a large and sustained suction is delivered through the patch pipette. The activities of at least two types of ion channels are found on the <u>E. coli</u> surface to date (Martinac et al., 1987). One is voltage-activated and the other is activated by voltage and by pressure. The activation of the pressure-sensitive channel at constant voltage is shown in Fig. 9B. This channel has been studied in greater detail as described below.

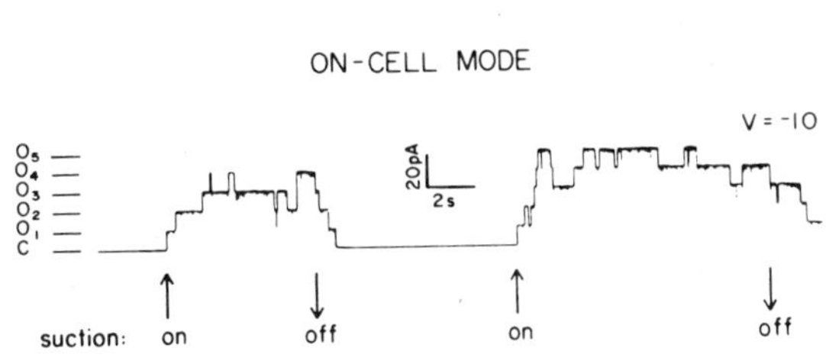

Fig. 9B. A recording segment showing the activation of pressure-sensitive ion channels in a giant spheroplast of <u>E. coli</u>. Channels were activated when suction was applied to the recording pipette, and terminated after release of the suction (arrows). The voltage imposed on the patch was -10 mV (from Martinac et al., 1987).

Suction of a few centimeter mercury opens this channels (Fig. 9A). The opening probability of these channels plotted against the suction can be fitted to a curve according to the Boltzmann distribution. These channels are also voltage sensitive. Depolarization shifts the Boltzmann

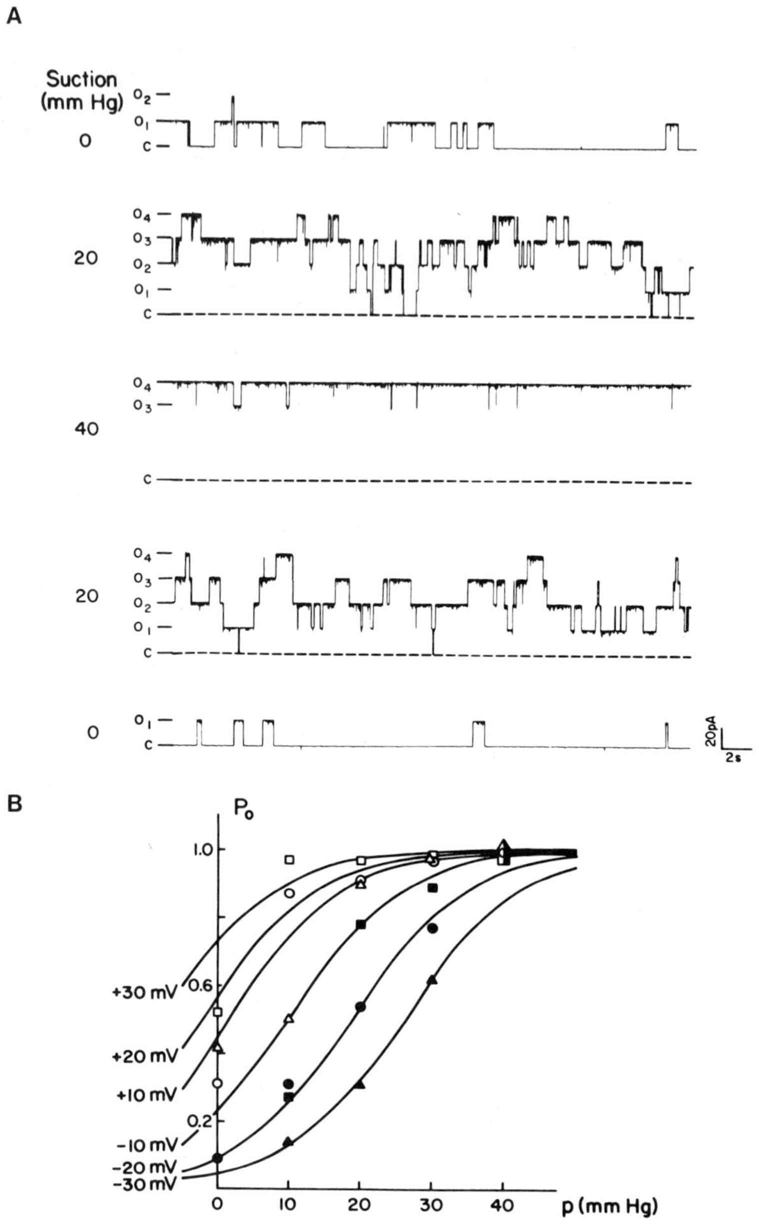

Fig. 10. A) Chart records from an on-cell patch showing activation of pressure-sensitive ion channels by suction of tens of millimeters mercury. Four channels were active here, and the channel activity changed reversibly upon a change in the negative pressure. The labels c and o_n correspond to closed and open states of channels (n = number of open channels). The imposed membrane voltage was −20 mV. B) The single channel opening probability P_O vs. negative pressure p was fitted to a Boltzmann distribution. An e-fold change in P_O for every 8mm Hg was calculated for data shown. Note that at depolarizing voltages, less pressure was necessary to activate the channels (from Martinac et al., 1987).

curve to the left, i.e. making the channel more likely to open at a
given pressure (Fig. 10B). Pressure exerted in either direction perpendicular to the membrane can open the channels. This channel has a large
unit conductance. In 300 mM salt, the conductance is about 970 pS. The
channel has little selectivity, although it slightly favors anions over
cations (Fig. 11). It passes inorganic as well as organic ions including
ions as large as glutamate. We found that the kinetic behavior of this
channel depends strongly on the species of ions in the solution. For
example, replacing K^+ with Na^+ greatly reduces the pressure sensitivity
and the mean open time. All permeant ions including protons appear to
affect the channel behavior suggesting a direct interaction between the
permeating ions and the gate.

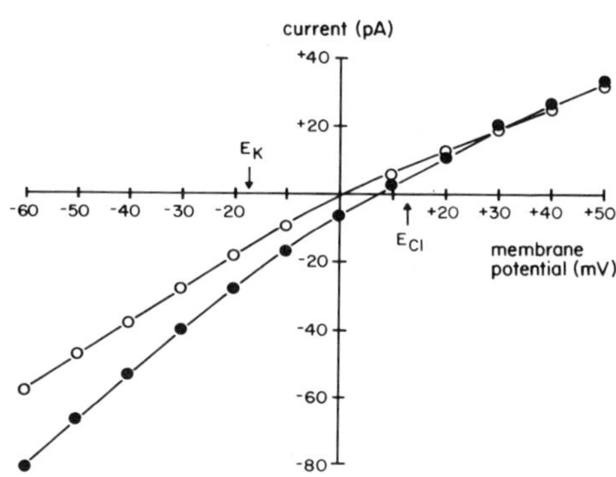

Fig. 11. Current-voltage relationship for the pressure-sensitive
channel gives a conductance of the
channel between 650 and 970 pS in
a symmetrical bath and pipette
solutions containing 200 mM KCl,
40 mM $MgCl_2$, 10 mM $CaCl_2$, 0.1 mM
EDTA, and 5 mM HEPES-KOH, pH 7.2
(open circles). Changing the KCl
concentration in the bath to 400
mM shifted the reversal potential
by about 8 mV in the positive
direction (filled circles). The
equilibrium potentials for Cl^- and
K^+ are +13 mV and -17 mV respectively (arrows) (from Martinac et
al., 1987).

     The location of this interesting channel is not yet clear. A Gram-
negative bacterium has two membranes separated by the periplasmic space
in which lies the peptidoglycan layer (the cell wall). It is not yet
clear whether the gigaohm seal is formed on the outer or the inner membrane of the giant spheroplast. The major proteins in the outer membrane
are porins which are nonselective channels of large conductances. Porins
have been purified and reconstituted into planar lipid bilayers (Benz
et al., 1978; Schindler and Rosenbusch, 1978). These reconstituted
porins are different from the pressure-sensitive channels in that all
porins so far reported show increased channel closure with voltage
(Benz, 1986), whereas pressure-sensitive channels we discovered are
activated more with positive voltage. Furthermore, no pressure sensitivity of porin activities has been reported. The major porins in E.
coli are the products of the ompF and the ompC genes. Mutants with both
genes deleted have been examined and were found to still have the
pressure-sensitive channels (Martinac, Delcour, Buechner, Adler and Kung,
unpublished).

     The large unit conductance and small cell volume make it unlikely
that these pressure-sensitive channels are used to regulate the membrane
potential. On the other hand, E. coli regulates both its cytoplasmic and
its periplasmic space upon changes of external osmolarity (Britten and
McClure, 1962). This channel may well play a role in osmoregulation.

CONCLUSION

The work reviewed here is preliminary, especially those on the channels of yeast and E. coli. The findings, however, lead to some interesting questions, some of which are listed below.

1. Most of the ion channels in Paramecium function in membrane excitation and therefore motile behavior. Do the channels in E. coli also govern motile behavior? These bacteria are flagellated and can perform chemotaxis (Adler, 1969) and osmotaxis (Li, Kung and Adler, unpublished). Are ion channels used in the transduction of chemical or osmotic stimuli into the motile responses? What are the functions of ion channels in yeast, a sessile, walled cell? Are they used in slow "behaviors" such as budding and mating, or in housekeeping fuctions such as membrane-potential regulation?

2. Within the last year, stretch- or pressure-gated channels have been reported in muscle cells, blood cells, secretory cells, lens cells, and eggs. They are found in plants, molluscs, mammals and now yeast and E. coli. Is there a universal mechanism to this mode of channel gating by mechanical forces or are there more than one mechanisms?

3. Channels are traditionally classified through their functions and not their structures. Thus we sort them by their permeant ions (Ca^{++}, Na^+, K^+ channel etc.), by their gates (V-sensitive, ligand-sensitive, Ca-activated etc.) or by their unit conductance (small, large or maxi). Structural information has only begun to be obtained. The few channels whose primary sequences are now known are unrelated. Based on amino acid sequence, the structural models of the Na^+ channel and the acetylcholine receptor have transmembrane alpha helices probably with pores lined with charged sectors of amphipathic helices, each contributed by one subunit or domain. Porins, on the other hand, are envisioned to be barrels of folded beta sheets (Benz, 1985). Into which categories do the microbial channels fall. Are there new categories? Are the channels in different categories analogous, i.e. fundamentally different structures independently evolved to serve a similar function such as the wings of insects, birds and bats? Or do they all derive from one promordial channel which may still be found in microbes?

4. We chose to study the three microbes reviewed here out of millions possible candidates because they have been studied intensively and extensively by geneticists and molecular biologists. The progress of yeast and E. coli as systems to advance molecular understanding is unquestioned. Transformation of ciliates (Tondravi and Yao, 1986) make it hopeful to study channel molecules in Paramecium with recombinant DNA technology in the near future. Cloning and expressing foreign channel genes into yeast, studying them through site-directed mutagenesis and making channel proteins in fermentation scales are now being explored. Besides these popular ventures, can we study synthesis, assembly, deployment, turnover, mediation and modulation of ion channels using these microbes? They have been used to reveal entire enzymatic cascades in many physiological and cell-biological processes through genetic and biochemical analyses. Much of modern textbooks consists of pathways and processes such as DNA synthesis and recombination, enzyme inductions, protein synthesis, lipid synthesis, secretion pathway etc. understood by way of microbes. Can a study of microbial channel be similarly helpful in advancing our knowledge of channels in general?

Acknowledgement

     We thank Pat Hanson and Leslie Rabas for technical assistance.
Supported by grant from NIH GM-22714 and GM-36386.

References

Adler, J., 1969, Chemoreceptors in bacteria, Science, 166:1588-1597.
Benz, R., Janko, K., Boos, W. and Läuger, P., 1978, Formation of large
     ion-permeable membrane channels by the matrix protein (porin) of
     Escherichia coli, Biochim. Biophys. Acta, 511:305-319.
Benz, R., 1985, Porin from bacteria and mitochondrial outer membranes,
     CRC Crit. Rev. Biochem., 19:145-190.
Benz, R., 1986, Analysis and chemical modification of bacterial porins,
     in: "Ion Channel Reconstitution," C. Miller, ed., Plenum Press,
     New York.
Britten, R.J. and McClure, F.T., 1962, The amino acid pool of Escherichia
     coli, Bact. Rev., 26:292-335.
Eckert, R., 1972, Bioelectric control of ciliary activity, Science,
     176:473-481.
Ehrlich, B.E., Finkelstein, A., Forte, M. and Kung, C., 1984, Voltage-
     dependent calcium channels from Paramecium cilia incorporated into
     planar lipid bilayers, Science, 225:427-428.
Fujita, N., Nelson, N., Fox, T.D., Claudio, T., Lindstrom, J., Riezman, H.
     and Hess, G.P., 1986, Biosynthesis of the Torpedo californica
     acetylcholine receptor subunit in yeast,Science, 231:1284-1287.
Fujita, N., Sweet, M.T., Fox, T.D., Nelson, N., Claudio, T., Lindstrom,
     J.M. and Hess, G.,1987, Expression of cDNA for acetylcholine receptor
     subunits in yeast cell plasma membrane, Biochem. Soc. Symp., 52: (in
     press).
Gustin, M.C., Martinac, B., Saimi, Y., Culbertson, M.R. and Kung, C.,
     1986, Ion channels in yeast, Science, 233:1195-1197.
Gustin, M.C., Zhou, X.L., Martinac, B., Culbertson, M.R. and Kung, C.,
     1987, Stretch-activated cation channel in yeast, Biophys. J., 51:251a.
Haga, N. and Hiwatashi, K.,1982, A soluble gene product controlling
     membrane excitability in Paramecium, Cell Biol. Int. Rep., 6:295-300.
Haga, N., Saimi, Y., Takahashi, M. and Kung, C., 1983, Intra and inter-
     specific complementation of membrane-inexcitable mutants of Paramecium,
     J. Cell Biol., 97:378-382.
Haga, N., Forte, M., Ramanathan, R., Hennessey, T., Takahashi, M. and
     Kung, C., 1984, Characterization and purification of a soluble protein
     controlling Ca-channel activity in Paramecium, J. Cell Biol., 39:71-78.
Hammil, O.P., Marty, A., Neher, E., Sakmann, B. and Sigworth, F.J., 1981,
     Improved patch-clamp techniques for high-resolution current recording
     from cells and cell-free membrane patches, Pflügers Arch., 391:85-100.
Hinrichsen, R.D. and Saimi, Y. 1984, A mutation that alters properties of
     the calcium channel in Paramecium tetraurelia, J. Physiol.,
     351:397-410.
Hinrichsen, R.D., Amberger, E., Saimi, Y., Burgess-Cassler, A. and Kung,
     C., 1985a, Genetic analysis of mutants with a reduced Ca^{++}-dependent
     K^+ current in Paramecium tetraurelia, Genetics, 111:433-445.
Hinrichsen, R.D., Saimi, Y., Ramanathan, R., Burgess-Cassler, A. and Kung,
     C., 1985b, A genetic and biochemical analysis of behavior in
     Paramecium, in: "Sensing and Responses in Microorganisms," M. Eisenbach
     and M. Balaban, eds., Elsevier Sci. Pub., New York.
Kamada, T. and Kinosita, H., 1940, Calcium-potassium factor in ciliary
     reversal of Paramecium, Jap. Acad. Proc., 16:125-130.

Kinosita, H., Dryl, S. and Naitoh, Y. 1964a, Changes in the membrane potential and the response to stimuli in Paramecium, J. Fac. Sci. Tokyo Univ. (Sect. IV), 10:291-301.

Kinosita, H., Dryl, S. and Naitoh, Y. 1964b, Relation between the magnitude of membrane potential and ciliary activity in Paramecium, J. Fac. Sci. Tokyo Univ. (Sect IV), 10:303-309.

Klee, C.B., Crouch, T.H. and Richman, P.G. 1980, Calmodulin, Ann. Rev. Biochem., 49:489-515.

Kretsinger, R.H. 1980, Structure and evolution of calcium-modulated proteins, CRC Crit. Rev. Biochem., 8:119-174.

Kung, C., 1971a, Genic mutations with altered system of excitation in Paramecium aurelia. I. Phenotypes of the behavioral mutants, Z. Vergl. Physiol., 71:142-164.

Kung, C., 1971b, Genic mutations with altered system of excitation in Paramecium aurelia. II. Mutagenesis, screening and genetic analysis of the mutants, Genetics, 69:29-45.

Kung, C. and Eckert, R. 1972, Genetic modification of electric properties in an excitable membrane, Proc. Natl. Acad. Sci. USA, 69:93-97.

Kung, C. and Saimi, Y., 1982, The physiological basis of taxes in Paramecium, Ann. Rev. Physiol., 44:519-534.

Martinac, B., Saimi, Y., Gustin, M. and Kung, C., 1986, Single-channel recording in Paramecium, Biophys. J., 49:167a.

Martinac, B., Buechner, M., Delcour, A.H., Adler, J. and Kung, C., 1987, A pressure-sensitive ion channel in Escherichia coli, Proc. Natl. Acad. Sci. USA, (in press).

Naitoh, Y., 1968, Ionic control of the reversal response of cilia in Paramecium caudatum. A calcium hypothesis, J. Gen. Physiol., 51:85-103.

Naitoh, Y. and Eckert, R., 1969, Ionic mechanisms controlling behavioral responses in Paramecium to mechanical stimulation, Science, 164:963-965.

Oertel, D., Schein, S.J. and Kung, C., 1977, Separation of membrane currents using a Paramecium mutant, Nature, 268:120-124.

Ramanathan, R., Saimi, Y., Hinrichsen, R.D., Burgess-Cassler, A. and Kung, C., 1986, A genetic dissection of ion-channel functions in Paramecium, in: "Paramecium," H.D. Gertz, ed., Springer Verlag, Heidelberg, (in press).

Richard, E.A., Saimi, Y. and Kung, C., 1986, A mutation that increases a novel calcium-activated potassium conductance of Paramecium tetraurelia, J. Membrane Biol., 91:173-181.

Ruthe, H.-J. and Adler, J., 1985, Fusion of bacterial spheroplasts by electric fields, Biochim. Biophys. Acta, 819:105-113.

Saimi, Y., Hinrichsen, R.D., Forte, M. and Kung, C., 1983, Mutant analysis shows that the calcium induced potassium current shuts off one type of excitation in Paramecium, Proc. Natl. Acad. Sci. USA, 80:5112-5116.

Schaefer, W.H., Lukas, T.J., Blair, I.A., Schultz, J.E. and Watterson, D.M., 1987a, Amino acid sequence of a novel calmodulin from Paramecium tetraurelia that contains dimethyl-lysine in the first domain, J. Biol. Chem., (in press).

Schaefer, W.H., Hinrichsen, R.D., Burgess-Cassler, A., Kung, C., Blair, I.A. and Watterson, D.M., 1987b, A mutant Paramecium with a defective calcium-dependent potassium channel has an altered calmodulin: a non-lethal selective alteration in calmodulin regulation, Proc. Natl. Acad. Sci. USA, (in press).

Schindler, H. and Rosenbusch, J., 1978, Matrix protein from Escherichia coli outer membrane forms voltage-controlled channels in lipid bilayers, Proc. Natl. Acad. Sci. USA, 75:3751-3755.

Tondravi, M.T. and Yao, M.-C., 1986, Transformation of Tetrahymena thermophila by microinjection of ribosomal RNA genes. Proc. Natl. Acad. Sci. USA, 83:4369-4373.

Speakers at the Symposium on Ion Channel Modulation, UCLA, February 27 - March 1, 1987.
front row: Anton Hermann, Joy Umbach, Stan Kater, Irwin Levitan, Kathy Dunlap, Rudolfo Llinas, Susumu Hagiwara, Richard Orkand, Dieter Lux, Martin Morad, Barbara Ehrlich, Laurinda Jaffe; second row: John Chad, George Augustine, Paul Brehm, David Armstrong, Erwin Neher, John Connor, Steve Smith, Len Kaczmarek, Susan DeRiemer, Jochen Deitmer, Sergei Mironov; third row: Alan Grinnell, Lou Byerly, Meyer Jackson, Doug Tillotson, Mahlon Kriebel, Boris Martinac, Gordon Fain, Julio Vergara, Paul O'Lague; fourth row: Aaron Fox, Gary Westerbrook, Vladimir Brezina.